Petroleum Geochemistry
and Geology

A Series of Books in Geology

Editor: James Gilluly

Petroleum Geochemistry and Geology

John M. Hunt

Woods Hole Oceanographic Institution
Woods Hole, Massachusetts

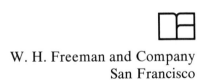

W. H. Freeman and Company
San Francisco

Sponsoring Editor: John H. Staples
Project Editor: Pearl C. Vapnek
Manuscript Editor: Linda Purrington
Designer: Perry Smith
Production Coordinator: Linda Jupiter
Illustration Coordinator: Batyah Janowski
Compositor: Typothetae Inc.
Printer and Binder: The Maple-Vail Book Manufacturing Group

Library of Congress Cataloging in Publication Data

Hunt, John Meacham.
　Petroleum geochemistry and geology.

　(A Series of books in geology)
　Bibliography: p.
　Includes index.
　1. Petroleum—Geology.　2. Gas, Natural—Geology.
3. Geochemical prospecting.　I. Title.
TN870.5.H86　　　　553′.28　　　　79–1281
ISBN 0–7167–1005–6

Printed in the United States of America

3 4 5 6 7 8 9　MP　0 8 9 8 7 6 5 4 3 2

To those pioneering scientists who were in the
Geological Research Section of the former
Carter Research Laboratory of Tulsa, Oklahoma,
in the late 1940s and early 1950s.

Contents

Foreword

The continually rising standards of petroleum geoscience and oil search require an ever greater knowledge of the fundamental nature of petroleum, its origin, and its behavior in the rocks of the earth's crust. In the early stages of the petroleum era, most oil men were not much concerned about the *how, why,* and *when* of oil and gas; all they cared to know was *where.* Those years are gone, and it is now more and more recognized that the how, why, and when of petroleum are critical keys to where it should be sought.

Someone has said that we will never know the answers to the problems of the origin, migration, accumulation, and preservation of petroleum until the last drop is found. Perhaps this is so; but equally certain is it that we will not come close to finding this last drop until we know more of these answers.

With all the current talk of oil shortages and "running out of oil and gas," we still probably have not used up half of the earth's petroleum accumulations, and it is even questionable whether we have yet discovered half of them. It is a fact, however, that we have found the easiest, most accessible, and least expensive portion, and that the remainder will only be found and recovered through more sophisticated methods than those of the past. And these methods will in a large part be based on better knowledge of the chemical and physical properties of petroleum and the principles of chemistry and physics that control its origin, its movements, its concentration into commercial deposits, its preservation, and its ultimate recovery from the rocks. Moreover, this knowledge will benefit us not only as regards conventional oil and gas deposits but also with respect to alternative hydrocarbon sources such as asphalt sands, oil shales, coal gas, gas-charged waters, and the like.

It is therefore a great boon to petroleum geologists, geochemists, geophysicists, engineers, teachers, and all others who are concerned with the finding and recovery of petroleum that one of our most widely experienced and greatest contributors to petroleum geochemistry has taken time from his busy career to put together this monumental volume, *Petroleum Geochemistry and Geology.*

John M. Hunt received his Ph.D. in chemistry from Pennsylvania State University in 1946 and, after a year on the faculty there, spent the next sixteen years at the Standard Oil Company of New Jersey (now Exxon) exploration research laboratory in Tulsa, Oklahoma, where from 1956 to 1963 he was Head of Geochemical Research. In this position, he was brought into close contact with the manifold petroleum research and exploration activities of a major petroleum company. In 1964, he left industry to become Chairman of the Department of Chemistry and Geology at Woods Hole Oceanographic Institution, 1964–1967, and Chairman of the Department of Chemistry, 1967–1974. He is presently a Senior Scientist at Woods Hole.

Throughout his life career, Dr. Hunt has been vitally concerned with research and the application of research on the origin, migration, and emplacement of petroleum. He has published numerous articles, alone and in collaboration with others, many of which have marked outstanding advances in our knowledge of this topic. Among these are his studies of the composition of crude oil in relation to stratigraphy in Wyoming (1953), Hunt and Jamieson on source rocks of petroleum (1956), Kidwell and Hunt on the migration of oil in recent sediments in Venezuela (1958), Forsman and Hunt on the kerogen in sedimentary rocks (1958), Dunton and Hunt on the distribution of low-molecular-weight hydrocarbons in Recent and ancient sediments (1962), the composition and origin of Uinta Basin bitumens (1963), the origin of petroleum in carbonate rocks (1967), Dickey and Hunt on prospecting for stratigraphic traps (1972), and his studies of light hydrocarbons in deep sea drilling samples (1974–1978).

Dr. Hunt's professional career with Exxon and at Woods Hole is moreover ornamented by numerous excursions into fringe activities to both. He was a Distinguished Lecturer of the American Association of Petroleum Geologists (AAPG) in 1964 and has been a Lecturer on their Continuing Education Program since it started. He has been an Associate Editor of the Bulletin of the AAPG since 1966. He was Chairman of the JOIDES Advisory Panel on Organic Geochemistry for many years and is currently a member of it and of the JOIDES Panel on Passive Margins. He was Chief Scientist of a Red Sea expedition in 1966 and a Black Sea expedition in 1969, the results of the latter being published in AAPG Memoir No. 20. He has lectured on geochemistry in various countries all over the world, and his works in this field are among those from this country frequently quoted in foreign publications.

The scope of the book is immense, and it is truly a happy blend of chemistry and geology as related to petroleum. Following an introductory section (Part I) on carbon and the composition of petroleum, Part II of the book consists of three extensive chapters on how oil forms, how gas forms, and how they both migrate and accumulate. Part III is a two-chapter discussion of source and reservoir rocks. Part IV, titled "Applications," deals with seeps and surface prospecting, subsurface prospecting, crude oil correlation, and prospect evaluation. Abundant tables and figures, chapter summaries, and copious literature references at the end of the book add greatly to the clarity and usefulness of the work.

The history of scientific progress in exploration for petroleum is replete with examples of new approaches and new techniques that were slow to catch on at first but that, once accepted, were carried even too far in the enthusiasm that attended their early successes. The anticlinal theory, once established, long dominated exploration to the exclusion of many nonanticlinal trap prospects that we now know exist. Micropaleontology, heavy minerals, the electrical log, the reflection seismograph, the air-borne magnetometer, clay–mineral transformations, the stratigraphic trap, the new global tectonics, the bright spot, the thermal window—all are examples of worthy and once new concepts that have played a very helpful role in petroleum exploration, and continue to do so, but in the flush of victory have often been carried too far. Organic geochemistry is relatively young as regards widespread application to petroleum exploration, but its contribution has already been phenomenal. However, it too must be used with discretion and understanding, or we may run the risk of prejudicing its most effective utilization.

In the careful reading of Dr. Hunt's book, I think one cannot help but realize the variety and complexity of the problems involved in the geochemical approach, the uncertainties and unknowns that still remain, and the differences of opinion that exist. If we are to get the most in exploration results out of the great potential contributions of organic geochemistry, we need the benefit of the experience, learning, and stimulating views of several leaders in this field, each of whom may have somewhat different backgrounds of geochemical experience. Moreover, geochemistry must also be melded and integrated into proper balance, as Dr. Hunt has done, with the contributions of geology, geophysics, and other branches of geoscience, each of which plays its own important role in exploration for petroleum.

I trust that all who are interested in petroleum—the second most abundant fluid in the earth's crust after water—whether for the sake of pure science or for commercial exploration, whether in academia or as professional, practicing explorationists, will take occasion to read, study, and ponder this outstanding volume.

January 1979   Hollis D. Hedberg
                                                        Professor of Geology (Emeritus)
                                                        Princeton University

Preface

This book was written for students who have had the basic courses in geology and chemistry and also for oil company operating personnel who are interested in the application of geochemistry to petroleum exploration. It is intended to be used both as a text and as a reference book. It discusses both the geochemistry and the geology of petroleum, but the emphasis is on the former.

Thirty years ago, petroleum geochemistry was limited to a few studies by major oil companies on such subjects as surface prospecting, crude oil correlation, and source rock identification. Today, it is a highly diversified applied science with a variety of geochemical concepts and techniques playing an important role in exploration decisions. The objective of this book is to explain the basic principles of petroleum geochemistry and to show how this information can be effectively integrated with geology and geophysics in the search for oil and gas.

The outline of this book and the scope of subjects is written to be easily understood by the geologist as well as by the chemist. Each geochemical concept, such as carbon isotopes, is explained in detail prior to discussions on its application. The composition and uses of petroleum are presented at the beginning of the book, so that readers will fully understand the subject with which they are dealing.

There is a worldwide trend toward the simplification of measurement. The International System of Units (SI), sometimes referred to as the *metric system,* has been adopted by over 30 countries and is destined to become universal in science and commerce. Both SI and English units of measurement are given in this book (with round numbers given first), so that the reader may become accustomed to the SI system. Some conversion factors for SI and English units are given in the Appendix.

This book evolved from notes used in continuing education courses given to exploration operating personnel at industrial and university seminars. I have participated in such courses since the early 1960s, when the Jersey Production Research Company (formerly Carter Research Laboratory) of Tulsa, Oklahoma, held a school for the worldwide affiliates of Exxon. The late A. I. Levorsen, who was one of the outstanding speakers on our early faculty, first suggested that I write this book. I also was encouraged by the comments of students at courses given in major cities of the United States, Canada, South America, Europe, Africa, and the Middle East. In most of the courses, I kept notes on the frequently asked questions, and these have been answered here in detail, insofar as possible. I also was able to include some anecdotes from my visits to field operations in the United States, Canada, Venezuela, and the Soviet Union.

Today, petroleum geochemistry is a rapidly changing field. This book will provide a background for understanding the basic concepts and principles, but readers are encouraged to watch for new developments through the literature and continuing education courses. The literature on petroleum geochemistry is increasing so fast that I was not able to quote all the important papers in this field. I do want to thank my many friends in geochemistry who sent me preprints of their papers prior to publication and thereby enabled me to quote some recent references.

I am particularly grateful to the many reviewers who generously provided their time and expertise. Hollis Hedberg and James Gilluly were the geologists who made detailed comments on the entire manuscript. Jean Whelan reviewed all of the chemistry, and Brian Hitchon commented on the geochemistry. Chapter 6, on migration, was reviewed by Philip Low, Parke Dickey, Peter Gretener, and Gerard Lijmbach, who also reviewed Chapter 5.

Others who provided valuable comments on parts of the manuscript were A. O. Woodford, Thane McCulloh, T. P. Goldstein, Oliver Zafiriou, and K. O. Emery.

I am also grateful to my wife, Phyllis Laking, who handled references, permission letters, indexing, and many other time-consuming jobs. Her experience in previously publishing her own book, *The Black Sea—A Bibliography* (Woods Hole Oceanographic Institution, 1974) was valuable in handling this text.

Special thanks go to Sharon Callahan and Julie Kertyzak, who listened to endless numbers of tapes while typing the manuscript, and to Christine Johnson, who typed some of the early drafts.

January 1979   John M. Hunt

Abbreviations Used in Text

Å	angstrom (1×10^{-10}m)
°API	degrees API gravity
Ar	aromatic
ASTM	American Society for Testing and Materials
bbl	barrel
C_1	methane
C_2	ethane
C_{2+}	ethane, propane, butanes, and pentanes
C_3	propane
C_4	butanes
C_7	heptanes
C_4–C_7	butanes, pentanes, hexanes, and heptanes
C_{11+}	all hydrocarbons containing 11 or more carbon atoms
C_{15+}	all hydrocarbons containing 15 or more carbon atoms
^{12}C	stable isotope of carbon with an atomic mass of 12
^{13}C	stable isotope of carbon with an atomic mass of 13
C_{org} (or C_o, or C_T)	total organic carbon
C_{eff}	effective carbon
C_R	carbon residue (nonvolatile organic carbon)
CI	correlation index
cm	centimeter
COST	Coastal Offshore Stratigraphic Test
CPI	carbon preference index
d.a.f.	dry, ash free

DSDP	Deep Sea Drilling Project
DST	drill-stem test
E_a	activation energy
Eh	redox potential—a measure of the oxidizing or reducing intensity of the environment
EPR (ESR)	electron paramagnetic spin resonance
FID	flame ionization detector
G	specific gravity
GC	gas chromatography
GCMS	gas chromatography—mass spectrometry
GOR	gas—oil ratio
GPC	gel permeation chromatography
HC	hydrocarbon
IPOD	International Program of Ocean Drilling
IR	infrared
JOIDES	Joint Oceanographic Institutions for Deep Earth Sampling
kcal	1,000 calories
kHz	kilohertz (1,000 cycles/sec frequency)
kPa	kilopascal
LNG	liquefied natural gas (methane, ethane)
LPG	liquefied petroleum gas (propane, butanes, pentanes)
ls	limestone
MCF	thousand cubic feet
md	millidarcy
mg	milligram
μg	microgram
mi	mile
ml	milliliter
mm	millimeter
MP	melting point
MPa	megapascal
MT	metric ton (Mg, megagram in SI units)
MW	molecular weight
N	naphthene
NBS	National Bureau of Standards
n—d—M	refractive index—density—molecular weight
ng	nanogram
nm	nanometer
NMR	nuclear magnetic resonance
NSO	nitrogen, sulfur, oxygen

OCS	Outer Continental Shelf
OEP	odd–even predominance
OM	organic matter
P	paraffin
Pa	pascal
PAH	polycyclic aromatic hydrocarbon
PDB	Peedee belemnite (carbon isotope standard)
PF	pyrolysis–fluorescence
pH	the negative logarithm of the hydrogen ion concentration; a measure of the acidity or alkalinity of a solution (acids, less than 7; bases, more than 7)
ppb	parts per billion
ppm	parts per million
ppt	parts per thousand
psi	pounds per square inch
psia	pounds per square inch absolute
PVT	pressure–volume–temperature
R_a	reflectance in air
R_o	reflectance in oil immersion
sh	shale
SI	international system of units
ss	sandstone
STP	standard temperature and pressure, 60°F (15.6°C) and 760 torr (133.3 Pa)
S.U.	Saybolt Universal (a viscosity measurement in seconds with a Saybolt viscosimeter)
T	temperature
TAI	thermal alteration index
TCD	thermal conductivity detector
TD	total depth
TG	thermal gravimetry
TLC	thin-layer chromatography
UV	ultraviolet
VI	viscosity index
‰	parts per thousand

The Geologic Time Scale

Era	North America — Period	North America — Epoch	North America — Age	Europe — Period	Europe — Epoch	Europe — Age	Approx. Age 10^6 Years
Cenozoic	Quaternary	Recent / Pleistocene		Neogene	Holocene / Pleistocene		1.6
		Pliocene / Miocene			Pliocene / Miocene		5
	Tertiary	Oligocene	Jacksonian	Paleogene	Oligocene	Chattian	23
		Eocene	Claibornian / Wilcoxian		Eocene	Bartonian	37
		Paleocene	Midwayan		Paleocene	Danian	53
	Cretaceous	Upper	Maastrichtian / Senonian / Turonian / Cenomanian	Cretaceous	Upper	Maastrichtian / Senonian / Turonian / Cenomanian	65
Mesozoic		Lower	Albian / Aptian / Neocomian		Lower	Albian / Aptian / Neocomian	100
	Jurassic	Upper / Middle / Lower	Kimmeridgian / Bathonian / Toarcian	Jurassic	Upper / Middle / Lower	Malm / Dogger / Lias	136
	Triassic	Upper / Middle / Lower	Keuper / Anisian / Scythian	Triassic	Upper / Middle / Lower	Keuper / Anisian / Scythian	190

Era	System	Series	Stages	Period	Series	Stages	Age (Ma)
Paleozoic	Permian	Upper / Lower	Ochoan, Guadalupian, Leonardian, Wolfcampian	Permian	Upper / Lower	Zechstein, Rotliegendes	230
	Pennsylvanian	Upper / Middle / Lower	Virgillian, Missourian, Desmoinesian, Atokan, Morrowan	Carboniferous	Upper	Stephanian, Westphalian, Namurian	280
	Mississippian	Upper / Middle / Lower	Chesterian, Meramecian, Osagean, Kinderhookian	Carboniferous	Lower	Viséan, Tournaisian	325
	Devonian	Upper / Middle / Lower	Chautauquan, Senecan, Erian, Onesquethawan, Oriskanian, Helderbergian	Devonian	Upper / Middle / Lower	Famennian, Frasnian, Givetian, Couvinian, Siegenian, Gedinnian	360
	Silurian	Upper / Middle / Lower	Cayugan, Niagaran, Medinan	Silurian	Upper / Lower	Ludlovian, Wenlockian, Llandoverian, Valentian	400
	Ordovician	Upper / Middle / Lower	Cincinnatian, Champlainian, Canadian	Ordovician	Upper / Middle / Lower	Ashgillian, Caradocian, Arenigian	435
	Cambrian	Upper / Middle / Lower	Croixan, Albertan, Waucoban	Cambrian	Upper / Middle / Lower	Tuorian, Amgan, Aldanian	500
Precambrian	Proterozoic, Archeozoic						570

I

INTRODUCTION

Esso Chemical Company, Inc.

1

The Development of Petroleum Geochemistry and Geology

The science of petroleum geochemistry is the application of chemical principles to the study of the origin, migration, accumulation, and alteration of petroleum, and the use of this knowledge in exploration and recovery of oil, gas, and related bitumens. Although petroleum has been known since ancient times, only in this century has there developed the technological capability of obtaining the enormous quantities of this fossil fuel now required to meet the energy demands of the world's expanding economy. The world's requirement for petroleum is now about 60 million barrels (8 million metric tons) per day; it is estimated that 120 million barrels (16 million metric tons) per day will be required in the year 2000. To find and produce such a vast amount of oil has required, and will require, the ingenuity and hard work of many people—the petroleum exploration geologists, geophysicists, and geochemists and the drilling and production engineers.

There is no clear evidence of when geological or geochemical principles were first applied to the search for oil. Natural seeps of oil and gas have been known since the beginning of recorded history, and hand-dug wells were common on the sites of such seeps. Petroleum was a frequent by-product of drilling for salt water. Confucius mentioned wells a few hundred meters deep in 600 BC (Owen 1975, p. 2). Burning gas was used in some salt works to evaporate the brine. Chinese drilling tools reached depths of about 1,000 meters by the year 1132. By the end of the eighteenth century, the Yenangyaung oil field in Burma had over 500 wells and produced about 40,000 tons of oil annually. The spectacular flows of oil and gas at Baku resulted in early development of the petroleum industry in the USSR. In 1870, the annual production at Baku reached about 28,000 tons. Colonel Edwin L. Drake is credited with starting the American oil industry on its spectacular career by drilling near the Titusville, Pennsylvania, seep in 1859.

By 1871, 700,000 tons of oil, 91 percent of the world's production, was coming from the Pennsylvania area opened by the Drake well. Annual production at Baku steadily increased to nearly 4 million tons in 1890, almost equal to the production in Pennsylvania and New York that year (Owen 1975, p. 101).

These early exploration ventures were carried out by wildcatters with little or no geological knowledge. Eventually, geological principles were brought into play. The first and foremost was the "anticlinal theory." To state it simply, the theory was that because oil is lighter than water it will seek the highest part of an underground structural fold. Thus, an anticline is a more favorable place to drill for oil than a syncline.

The first statement of this concept was made by T. Sterry Hunt, geologist and chemist of the Geological Survey of Canada, in a public lecture in Montreal in March 1861. Subsequently, he published the details of his talk in the *Canadian Naturalist and Geologist* (Hunt 1861, p. 249):

> These wells occur along the line of a low, broad, anticlinal axis, which runs nearly east and west through the western peninsula of Canada and brings to the surface in Enniskillen the shales and limestones of the Hamilton Group, which are there covered with a few feet of clay. The oil doubtless rises from the Corniferous Limestone, which as we have seen contains petroleum; this, being lighter than the water which permeates at the same time the porous strata, rises to the higher portion of the formation, which is the crest of the anticlinal axis where the petroleum of a considerable area accumulates and slowly finds its way to the surface through vertical fissures in the overlying Hamilton shales, giving rise to the oil springs of the region.

On page 250 of the same report, he states, "A well yielding considerable quantities of petroleum is said to occur in the Township of Dereham, about a quarter of a mile S.W. of Tilsonburg, and we may reasonably expect to find others along the line of the anticlinal or of the folds which are subordinate to it."

While Hunt was describing the anticline at Enniskillen, Professor E. B. Andrews of Marietta College, Ohio, was writing about the anticlines of western Virginia and southeastern Ohio (Andrews 1861). His paper describes several anticlines with oil fields on the crest, typical of which was Cow Creek, Virginia (now West Virginia). At the anticlinal line marked A in Figure 1-1 are oil and gas springs. Andrews felt that the tensional forces at the crest of anticlines developed fissures that filled with oil during long geological ages. He writes (p. 92), "In the broken rocks, as found along the central line of a great uplift, we meet with the largest quantity of oil."

This basic principle of looking for petroleum high on a structure is still the first criterion of exploration in rank wildcat areas. However, the surface geological mapping has long since been supplemented by the geophysical mapping of subsurface structures.

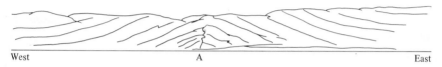

West   A   East

Figure 1-1
Anticlinal section on Cow Creek, Virginia (now West Virginia), drawn by Professor E. B.
Andrews of Marietta College in 1861. Oil and gas springs occur at the crest (A). [Andrews
1861]

As the oil-producing areas began to spread through the United States and
Canada in the early 1900s, it was soon realized that there is oil in many
synclines, as well as in anticlines. In fact, it could be found under a variety of
geological conditions not explained by the anticlinal theory. For a while, the
seemingly erratic nature of oil accumulations made it difficult for geologists to
convince drillers of the importance of using geological principles in locating
well sites. Some of the early, disillusioned drillers felt that the best way to drill
a "dry hole" was to employ a geologist. During these difficult times, the
nature of petroleum was clearly expressed by a Pennsylvania judge in an early
court decision when he said, "Oil is a fugacious mineral." What he meant was
that oil could move from its point of origin, thereby making it difficult legally
to define its geographic boundaries.

The discovery of oil beneath large anticlinal structures in Kansas, Oklahoma, and California about the time of World War I brought about a
resurgence of geological structural prospecting, with geologists firmly in
control of exploration decisions. The successful application of the reflection
seismograph to subsurface structural mapping in the 1920s further strengthened faith in the anticlinal theory. Then, in the middle 1930s, the great East
Texas pool was found. The discovery of this stratigraphically trapped oil and
gas reservoir made drillers and geologists alike realize that finding oil requires
a knowledge of all available principles of earth science. No longer could
exploration geologists simply look for folded structures. They had to
understand sedimentation, stratigraphy, paleontology, geochemistry, miner-
alogy, petrology, geomorphology, and historical geology. The day of the
wildcatter with a nose for oil was waning, and fields were being found by
intensive detailed studies using all scientific data available.

Even with all these new approaches, the search for oil continues to be a
high-risk venture. Today only 1 rank wildcat in 10 discovers a good show of
oil. About 1 in 50 discovers a field with reserves in excess of 1 million barrels
of oil or 6×10^9 cubic feet of gas. In view of these odds, it is no wonder that
the U.S. oil industry has been unable to meet the energy demands of its
expanding population. The exploration geologist must be a cautious optimist.
There is no room in this industry for conservatism. Periodic predictions that

the world is running out of oil must be offset by the injection of new ideas, new concepts, and new approaches to exploration. Structural mapping has given way to stratigraphic plays—reefs, salt domes, unconformities, updip pinchouts, lithofacies mapping, hinge lines, deltaic deposits, nonmarine sediments, and, ultimately, the application of all these concepts to offshore drilling.

H. D. Hedberg (1954, p. 1724) summarized the philosophy of oil finding thus:

> For geological exploration to have effective results, economically operative leases and concessions must be acquired in the regions of generally favorable geology. The definition of such regions should be geological, but should not be restricted by geological overassurance—an exaggerated feeling of omniscience on the part of the geologist. The evaluation of most regions will usually require drilling, and often this will have to be very extensive before a reasonably conclusive result is obtained. It should not be hampered by excessive geological conservatism; at the same time, it should never be haphazard and should always have the benefit of full geological advice.

As oil became more difficult to find, it was clear that the geologist required an understanding of the geochemistry of petroleum. What is the composition of petroleum? How does it originate, and how does it migrate in the subsurface? How does it change with depth, temperature, and pressure? How can we use such knowledge to help us find commercial accumulations?

Although the Canadian, T. Sterry Hunt, has been called the world's first real authority in petroleum geology (Owen 1975, p. 54), he was also a scientist of broad knowledge who applied chemical concepts to his search for oil. His first statement on the origin of petroleum supported the theory of Karl G. Bischof, professor of chemistry at Bonn, that petroleum originated by the slow decomposition of organic matter. Hunt elaborated on this theory by defining the lower forms of marine life as the probable sources of petroleum (Hunt 1863, p. 527):

> In the Paleozoic rocks of North America, the organic matter which has yielded the bitumen must be derived either from a marine vegetation or from the remains of marine animals. These, especially the lower forms, differ but little in elementary composition from plants and may as readily yield bitumen by their change. The transformation by which organic matters may be converted into bitumen does not differ very greatly from that which produces the more bituminous coals—to some of which indeed, certain of the asphaltums approach very closely in composition. The true petroleums retain a larger proportion of hydrogen and result from a change, under conditions as yet but imperfectly understood, by which a greater proportion of hydrogen is retained in combination.

While some geologists were speculating on the origin of petroleum, others were turning their attention to practical uses of geochemistry, such as the carbon ratio theory. This theory, which relates the metamorphism of coal to the occurrence of petroleum, gradually evolved in the late nineteenth century in a series of papers by H. D. Rogers. However, it remained for David White of the U.S. Geological Survey (USGS) to state the concept clearly in a classic paper delivered as his presidential address to the Washington Academy of Sciences (White 1915). He showed that the oil fields of the eastern United States died out where the coals had a fixed carbon content (nonvolatile carbon) of 60 percent. Gas pools disappeared beyond fixed carbon values of 65–70 percent. The theory showed for the first time that the accumulation of oil and gas in the earth is limited by metamorphism. It is still useful in a modified form today, as will be discussed later.

As drilling spread west and south across the United States, many wells favorably located structurally encountered no oil but good porosity. Naturally the question arose, "Is the absence of oil caused by a lack of source materials?" The concept of bituminous shales as source beds for the Pennsylvania oil accumulations was proposed as early as 1860 by state geologist John Newberry. Later, Newberry and others emphasized that in Ohio and Kentucky the oil should be sought where sands are in contact with the Ohio black shale.

The principle that petroleum originates in bituminous source beds and migrates into reservoir beds was well established by 1926. That year, the American Petroleum Institute cooperated with the U.S. Geological Survey in supporting a detailed study aimed at developing diagnostic criteria for recognizing petroleum source beds. Parker Trask (1942) undertook this formidable task and ultimately published his analyses of 35,000 cuttings and core samples from oil-producing areas of the United States. Trask's studies did not result in a technique for the positive identification of source rocks, but he did learn a great deal about the geochemistry of sediments that are associated with oil-producing horizons.

Meanwhile, the workers in the field were still searching for a magic black box that would pinpoint an oil accumulation. The first possibility of a direct, surface prospecting method appeared in a patent filed in Germany and the United States in 1929 entitled "Method and Apparatus for Detecting the Presence of Profitable Deposits in the Earth." G. Laubmeyer, the inventor, planned to assay soil gas for hydrocarbons as an indicator of petroleum accumulations in the subsurface. About the same time, a young Russian nuclear physicist, V. A. Sokolov, also was working on surface prospecting methods. In 1930, he and a co-worker, M. G. Gurevitch, devised an apparatus for measuring the radon and thoron liberated during the radioactive disintegration of the uranium series elements (Sokolov 1933). It occurred to Sokolov that if microgas seeps exist over oil reservoirs they could alter the adsorption of radioactive gases present at the surface and that his instrument would show this up. He made tests over the Grozny and Baku fields and did find some surface evidence that subsurface petroleum deposits could be recognized. Sokolov then combined the radioactive survey with direct sensitive methods of analysis for methane and heavier hydrocarbons. This led to an intense program of surface geochemical prospecting by the Russians over the next several decades.

The pioneering work of Sokolov and Laubmeyer stimulated geochemists in the United States, notably Rosaire (1938) and Horvitz, to organize a company to perform surface prospecting for oil companies that did not have a geochemical staff. Major oil companies, meanwhile, started geochemical groups in their research laboratories to investigate surface prospecting. These studies were not able to pinpoint petroleum accumulations except in unusual circumstances, but they did have some value as a regional tool. A detailed discussion of surface prospecting is in Chapter 9.

Historically, surface prospecting was the major impetus to the development of all geochemical prospecting techniques. The early geochemists soon recognized that subsurface as well as surface methods could be employed in exploration, and today there is more research on well cuttings, cores, and subsurface fluids than on surface prospecting.

Many successful operating geologists had a chemical understanding of the oil they were looking for. This naturally helped them make more intelligent decisions regarding oil leases. One of the foremost field geologists of these early years was Wallace E. Pratt, who became chief geologist and director of the Humble Oil and Refining Company. In 1941, Pratt gave a series of lectures at the University of Kansas that showed remarkable insight into principles of the origin, migration, and accumulation of petroleum. Many of the concepts he proposed in those lectures have been shown to be correct by subsequent research and exploration programs. His most important statement for that time was in regard to the ubiquity of hydrocarbons (Pratt 1943, pp. 9–10):

> Petroleum is an inevitable result of fundamental earth processes, processes so typical that they have been repeated in each successive cycle of earth history. I believe that oil in the earth is far more abundant and far more widely distributed than is generally realized. Oil is a normal constituent of unmetamorphosed marine rocks of nearshore origin. Rocks of this character comprise nearly 40 percent of the total land surface of the earth. Oil is a creature of the direct action of common earth forces on common earth materials.

Today we recognize that hydrocarbons are found in practically all sedimentary rocks, so that no unmetamorphosed section can be condemned until the quantity of hydrocarbons it contains has been evaluated. The real differences between oil-rich and oil-poor areas are more quantitative than qualitative.

Pratt also clearly described what is now considered the most logical theory of the origin of oil, namely, that small molecules are formed from large molecules through natural cracking at depth. In the Kansas lectures, he stated (1943, pp. 17, 18),

> The evolution of oil in the earth's crust through geologic time is marked by a decreasing molecule size and a concurrently increasing hydrogen-to-carbon ratio. Young oils are characteristically undersaturated; they are deficient in hydrogen and contain but little gas; they are asphaltic, made up of large or heavy molecules. Oil in geologically old rocks, on the other hand, is often paraffinic, with small, fully saturated molecules, and is typically associated with large volumes of free gas. These facts suggest a progressive natural cracking in nature's laboratory.

Pratt's application of geological and geochemical principles to field operations resulted in the growth of Humble Oil from a relatively small operator to the largest producer of oil in the United States.

Other countries soon recognized that finding oil was a complex problem that required the application of precise scientific knowledge. Although the USSR was late in developing its oil field technology, it had already acquired a vast amount of background data on the geochemistry of oil and related organic matter. Russia's father of geochemistry, V. I. Vernadskii, was the real inspiration for the development of the vast mineral resources of the USSR. In his book, *Outlines of Geochemistry* (1934), which has been translated into many languages, he reiterates the importance of organic matter in the origin of oil: "On the surface of the earth, there is not a chemical force that acts more constantly, and therefore more powerfully in its final consequences, than living organisms taken as a whole—the chemical phenomena of the biosphere have lasted throughout the course of all geologic history." Vernadskii (1934, pp. 152–153) was one of the first to state clearly that the hydrocarbon gases (largely methane) originate from all types of organic matter and can have sources independent of oil: "A part of the gases are genetically related to oil fields, namely, the gas phase of the oil. Other gases are related to the disseminated organic matter in the sedimentary rock. Its origin may be expressed schematically as follows: Marine life to marine muds to sedimentary rocks to gases." Vernadskii rejected the inorganic (carbide) hypothesis of oil formation: "The general features of oil genesis are clear. We should consider oils as sedimentary minerals genetically related to organic matter. Organisms are undoubtedly the source material of oils—oils cannot contain any significant amounts of juvenile (primordial) hydrocarbons" (Vernadskii 1934, pp. 152–153). Vernadskii also recognized that the total mass of dispersed oil in nonreservoir rocks such as shales far exceeded that in reservoir accumulations.

The USSR has many geochemists engaged in prospecting for oil. Although the Soviet oil industry started late, it has made remarkable progress in recent decades, and Soviet discoveries of new reserves are still on a rising trend compared to a falling trend for the 48 conterminous states of the United States.

Many other countries have increased both their geochemical research and exploration use of petroleum geochemistry in the last decade. Rapid developments in analytical techniques have enabled detailed studies to be made on the distribution of all types of hydrocarbons throughout a sedimentary basin. Source rock identification, crude oil correlation, oil-source rock correlation, basin evaluation from geochemical data, and the recognition of organic diagenesis and metamorphism have become accepted routine geochemical techniques of many major oil companies. Even smaller companies are looking closely at such techniques as mudlogs and cuttings gas analyses recorded down well bores to evaluate petroleum source bed characteristics, as well as to detect oil in reservoirs. A most encouraging trend has been the realization by petroleum geologists that having a good structural or stratigraphic trap with satisfactory reservoir characteristics is of no value if conditions were not right for hydrocarbons to form and accumulate. Knowing the origin, migration, and accumulation of petroleum in a basin is of importance for even a semiquantitative evaluation of the oil potential of a prospect.

The ensuing chapters provide a background in petroleum geochemistry for geologists and show how to use geochemical concepts and techniques to assist in locating new oil and gas deposits.

SUMMARY

1. Petroleum geochemistry involves the application of chemical principles to the study of the origin, migration, accumulation, and alteration of petroleum, and the use of this knowledge in exploration and recovery of oil, gas, and related bitumens.

2. The carbon ratio theory of the late nineteenth century was the first geochemical concept that related oil and gas accumulations to metamorphism. Oil fields changed to gas where the fixed carbon content of coals exceeded 60 percent, and gas fields disappeared at values beyond 65–70 percent.

3. Surface geochemical prospecting for gas migrating vertically over petroleum accumulations was a major impetus to the early development of geochemical groups within major oil companies.

4. The concept of oil originating from the organic matter of bituminous shales and migrating into sands first developed from field observations of operating geologists in the late nineteenth century.

SUPPLEMENTARY READING

Owen, E. W. 1975. Trek of the oil finders: A history of exploration for petroleum: Amer. Assoc. Petrol. Geol. Memoir 6. Tulsa, Okla.: American Association of Petroleum Geologists.

2

Carbon and the Origin of Life

Carbon (from *carbo*, meaning "charcoal") is in the fourth group of the periodic table of elements, which means it has four electrons in its outermost electron shell. Carbon is unusual in that it forms strong carbon-carbon bonds, which remain strong when the carbon groups combine with other elements. The most stable elements, or combinations of elements, are those that contain eight electrons (an octet) in the outer shell. Carbon assumes this configuration by forming covalent bonds; that is, by sharing electrons with itself and other elements. For example, carbon is readily reduced with hydrogen, or oxidized with oxygen, to form the two most common carbon compounds in the earth's crust, methane and carbon dioxide.

$$\cdot \overset{\cdot}{\underset{\cdot}{C}} \cdot$$

$$\begin{array}{c} H \\ H\!:\!\overset{\cdot\cdot}{\underset{\cdot\cdot}{C}}\!:\!H \\ H \end{array}$$
Methane

$$:\!\overset{\cdot\cdot}{\underset{\cdot\cdot}{O}}\!:\!:\!C\!:\!:\!\overset{\cdot\cdot}{\underset{\cdot\cdot}{O}}\!:$$
Carbon dioxide

The carbon atoms form an octet of electrons around them by sharing one electron from each of four hydrogen atoms or by sharing two electrons from each of two oxygen atoms.

The real uniqueness of carbon, which enables it to be the basic element of all life, lies in its ability to combine with itself to form long carbon chains, rings, and complex, bridged structures. Only one other element, silicon, also with a valence bond of 4, can do this. Silicon chains can be made in the laboratory, but they do not exist in nature for the following reasons: (1) The Si—Si bond energy of 53 kcal/mole is much weaker than the C—C bond energy of 83 kcal/mole; (2) the outermost electron shell of silicon is readily attacked by water, oxygen, or ammonia, so silicon chains are unstable in the presence of these compounds; and (3) silicon is unable to form double bonds with oxygen to yield an SiO_2 monomer in the same manner that carbon forms CO_2 gas. Silicon oxides exist only as high-molecular-weight solid polymers. These crystalline solids do not circulate through the hydrosphere and biosphere, as does CO_2.

Carbon has been the basic structure of all life as we know it since the beginning of life on earth. Consequently, the chemistry of carbon is often referred to as *organic* chemistry, whereas the chemistry of all other elements is called *inorganic* chemistry. The 100-plus elements other than carbon combine with each other to form about 70,000 inorganic compounds, whereas carbon combines with itself and the others to form about 4 million organic compounds. Carbon is in the food we eat, the air we breathe, the clothes we wear, the houses we build, the fuel to heat those houses and power our cars, trains, airplanes—carbon is the most ubiquitous element on earth; it is everywhere in the crust.

THE PRIMITIVE EARTH

Since oil is organic in origin, as will be discussed later, it is important to understand the origin and development of life on earth. Speculations about the prevalence of Precambrian oil really depend on the extent to which organic matter was formed and preserved in Precambrian times.

The earth is believed to be as old as the oldest known meteorites and terrestrial lead, about 4.7×10^9 years (Patterson 1956). At first, the earth was probably composed of about 90 percent iron, oxygen, silicon, and magnesium and 10 percent all other natural elements. One model proposes that the earth heated up during this first billion years because of the impact energy of falling planetesimals, compression of the earth due to gravity, and the disintegration of radioactive elements. The rise in temperature caused the iron to melt and sink to the center, while the lighter material floated to the surface. In effect, this converted the earth from a relatively homogeneous body to a heterogeneous layered body with a dense iron core, a mantle of original body, and a surface crust of light material, as shown in Figure 2-1. In commenting on this model, Press and Siever (1978, pp. 12–13) state, "Differentiation is perhaps the most significant event in the history of the earth. It led to the formation of a crust and eventually the continents. Differentiation probably initiated the escape of gases from the interior, which eventually led to the formation of the atmosphere and oceans." Although the details of this model are speculative, there is general agreement that during this first billion years the earth underwent cataclysmic changes that eliminated the original crust. No earth

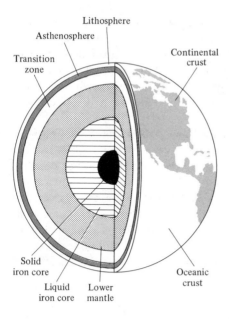

Lithosphere
Asthenosphere
Transition zone
Continental crust
Solid iron core
Liquid iron core
Lower mantle
Oceanic crust

Figure 2-1
The earth is density zoned, with the great-est density in the center iron core and decreasing densities toward the surface to a crust of light rock. [From *Earth,* 2nd ed., by F. Press and R. Siever. W. H. Freeman and Company. Copyright © 1978.]

rocks have been found older than 3.7×10^9 years, whereas moon rocks range in age from 3.1 to 4.5×10^9 years.

The evolution of the atmosphere is more controversial. One model (Holland 1962) proposes that the earliest atmosphere was reducing; that is, it contained hydrogen and some water vapor, but no oxygen. It also may have contained nitrogen, methane, and ammonia in lesser amounts. As the earth heated up and lighter materials came to the surface, enormous quantities of water vapor, carbon dioxide, nitrogen, and hydrogen were contributed to the atmosphere by volcanism. The hydrogen gradually diffused into outer space, and the water vapor condensed leaving nitrogen and carbon dioxide as the major atmospheric components. By about 3.7×10^9 years ago, there was enough carbon dioxide in the atmosphere to cause extensive chemical weathering because of the higher acidity of surface waters. This weathering caused dissolution of enough silica to form large chert and quartz deposits in the Precambrian. Other evidence, such as the results of carbon isotope analyses of Precambrian sediments, supports the model of a higher propor-tion of carbon dioxide in the Precambrian atmosphere than today's (Galimov et al. 1968).

In an atmosphere mainly of nitrogen and carbon dioxide, life was limited to unicellular organisms that could live under reducing conditions, such as the sulfate-reducing bacteria that are found today in anaerobic, stagnant waters. The development of life as we know it now did not occur until oxygen became an important atmospheric component.

The geochemical evidence supports the idea that a nonoxidizing atmosphere existed until about 2×10^9 years ago. Extensive banded iron formations of the Lake Superior type, which consist of alternate layers of iron and silica, are unique to the Precambrian, none being younger than about 1.7×10^9 years (Govett 1966). These formations, which occur on all continents, imply a nonoxidizing atmosphere under which large quantities of iron could be transported in the soluble ferrous state to sites of deposition. Examples are the Soudan Formation of North America (over 2.5×10^9 years old), the Dharwar Formation of India (about 2.5×10^9 years), the Krivoi Rog of Russia (about 2.1×10^9 years), and deposits of the Huronian period of North America (1.7 to 2.5×10^9 years old).

Other evidence for a nonoxidizing atmosphere is the presence of uraniferous conglomerates and detrital uraninite prior to about 2×10^9 years ago. Holland (1975) estimates the partial pressure of oxygen at that time to have been no more than about 0.004 atmospheres. The present-day partial pressure of oxygen is 0.2 atmospheres and of carbon dioxide is 0.0003 atmospheres.

Red beds—sediments with the grains coated with ferric oxide—first appear in abundance about 1.8 to 2×10^9 years ago. Their appearance marks the beginning of a significant oxygen increase in the atmosphere. Photosynthesis by primitive organisms produced large amounts of oxygen. In photosynthesis, plants that contain chlorophyll convert carbon dioxide and water into sugars (carbohydrates) needed for biological energy, releasing oxygen as follows:

$$CO_2 \;+\; H_2O \xrightarrow[\text{Sunlight}]{\text{Chlorophyll}} CH_2O \;+\; O_2$$

CO₂	H₂O		CH₂O	O₂
Carbon Dioxide	Water		Sugars	Oxygen

By this process, every atom of carbon that goes through photosynthesis produces one molecule of oxygen. Some of this oxygen has been utilized to form carbonates, oxides, and sulfates of the crust.

PRIMITIVE LIFE

Evidence for the earliest suggestion of life occurs in the Swaziland Supergroup near Barberton, South Africa. In rocks with an age ranging from 3.1 to over 3.3×10^9 years there are bacterium-like rods and algal-like spheroids (Schopf 1970). The latter are similar to modern blue-green algae in morphology. Their algal-like structure suggests that oxygen-producing green plant photosynthesis may have evolved by this time.

These early organisms are called *procaryote* because their genetic material is disarranged in the cell nucleus and they are asexual. The morphology of such organisms would change very little over geologic time because of their inability to mutate. Precambrian and Recent forms are thus similar. Layered mats of blue-green algae, or *stromatolites,* first appeared about 3.1×10^9 years ago. They are comparable in gross morphology to the layered bioherms of modern blue-green algal communities. By 2.3×10^9 years ago, the stromatolites were widespread. Diverse morphologically complex microorganisms also were relatively abundant (Schopf 1975).

A most significant step in evolutionary history was the origin of *eucaryote* organisms. These have chromosomes and nucleated cells comparable to those of all higher life. The early eucaryotes (1 to 1.5×10^9 years ago) were asexual, incapable of genetic variability. Around 0.8 to 1×10^9 years ago, the sexual eucaryotes appeared. According to Schopf et al. (1973), they were the "evolutionary trigger" that caused an explosive increase in both the diversity and the evolutionary development of life. Within a few hundred million

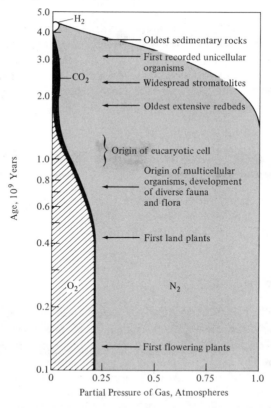

Figure 2-2
The evolution of atmosphere and life.

years, megascopic green, red, and brown algae were populating the world's oceans. Brown algae of Wendian age (680 to 570 million years ago) have been identified by Gnilovskaja (1971) in the USSR and green algae in the late Precambrian of Norway by Spjeldnaes (1963). Other examples are cited by Schopf et al. (1973).

These multicellular organisms allowed widespread development of fauna and flora in the Early Cambrian. Cambrian strata contains at least 1,200 different kinds of life, including brachiopods, gastropods, calcareous sponges, algae, worms, and complex trilobites up to 5 kilograms (11 lb) in weight.

Life was still limited to the lakes, rivers, and oceans. Not until the Late Silurian were the land surfaces invaded by plants. Later sporadic increases and decreases in organic productivity are indicated by variations in the organic carbon content of Phanerozoic sediments (Ronov 1958) and in phytoplankton abundance (Tappan 1974).

In summary, the geochemical and biological evidence suggests a history as shown in Figure 2-2: hydrogen initially dominant, but lost early; nitrogen, which is less reactive than the other gases, building up early in geologic time; carbon dioxide building up and then decreasing with a decreasing rate of juvenile and recycled carbon dioxide degassing; oxygen increasing as plants spread over the earth. Oxygen levels have probably stabilized since Late Mesozoic (Holland 1973). Diversity of organisms was limited until the first sexual eucaryotes appeared. Then, by 0.6×10^9 years ago, both fauna and flora developed diversely.

PETROLEUM POTENTIAL OF PRECAMBRIAN ROCKS

In 1965, Grover Murray pointed out that petroleum geologists were overlooking large, potentially favorable areas by failing to explore unmetamorphosed Precambrian strata. Murray's paper was stimulated by the discovery of indigenous hydrocarbons in the Precambrian of central Australia, but reports of indigenous hydrocarbons in the late Precambrian of the Russian and Siberian platform had preceded this. In 1962, the Russians discovered the giant Markovo field north of Irkutsk, a multistratal deposit of gas and condensate in the Precambrian Riphean and Wendian formations, and oil in the Lower Cambrian. The distribution of oil and gas in these sediments is indicated in Table 2-1 according to Vassoevich (1971). Deepest production is from a small gas accumulation in terrigenous sands and gravels lying on gneisses and schists. A large gas condensate production is from the Markovo horizon, where 19 to 29 meters of poorly sorted sands are overlain by 70 m of dolomitized argillites. Cracks in these brittle argillites appear to allow hydrocarbon communication with the major producing horizon, the Parfenovo, at the base of the Wendian (Kotel'nikov and Florenskaya 1974).

Table 2-1

Distribution of Oil and Gas in Cambrian-Proterozoic Rocks of Irkutsk, USSR

Age, 10^6 Years	Thickness	Lithology	Production	Percent C_1–C_4 HC in gas[b]			
				C_1	C_2	C_3	C_4
Lower Cambrian	270–1350 m	Interbedded dolomites, limestones,					
570		evaporites	Oil	72	13	8	
Wendian	380–600 m	Dolomites, limestones,	Gas conden- sate and				
680[a]		shales, sands	trace oil	74	8	3	
Upper Riphean	0–300 m	Argillites, sands, shales	Gas conden- sate (Markovo)	72	11	7	5
950			Gas	81	11	4	2

[a]Dated by glauconites.
[b]C_1 = methane, C_2 = ethane, C_3 = propane, C_4 = butanes.
Source: From Vassoevich et al. 1971.

The Parfenovo sands are 16 to 37 m thick with a permeability of 0.5 to 30 millidarcys. The cap rock for this gas condensate is a thick dolomitic clay containing smectite, which in turn is overlain by 250 m of fine-grained dolomite. A small oil accumulation is in fissured, cavernous algal limestones of the basal Cambrian.

The fact that the produced gas is wet instead of dry indicates that the source and reservoir beds have not been subjected to high temperatures as in the deep, organic, metamorphosed dry gas facies of the Western Canada Basin.* Vassoevich et al. (1971) give many other examples of Precambrian hydrocarbons and show a map containing nine potential Precambrian oil and gas basins in Eurasia, Africa, and Australia.

Clearly, unmetamorphosed Precambrian sediments should not be ignored. Nevertheless, our knowledge of the origin and development of life and the distribution of organic matter in the Precambrian eras does help in evaluating those Precambrian deposits that might contain petroleum. Life originated at least 3×10^9 years ago (Figure 2-2). If we assume that the bulk of all petroleum is organic in origin, as will be shown later, then there can be little or no petroleum from rocks older than this. In the period from 3×10^9 to 1×10^9 years ago, life was limited to a few, very primitive forms. Sometime between 0.6 and 1×10^9 years ago, the world population exploded to literally thousands of species of plants and animals. This period most likely marks the beginning of oil-forming sediments on the scale observed in the Phanerozoic.

*Wet gas contains more than 0.3 gallon condensible hydrocarbons, such as gasoline, per 1,000 ft³; dry gas contains less than 0.1 gallon. Gases high in the ethane-plus fraction are generally wet.

The increase of oxygen in the atmosphere and the origin of multicellular organisms in the period around 0.7 to 1.0×10^9 years ago is coincident with the marked increase in life on earth. Ronov et al. (1973) show a large increase in the average organic carbon content of slates and clays of the USSR and North America from the Precambrian to the Phanerozoic. At least one sediment of this age, the Michigan Nonesuch Shale, has a hydrocarbon content similar to that of good Phanerozoic source rocks.

Data on the organic carbon and hydrocarbon content of Precambrian and Phanerozoic cherts and shales are shown in Table 2-2. The rocks dating from 1.9 to 3.3×10^9 years ago all have low hydrocarbon contents, and all but the Soudan Shale have low carbon contents. The hydrocarbon contents are so low as to preclude the possibility of these rocks being sources of oil or gas. If they once contained more hydrocarbons, they probably were reduced to this low level by metamorphism. Smith et al. (1970) have presented evidence that even these hydrocarbon yields may be contaminants that have entered the rocks since their origin.

In contrast, it has been known for some time that both liquid and solid hydrocarbons of unquestionably indigenous origin are associated with the black Nonesuch Shale in the Keweenawan series of the Lake Superior region (Butler and Burbank 1929). These shales are overlain and underlain by red beds that show no evidence of hydrocarbons. Individual analyses of the Nonesuch vary somewhat, but all show a complete range of hydrocarbons typical of crude oil, as shown in Table 2-3. The compounds shown in this table are discussed in more detail in Chapter 3.

Table 2-2
Organic Carbon and Hydrocarbons in Precambrian and Phanerozoic Sediments

Sample	Age in 10^9 years	Weight percent C_{org}	Alkane hydrocarbons $\mu g/g$
Precambrian			
Theespruit chert	>3.3	0.2–1.0%	
Onverwacht chert	3.3	0.05–1.0	0.05
Figtree chert	3.1	0.02–0.14	0.015
Soudan shale	2.5	3.0	0.02–0.5
Gunflint chert	1.9	0.07–0.37	0.03
Nonesuch shale	1.0	0.8–1.2	620
Pertatataka shale	0.8	0.9	∿500
USSR shales (33)[a]	∿0.8	0.3	—
Phanerozoic			
USSR Cenozoic shales	∿0.04	0.94	∿180
North American shales (800)[a]	∿0.1	1.2	300

[a]Number of samples.
Source: Data from Dungworth and Schwartz 1974; Kvenvolden 1972; Smith et al. 1970; Oro and Nooner 1967; Siller et al. 1963; Hunt 1963b; and Ronov 1958.

Table 2-3
Hydrocarbons Extracted from Nonesuch Shale

C_1 to C_8 hydrocarbons	$\mu g/g$	C_{15+} hydrocarbons and non-HC	$\mu g/g$
Methane + ethane	0.232	Normal paraffins +	
Propane	0.365	isoparaffins	479
Isobutane	0.058	1-Ring naphthenes	100
Butane	0.356	2-Ring naphthenes	54
Isopentane	0.040	3-Ring naphthenes	41
n-Pentane	0.134	4-Ring naphthenes	45
Cyclopentane	0.004	Alkylbenzenes	11
2-Methylpentane	0.018	Indanes	5
3-Methylpentane	0.006	Indenes	8
n-Hexane	0.095	Naphthalenes	7
Methylcyclopentane	0.002	Acenaphthylenes	23
Cyclohexane	0.021	Acenaphthenes	14
2-Methylhexane	0.009	Phenanthrenes	16
3-Methylhexane	0.014	Pyrene	0
Dimethylcyclopentanes +		Chrysene	6
2,2,4-Trimethylpentane	0.064	Benzothiophenes	7
n-Heptane	0.092	Dibenzothiophenes	17
Iso-octanes	0.082	Thiophenophenanthrene	7
n-Octane	0.194		
Benzene	0.024	Asphalt	100
Toluene	0.125	Kerogen	8,560
Total	1.935	Total	9,500

Note: Using the technique of Dunton and Hunt (1962) for C_1–C_8 hydrocarbons (detection limit is 0.001 $\mu g/g$); C_{15+} range analyzed on mass spectrometer. Normal and isoparaffins about equal above C_{15+}; $\mu g/g$ = micrograms of hydrocarbon per gram of dried shale.

The Nonesuch Shale differs from Phanerozoic source rocks, however, in having only 1 $\mu g/g$ C_4–C_8 (gasoline range) hydrocarbons, compared to more typical values from 5 to several hundred $\mu g/g$ for known source shales. Table 2-2 also shows that Precambrian shales from the Russian platform have a third of the organic carbon of the average USSR Cenozoic sediments.

In summary, the sparse data available indicate that the probability of finding favorable source–reservoir relationships in rocks older than about 1×10^9 years ago is remote. However, unmetamorphosed rocks younger than this may have good source potential. There is no biogeochemical reason for discounting them, and they certainly appear to be sources for large oil and gas accumulations in the USSR. The problem of source potential in late Precambrian–Cambrian rocks is as much related to metamorphism and tectonic activity as to original organic content. Rocks this old are frequently "burned out" in terms of their hydrocarbon generation, as will be discussed later, and many petroleum accumulations that must have been formed early have been destroyed through erosion and leakage of reservoirs over geologic time. Petroleum will continue to be found in the Precambrian both in unmetamorphosed, relatively undisturbed sequences and in sediments where there is fluid communication with younger source beds. The fact that much of these sediments are untested suggests that large accumulations remain to be found. However, the geologist should recognize that the dry hole odds are somewhat greater than average, for the reasons already mentioned.

INVENTORY OF CARBON IN SEDIMENTARY ROCKS

Carbon is cycled through the biosphere by photosynthesis and oxidation.

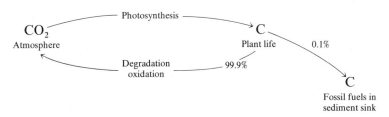

Plants (phytoplankton) utilize CO_2 to form the carbon of their cells, and animals (zooplankton) eat the plants and give off the excess carbon as CO_2. Dead organisms are microbially or chemically oxidized to CO_2. Out of this cycle, about 0.1 percent of the carbon is buried with sediments (Ryther 1970). Since the beginning of life, about 1 part in 11,000 of this 0.1 percent carbon has become a commercial petroleum accumulation that has survived to the present.

Carbon exists in sedimentary rocks in two forms: as reduced carbon, principally organic matter formed by biological processes over geologic time, and as oxidized carbon, principally carbonate. Table 2-4 shows the distribution of these forms of carbon in the earth's sedimentary rocks (Hunt 1972, 1977a). The quantity in sedimentary rocks was determined by using the mass of rocks in the earth's crust as calculated by Ronov and Yaroshevsky (1969) and by using organic carbon data from the general literature. Asphalt data from the literature have also been used. The asphalt in nonreservoir rocks is defined as the largely nonhydrocarbon fraction soluble in lipid solvents such as benzene. The figures in Table 2-4 must be considered approximate, since there are inherent errors in all such calculations. For example, Strakhov (1974) has indicated that Ronov's figure for volume of Phanerozoic sediments for the Russian platform is low. Some figures, such as the mass of coal, are reasonably well documented, but others, such as the carbonate content of sands worldwide, are at best approximations.

Table 2-4

Carbon in 10^{18} Grams in Sedimentary Rocks

	Organic carbon (reduced)	Carbonate carbon (oxidized)
All Sediments	Insoluble Organic Matter	
Clays and shales	8,900	9,300
Carbonates	1,800	51,100
Sands	1,300	3,900
Coal beds thicker than 4.6 meters (15 feet)	15	
Nonreservoir rocks	Soluble Organic Matter	
Asphalt	275	
Petroleum	265	
Reservoir rocks		
Asphalt	0.5	
Petroleum	1.1	
Total	$\sim 13,000$	$\sim 64,000$

The data do indicate that about 17 percent of the carbon in existing sediments is reduced and the rest oxidized. To determine this ratio over all geologic time, the carbon of metamorphic rocks and of sedimentary rocks being recycled as igneous rocks would have to be added.

Of particular interest to the petroleum geologist is the ratio of organic carbon in sedimentary rocks to that in the petroleum of reservoir and nonreservoir rocks. The data clearly show that the whole process of origin, migration, and accumulation of oil is extremely inefficient. If we assume that most oil forms from organic matter, then only 2 percent of the carbon in sedimentary rocks eventually becomes the carbon of petroleum, and less than 0.5 percent of the latter finds its way to a reservoir accumulation that survives through geologic time. This is an efficiency of 0.01 percent from starting material to product. Although sedimentary basins vary, it is clear that nonreservoir rocks in general give up only a trivial amount of their oil potential to fill associated reservoirs. The ratio shown here between source and reservoir oil is 240/1, but in oil-producing basins it generally varies from 10/1 to 100/1. Vassoevich et al. (1967) also report that the dispersed micro-oil far exceeds the petroleum in reservoirs in all Russian areas studied.

A large amount of hydrogen is available for forming petroleum from the organic matter. Most petroleum geochemists believe that oil originates from the gradual cracking of organic matter as it is buried deeper in the earth (Tissot et al. 1971). The formation of petroleum involves an exchange of hydrogen between the very large, unstable organic molecules and the small, newly formed hydrocarbon molecules. McIver (1967) found the organic matter in sediments of oil-producing basins to contain about 6 percent hydrogen. If the $10,700 \times 10^{18}$ grams of organic carbon in clays and carbonates (Table 2-4) is multiplied by the weight ratio of hydrogen to carbon in sediments (6/82), there is, stoichiometrically, 800×10^{18} grams of hydrogen available to form oil. This is about 6,000 times more than the hydrogen in all the petroleum accumulations of all reservoir rocks of the world. It is 13 times as much as the hydrogen in the petroleum and asphalt of both reservoir and nonreservoir rocks. As sediments are buried deeper and the organic matter is gradually metamorphosed to graphite, part of this large quantity of hydrogen is released and utilized to form methane and the related gases. These are the principal hydrocarbons in reservoirs deeper than 20,000 feet (6,096 meters).

THE ISOTOPES OF CARBON

Practically all the mass of an atom is in the dense nucleus as protons with a positive electrical charge of $+1$ and as neutrons that are electrically neutral. The protons are balanced by an equal number of electrons, with charges of -1, orbiting around the nucleus. The number of protons is unique for each element and is called the *atomic number*. The sum of the masses of protons and neutrons is the *atomic weight*. Protons and neutrons have the same mass.

Atoms whose nuclei contain the same number of protons but a different number of neutrons are called *isotopes*. All carbon atoms have six protons, but there are three carbon isotopes containing 6, 7, and 8 neutrons, giving atomic masses respectively of 12, 13, and 14. The distribution of these three isotopes in the biosphere is shown in Table 2-5. Carbon-12 and -13 are the original forms of carbon in the earth. Carbon-14 is formed from the bombardment of atmospheric nitrogen with neutrons produced by cosmic radiation and enters the biosphere as CO_2.

The isotopes ^{12}C and ^{13}C are stable, but ^{14}C decays to ^{14}N. One neutron (n) in a ^{14}C atom spontaneously decays, giving off an electron (β particle) and

Table 2-5
Distribution of ^{12}C, ^{13}C, and ^{14}C in the Biosphere

Symbol	Protons	Neutrons	Atomic mass	Weight percent of carbon in biosphere
^{12}C	6	6	12	98.89%
^{13}C	6	7	13	1.11
^{14}C	6	8	14	1×10^{-11}

leaving a new proton (p). This produces ^{14}N, since nitrogen atoms have seven protons in their nuclei. The reaction is ^{14}C (6 p + 8 n) $- \beta \rightarrow$ ^{14}N (7 p + 7 n). A mass of ^{14}C atoms disintegrates at a fixed rate such that half of the mass is changed from carbon to nitrogen in 5,570 years (the half-life period of ^{14}C). Consequently, the age of a carbon-containing substance can be determined by measuring its output of β particles. The technique is only good for about five half-lives (30,000 years) because β-particle emission becomes too low in older materials to distinguish it from "background noise." The isotope ^{14}C can be used to distinguish marsh gas methane from methane seeping from a petroleum accumulation, since the latter is too old to contain any ^{14}C. It also has been used to date hydrocarbons found in very recent sediments.

The isotope ^{13}C is distributed through sediments of all geological ages, in contrast to ^{14}C, which is limited to very young sediments. While it cannot be used for dating, ^{13}C can solve many geochemical problems, because its difference in mass relative to ^{12}C results in fractionation by both biological and physical processes. The ratio of ^{13}C to ^{12}C is determined on an isotope-ratio-mass spectrometer, using the following equation to calculate the ratio difference (δ), in parts per thousand (parts per mil, or ‰), relative to a standard.

$$\delta^{13}C = \left| \frac{(^{13}C/^{12}C) \text{ sample}}{(^{13}C/^{12}C) \text{ standard}} - 1 \right| \times 1000$$

The standard that has been used most widely in the literature over the years is a belemnite from the Peedee Formation in South Carolina. Table 2-6 shows a comparison of the actual ^{13}C content as a percent of total carbon for three materials. Modern isotope-mass spectrometry can determine the ^{13}C content with a precision of 1 part in 10,000. As shown in Table 2-6, it is unwieldy to compare these numbers in the fourth decimal place, but it is easy to understand them in terms of the reference standard.

The limestone contains 5 ‰ more ^{13}C than PDB (Peedee belemnite), whereas the lipids contain 30 ‰ less ^{13}C than PDB. As a general rule, reduced carbon—that is, the carbon of methane, petroleum, coal, and organic matter—is light, with less ^{13}C than PDB. Oxidized carbon, as in carbonate rocks, is heavy, containing as much or more ^{13}C than PDB.

Table 2-6
Variation in Carbon-13 in Natural Material

	Percent of carbon-13	$\delta^{13}C$ ‰ Relative to PDB
Peedee belemnite (PDB)	1.1112	0
A typical limestone	1.1162	+5
Plankton lipids	1.0862	−25

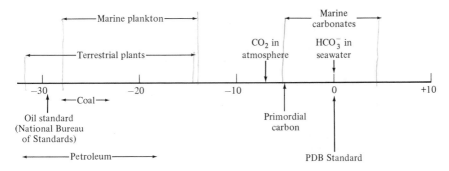

Figure 2-3
The range in the carbon-13 content of carbon reservoirs (in parts per thousand relative to the Peedee belemnite standard).

Much ^{13}C data of petroleum is in the Russian literature. The Russian papers frequently show $\delta^{13}C$ in parts per hundred (percent, %) relative to PDB (Sidorenko et al. 1972). Their numbers should be multiplied by 10 to compare with U.S. data using the PDB standard.

The common $\delta^{13}C$ ranges of some carbon reservoirs on earth are shown in Figure 2-3 relative to PDB. The bicarbonate of seawater is about the same as PDB. The CO_2 in the atmosphere has 7 ‰ less ^{13}C than PDB. The U.S. National Bureau of Standards has an oil standard with a $\delta^{13}C$ of -29.4 ‰ on the PDB scale. Fossil organic matter falls in the range covered by marine plankton and terrestrial plants. Methane formed microbiologically ranges in $\delta^{13}C$ from about -55 ‰ to -85 ‰, while methane formed by thermal cracking at depth ranges from about -25 ‰ to -60 ‰ (Dubrova and Nesmelova 1968; Frank et al. 1974).

The value for primordial carbon was calculated by using the data on the mass of carbon in the crust of the earth from Hunt (1972) and the $\delta^{13}C$ data from Hoefs (1969). Since the quantity of ^{13}C relative to ^{12}C in the earth's crust has probably not changed over geologic time, it is possible to estimate primordial $\delta^{13}C$ values from the mean of various carbon reservoirs assuming that there is no exchange of carbon between these reservoirs and the mantle. This is shown in Table 2-7. The mean $\delta^{13}C$ from this estimate is -5 ‰. This value is close to the mean for diamonds, carbonatites (Degens 1969), volcanic CO_2, and gas inclusions in igneous rocks (Galimov 1968).

Biological activity causes the largest fractionation of carbon isotopes. It is the principal cause of the variations in $\delta^{13}C$ in Table 2-7. Carbonates forming from the HCO_3^- of seawater have $\delta^{13}C$ values near 0. Marine organisms utilize dissolved CO_2 rather than HCO_3^- to build cellular material. Deuser et al. (1968) showed that dissolved CO_2 varies from -7 ‰ in warm water to -9 ‰ in cold water. Park and Epstein (1960) showed that photosynthesis depletes ^{13}C in going from the carbon source to the cell from 6 to 19 ‰ so that

Table 2-7
Balance of Carbon-13 in Earth's Crust

	Total carbon mass in 10^{21} grams	$\delta^{13}C_{PDB}$
Igneous rocks	7.2	−6
Metamorphic rocks		
C_{org}	3.5	−27
C_{CO_3}	2.6	−2
Sedimentary rocks		
C_{org}	12.0	−27
C_{CO_3}	64.3	0
Petroleum	0.2	−25
Mean (primordial carbon)		$\sim$ −5

the range of $\delta^{13}C$ for marine plankton is from $-13\,\%_{00}$ to $-28\,\%_{00}$. This is the sum of the 7 to 9 $\%_{00}$ fractionation between HCO_3^- and dissolved CO_2, plus the photosynthesis fractionation between CO_2 and the cell of 6 to 19 $\%_{00}$. Craig (1953) reported $\delta^{13}C$ of $-13\,\%_{00}$ for plankton from warm Gulf Stream waters, and Sacket et al. (1965) reported $-28\,\%_{00}$ for plankton from cold waters of the southern South Atlantic. Since no known inorganic process will produce such a large fractionation under geological conditions, a deficiency in ^{13}C has been used to indicate that a particular sample has passed through the photosynthesis process. For example, Rankama (1954) concluded that disseminated carbon in Precambrian schists and slates (aged 2.5×10^9 years) was of biogenic origin, since it had a $\delta^{13}C$ of $-28\,\%_{00}$ relative to PDB.

Organic carbon that coexists with carbonate materials in Phanerozoic sediments averages about 25 $\%_{00}$ less ^{13}C than the carbonate carbon, because of the photosynthesis process. A study of 53 coexisting organic carbon–carbonate pairs in Precambrian sediments showed a difference of about 26 $\%_{00}$ (Eichmann and Schidlowski 1975). This indicates that fractionation of carbon isotopes by photosynthesis has not changed since the early Precambrian around 3.1×10^9 years ago.

The use of $^{13}C/^{12}C$ ratios in solving geochemical and geological problems will be discussed in more detail in subsequent chapters.

SUMMARY

1. Life originated on earth between 3.1 and 3.3×10^9 years ago, but the most explosive increase in diversity and evolutionary development occurred between 0.8 to 1×10^9 years ago.

2. Unmetamorphosed Precambrian rocks younger than 1×10^9 years in relatively undisturbed regions are prospective for petroleum accumulations although the dry hole odds are greater than for Late Phanerozoic sediments.

3. The ratio of carbonate carbon to organic carbon in sedimentary rocks is about 5 to 1. The ratio of petroleum carbon in fine-grained nonreservoir rocks to petroleum carbon in commercial reservoirs is about 240 to 1. The ratio of organic carbon in sedimentary rocks to petroleum carbon is about 11,000 to 1.

4. Primordial carbon has undergone isotope fractionation on earth, with the heavier isotope ^{13}C concentrating in carbonate carbon and the lighter isotope ^{12}C in organic carbon. Organic carbon contains about 27 parts per thousand less ^{13}C than does carbonate carbon. Petroleum $\delta^{13}C$ values are in the range of organic carbon.

5. Petroleum originated in the last 1×10^9 years from a very small fraction of the biological life that has inhabited the earth.

SUPPLEMENTARY READING

Cloud, P. 1970. *Adventures in earth history.* San Francisco: W. H. Freeman and Company. 992 p.

Faure, G. 1977. *Principles of isotope geology.* New York: Wiley. 464 p.

Hoefs, J. 1973. *Stable isotope geochemistry.* New York: Springer-Verlag. 140 p.

Schopf, J. W. 1975. Precambrian Paleobiology: Problems and Perspectives. *Annual Review of Earth and Planetary Sciences, 3,* 213–249.

3

Petroleum and Its Products

Petroleum is a form of *bitumen* composed principally of hydrocarbons and existing in the gaseous or liquid state in its natural reservoir.* The word *petroleum* originates from the Latin *petra* ("rock") and *oleum* ("oil"). In common usage, it has come to mean any hydrocarbon mixture that can be produced through a drill pipe. Thus some of the Duchesne oils produced in the Uinta Basin, Utah, come to the surface as liquids at their reservoir temperature of about 93°C (200°F), but soon cool to solids. The principal forms of petroleum are *natural gas,* which does not condense at standard temperature and pressure (STP = 760 mm Hg, 60°F or 15.6°C), *condensate,* which is gaseous in the ground but condenses at the surface; and *crude oil,* the liquid part of petroleum.

ELEMENTAL COMPOSITION

Petroleum is composed almost entirely of the elements hydrogen and carbon, in the ratio of about 1.85 hydrogen atoms to one carbon atom. The minor elements sulfur, nitrogen, and oxygen constitute less than 3 percent of most petroleum. Traces of phosphorus and heavy metals such as vanadium and nickel are also present. Table 3-1 compares the elemental composition of oil with that of natural asphalt and the dispersed organic matter (*kerogen*) in sedimentary rocks. In going from the kerogen to the asphalt to the oil, there is a marked increase in hydrogen and a corresponding decrease in sulfur,

---

*Definitions of the various bitumens and their fractions are given in the Glossary at the end of the book.

Table 3-1
Elemental Composition of Natural
Materials (percent)

	Oil	Asphalt	Kerogen
Carbon	84.5	84	79
Hydrogen	13	10	6
Sulfur	1.5	3	5
Nitrogen	0.5	1	2
Oxygen	0.5	2	8
	100	100	100

nitrogen, and oxygen relative to carbon. The process of petroleum formation, as will be discussed later, involves this generation and accumulation of the lighter molecules rich in hydrogen from kerogen.

Since hydrogen is a much lighter element than the others in Table 3-1, oils with a higher hydrogen content have lower specific gravities. Thus a Pennsylvania crude with a hydrogen content of 14.2 percent has a specific gravity of 0.862 (33° API) compared to a Coalinga, California, crude with 11.7 percent hydrogen and a specific gravity of 0.951 (17° API). The elemental analysis in Table 3-1 is about the average for oils worldwide. Some oils have much higher contents of nitrogen, sulfur, and oxygen (NSO) than are shown here. Erdman (1962) analyzed a group of oils of varying origin and age with higher than normal nonhydrocarbon content. The data are shown in Table 3-2. The highest values here are 7.9 percent sulfur in a Wafra crude, 1.6 percent nitrogen in a West Texas crude, and 1.8 percent oxygen in a Venezuela crude. Some of these are light crudes, yet they all have a relatively high NSO concentration.

The elements carbon and hydrogen are combined as hydrocarbons that vary both in the size and type of the molecule. Differences in the physical and chemical properties of petroleum are due to the variations in the distribution of the different sizes and types of hydrocarbons.

MOLECULAR SIZE VARIATION

The smallest molecule in petroleum is methane with a molecular weight of 16. The largest molecules are the asphaltenes with molecular weights in the thousands. Between these two extremes, there are hundreds of compounds from simple to very complex structures. Hydrocarbons form homologous series; that is, families of molecules whose members have similar properties and differ in size by a CH_2 group. The formula for the paraffin series is C_nH_{2n+2}, where n is any number from 1 to about 60. As the molecular size

Table 3-2

Analysis of Various Crude Oils for Nitrogen, Sulfur, and Oxygen

Crude oil	Source	Age	Gravity, °API	Percent in crude of		
				Nitrogen	Oxygen	Sulfur
Ellenburger	W. Texas	Ordovician	40.3	0.018	0.082	0.19
N. Ward Estes	W. Texas	Permian	35.3	0.16	0.76	1.23
Baxterville	Mississippi	Upper Cretaceous	16.0	0.18	0.28	3.05
Vinton	Louisiana	Miocene	18.8	0.20	0.44	0.09
Merey	Venezuela	Lower Oligocene	17.6	0.39	1.82	1.64
La Paz	Venezuela	Paleocene	24.0	0.44	0.97	2.56
Frio	S. Texas	Oligocene	25.0	0.49	0.66	0.26
Wafra No. 17	Neutral Zone	Middle Cretaceous	22.8	0.55	1.24	7.90
Wilmington	California	Miocene	19.4	0.65	0.51	1.51
N. Belridge	California	Pliocene	13.8	0.77	1.45	1.14
Raudhatain	Kuwait	Middle Cretaceous	31.2	0.82	1.81	7.70
Wafra No. A1	Neutral Zone	Middle Cretaceous	22.7	1.03	1.51	7.80
McElroy	W. Texas	Permian	31.5	1.61	0.89	2.60

Source: From Erdman 1962.

increases, the individual members go from gases to liquids to solids. In the paraffin series, *n* equals 1 to 4 for gases, 5 to 16 for liquids, and above 16 for solids for the straight-chain paraffins.

Petroleum is separated into its various molecular sizes by distillation. A typical refining tower will yield products from the smallest to the largest size, of molecule, as follows: gas, gasoline, kerosine, light gas oil (diesel fuel), heavy gas oil, lubricating oil, and residuum. These molecular size groups are discussed in more detail in the section on the uses of petroleum.

MOLECULAR TYPE VARIATION

Hydrocarbon molecules occur in different structural forms with the following names: *Alkanes* are open-chain molecules with single bonds between carbon atoms, *cycloalkanes* are alkane rings, *alkenes* contain one or more double bonds between carbon atoms, and *arenes* are hydrocarbons with one or more benzene rings. Most petroleum geologists and engineers are more familiar with the terms *paraffins* for alkanes, and *naphthenes* or *cycloparaffins* for cycloalkanes, *olefins* for alkenes, and *aromatics* for arenes. Consequently, these terms will be used in this text.

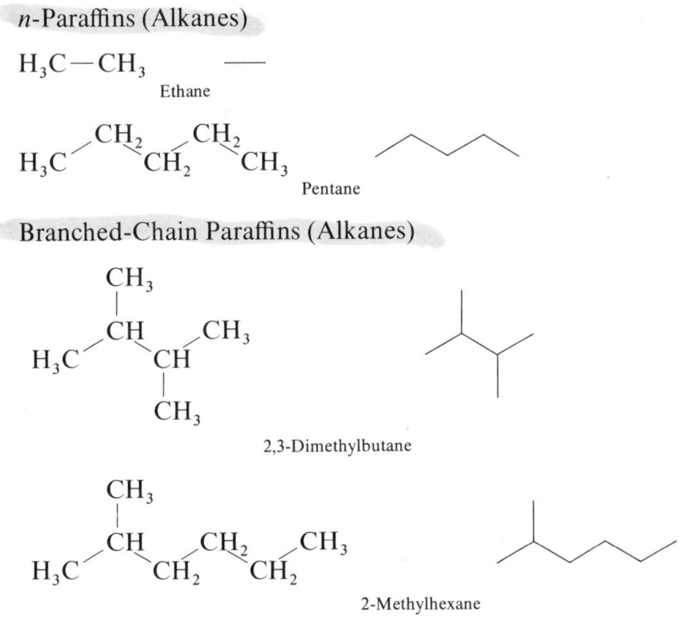

n-Paraffins (Alkanes)

$$H_3C — CH_3 \qquad \text{———}$$

Ethane

Pentane

Branched-Chain Paraffins (Alkanes)

2,3-Dimethylbutane

2-Methylhexane

Figure 3-1
Hydrocarbon formulas for normal and branched-chain alkanes. Conventional formula is on the left; skeletal formula is on the right.

In discussing molecular structures, both conventional and shorthand skeletal formulas will be used, as shown in Figures 3-1 and 3-2. Figure 3-1 shows these structures for normal and branched-chain paraffins. Figure 3-2 shows an olefin, a naphthene (cycloparaffin), and aromatic hydrocarbons. In the skeletal formulas shown on the right, a carbon atom with enough hydrogen atoms to give a total of four bonds is implied at each corner or end of the structures.

Paraffins (C_nH_{2n+2})

The paraffin-type hydrocarbons are the second most common constituents of crude oil next to naphthenes. Paraffins dominate the gasoline fraction of crude oil, and they are the principal hydrocarbons in the oldest, most deeply buried reservoirs. The terms *saturated* and *aliphatic* hydrocarbons are also used for this group. Straight-chain paraffins shown in the first two examples of Figure 3-1 are called *normal paraffins,* or *n*-paraffins. The normal paraffins form an homologous series. In organic chemistry, an homologous series is a series of compounds in which each member differs from the next member by a constant amount. The *n*-paraffins on the left side of Figure 3-3 form an

Olefin (Alkene)

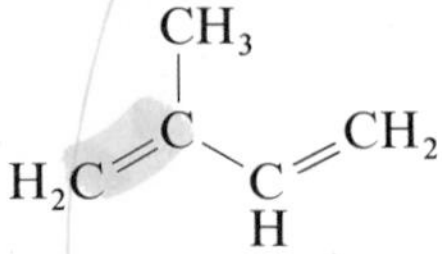

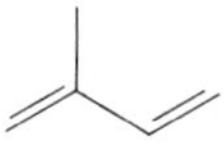

Isoprene

Naphthene (Cycloalkane)

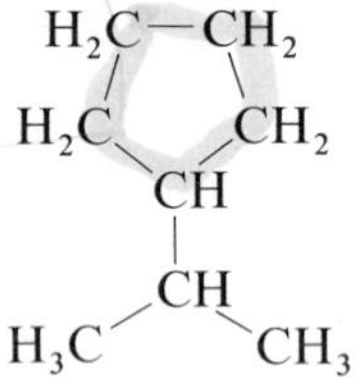

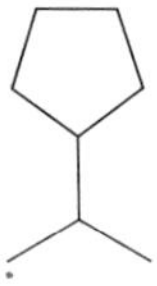

Isopropylcyclopentane

Figure 3-2
Formulas for an olefin, a naphthene, and three aromatic hydrocarbons.

Aromatics (Arenes)

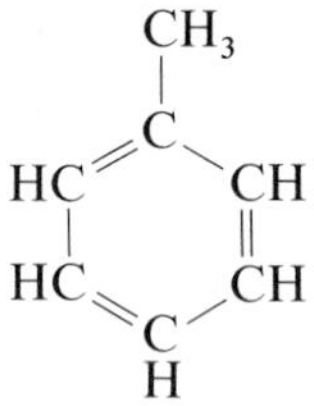

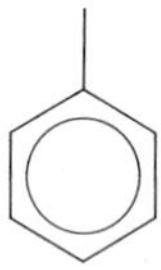

Toluene

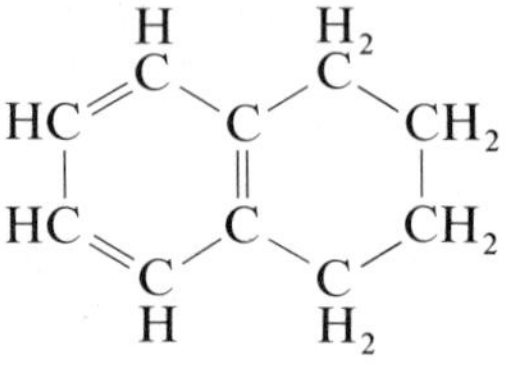

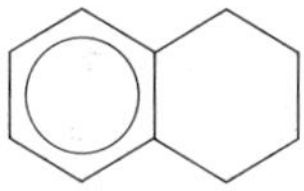

Tetralin

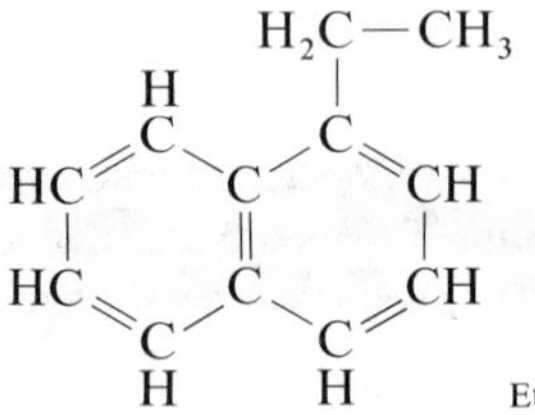

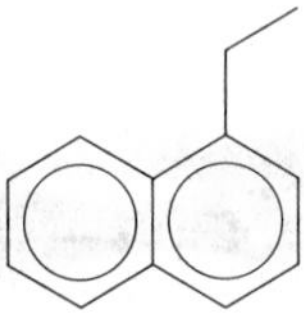

Ethylnaphthalene

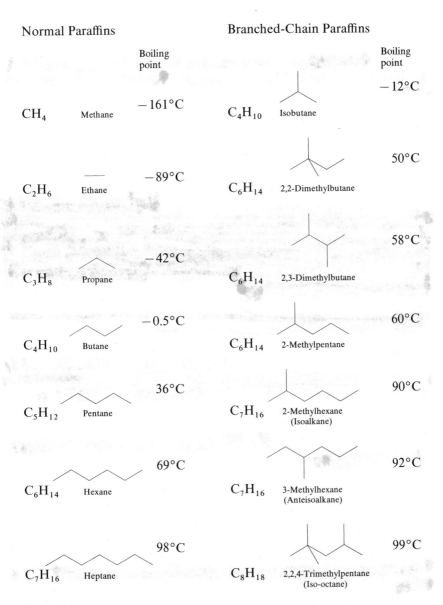

Normal Paraffins

Boiling point

CH$_4$ Methane − 161°C

C$_2$H$_6$ Ethane − 89°C

C$_3$H$_8$ Propane − 42°C

C$_4$H$_{10}$ Butane − 0.5°C

C$_5$H$_{12}$ Pentane 36°C

C$_6$H$_{14}$ Hexane 69°C

C$_7$H$_{16}$ Heptane 98°C

Branched-Chain Paraffins

Boiling point

C$_4$H$_{10}$ Isobutane − 12°C

C$_6$H$_{14}$ 2,2-Dimethylbutane 50°C

C$_6$H$_{14}$ 2,3-Dimethylbutane 58°C

C$_6$H$_{14}$ 2-Methylpentane 60°C

C$_7$H$_{16}$ 2-Methylhexane (Isoalkane) 90°C

C$_7$H$_{16}$ 3-Methylhexane (Anteisoalkane) 92°C

C$_8$H$_{18}$ 2,2,4-Trimethylpentane (Iso-octane) 99°C

Figure 3-3
Skeletal formulas for some normal paraffin and branched-chain paraffin hydrocarbons in petroleum. Iso-octane, at bottom of figure, is the standard for motor fuel octane ratings.

homologous series, since each hydrocarbon differs from the succeeding member by one carbon and two hydrogen atoms. The members of the series are called *homologs*. Since a normal paraffin is defined as a straight chain, there are a limited number of them in crude oil—usually less than 60 ($n = 1$ to 60).* This makes them the most easily identified compounds in petroleum. All other molecular types include hundreds of different molecules, so identification is much more difficult.

The word *paraffin* is derived from Latin *parum affinis,* which means "of slight affinity." The normal paraffins are relatively inert with strong acids, bases, and oxidizing agents. Sulfuric acid, for example, is used to clean normal paraffins of other hydrocarbon-type impurities so they may be used medicinally and as coatings for food containers. Plants began to synthesize paraffin waxes early in geologic history as coatings for seeds, spores, leaves, and other cells that had to be protected from weathering. As sediments undergo diagenesis, much of the original organic matter is altered or destroyed, but the paraffin coatings last unless the rock is subjected to high-temperature metamorphism. Plants growing in dry desert areas form particularly hard waxes as coatings in order to minimize loss of water. Consequently, many ancient sediments deposited in desert areas contain a preponderance of the paraffin-type hydrocarbons.

In addition to straight chains, the paraffins can form branched-chain paraffins, as shown in Figure 3-3. Whereas only about 60 structures of straight-chain normal paraffins exist in petroleum, it is theoretically possible to have over a million branched-chain structures, as shown in Table 3-3. This table lists the number of possible isomers representing different kinds of branching all containing the same number of carbon atoms and corresponding to the formula C_nH_{2n+2}. Isomers are different compounds with the same molecular formula.

Fortunately, since crude oil is derived from a finite number of structures in living things, it is not as complex as the theoretical number of isomers seems to indicate. However, since a number of these isomers can form through the cracking and rearrangement of organic structures over geologic time and since there are an equally large number of isomers possible with naphthenes and aromatics, it is obvious that petroleum is very complex in composition.

About 600 individual hydrocarbons have been identified in petroleum to date. The American Petroleum Institute supported a study on the hydrocarbon constituents of petroleum under the direction of F. D. Rossini (1960) and, later, B. J. Mair (1967). As of 1967, 295 individual hydrocarbons were isolated from Ponca City crude, or about 60 percent of the total crude. The remaining 40 percent undoubtedly consists of thousands of compounds,

---

*A few very waxy crudes, such as from the Altamount and Bluebell fields of the Uinta Basin, Utah, contain traces of paraffin chains with up to 200 carbon atoms.

Table 3-3
Possible Number of Paraffin Isomers
for Each Size of Molecule

Size	Isomers	Size	Isomers
C_1, C_2, C_3	1 each	C_{10}	75
C_4	2	C_{11}	159
C_5	3	C_{12}	355
C_6	5	C_{13}	802
C_7	9	C_{15}	4,347
C_8	18	C_{18}	60,523
C_9	35	C_{25}	36,797,588

many of which will never be identified. Most of the hydrocarbons identified are in the lower molecular-weight range, from C_1 to about C_{20}. The correlation of crude oils can be made on individual hydrocarbons in this lower molecular-weight range, but in the higher ranges it is usually made by groups of compounds.

The boiling point of a normal paraffin is slightly higher than that of any isoparaffins with the same molecular formula. Thus normal heptane boils higher than its two isomers shown in Figure 3-3.

Paraffins most commonly synthesized by plants are the normal alkanes and the 2- or 3-methyl isomers. The 2-methylalkanes are sometimes referred to as the *isoalkanes* and the 3-methyl as the *anteisoalkanes*. Examples of these are shown on the lower right side of Figure 3-3.

In paraffin hydrocarbons, the covalent bonds of the carbon atom are normally at an angle of 109.5°. This is the angle of C—H bonds in methane.

Naphthenes or Cycloparaffins (C_nH_{2n})

The cycloparaffins that are formed by joining the carbon atoms in a ring are the most common molecular structures in petroleum. Naphthene rings (Figure 3-4) generally contain 5 or 6 carbon atoms, because in these ring sizes the carbon-carbon bond angles approach 109.5°. The 5-membered cyclopentane ring has bond angles of 108°. The carbon atoms lie in a plane, and the ring is not strained. The 6-membered cyclohexane ring, however, would form valence angles of 120° if it were in a plane. In order to eliminate this strain, the cyclohexane ring is actually a puckered configuration not in a plane. It is theoretically possible to form rings with more than 6 carbon atoms by warping the ring further. A few cycloheptanes (C_7H_{14}) have been identified in petroleum, but no rings smaller than C_5 or larger than C_7 have been found. A few rings are formed by living things outside this range, but they have not been identified in petroleum.

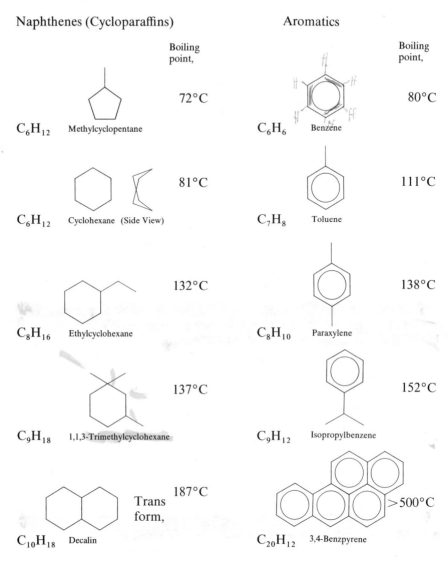

Figure 3-4
Skeletal formulas for cycloparaffin (naphthene) and aromatic hydrocarbons in petroleum.
All C_6 rings are warped. Cyclohexane is shaped like a twisted boat.

The average crude oil contains about 50 percent naphthenes with the
quantities increasing in the heavier fractions and decreasing in the lighter
fractions. In the heavier fractions, the naphthenes tend to fuse into polycyclic
rings; that is, a group of rings in which two or more carbon atoms are shared
between the rings. Decalin, in Figure 3-4, is an example. The most common

naphthenes are methylcyclopentane and methylcyclohexane, which together represent 2 percent or more of the average crude.

The naphthenes and paraffins are also referred to as *saturated hydrocarbons,* because all available carbon bonds are saturated with hydrogen. If hydrogen is removed from a paraffin, it will form one, two, or three double bonds, depending on whether 2, 4, or 6 hydrogen atoms are removed. Removing hydrogen from naphthenes forms either cycloolefins or aromatics. During World War II, methylcyclohexane was concentrated in refinery runs and stripped of half its hydrogen to form toluene, the starting material for trinitrotoluene (TNT).

Aromatic Hydrocarbons (C_nH_{2n-6})

The term *aromatic hydrocarbons* originated when some early, pleasant-smelling compounds such as cymene were isolated from natural fragrant oils. However, most hydrocarbons have very little odor in the pure state. The strong odor of petroleum is due to the nonhydrocarbons. All aromatic hydrocarbons contain at least one benzene ring. This is a flat 6-carbon ring (top of Figure 3-4) in which the fourth bond of each carbon atom is shared throughout the ring. For simplicity, the ring is shown with an inner circle, which indicates that the fourth bond's unpaired electrons are constantly oscillating between all carbon atoms in the ring. The aromatics are unsaturated hydrocarbons that will react to add hydrogen or other elements to the ring. The aromatics rarely amount to more than 15 percent of a total crude oil. They tend to be concentrated in the heavy fractions of petroleum, such as gas oil, lubricating oil, and residuum, where the quantity often exceeds 50 percent. Toluene (Figure 3-4) and metaxylene are the most common aromatic hydrocarbons in petroleum. Aromatics have the highest octane ratings of the hydrocarbon types, so they are valuable in gasoline blends. However, they are undesirable in the lubricating oil range, because they have the highest change in viscosity with temperature of all the hydrocarbons.

The heavy gas oil, lubricating oil, and residuum of petroleum contain increasing amounts of polycyclic (condensed-ring) hydrocarbons (Figure 3-4). McKay and Latham (1973) found that the 335–530°C boiling range of Recluse, Wyoming, crude contains four to eight condensed ring systems, and the higher boiling ranges appear to have even larger systems of condensed rings. It is probable that the number of rings in polycyclic molecules of residuum increase continuously up to the size of asphaltene particles (Figure 3-15).

Polycyclic aromatic hydrocarbons in natural products are being studied more intensively, because some of them, such as 3,4-benzpyrene, 3,4-

benzphenanthrene, and 1,2,3,4,-dibenzphenanthrene, are potent carcinogens. Besides being found in coal tars and petroleum, they are a common constituent of the burning of most organic material. The benzpyrene in coal smoke is about 300 mg/kg, in petroleum 0.5 to 2 mg/kg, in polluted urban air 100 μg/1000 m^3, in smoke from 100 cigarettes 10 to 15 μg, and in smoked meats 2 to 10 μg/kg. Practically all smoked or burned foods, such as charcoal-broiled meats, contain carcinogens such as benzpyrene.

Olefin Hydrocarbons (C_nH_{2n-2})

Olefin hydrocarbons contain double bonds between two or more carbon atoms, as shown in Figure 3-5. This causes them to be very reactive, compared to the other hydrocarbon types. The unsaturated state of the olefin is much more unstable than that of the aromatics. If hydrogen or other elements are not available to react with the unsaturation, some olefins will react with themselves to form high-molecular-weight polymers.

Many hydrocarbons formed by plants and animals are olefins. Ethylene (C_2H_4) is the major gas formed by the ripening of fruits and vegetables. Apples, pears, tomatoes, and corn all yield ethylene on ripening. In fact, ethylene is now used to control the ripening of bananas when they are ready to be marketed.

Fish oils and vegetable oils are high in olefins, which are believed to be useful in controlling the deposition of fatty material on human arteries. Peanut oil, olive oil, fishliver oil, and wheat germ oil all contain some olefinic hydrocarbons. Squalene, a natural component of human tissues that is an intermediate in the biosynthesis of cholesterol, is an olefin. Vitamin A and many pigments, such as the orange of carrots and the red of tomatoes, are olefins.

Olefins are uncommon in crude oil because they are readily reduced to paraffins with hydrogen or to thiols with hydrogen sulfide in the sediments. The two reactions are as follows:

$$
H_2C{=}CH{-}CH_3 \xrightarrow[H_2S]{H_2}
\begin{cases}
H_3C{-}CH_2{-}CH_3 & \text{Propane} \\[2ex]
\overset{\displaystyle SH}{\underset{}{H_3C{-}\underset{|}{C}{-}CH_3}} & \text{2-Propylthiol}
\end{cases}
$$

Propylene

Isoprene is a diolefin whose basic structure is one of the most important in nature (Figure 3-5). The isoprene structure appears to have been formed by the first photosynthetic organisms about 3.3×10^9 years ago. It is the basic building block for many hydrocarbon structures in living things, including the terpenes, rubber, most pigments, vitamin A, and the sterols. It is the precursor of essential oils of flowers, fruits, seeds, and leaves. Of the biological structures that have formed the hydrocarbons of recent sediments, isoprene is undoubtedly the most important.

Although olefins are the most common hydrocarbons of living things, they are found only in traces in a few petroleums, because they are readily reduced or polymerized to alkanes early in diagenesis. Olefins are formed in refinery processes, where they are major materials for petrochemicals.

Nitrogen, Sulfur, and Oxygen Compounds (Asphaltics)

The fifth molecular type are the nonhydrocarbons; that is, compounds containing atoms of nitrogen, sulfur, or oxygen in the molecule. Although these elements are present in small amounts, they disproportionately increase the nonhydrocarbon fraction of a crude oil by being incorporated in the molecules. For example, if an asphalt were composed of a single compound having the formula $C_{30}H_{60}S$, it would contain by weight 80 percent carbon, 13 percent hydrogen, and 7 percent sulfur. Yet there would not be a single hydrocarbon in this asphalt. Most crude oil residua contain a high percentage of nonhydrocarbon compounds.

There are small amounts of nonhydrocarbons scattered through the entire boiling range of crude oil, and a few of these are listed in Figure 3-5. Sulfur compounds include thiols, sulfides, thiophenes, and benzothiophenes. Nitrogen compounds include pyrroles, indoles, pyridines, quinolines, and carbazoles. Oxygen compounds are mainly chain or ring acids, as shown in Figure 3-5, where R equals a straight or branched paraffin chain. Carboxylic (chain or ring) acids and phenols represent 3.5 percent of the Midway Sunset, California, crude oil. Seifert and Teeter (1970) have identified 40 classes of carboxylic acids, including some 200 compounds, in this oil. Rall et al. (1972) identified 13 classes of sulfur compounds, including 176 individual structures in four crude oils.

Molecules in the high boiling ranges of petroleum frequently contain more than one of the molecular types previously discussed. To avoid misunderstanding, a molecule is called *aromatic* if it contains at least one aromatic ring. It is called *naphthenic* if it contains at least one cycloparaffin ring, and *paraffinic* if it does not contain either an aromatic or cycloparaffin ring. For example, a combination of an aromatic ring with the other two types would be called an *alkylaromatic* and *cycloalkylaromatic*.

Olefins

C_2H_4 Ethylene

C_4H_8 Butylene

C_5H_8 Isoprene

C_7H_{12} Methylcyclohexene

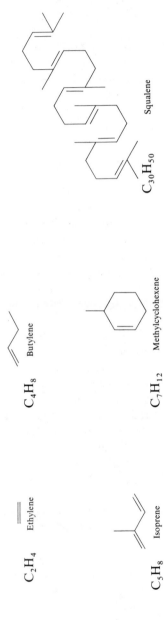

$C_{30}H_{50}$ Squalene

Nitrogen, Sulfur, and Oxygen Compounds

C_4H_9SH    Isobutylthiol

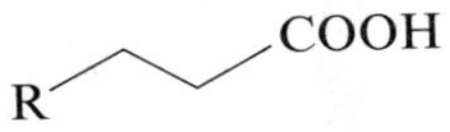

R—COOH   Fatty acid

C_6H_8S 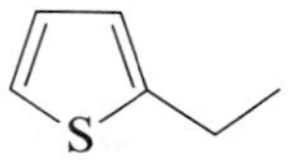   2-Ethylthiophene

C_8H_6O 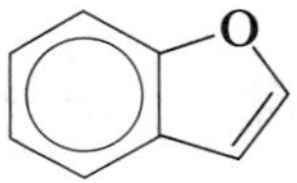   Benzofuran

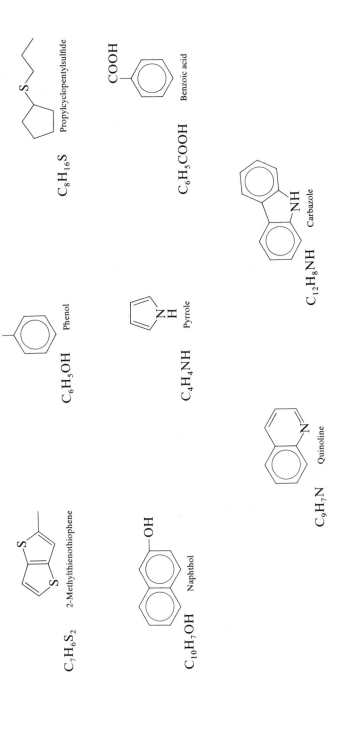

Figure 3-5
Skeletal formulas for olefin hydrocarbons and for nitrogen, sulfur, and oxygen compounds. Squalene, the precursor of sterols, is formed from six isoprene units. Heterocyclics are compounds with either N, S, or O in the ring and are common in the residuum of petroleum.

COMPOSITION AND USES OF PETROLEUM

Distillation is the principal method for separating crude oil into useful products. When promoters were trying to raise money to drill the Drake well, they submitted a sample of the oil from the Titusville seep to Professor Benjamin Silliman of Yale to determine its value. Silliman placed it in a distillation flask and boiled off eight fractions, each of which he described in detail. His results showed that the seep would make an illuminating oil that would be superior to most oils then available. This assured the financing of the Drake well.

Today a modern refinery distills thousands of barrels of oil a day through continuously operating distillation towers that are based on the same principle as Silliman's distillation flask. A refinery tower is equivalent to a series of individual distillation flasks, in which the distillate from the first flask is condensed in the second flask and redistilled to produce a distillate for the third flask. Instead of flasks, there are condensation plates in a tower, as shown in Figure 3-6. The vapor distilled from one of the chambers rises to the chamber above and passes through the condensed liquid of the overlying

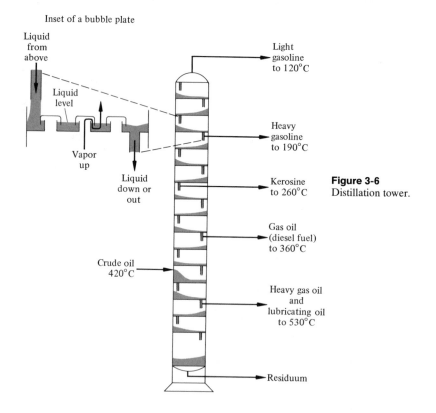

Inset of a bubble plate

Liquid from above

Liquid level

Vapor up

Liquid down or out

Crude oil 420°C

Light gasoline to 120°C

Heavy gasoline to 190°C

Kerosine to 260°C

Gas oil (diesel fuel) to 360°C

Heavy gas oil and lubricating oil to 530°C

Residuum

Figure 3-6
Distillation tower.

chamber, as shown in the inset of Figure 3-6. Each overlying chamber in the tower condenses successively lighter and smaller molecules, until only the light gasoline escapes from the top. At the bottom of the tower, the molecules that are so large and heavy that they cannot penetrate through the first plate as gases end up in the residuum. Refining towers may have different internal designs for condensing the vapors, but the efficiency of all of them is measured in terms of the number of plates, each bubble plate being the equivalent of the original distillation flask.

The refining tower in Figure 3-6 is run continuously by taking products out at various levels in the tower while continuously introducing fresh crude oil. The boiling ranges for the various crude oil fractions shown are for a typical Gulf Coast refinery. Refineries in other areas will show some variation in products and boiling ranges.

The composition of a typical 35° API-gravity oil is shown in Table 3-4. The molecular types can vary considerably from the figures shown. The average oil tends to have more paraffins in the gasoline fraction and more aromatics and asphaltics in the residuum. However, a highly paraffinic oil will have waxes predominating over asphaltic compounds in the residuum. The density, or °API gravity, of an oil varies with both the size and types of molecules. Since the element carbon is heavier than hydrogen, the density of hydrocarbons generally increases with a decreasing ratio of hydrogen to carbon atoms. Thus, in Table 3-5, normal hexane, cyclohexane, and benzene have increasing density (decreasing °API) as the H/C ratio decreases. A

Table 3-4
Composition of a 35° API Gravity
Crude Oil

Molecular size	Volume percent
Gasoline (C_5 to C_{10})	27
Kerosine (C_{11} to C_{13})	13
Diesel fuel (C_{14} to C_{18})	12
Heavy gas oil (C_{19} to C_{25})	10
Lubricating oil (C_{26} to C_{40})	20
Residuum ($>C_{40}$)	18
Total	100

Molecular type	Weight percent
Paraffins	25
Naphthenes	50
Aromatics	17
Asphaltics	8
Total	100

Table 3-5
Change in Hydrocarbon Gravity with Molecular Type

Hydrocarbon	Molecular type	Formula	H/C atomic ratio	Gravity	
				Density, d_4^{20}	°API
n-Hexane	Paraffin	C_6H_{14}	2.3	0.6594	82
Cyclohexane	Naphthene	C_6H_{12}	2.0	0.7786	50
Benzene	Aromatic	C_6H_6	1.0	0.8790	29

highly paraffinic oil will be lighter than an aromatic or asphaltic oil with the same molecular size distribution. However, different molecular sizes have a larger effect on gravity than do different molecular types. An oil with 50 percent gasoline is always lighter than one with 50 percent lubricating oil and residuum, irrespective of molecular type distribution. The greater density of the large molecules outweighs any type differences.

Figure 3-7 shows the distribution of various hydrocarbon types with boiling range (molecular size) in a naphthenic crude oil. The light gasoline fraction of the oil is dominated by the normal, iso-, and cycloparaffins because there are only two aromatics, benzene and toluene, that boil below 130°C (266°F). Moving from gasoline into the heavier fractions of oil, there is a marked increase in the aromatic content of kerosine, after which the aromatic content increases slowly until reaching the heavy lubricating-oil range. Aromatics and NSO compounds represent about 75 percent of the residuum.

Even though normal paraffins and isoparaffins decrease in the heavier fractions, there are both straight and branched carbon chains attached to both the naphthene and aromatic rings. Consequently, cracking of these heavier fractions of oil will release large quantities of paraffins.

The resins, waxes, and asphaltenes are not shown in Figure 3-7, because they are not distillation products. They are separated from the heavy lubricating oil and residuum by extraction with solvents. The resin fraction is defined as the propane-insoluble, pentane-soluble fraction, while the asphaltenes are pentane insoluble, benzene soluble.

As previously mentioned, Rossini and co-workers (1960) found that about 60 percent of the Ponca City crude oil was composed of only 295 compounds. Although hundreds of compounds are possible in the range up to C_{15}, only certain hydrocarbons are abundant. Bestougeff (1967) has proposed designating these more common hydrocarbons as "predominant constituents," several of which are listed in Table 3-6. It is interesting that the 2- and 3-methylalkanes and pristane are the dominant isoparaffins. These are the same types of structures synthesized by living organisms. Among the biologically produced alkanes, the 2- and 3-methyl isomers usually dominate over all other structures except the normal alkanes.

The cycloparaffin rings with a single methyl group attached are more common than the unsubstituted rings. It is not unusual to find 20 times as much methylcyclopentanes as cyclopentane in crude oil. The methyl-substituted aromatics (toluene and ethylbenzene) are also in higher concentration than the unsubstituted benzene ring.

Since World War II, the products from petroleum have dominated not only the energy market but also the chemical industry. Petroleum has been the true Aladdin's lamp in providing a myriad of products that have greatly advanced our civilization. Petroleum is the original raw material for an

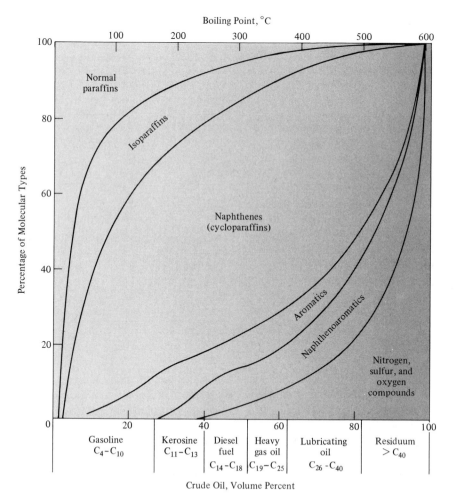

Figure 3-7
Chemical composition of a crude oil.

Table 3-6
Predominant Constituents of Petroleum (1 to 3 percent range, maximum concentration)

Hydrocarbon	Formula	Maximum weight percent in crude oil
Normal paraffins		
Pentane	C_5H_{12}	3.2
Hexane	C_6H_{14}	2.6
Heptane	C_7H_{16}	2.5
Octane	C_8H_{18}	2.0
Nonane	C_9H_{20}	1.8
Decane	$C_{10}H_{22}$	1.8
Branched-chain paraffins		
2-Methylpentane	C_6H_{14}	1.2
3-Methylpentane	C_6H_{14}	0.9
2-Methylhexane	C_7H_{16}	1.1
3-Methylhexane	C_7H_{16}	0.9
2-Methylheptane	C_8H_{18}	1.0
Pristane (isoprenoid)	$C_{19}H_{40}$	1.1
Cycloparaffins (naphthenes)		
Methylcyclopentane	C_6H_{12}	2.4
Cyclohexane	C_6H_{12}	1.4
Methylcyclohexane	C_7H_{14}	2.8
1,2-Dimethylcyclopentane	C_7H_{14}	1.2
1,3-Dimethylcyclopentane	C_7H_{14}	1.0
1-3,Dimethylcyclohexane	C_8H_{16}	0.9
Aromatics		
Benzene	C_6H_6	1.0
Toluene	C_7H_8	1.8
Ethylbenzene	C_8H_{10}	1.6
m-Xylene	C_8H_{10}	1.0

Source: Data primarily from Bestougeff 1967.

estimated 7,000 end-use chemicals. Since the energy market is enormous compared to the chemicals market, it takes less than 3 percent of the output of oil refineries to produce more than two-thirds of the organic chemicals used in the United States. The most useful hydrocarbons are methane (CH_4), ethylene (C_2H_4), propylene (C_3H_6), butylene (C_4H_8), and benzene (C_6H_6). They supply more than half the synthetic fibers and plastics and two-thirds of the synthetic rubber, soaps, and detergents. They are raw materials for cosmetics, drugs, paints, and agricultural chemicals. The composition and uses of various products are discussed in the following sections.

Gases

Gas at the wellhead usually consists of methane (CH_4) with decreasing amounts of the heavier hydrocarbons, sometimes including traces as high as nonane (C_9H_{20}). The principle nonhydrocarbon gases are nitrogen, carbon dioxide, and hydrogen sulfide. Small amounts of helium are also found in some gases. Dry gas is predominantly methane and ethane, whereas wet gas may contain 50 percent or more of propane and butanes. If the gas cap on an oil accumulation has a high content of wet gas, the oil will contain more gasoline than a field with a dry gascap. Normal butane usually predominates over isobutane in the older, more deeply buried gases. If the carbon dioxide content of a gas is high, it may be used to manufacture "dry ice," provided that the field is near a suitable market for the ice.

Restrictions on the emission of sulfur gases to the atmosphere have hastened the construction of plants to recover H_2S from natural and refinery gases. Hydrogen sulfide is one of the most poisonous gases known. A 0.1 percent concentration in air is fatal in less than 30 minutes. Drillers have died from sniffing H_2S at the wellhead. At the refinery, it is converted to sulfur as follows:

$$2\,H_2S + 3\,O_2 \rightarrow 2\,SO_2 + 2\,H_2O$$
$$2\,H_2S + SO_2 \rightarrow 3\,S + 2\,H_2O$$

The sulfur is used to manufacture sulfuric acid and other sulfur products.

Gases are taking an increasing share of the energy market from oil, and this may continue, because gas is the cleanest fossil fuel. Also, many countries now require that wellhead gas be saved rather than flared, so processing plants are being built in the areas of major oil fields. Most gas used to be transported to the consumer by pipeline, but increasing amounts are now being carried as liquids in tankers. Liquefied natural gas (LNG) is primarily methane with a boiling point of $-161°C$ ($-258°F$). Liquefied petroleum gas (LPG) is largely propane and butane. It can be liquefied under pressure at room temperature. The bottled-gas tanks used on farms and to run some automobiles are at pressures of about 200 psi (pounds per square inch), or 585 kg/cm² (kilograms per square centimeter). Because of the costs of maintaining the low temperature, LNG is much more expensive to process and transport than LPG. Consequently, LNG shipped by tanker will continue to be used primarily for peak gas demands in urban areas until the costs become more competitive.

Ethylene, propylene, and butylene (Figure 3-5) are olefins, not present in natural gas but formed in the refinery by cracking the gas oils and heavier hydrocarbons to make gasoline. High-temperature (700–900°C or 1,292–1,652°F), low-pressure (5 psi, or 14.6 kg/cm²) vapor-phase cracking favors production of the olefins. Some of the principal products made from petroleum gases are shown in Table 3-7.

Gasoline

Gasoline is composed of hydrocarbons mainly ranging from C_5 to C_{10}. From the time of the Drake well until the advent of the automobile, gasoline had no value and was usually discarded. Automobiles required an enormous increase in gasoline production. Since crude oils contain only 10–40 percent gasoline, the cracking process, which involved breaking large molecules into gasoline-

Table 3-7
Products of Refinery Gases

Refinery gas	Intermediate chemical	Product
Methane, CH_4	Formaldehyde	
	Bakelite plastics	Electrical fixtures
	Urea plastics	Buttons
	Melamine plastics	Dinnerware
	Polyformaldehyde	Gear wheels
	Pentaerythritol	Explosives
	Methylmethacrylate	Plexiglas
	Methanol	Antifreeze
	Chlorodifluoromethane	Teflon
Ethylene, C_2H_4	Polyethylene	Plastic wrapping, bottles, pipes
	Vinyl chloride	Plastic toys
	Vinyl acetylene	Neoprene rubber
	Vinyl acetate	Vinyl paints
	Acrylonitrile	Acrylic fiber, clothes, carpets
	Acetic anhydride	Acetate rayon, clothes, aspirin, sulfa drugs
	Ethylene glycol	Dacron clothing, orlon clothing, antifreeze
	Ethylene dichloride	Saran plastic, industrial solvents
Propylene, C_3H_6	Polypropylene	Plastic containers, suitcases
	Bisphenol	Epoxy adhesives
	Dodecylbenzene	Anionic detergents
	Propylene glycol	Cellophane film, glass fiber resins
	Nonylphenol	Nonionic detergents
	Glycerol	Alkyd resin paints
Butylene, C_4H_8	Butyl rubber	Tires
	Nitrile rubber	Hoses
	Hexamethylenediamine	Nylon

sized ones at high temperatures, was developed. Also, molecules smaller than the gasoline range were polymerized to the larger gasoline size. Combined cracking and polymerization would yield as much as 70 percent gasoline from a barrel of crude. Later, more esoteric processes were developed, such as cyclization of paraffins to form naphthenes and dehydrogenation of naphthenes to form aromatics. These processes reform the molecules. Today a refinery operation can be shifted to produce almost any molecular type and size range of hydrocarbons from a single crude feed stock, although each operation naturally adds to the cost of the product. Examples of these processes are shown in Figure 3-8.

Reforming hydrocarbon molecules is particularly important in developing high-octane gasolines without the use of such additives as tetraethyl lead. When gasoline was first used in a spark ignition engine, it was found that the initial explosion in the cylinder was followed by secondary explosions that

Cracking

$$C_{30}H_{60} \rightarrow CH_4 + C_2H_4 + C_2H_6 + C_3H_6 + C_7H_8 + C_7H_{14} + C_8H_{18}$$

Gas oil   Gases   Gasoline

Polymerization

$$C_3H_6 + C_4H_8 \rightarrow C_7H_{14} \xrightarrow{H_2} C_7H_{16}$$

Propylene + Butylene   Heptene   Heptane

Alkylation

$$C_3H_6 + C_4H_{10} \rightarrow C_7H_{16}$$

Propylene + Butane   Heptane

Reforming: Dehydroisomerization

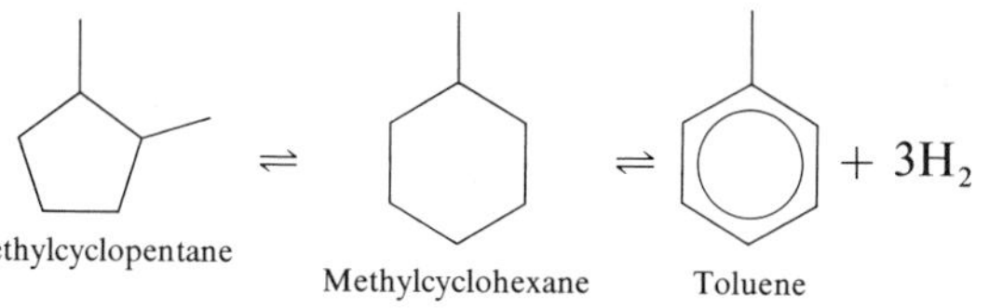

Dimethylcyclopentane   Methylcyclohexane   Toluene $+\ 3H_2$

Reforming: Dehydrocyclization

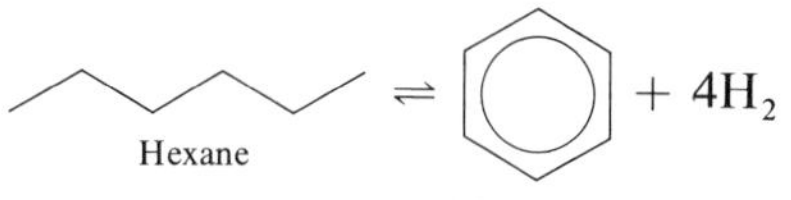

Hexane   Benzene $+\ 4H_2$

Figure 3-8
Refinery processes for rebuilding hydrocarbons.

caused the engine to knock badly. Since the cause of knock was not understood, the chemists initially tried the approach of adding every chemical in the laboratory to gasoline in hopes of reducing engine knock. The ultimate antiknock chemical was $Pb(C_2H_5)_4$. Adding 1 to 3 ml (0.03 to 0.1 oz) to a gallon of fuel caused the knock to disappear. Meanwhile, engines were being tested with individual hydrocarbons rather than mixtures. Iso-octane (2,2,4-trimethylpentane) caused the least knock of the first pure hydrocarbons tested, so it was given a rating of 100. Normal heptane, which caused considerable knock, was rated 0. The knock characteristics of all hydrocarbons and fuels were rated relative to the *n*-heptane–iso-octane scale. It was later learned that structural groups that retarded oxidation had the least knock (Livingston 1951). Thus, ring compounds (aromatics and naphthenes) and highly branched isoparaffins would not oxidize until the temperatures were high enough for complete combustion. Long-chain paraffins would start oxidizing at a lower temperature, and later combustion would cause knock.

A comparison of standard oxidation temperatures and octane numbers is shown in Table 3-8. The increase in oxidation temperature and octane number with increased branching of the chain is apparent.

A visual picture of the effect of structure on octane number is shown in Figure 3-9. Four of the structures on the left are the same as those listed in Table 3-8. As a straight chain becomes branched, the octane rating is raised. For example, the first four hydrocarbons on the left of Figure 3-9 have exactly the same formula, C_8H_{18}. They differ in that the first is a straight chain, and each succeeding hydrocarbon has an additional side chain that is taken off the end. These four hydrocarbons achieve an octane rating range from -19 to 100 simply by altering the structure to make it more compact. The fifth hydrocarbon on the left, 2,2,3-trimethylbutane, is even more compact than iso-octane and has a higher octane rating.

The naphthenes and aromatics generally have higher octane numbers than paraffins, because their rings are basically compact. Adding long chains to the ring will lower the octane number, as shown in the examples to the right of

Table 3-8

Comparison of Knock Resistance and Oxidation Temperature of Hydrocarbons

Hydrocarbon (C_8H_{18})	Oxidation temperature °C	Research octane number
Normal octane	265	-19
Normal heptane	275	0
3-Methylheptane	295	27
2,4-Dimethylhexane	320	65
2,2,4-Trimethylpentane	465	100

Source: From Livingston 1951.

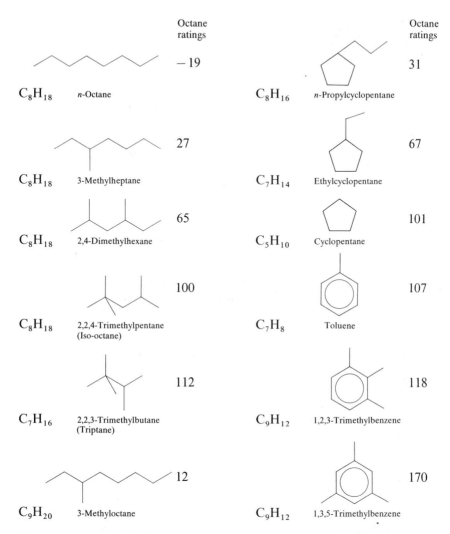

Figure 3-9
Variation in octane rating with hydrocarbon structure. The numbers to the right of each formula are the research octane ratings.

Figure 3-9. The three cyclopentanes have increasing octane numbers as the side chain is shortened. Making the molecules symmetrical, as in 1,3,5-trimethylbenzene increases the octane number above the unsymmetrical 1,2,3 compound. In order to make high-octane fuels without additives, it is necessary to convert both paraffins and alkylnaphthenes to aromatics, as shown in Figure 3-8. This is the direction refinery operations will take as tetraethyl lead is phased out.

Table 3-9
Products from Benzene

Starting products	Intermediate products	Final products
C_8H_{10}	Styrene	Polystyrene plastics / Styrene-butadiene rubber
$C_6H_5SO_3H$ → Phenol	2,4-Dichlorophenol → / Salicylic acid →	2,4-D weedkiller / Aspirin
C_6H_5Cl		DDT insecticides
$C_6H_5NO_2$ → Aniline	Acetanilide →	Aniline dyes / Analgesic drugs / Dye solvents
$C_6H_4Cl_2$	p-Dichlorobenzene → / o-Dichlorobenzene →	Insecticides / Industrial solvents
C_6H_{12}	Caprolactam → / Adipic acid →	Nylon 6 fibers / Nylon 66 fibers / Polyurethane foams
C_6Cl_6		Insecticides
$C_4H_2O_3$	$C_4H_4O_2N_2$ →	Agricultural chemicals
$C_{18}H_{30}$	$C_{18}H_{31}O_4SNa$ →	Anionic detergents

One of the most useful hydrocarbons in the gasoline range is benzene, which is a major raw material for a host of petrochemicals, as shown in Table 3-9. Benzene is produced by reforming methycyclopentane, cyclohexane, and some hexanes over a platinum catalyst (platforming). As previously mentioned, only a small percentage of the gasoline output is required for this purpose.

Kerosine

Kerosine replaced whale oil in the lamps of the world during the late nineteenth century. It, in turn, was replaced by the gas lamp and electric light. In order to prevent kerosine from being diluted with the more dangerous gasoline, the *flash point* test was developed. The flash point is the temperature to which an oil can be heated before its vapors will flash when a flame is passed over the oil. Flash points are still determined for crude oils because they indicate the temperature below which an oil can be handled without danger of fire.

The first high-temperature cracking process was developed to increase the yield of kerosine by cracking the heavier fractions of crude oil. As the use of gasoline increased and of kerosine diminished, the latter was cracked to make gasoline. The increased use of both jet and diesel fuels has reversed this trend, so that kerosine and light gas oils are now in heavy demand.

The kerosine fraction of crude oil (C_{11} to C_{13}) is the first to show an appreciable increase in the cyclic hydrocarbons that dominate the heavier fractions of crudes. The aromatics in kerosine range from 10 to 40 percent, considerably higher than for the total crude. Sachanen (1945) reports kerosine fractions of some California, Mexico, and Borneo crudes ranging from 25 to 40 percent aromatics. Condensed bicyclic naphthenes and aromatics, such as tetralin and naphthalene, are common in this range. Naphthenic acids, phenols, and thiophenes are among the nonhydrocarbons of kerosine.

Gas Oil

Light gas oils (C_{14} to C_{18}) are used in both jet fuels and diesel fuels. A diesel engine is a compression ignition engine because hot compressed air ignites the fuel. Fuel requirements for high thermodynamic efficiencies are exactly the opposite of the gasoline engine. The long-chain paraffin hydrocarbons that knock badly in spark ignition are the best fuels for the diesel engine. Cetane, normal hexadecane ($C_{16}H_{34}$), is the standard for the diesel engine, just as iso-octane is for the gasoline engine. Branched and cyclic hydrocarbons have low cetane numbers or high octane numbers, whereas long-chain hydrocarbons have high cetane numbers and low octane numbers. Consequently, a paraffin-base Pennsylvania crude oil will yield an excellent diesel fuel, but a poor gasoline; an aromatic-base California crude will yield an excellent gasoline but a poor diesel fuel.

The composition of the gas oil fraction is known in terms of the grouping of molecular types; however, it is so complex that only a few hydrocarbons have been identified in its total range (C_{14}–C_{25}). Sergienko (1964) has published a comprehensive survey of the composition, properties, and reactions of hydrocarbons in the gas oil, lube oil, and residuum ranges. More recently, Robinson (1971) published a detailed mass spectra analysis of the gas oil–lube oil fraction of a West Texas crude (Table 3-10). Robinson was able to identify 19 molecular compound types; 4 saturated, 12 aromatic hydrocarbon, and 3 nonhydrocarbon types. The carbon number ranges (Table 3-10) are only approximate, as there is considerable overlap. These data show a marked decrease in paraffins (saturates) and an increase in aromatics in going from the diesel oil through the lube oil range. There is also an increase in the hydrocarbons with three or more condensed rings, both in the paraffins and aromatics, in going to the higher boiling fractions. Single-ring aromatics (benzenes) decrease in the same direction. The complexity of the higher range is apparent from the increase in unidentified aromatics. The sharp increase in cycloparaffins having three or more condensed rings in the lubricating oil is attributed to steranes. These are the diagenetic derivatives of sterols, such as cholesterol, formed by living things.

Table 3-10

Composition of a West Texas Gas Oil to Lubricating Oil Fraction

Hydrocarbon range	Weight percent in each fraction		
	Diesel oil	Heavy gas oil	Lube oil
	C_{15}–C_{18}	C_{19}–C_{25}	C_{26}–C_{40}
Paraffins	31.0	20.0	10.9
Cycloparaffins, noncondensed	23.7	18.6	19.5
2-Ring condensed	10.2	9.0	8.0
3-Ring + condensed	6.1	6.4	10.9
Total saturates	71.0	54.0	49.3
Aromatics			
Benzenes	6.3	4.6	3.8
Naphthenebenzenes	4.3	4.0	3.5
Dinaphthenebenzenes	4.4	5.2	5.3
Naphthalenes	5.1	2.7	1.3
Acenaphthenes, dibenzofurans	2.7	5.0	3.7
Fluorenes	1.5	4.5	4.6
Phenanthrenes	0.7	3.4	2.2
Naphthenephenanthrenes	0.2	1.8	2.7
Pyrenes	0.4	2.5	2.2
Chrysenes	0.0	1.2	1.5
Perylenes	0.0	0.5	1.4
Dibenzanthracenes	0.0	0.2	0.7
Benzothiophenes	2.6	3.3	2.8
Dibenzothiophenes	0.8	4.4	2.4
Naphthobenzothiophenes	0.0	1.0	1.9
Unidentified aromatics	0.0	1.7	10.7
Total aromatics	29.0	46.0	50.7
Total all compounds	100	100	100

Source: From Robinson 1971.

The nonhydrocarbons in Table 3-10 are the sulfur compounds shown below:

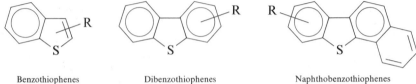

Benzothiophenes   Dibenzothiophenes   Naphthobenzothiophenes

The letter *R* represents any substituent group, such as a paraffin chain. The unidentified aromatics probably include three or more fused aromatic rings with one or more nitrogen, sulfur, or oxygen atoms in the rings.

Lubricating Oils and Waxes

Lubricating oil normally ranges from about C_{26} to C_{40}, but it can go as low as C_{20} and as high as C_{50}, depending on the distillation process. This range contains the normal paraffin waxes (C_{22}–C_{40}) and some asphaltics (NSO compounds). Highly paraffinic crude oils frequently have a high wax content in this range and a correspondingly high *pour point*. Pour point is determined by heating an oil in a tube to 46°C (115°F) to dissolve all the wax and then gradually cooling it in a bath that is held about 11°C (20°F) below the estimated pour point. The temperature at which the oil will not flow when the tube is horizontal is the pour point. Crude oils have pour points ranging from about −57 to 43°C (−70 to 110°F). The pour point is raised by straight-chain hydrocarbons and lowered by branched-chain hydrocarbons, cyclic compounds, and asphaltic substances. Pour points of lubricating oils are lowered by removing the waxes by solvent extraction with liquid propane or with ketones such as methylethylketone.

The paraffinicity of a crude and the volume of its +300°C (+572°F) fraction determines its wax content. Some typical wax analyses of crude oils reported by Nelson (1962) are shown in Table 3-11.

The waxes extracted from lubricating oil are not all normal paraffins. For example, Levy et al. (1961) cites a wax containing 39 percent *n*-paraffins, 32 percent isoparaffins, 27 percent naphthenes, and 1 percent aromatics. The isoparaffins and cyclic hydrocarbons tend to lower the melting point. Levy identified 67 compounds in a commercial paraffin wax with a melting point of 53°C (128°F). The principal groups are shown in Table 3-12. The normal paraffins peak at C_{26}. It is interesting that the 2- and 3-methylalkanes are the dominant branched hydrocarbons. They are also the principal branched paraffins synthesized by living things.

An important property of good lubricating oils is the change in viscosity with temperature, or *viscosity index* (VI). This index is a series of numbers ranging from 0 to 100. A VI of 100 indicates an oil does not tend to become

Table 3-11
Typical Wax Analyses of Crude Oils

Crude oil	Gravity °API	Weight percent wax in	
		Total crude	371°–482°C (700°–900°F) fraction
Coldwater, Michigan	50	2.9	12
Rodessa, Texas	43	6.6	19
San Joaquín, Venezuela	44	13	45
Bahia, Brazil	40	15	32
Esquina, Venezuela	41	18	55
Taman, Venezuela	38	24	51

Table 3-12
Composition of Commercial Paraffin Wax

Molecular type	Weight percent
Normal paraffins (C_{20}–C_{33})	79.1
Branched paraffins	
2-Methylalkanes	6.8
3-Methylalkanes	3.5
Other branched paraffins	1.3
Naphthenes	
C_5 rings	3.9
C_6 rings	5.2
Aromatics	0.2
Total	100

viscous at low temperatures or become thin at high temperatures. Paraffin-base lubricating oils containing long-chain hydrocarbons have a VI of nearly 100, whereas naphthene-base oils composed of rings have VIs around 40, with the more naphthenic aromatic oils going as low as 0. Before oil additives were widely used, the best lubricating oils came from the paraffin-base Pennsylvania crudes while the worst lube oils came from the more aromatic California crudes. For years, Pennsylvania crude oils sold at a premium because they had a higher VI and better lubricating qualities than other available oils. The introduction of additives that raised the VI of both Mid-Continent and California oils made them more competitive.

The color and odor of crude oil is largely caused by the nitrogen, sulfur, and oxygen (NSO) compounds that are concentrated in the lubricating oil and residuum fractions. Most of the pure hydrocarbons in petroleum are colorless and odorless, but traces of the NSO compounds can impart a host of colors and odors. Snyder (1969) made a detailed analysis of the kerosine through heavy lubricating oil fraction (C_{12} to C_{40}) of a California crude oil by high-resolution mass spectrometry. He found a wide range of NSO compound types present, as shown in Table 3-13. He was analyzing specifically for nitrogen and oxygen compounds, but some of the structures identified also contained sulfur. Snyder's data clearly show that both the oxygen and nitrogen compounds increase considerably in the heavy fractions. The heavy lubricating oil contains seven times the weight of heterocompounds as the kerosine to light gas oil fraction. Every individual compound group increases with molecular size. Also, there are about five times as many aromatic heterogroups in the heaviest fraction, compared to paraffin groups. This increase in both aromatics and NSO compounds continues into the residuum, where the heaviest fraction may have no hydrocarbons, as previously shown

Table 3-13
Heterocompounds of a California Kerosine to Lubricating Oil Fraction

	Weight percent in each fraction		
	Kerosine to light gas oil	Heavy gas oil to light lube oil	Heavy lube oil
Hydrocarbon range	$C_{12}-C_{22}$	$C_{23}-C_{30}$	$C_{31}-C_{40}$
Paraffin heterocompounds[a]			
Carboxylic acids	0.5	1.74	1.35
Other oxygen compounds	0.5	0.72	2.30
Total paraffin heterocompounds	1.0	2.5	3.6
Aromatic heterocompounds			
Indoles	0.07	0.69	1.15
Carbazoles	0.28	3.90	5.86
Pyridines	0.35	0.66	1.3
Quinolines	0.24	2.0	3.6
Pyridones, quinolones	0.2	1.2	2.0
Furanes	0.6	1.2	2.0
Phenols	0.3	0.97	1.4
Total aromatic heterocompounds	2.0	10.6	17.3
Total heterocompounds	3.0	13	21

[a]Compounds with one or more nitrogen, oxygen, or sulfur atoms.
Source: From Snyder 1969.

in Figure 3-7. Rall et al. (1972) and Coleman et al. (1973) have shown that the sulfur content of most crude oil fractions increases in the heavy fractions (Figure 3-10), as do nitrogen and oxygen. Thiols and chain and cyclic sulfides are in the lower ranges, with thienothiophenes, thiaindanes, and benzothiophenes in the higher ranges.

Hydrodesulfurization is a process for removing the sulfur and nitrogen from crude oil. The organic sulfur compounds are decomposed to hydrogen sulfide and a hydrocarbon as follows:

$$\begin{array}{c} HC \!\!-\!\!\!-\!\!\!-\!\! CH \\ \| \qquad \| \\ HC \diagdown_{\textstyle S}\diagup CH \end{array} + 4\,H_2 \rightarrow C_4H_{10} + H_2S$$

Strong acids and bases are also used to clean up oil fractions. A lubricating oil fraction can be made completely colorless and odorless by treating with fuming sulfuric acid until no further reaction occurs. This is the way that white oil or Nujol, a laxative, is prepared.

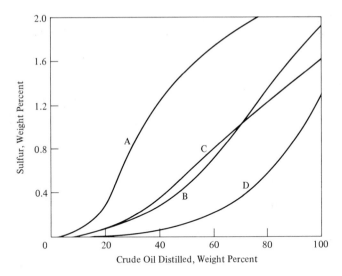

Figure 3-10
Increase in the sulfur content of the high-molecular-weight fractions
of crude oil: A, Huntington Beach, Calif.; B, Maljamar, New Mexico;
C, Shiells Canyon, Calif.; D, Schuler, Ark. [Rall et al. 1972]

Table 3-14
Resins and Asphaltenes in Crude Oils

Crude oil	Source	Weight percent	
		Resins	Asphaltenes
Ellenburger	W. Texas	4.2	0.24
Ragusa	Sicily	9	0.28
Grozni	USSR	8	1.5
Karami	China	14	1.8
Wilmington	California	14	5
N. Belridge	California	18	5
Khaudag	USSR	33	8
Belaim	Egypt	20	13
Boscan	Venezuela	29	18
Athabasca	Canada	24	19

Source: From Erdman 1962; Sergienko 1964, p. 387.

Residuum: Resins, Asphaltenes, and Waxes

The most complex and least understood fraction of petroleum is the residuum. The principal constituents are some of the very heavy oils, resins, asphaltenes, and high-molecular-weight waxes. The wax fraction in most residua is about half that in the lube oil fraction. Treatment of the residuum with liquid propane at temperatures not exceeding 21°C (70°F) precipitates the resins and asphaltenes. This fraction is then treated with normal pentane, which dissolves the resins and precipitates the asphaltenes as shown below:

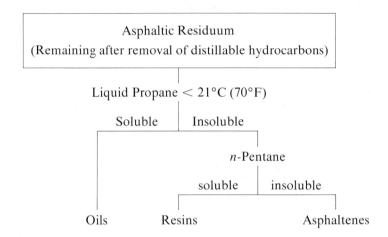

The asphaltenes are dark brown to black amorphous solids. The resins may be light to dark colored, thick, viscous substances to amorphous solids. The percent resins and asphaltenes in a variety of crude oils is shown in Table 3-14. These yields are after the removal of the waxes and the hydrocarbon oils soluble in liquid propane. The resins always exceed the asphaltenes, although there is considerable variation in the ratio. The resins and the asphaltenes contain about half of the total nitrogen and sulfur in crude oil. Most of this is in the form of heteromolecules, which are condensed to both aromatic and naphthene rings. In going from oils to resins to asphaltenes, there are increases in molecular weight, in aromaticity, and in nitrogen, oxygen, and sulfur compounds. Heavy crude oils invariably have more nitrogen and sulfur, as shown in Figure 3-11 and 3-12. These show the approximate ranges for nitrogen and sulfur for a variety of total crudes. Oxygen follows the same trend, with residua frequently containing over 5 percent oxygen.

A simple, rapid way to compare the paraffinicity or aromaticity of any fossil fuel or its fraction such as resins and asphaltenes is by showing the ratio of hydrogen to carbon atoms. As previously shown (Table 3-5), the paraffin, n-hexane, has 14 hydrogen atoms to 6 carbon atoms, whereas the aromatic

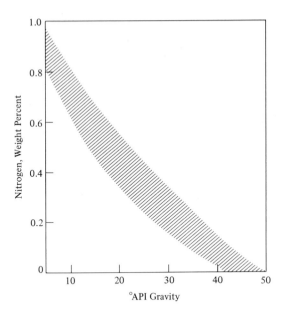

Figure 3-11
Variation in nitrogen content with °API gravity for crude
oils. [Nelson 1974]

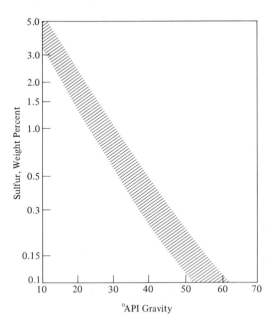

Figure 3-12
Variation in sulfur content with °API gravity for crude
oils. [Nelson 1972]

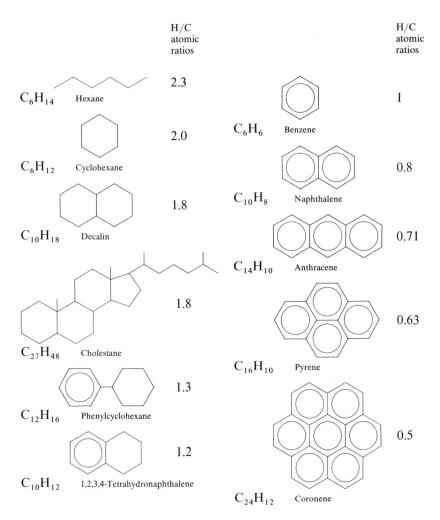

H/C
atomic
ratios

2.3

C_6H_{14} Hexane

2.0

C_6H_{12} Cyclohexane

1.8

$C_{10}H_{18}$ Decalin

1.8

$C_{27}H_{48}$ Cholestane

1.3

$C_{12}H_{16}$ Phenylcyclohexane

1.2

$C_{10}H_{12}$ 1,2,3,4-Tetrahydronaphthalene

H/C
atomic
ratios

1

C_6H_6 Benzene

0.8

$C_{10}H_8$ Naphthalene

0.71

$C_{14}H_{10}$ Anthracene

0.63

$C_{16}H_{10}$ Pyrene

0.5

$C_{24}H_{12}$ Coronene

Figure 3-13
Hydrogen to carbon (H/C) atomic ratios shown as numbers to right of structures. Ratio decreases with increasing cyclization and aromaticity.

benzene has 6 to 6. The H/C ratios are 2.3 and 1, respectively. As a hydrocarbon becomes more compact, with more condensed aromatic rings having less hydrogen, the H/C ratio continues to drop. Figure 3-13 shows several hydrocarbon structures with their H:C ratios. Condensed saturated rings, such as the cycloparaffins decalin and cholestane, have lower ratios compared to a single ring. The lowest ratios occur with condensed aromatic rings such as anthracene, pyrene, and coronene. For comparison, a typical oil, an asphalt, and a coal have H/C values of 1.85, 1.5, and 0.6 respectively.

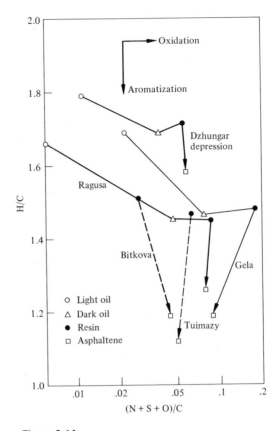

Figure 3-14
Atomic ratios of oil fractions. The NSO/C ratio
increases with oxidation. The H/C ratio decreases with
loss of more volatile components and aromatization.

In Figure 3-14, the H/C and NSO/C ratios for the colorless oils, dark oils, resins, and asphaltenes from the residuum of a Ragusa crude oil and a natural asphalt from Gela, Sicily, are plotted (Colombo and Sironi 1959). Also plotted are the resin and asphaltene data for two crude oils from the USSR (Sergienko 1964).

A line is drawn between light oils, dark oils, resins, and asphaltenes of each sample for clarification. The H/C ratio decreases from light to dark oils, is constant from dark oils to resins, and drops sharply from resins to asphaltenes. Colorless oils are the most paraffinic. Dark oils and resins are similar in aromaticity and paraffinicity, while asphaltenes are the most aromatic of the fractions. The NSO/C ratio increases, except between resins and asphaltenes, where three samples decrease. The resins are highest of the fractions in oxygenated compounds, such as acids, while the asphaltenes are highest in sulfur compounds, so the balance of these two groups determines whether resins or asphaltenes have the highest NSO/C ratio. A sample of natural bitumens altered by weathering from the Dzhungar depression, USSR (Uspenskiy et al. 1961) also is shown for comparison. In this case, the four points represent seep oil; dark, weathered oil; asphalt; and asphaltite. All the Dzhungar bitumens are more paraffinic than the other examples, probably the source is more paraffinic.

The physical nature of resins and asphaltenes has been deduced from a combination of nuclear magnetic resonance (NMR) studies, x-ray diffraction, and molecular weight distributions. Asphaltic components such as asphaltenes exist in petroleum as colloidal particles dispersed in an oily medium. As the oily medium is removed by distillation, the particles become more concentrated, to form an asphalt. At standard temperatures, asphalts are highly viscous, appearing to be solids. At high temperatures, they behave as Newtonian liquids. The viscous behavior depends on the quantity and size of asphalt particles. Reerink and Lijzenga (1973) have shown that the molecular weight distribution of asphaltene particles in Kuwait bitumen is very wide, ranging from 2,000 to 200,000 depending on the method of preparation. Air blowing of bitumens widened the molecular weight distribution of the asphaltenes. The asphaltene molecules are flat ellipsoids with minor axes varying from 0.9 to 1.3 nm and major axes from 4 to 11 nm (a nanometer = 10^{-9} meters = 10 angstroms).

The probable structures of resins and asphaltenes have been worked out through combined NMR and x-ray studies (Winniford and Bersohn 1962; Wetmore et al. 1966; Yen 1974). Figure 3-15 shows the kinds of structures that can be drawn from these studies. An asphaltene molecule consists of 10 to 20 condensed aromatic and naphthenic rings with paraffin and naphthenic side chains. This includes long normal paraffin chains. Combinations of these molecules form particles as shown. In going from light oils to dark oils to resins to asphaltenes, the asphaltene particles increase in size, with a corresponding increase in molecular weight. The range in observed molecular weights is caused by the variation in dispersion and clustering of the asphaltene particles and micelles in the asphalt.

The condensed aromatic structures of asphaltenes also contain free radical sites. These sites contain highly reactive, unpaired electrons. Some of these sites in asphaltenes are capable of complexing metals, and Erdman (1962) has shown that the Boscan asphaltene is capable of complexing an additional 1,200 ppm (parts per million) of vanadium even though it already contains 4,000 ppm. Both vanadium and nickel appear to be complexed with large molecules in the lubricating oil-residuum range. The presence of these metals even in trace amounts has adverse effects on the refining and use of these fractions. Both vanadium and nickel tend to poison cracking catalysts. Gas

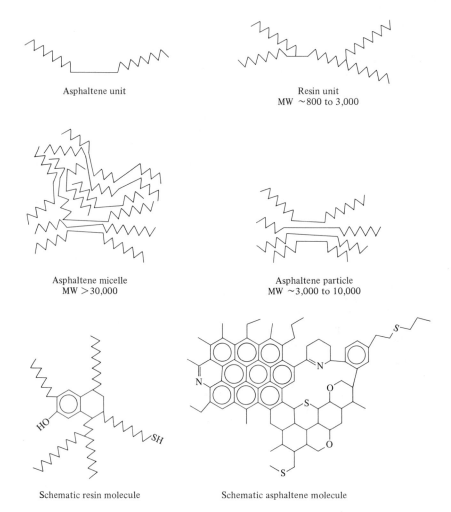

Asphaltene unit

Resin unit
MW ~800 to 3,000

Asphaltene micelle
MW >30,000

Asphaltene particle
MW ~3,000 to 10,000

Schematic resin molecule

Schematic asphaltene molecule

Figure 3-15
Structures of asphaltenes and resins. MW = molecular weight. Resins contain fewer condensed polycyclic rings than do asphaltenes. Note open area in asphaltene structure where heavy metals such as vanadium and nickel can be complexed with the asphaltene.

turbines corrode badly with high-vanadium fuels. Fortunately, most of the vanadium and nickel in the high-lubricating-oil residuum fraction is in the asphaltenes. Consequently, deasphalting is particularly important with high-vanadium crudes.

The principal use of asphalt residua is as a blend for making furnace oils and for road construction. Miscellaneous uses of asphalt include applications as binders, fillers, water-insulating materials, and adhesives in construction and manufacturing. In 1971, about 50 percent of the crude oils produced in the United States had 10 percent or more asphaltic constituents. Asphalt contents as high as 50 percent have been reported for Middle East and South American crudes. The world production of straight-run asphalt is about 2 percent by weight of the total petroleum production, according to Sergienko (1964).

The potential for petroleum asphalts as an energy source is much greater than its present use indicates. There are enormous worldwide reserves of heavy asphaltic crudes. In the early 1960s, about 40 percent of the world production of petroleum was in heavy crudes (Sergienko 1964). In many countries, there are large reserves of heavy crudes that are shut in (not producible) because of economics. Exploratory wells in wildcat areas frequently encounter thick deposits of viscous bitumen that are not commercial. For example, during the early 1960s several companies encountered thick bitumen zones between 2,000 and 3,000 meters (6,560 and 9,840 feet) below lake floor in the central part of Lake Maracaibo, Venezuela. As production methods improve and the use of asphalt increases, more of these heavy crude reserves will be produced.

SUMMARY

1. Petroleum has an average composition of 85 percent carbon, 13 percent hydrogen, and 2 percent sulfur, nitrogen, and oxygen.

2. Distillation separates petroleum into molecular groups of different sizes: gas C_1-C_4, gasoline C_5-C_{10}, kerosine $C_{11}-C_{13}$, light gas oil (diesel fuel) $C_{14}-C_{18}$, heavy gas oil $C_{19}-C_{25}$, lubricating oil $C_{26}-C_{40}$, residuum $> C_{40}$. High °API-gravity crudes have a high gasoline and low residuum content whereas low °API-gravity oils are low in gasoline and high in residuum.

3. The different types of hydrocarbon molecules in crude oil are paraffins (alkanes with single bonds between carbon atoms), naphthenes (cycloalkanes, cycloparaffins with carbon rings), olefins (alkenes with one or more double bonds between carbon atoms), and aromatics (arenes with one or more benzene rings). Olefins with one or more double bonds are found in refinery cracking products but not normally in crude oil. Lubricating oil and residua contain polycyclic ring hydrocarbons with multiple aromatic or mixed naphthene–aromatic rings fused together, two rings sharing one side in common.

4. Nonhydrocarbons in petroleum are composed predominately of carbon and hydrogen but also contain one or more of the elements nitrogen, sulfur, and oxygen (NSO). Small amounts of NSO compounds occur throughout petroleum, with the largest quantities in the lubricating oil and

residuum. Heterocyclic compounds containing N, S, or O in a ring are common in residuum.

5. Less than 3 percent of the output of oil refineries is used to produce more than two-thirds of the organic chemicals used in the United States for producing 7,000 end-use materials.

6. Petroleum waxes are extracted from the lubricating oil fraction and residuum of high-wax crudes. Wax is predominately *n*-paraffins and branched paraffins in the range to about C_{60}.

7. Petroleum asphalts are either straight-run residues from distilling crude oils or "blown" asphalts produced by air oxidation of crude residues. Asphalts contain heavy oils, resins, asphaltenes, and high-molecular-weight waxes. Asphaltenes are agglomerations of molecules with condensed aromatic and naphthenic rings and molecular weights in the thousands.

SUPPLEMENTARY READING

Nagy, B., and U. Colombo. 1967. *Fundamental aspects of petroleum geochemistry.* New York: Elsevier. 388 p.

Sergienko, S. R. 1964. *High molecular weight compounds in Petroleum.* Translated by Israel Program for Scientific Translation, Jerusalem, 1965. 440 p.

Sokolov, V. A., M. A. Bestougeff, and T. V. Tikhomolova. 1972. *Khimicheskii sostav neftei i prirodnykh gazov v svyazi s ikh proiskhozhdeniem.* [*Chemical composition of crude oils and natural gas in relation to their origin.*] Moscow: Nedra. 276 p.

II

ORIGIN AND MIGRATION

4

How Oil Forms

The organic theory of petroleum origin is based on the accumulation of hydrocarbons from living things and on the generation of hydrocarbons by the action of heat on biogenically formed organic matter. The inorganic hypothesis assumes oil forms from the reduction of primordial carbon or its oxidized form at elevated temperatures deep in the earth. The overwhelming geochemical and geological evidence from both sediment and petroleum studies of the past few decades clearly shows that most petroleum originated from organic matter buried with the sediments in a sedimentary basin. This chapter documents the evidence for organic origin. Although a few hydrocarbons in the crust may be derived from inorganic sources, the quantities are negligible compared to those from organic sources.

The hypothesis that petroleum is of inorganic origin is discussed in relation to the occurrence of hydrocarbons in igneous rocks, as this association is the basis for the concept.

THE ORIGIN OF HYDROCARBONS IN IGNEOUS AND METAMORPHIC ROCKS

The occurrence of various hydrocarbon gases, liquids, and bitumens in igneous rocks has long been known. A compilation of such occurrences in the USSR was published by Kudryavtzev (1959). One area studied in detail is the Khibina Massif, where hydrocarbons are found in the alkaline rocks and in associated Precambrian gneisses, phyllites, and schists. Careful geological studies of such occurrences indicate that most of them have originated from

organic carbon originally deposited with sediments that have since undergone extensive alteration. Hydrocarbon gases and bitumens can accumulate
in igneous and metamorphic rocks in at least five ways.

1. *Genesis from primordial carbon by inorganic processes.* The famous
Russian chemist, Dimitri Mendeleeff, who developed the periodic table, was
the first to propose the carbide hypothesis of petroleum genesis. Basically, this
hypothesis assumes that deep in the earth exist metal carbides that form
hydrocarbons on contact with hydrothermal solutions, as follows:

$$FeC_2 + 2H_2O \rightarrow HC\equiv CH + Fe(OH)_2$$

At elevated temperatures, the acetylene polymerizes to form benzene and a
complex mixture of hydrocarbons. Also,

$$Al_4C_3 + 12H_2O \rightarrow 3CH_4 + 4Al(OH)_3$$

In addition to the carbide hypothesis, the Fischer–Tropsch reaction has
been suggested as a possible inorganic source of hydrocarbons, since it occurs
readily at about 250°C (482°F) in the presence of a catalyst.

$$CO_2 + H_2 \rightarrow CO + H_2O$$

$$CO + 3H_2 \rightarrow CH_4 + H_2O$$

Although these reactions can occur in the laboratory, neither of them has
been demonstrated to occur on a large scale in nature. There may be small,
unusual accumulations of hydrocarbons in igneous rocks formed by these
processes, but most such occurrences can be attributed to one of the organic
sources discussed as follows.

2. *Genesis from organic carbon in sediments subjected to metamorphism.*
One fact that is frequently overlooked is that the rock matrix may have
originally been sedimentary. If it is assumed that at least half of the 48×10^{20}
grams of carbon in metamorphic and granitic rocks in the crust (Hunt 1972)
was once kerogen in sediments, then the quantity of organic carbon that
would be converted to methane and other hydrocarbons by catagenesis and
metamorphism could be 5×10^{20} grams. This total of carbon as hydrocarbons exceeds the mass of carbon in all presently known oil accumulations by
two orders of magnitude. If only a fraction of these hydrocarbons were
preserved in igneous and metamorphic rocks, it would readily explain most
known occurrences.

The high temperatures causing metamorphism would be expected to
produce some bitumens in addition to methane. When methane is heated
much above 1000°C (1832°F), it yields acetylene, which is thermodynami-

cally more stable. Subsequently, the acetylene could polymerize to form a complex bitumen.

3. *Distillation of hydrocarbons from sedimentary rocks of Precambrian and Phanerozoic Age by igneous intrusions.* The high temperatures accompanying igneous intrusion will distill hydrocarbons out of almost any sedimentary rock in the same manner that oil is distilled from oil shales on heating. Studies by both Hunt (1963b) and Baker and Claypool (1970) have shown notable changes in the hydrocarbons of sediments, because of the baking effect of an intrusion. Generally, the yield of hydrocarbons is low, because the heat is quickly dissipated in the surrounding sedimentary rocks.

When the igneous intrusion scavenges carbon from the surrounding sediments, there is a tendency to incorrectly define such carbon as having an inorganic origin. For example, the bitumens associated with dolorite intrusions in the Kola Peninsula of the USSR were thought to be inorganic until Belyayeva (1968) clarified their origin by carbon-13 analysis. She showed that the isotopic composition of carbon in both the sedimentary and igneous rocks was identical, both being in the range of carbon of organic origin.

BIOGENIC VS. ABIOGENIC HYDROCARBONS

The fractionation of carbon isotopes by biogenic activity has been used to prove a biogenic origin for carbon in igneous and metamorphic rocks. This can be only partly correct if the carbon has been recycled. For example, Eichmann and Schidlowski (1975) have demonstrated that the isotope fractionation factor governing biological fixation of inorganic carbon has been practically constant since about 3×10^9 years ago. This means that biogenic organic matter formed at that time with a $\delta^{13}C$ of -25 ‰ PDB could be subducted into the deep lithosphere or asthenosphere where it might undergo abiogenic reactions to form carbon compounds with a $\delta^{13}C$ of -25 ‰. The abiogenic reaction in the recycling could be misinterpreted as biogenic because of the $\delta^{13}C$ value of -25 ‰ of the source carbon in this reaction. It would actually be a biogenic synthesis followed by abiogenic.

The evidence for biogenic origin frequently has been strengthened by combining stereochemistry with isotope geochemistry. Many biologically formed organic compounds in solution are *optically active;* that is, they will rotate a beam of plane-polarized light. One molecule will rotate light to the right (dextrorotatory), another to the left (levorotatory). The two molecules will have the same atoms but are mirror images, or *enantiomers* of each other, as in the lactic acid of our muscle tissues. The two mirror images are identical in chemical and physical properties except for the direction in which they rotate plane-polarized light.   **(continued)**

BIOGENIC VS. ABIOGENIC HYDROCARBONS (continued)

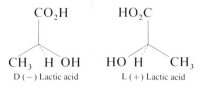

CO$_2$H   HO$_2$C

CH$_3$ H OH   HO H CH$_3$
D (−) Lactic acid   L (+) Lactic acid

The carbon atom in the center is called *asymmetric* because each of the four groups bonded to it is different. The tetrahedral arrangement contains no plane of symmetry and the molecules cannot be superimposed.

Biological synthesis generally forms only one enantiomer. Thus, enzymes made of only L-amino acids will synthesize only L-amino acids. In contrast, abiotic synthesis forms equal mixtures of D- and L-forms. Consequently, abiotically formed molecules are optically inactive, whereas biogenic molecules are optically active.

Cholesterol, the fatty deposit in arteries, has eight asymmetric carbon atoms. This means that abiotic synthesis could yield 2^8, or 256 different cholesterol structures. The enzymes that make cholesterol in humans are so specific that they make only 1 of these 256 possible structures.

Most crude oils contain optically active compounds, with the highest quantities in the youngest oils and the least in the most deeply buried oils. As organic compounds are buried in the earth, the thermal reactions form racemic mixtures so that optical activity gradually disappears with depth. The steranes and triterpanes containing optically active, condensed cyclo-alkane rings are partly converted to polycyclic aromatic hydrocarbons with no optical activity.

4. *Penetration of igneous rocks by hydrothermal solutions carrying organic matter from sediments.* When surface waters penetrate deep into the crust of the earth, they are converted to high-temperature hydrothermal solutions that cause considerable erosion of sedimentary sequences on returning to the surface. For example, there is evidence that the heavy metal deposits of the Red Sea are carried to the surface by recycled Red Sea water (Craig 1969). The stratified waters in the hot holes of the Red Sea are known to carry hydrocarbons considerably in excess of those in the overlying waters (Swinnerton and Linnenbom 1969). These hydrocarbons undoubtedly originate from the several thousand feet of underlying sediments.

Hydrothermal solutions are probably responsible for many bitumen deposits along fractures, fissures, and fault zones. They also destroy organic matter through oxidation by the sulfate ion. Germanov (1965) describes the removal of organic carbon by both solution and oxidation in the Kanzay ore deposits of Tadjik, USSR. He quotes the data of other Soviet authors that waters with temperatures up to 90°C (194°F) have been found to carry up to 417 mg of organic matter/ml of water.

5. *Migration of liquid and gaseous hydrocarbons to igneous and metamorphic rock reservoirs from sedimentary source beds.* This is the most common cause of hydrocarbon accumulations in igneous formations for essentially all petroleum produced from fractured or weathered igneous and metamorphic rocks, such as quartzite in Kansas, granite in Venezuela, basalt in Washington and California, and numerous schists in California, Morocco, and other countries. Thousands of seeps and solid hydrocarbon impregnations originate from igneous contacts with sedimentary rocks. Goldberg and Chernikov (1968) examined bitumens associated with the ultramafic rocks of the Siberian platform and concluded that they originated in the surrounding sedimentary rocks. Hedberg (1964) reviewed many occurrences of both hydrocarbon seepages and economic accumulations in all types of igneous and metamorphic rocks. His conclusion was that no appreciable amount of commercial liquid petroleum had an inorganic origin. This same conclusion was reached by N. A. Eremenko (1961) in his book *Geology of Oil and Gas.*

In summary, although a few unusual occurrences of hydrocarbons may have an inorganic origin, practically all economic petroleum accumulations originated from organic matter deposited with sedimentary rocks. Consequently, exploration programs should be designed with this basic concept.

MOLECULAR STRUCTURES OF LIVING ORGANISMS

All living things are formed from a few, simple, molecular building blocks that have changed relatively little over geologic time. The structures formed by these basic molecules contain widely varying amounts of carbon and hydrogen relative to oxygen, nitrogen, and sulfur. Consequently, although any organic matter may contribute to the formation of oil, there are valid reasons for believing that certain compounds are the principal precursors of petroleum, while others make up the mass of residual organic matter in sedimentary rocks.

A particular carbon structure or carbon skeleton (molecule minus H, O, N. and S) formed by a living organism and found in a Recent or ancient sediment or in a crude oil is called a *biological marker.* Such markers are useful in following reactions leading to petroleum.

Carbohydrates and Related Substances

The carbohydrates are polymers of monosaccharides, which are polyhydroxy-aldehydes and ketones, as shown in Figure 4-1. They provide the sweet taste in most food products. Glucose is in corn syrup and fructose in most fruits.

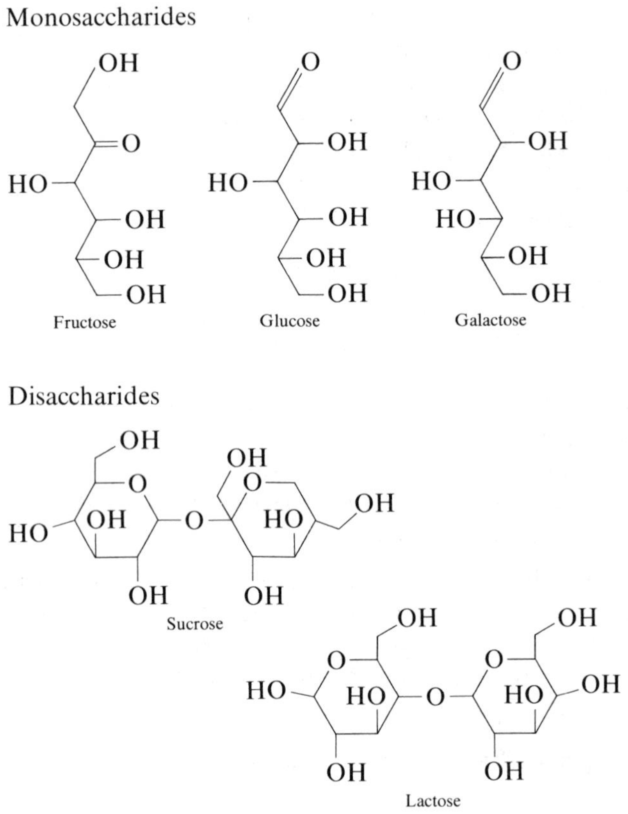

Monosaccharides

Fructose   Glucose   Galactose

Disaccharides

Sucrose

Lactose

Figure 4-1
Monosaccharides are the simple building blocks of all carbohydrates.
Starch and cellulose are polymers of glucose. Sucrose (table sugar)
is a combination of fructose (fruit sugar) and glucose. Lactose (milk
sugar) is a combination of galactose and glucose.

The combination of these two molecules forms the disaccharide sucrose,
common table sugar. Starch and cellulose are higher polymers of glucose, the
latter containing several thousand glucose units with the unit structure shown
in Figure 4-2.

Chitin, a polysaccharide that contains nitrogen in its structure (Figure 4-2),
is a common constituent of invertebrate animals. It is very prevalent in
Arthropoda, particularly crustaceans and insects. It forms the exoskeleton of
the body and appendages of crabs and lobsters, as well as the shells of
mollusks. It also forms the cell walls of some fungi.

Just as chitin was synthesized by organisms as a hard protective covering, so
lignin was eventually developed to give strength and rigidity to the cell wall.
Tissues composed only of cellulose are usually soft, with a higher percentage

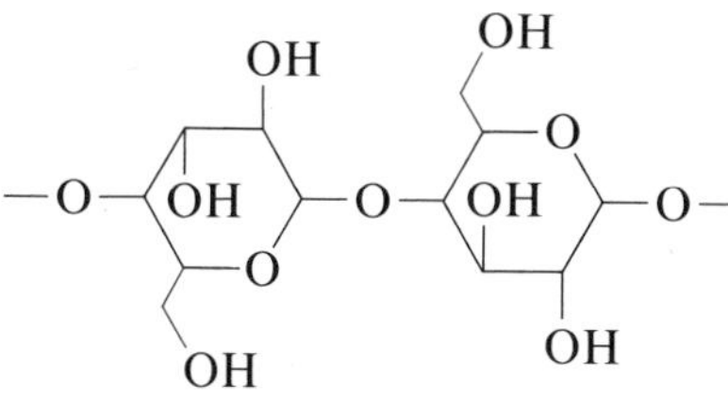

Cellulose unit

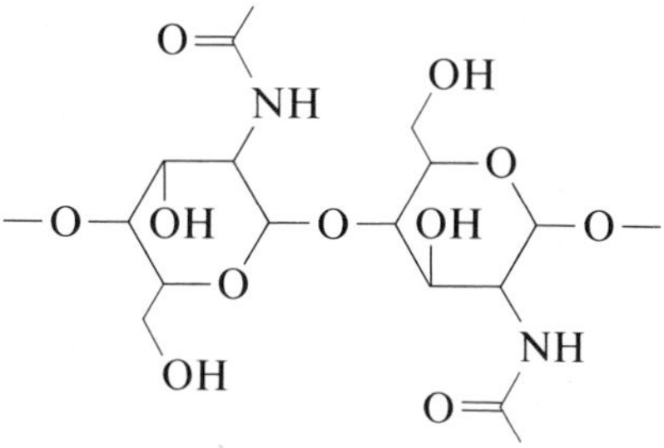

Chitin unit

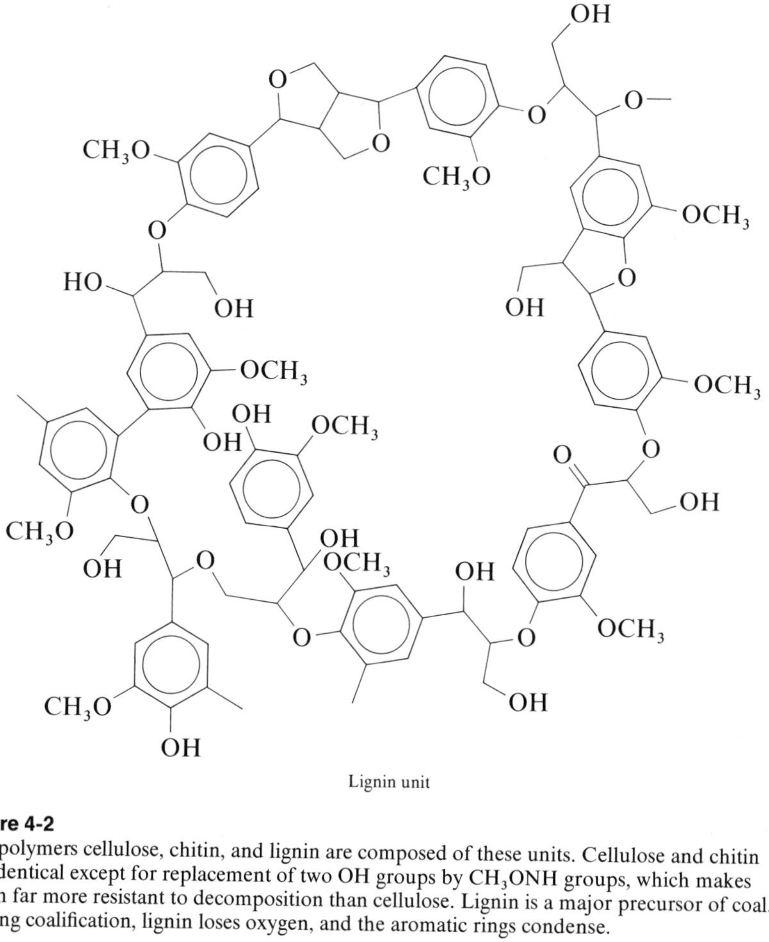

Lignin unit

Figure 4-2
The polymers cellulose, chitin, and lignin are composed of these units. Cellulose and chitin are identical except for replacement of two OH groups by CH_3ONH groups, which makes chitin far more resistant to decomposition than cellulose. Lignin is a major precursor of coal. During coalification, lignin loses oxygen, and the aromatic rings condense.

of water and a much lower breaking strength than those containing lignin. Lignin is a polyphenol built up from units, as shown in Figure 4-2, and is generally more resistant to decay and bacterial action than is cellulose. It is the most common natural product to contain aromatic rings. As lignin matures, these rings form the basic structure of humic coals.

Another group of natural substances also containing aromatic rings are the tannins, which are widely distributed in higher plants. Tannins are found in the bark of trees such as oaks and hemlocks; in unripe fruits, such as persimmons and plums; in the hulls of walnuts and hickory nuts; in seed coats; and in algae, fungi, and pathological plant growths. Plant galls generally contain 25–75 percent tannins. The tannins all contain polyhydroxyphenols or their derivatives, often in the form of complex condensed ring structures. Tannins appear to play a role in the healing process of plants and the lignification of plant cell walls.

The monosaccharides also combine with other molecules to form the glycosides, which may be represented by the formula sugar—O—R, where R represents the organic group attached to the sugar and O represents oxygen. The glycosides are widely distributed in nature, although usually present in small amounts. Amygdalin ($C_{20}H_{27}O_{11}N$) is the best known of the glycosides. It is obtained from the kernels of the bitter almond and also occurs in the kernels of many fruits, such as peaches, plums, and apricots. It consists of two glucose units linked with oxygen to a cyanide unit and benzene ring. Arbutin ($C_{12}H_{16}O_7$), another glycoside widely distributed in nature, consists of one glucose unit linked with oxygen to a phenol. Both of these molecules contain an aromatic ring, as shown below.

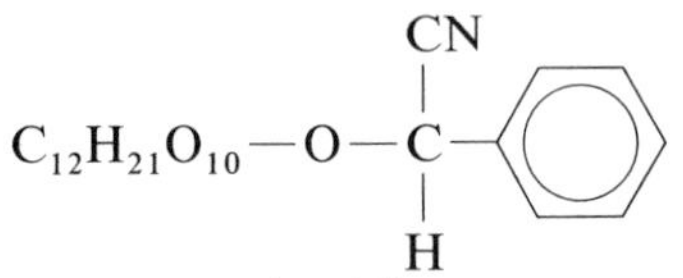

Amygdalin 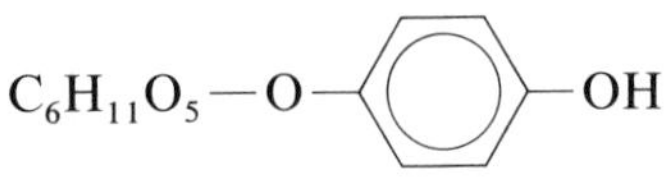   Arbutin

These are only typical examples of the glycosides, of which over 60 are known to occur naturally, many of which have aromatic rings and condensed cyclic rings related to the steroids shown in Figure 4-4.

Proteins

Proteins are high-molecular-weight polymers of amino acids. They constitute more than 50 percent of the dry weight of animals and account for most nitrogen compounds in living organisms. Twenty amino acids represent the building blocks of most proteins. Three of these—tyrosine, phenylalanine, and tryptophan—contain aromatic rings. The other 17 have straight and branched carbon chains, carbon–sulfur chains, and heterocyclic carbon–nitrogen rings. Some typical structures of amino acids are shown in Figure 4-3.

Proteins combine with other molecules to form most of the important compounds of life processes, such as the cytochromes, hemoglobins, enzymes, bacterial toxins, and antibodies. Proteins also form the structural components of marine organisms such as sponges and coral. Calcium carbonate shells form on protein templates. Most organic matter in shells is proteinaceous.

Amino acids are useful in following diagenetic processes in sediments. Abelson (1956) has shown that the thermally more stable amino acids, such as alanine, glycine, valine, proline, and the leucines, are present in the deeper, older sediments, whereas the thermally less stable amino acids are either absent or present in trace amounts. A more detailed discussion of the geochemistry of proteins, peptides, and amino acids has been published by Hare (1969), including a discussion of the possible use of amino acids in geochronology.

Lipids

The term *lipid* is derived from the Greek word meaning "fat." Lipids are biological substances insoluble in water, but soluble in fat solvents, such as ether, chloroform, or benzene. The lipids, essential oils, and plant pigments are more similar to petroleum in composition than are either the proteins or carbohydrates. In fact, many chemical structures in these groups can be recognized in crude oil.

The most common lipids are the animal fats and vegetable oils, which are formed by combining fatty acids with glycerol to form esters (glycerides), as shown below.

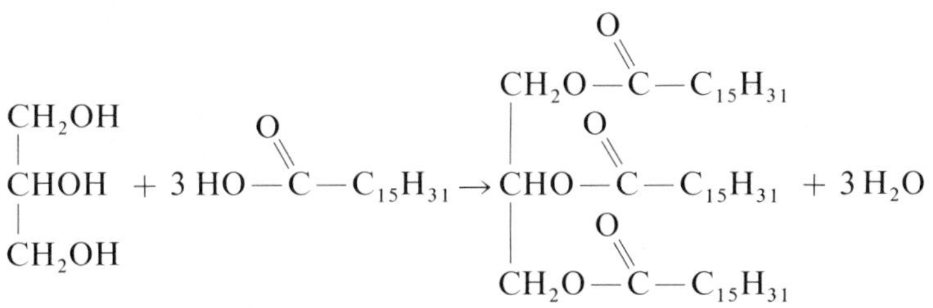

$$
\begin{array}{l}
CH_2OH \\
| \\
CHOH \\
| \\
CH_2OH
\end{array}
\ + \ 3\ HO{-}\overset{\displaystyle O}{\overset{\|}{C}}{-}C_{15}H_{31}
\ \rightarrow \
\begin{array}{l}
CH_2O{-}\overset{\displaystyle O}{\overset{\|}{C}}{-}C_{15}H_{31} \\
| \\
CHO{-}\overset{\displaystyle O}{\overset{\|}{C}}{-}C_{15}H_{31} \\
| \\
CH_2O{-}\overset{\displaystyle O}{\overset{\|}{C}}{-}C_{15}H_{31}
\end{array}
\ + \ 3\ H_2O
$$

Glycerol   Palmitic acid   Fat (Glyceride)

Animal fats, which are mostly solid at room temperature, are formed with fatty acids containing saturated paraffin chains, whereas the liquid vegetable oils have unsaturated olefin chains. In recent years, heart specialists have

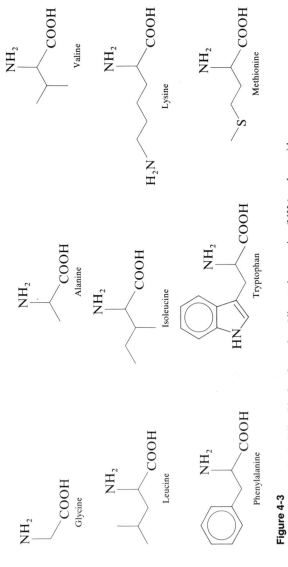

Figure 4-3
Amino acids are the building blocks of proteins. All contain an amino (NH_2) and an acid (COOH) group. All of these, except glycine and alanine, are essential in our diet.

encouraged the consumption of vegetable oils in preference to animal fats, since the former are believed to reduce solid fat deposition on arteries. Some typical fatty acids that form fats and oils are shown below. In fatty-acid nomenclature, 16:0 means a C_{16} acid with no double bonds, while 18:2 means a C_{18} acid with two double bonds (diolefin).

Palmitic	16:0	$C_{15}H_{31}COOH$	Linoleic	18:2	$C_{17}H_{31}COOH$
Stearic	18:0	$C_{17}H_{35}COOH$	Arachidonic	20:4	$C_{19}H_{31}COOH$
Oleic	18:1	$C_{17}H_{33}COOH$	Behenic	22:0	$C_{21}H_{43}COOH$

Palmitic and oleic acids are the most widespread of all fatty acids. They have been identified in most animal and vegetable fats. Stearic acid is dominant in animal fats, but only traces occur in vegetable fats. Butyric acid (4:0) occurs only in the milk fat of mammals.

Palmitoleic acid (16:1) is most abundant in marine fats. Oleic acid comprises more than 75 percent of the fatty acids of olive and almond oils and of many fruit and seed fats. It is also the chief depot fat of herbivorous animals. Linoleic acid occurs about as frequently as oleic acid in vegetable fats.

Many of the fatty acids have strong odors, whereas the fats and oils are relatively odorless. Fats and oils become rancid by hydrolysis, which releases the odorous acid.

Fats are stored in plants mainly in the seed, fruit, and spore; and in animals in the subcutaneous and inner muscular tissue and in the abdomen. Depot fats are mainly glycerides, whereas the fats deposited in active tissues, such as brain, liver, and kidney tend to be more complex. They are formed from alcohols other than glycerin. The most important of these are the phospholipids, the best known of which are the lecithins. Phospholipids can interact with both water-soluble proteins and insoluble lipids, so they are important in transporting materials in body fluids. They comprise half the lipids in blood plasma and are widely distributed in brain tissues. They also regulate the permeability of cell membranes.

The waxes of plant and animal origin are esters of high molecular weight, alcohols and fatty acids. They also include some of the uncombined higher alcohols and acids and saturated hydrocarbons. Beeswax, for example, is formed from paraffin alcohols containing 25 to 34 carbon atoms. Spermaceti, woolfat, applewax, and carnauba wax also are formed from higher alcohols.

The word *sterol* is derived from the Greek *stereos,* meaning "solid" and *-ol,* meaning "alcohol." Cholesterol (Figure 4-4) is the best known of the solid alcohols. It occurs in large amounts in brain and nerve tissues, as well as in the blood. Provitamin D and the sex pheromones are sterols. The bile of mammals is formed from acids containing sterol-like structures.

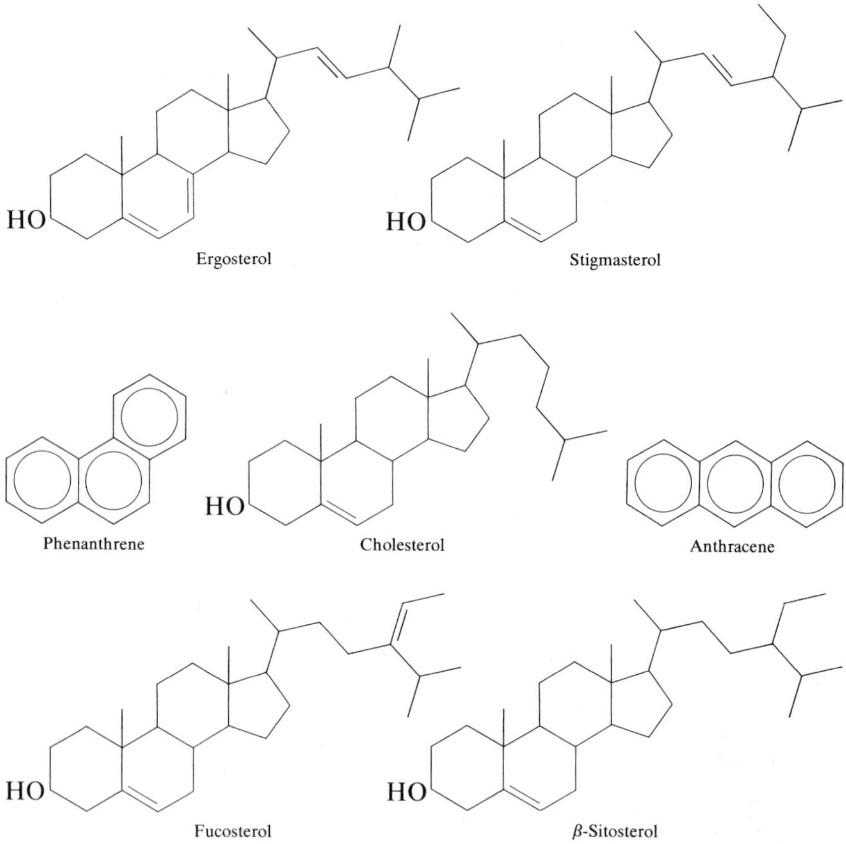

Figure 4-4
Some of the sterols that are precursors to steranes are here compared to the phenanthrene and anthracene structures. Sterols are optically active, and their maturation derivatives cause optical activity in the C_{25}–C_{35} range of crude oil. The sterol structure eventually matures to phenanthrenes, which are far more abundant in petroleum than the isomeric anthracenes that are not related to sterols.

The most important role of the insoluble lipids in body processes is in regulating the transfer of water-soluble substances, such as proteins. In addition, fats are essential in maintaining a healthy skin and as a stored food supply for pathogenic conditions. Fats are also used for body insulation and for controlling buoyancy in marine animals.

Fats can be hydrolyzed, and some can be decarboxylated to yield hydrocarbons, as will be discussed later. Consequently, both fatty acids and fatty alcohols are potential precursors to some of the hydrocarbons in petroleum.

Resins

The resins that bleed from the cut and damaged surfaces of tree trunks have one of the highest resistances to chemical and biological attack of all plant products. Trees also contain resins in their heartwood and on their leaf surfaces. Many of these resins are composed of unsaturated polycyclic acids that polymerize on exposure to air, forming a hard, tough layer over the wounded surface. Because of their resistance to decay over geologic time, resins containing perfectly preserved insects have been found in ancient sediments. Thomas (1969) has published a detailed survey of resins from the genus *Agathis,* a group of conifers found in land areas of the South Pacific. Three diterpene acids and their derivatives, whose structures are shown below, form the main constituents of the resins.

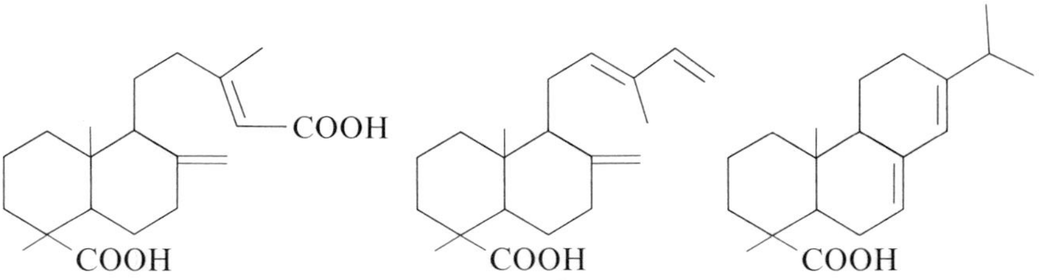

Agathic acid   Communic acid   Abietic acid

Small amounts of monoterpenes—mainly pinene—and some phenolic material are also present. The hardening of the resins on exposure is due to polymerization of the conjugated dienes. There also may be oxidation and cross linking of structures. Some 30 to 50 percent of the material remains unpolymerized, acting mainly as a plasticizer. This includes small amounts of hydrocarbons, methylesters, and acetates.

Resins are the progenitors of the resinite maceral of coal. Dehydrogenation yields nonlinear tricyclic hydrocarbons, comparable to those found in sediments and crude oil. Snowdon (1977) reported high concentrations of tricyclic diterpanes in the extracts of drill cuttings from a well in the Mackenzie Basin of northern Canada. Deroo et al. (1977) found tricyclic hydrocarbons, presumably derived from conifer resins, dominant in the cycloalkane fraction of a Cretaceous crude oil of the Western Canada Basin.

Essential Oils

In plants, those compounds that are volatile with steam and can be separated as an oily layer in the distillate are the essential oils. These include oils of wintergreen, clove, cinnamon, rose, orange, camphor, cedar, pine, and eucalyptus. Conifers contain essential oils in all their tissues, whereas in roses

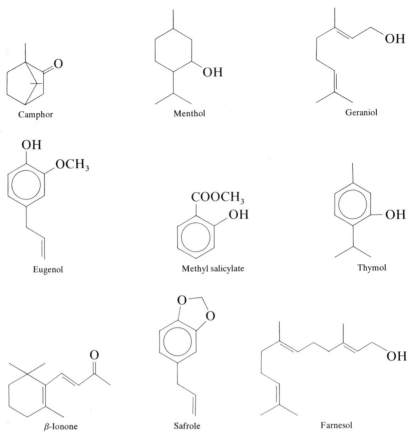

Figure 4-5
The fragrant odor of plants is caused by the essential oils shown here. These are
oxygenated derivatives of the terpenes (isoprenoids). They are readily reduced during
diagenesis to the hydrocarbons found in crude oil. For example, thymol would form
methylisopropylbenzene.

they occur mainly in the petals. The perfume industry extracts thousands of
tons of blossoms to obtain essential oils. It takes 8 tons of rose petals to yield 1
pound of attar of roses. The essential oils include a wide variety of chemical
structures, such as hydrocarbons, alcohols, ketones, aldehydes, acids, esters,
and organic bases. Most of these compounds contain only one oxygen atom to
ten or more carbon atoms, so they are very close to hydrocarbons in
composition.

Many of the more fragrant essential oils contain the 6-carbon ring
structure, which was first named *aromatic* because of this association with
pleasant odors. A. Kekulé, who was the first to describe the aromatic
structure, isolated it from such aromatic compounds as oil of bitter almonds.

Some of the typical essential oils are shown in Figure 4-5. Camphor, an important industrial chemical, is derived from the camphor tree, native to Taiwan. Menthol is the dominant oil in peppermint. Geraniol is the fragrant oil occurring in roses, geraniums, lavender, and other flowers. The violet odor is produced by β-ionone, which has a configuration similar to the isoprenoid carotene. Eugenol is the main constituent of oil of cloves. On oxidation, it is converted to vanillin, the common vanilla flavoring. Thymol occurs in many essential oils, particularly oil of thyme. Methylsalicylate is well known as the fragrant oil of wintergreen. It also has medicinal uses, since it is structurally related to aspirin. Farnesol is a fragrant oil found in orange blossoms and related plants. Safrole is the chief constituent of oil of sassafras.

The essential oils also include the terpene hydrocarbons, which are discussed in more detail later.

Plant and Animal Pigments

The most common pigment in plants is chlorophyll, the green coloring matter, which is involved in the photosynthesis process as previously discussed. Chlorophyll (Figure 4-6) is formed by joining four pyrrole rings to form the basic porphyrin structure and by then adding various carbon substituents to the ring, including the long phytol chain shown. A magnesium atom is chelated to the nitrogen in the center of the ring. Chlorophyll is in the plant chloroplasts and is usually accompanied by one or more of the yellow pigments, the carotenes and xanthophylls. Chlorophyll is hydrolyzed in soils and in the digestive system of animals to yield the long-chain alcohol phytol $C_{20}H_{39}OH$. Phytol can be further reduced to form phytane or oxidized and decarboxylated to form pristane, as shown in Figure 4-6. Both of these compounds occur in petroleum and are widespread in sediments. They are important biological markers.

Hemin is the red coloring matter of animal blood. It contains the same parent porphyrin ring structure as chlorophyll, but with iron in place of magnesium and with different carbon groups on the rings in place of phytol and the other substituents of chlorophyll.

Another major group of pigments are the carotenoids, which cause the light yellow to deep red colors of plants. These include hydrocarbons, alcohols, ketones, and carboxylic acids. A typical member is β-carotene, a hydrocarbon with the formula $C_{40}H_{56}$, as shown in Figure 4-7. The yellow of carrots and many fruits is due to β-carotene. Lycopene, the red coloring matter of tomatoes has a similar structure, but with the terminal rings open (Figure 4-7). Xanthophylls are yellow, oxygenated carotenoids. Lutein, the most widespread xanthophyll, is in both plant and animal material (for example, in egg yolks).

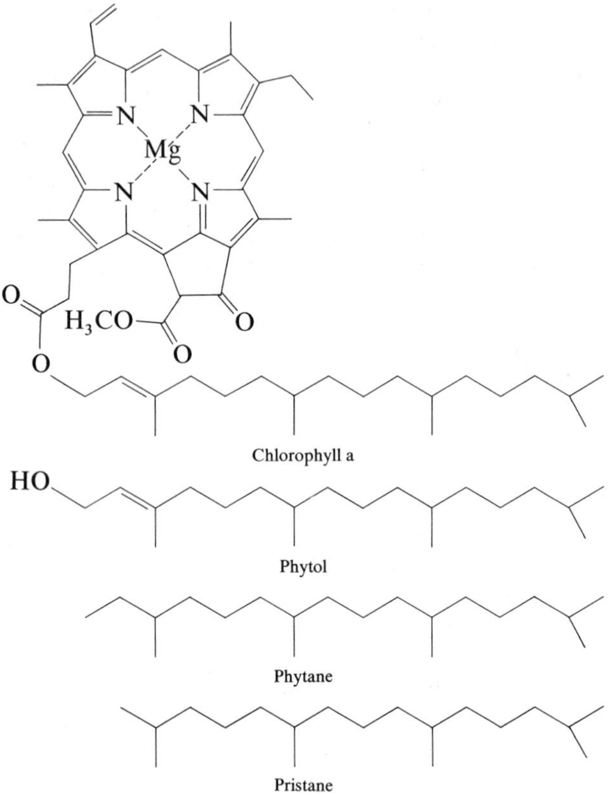

Figure 4-6
Chlorophyll and three of its derivatives: phytol, phytane, and pristane.

Carotenoids occur in all green plant tissue and are also found in some nonphotosynthetic bacteria and fungi. Vitamin A is an alcohol formed from the carotenoids.

Average Composition of Life Substances

The average chemical composition of the major constituents of living organisms is shown in Table 4-1, compared to petroleum. It is evident that the lipids can be converted to oil by the loss of a small amount of oxygen, whereas considerable oxygen would need to be removed to form hydrocarbons out of carbohydrates or lignin. Both oxygen and nitrogen would need to be removed from proteins. The ratio of carbon atoms to the N, S, and O atoms is approximately 1/1 in carbohydrates, 3/1 in proteins, and 10/1 in lipids.

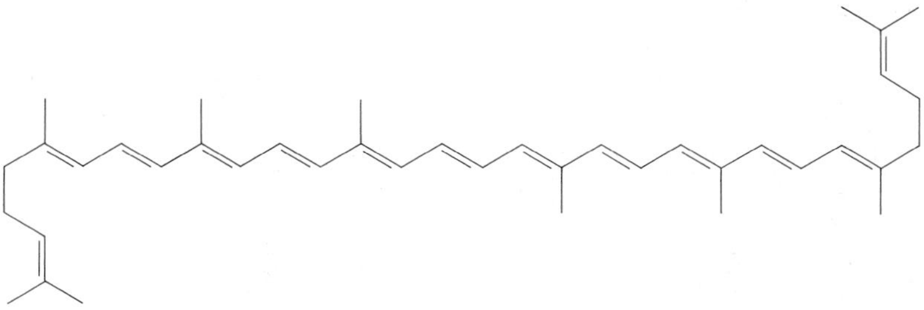

Lycopene

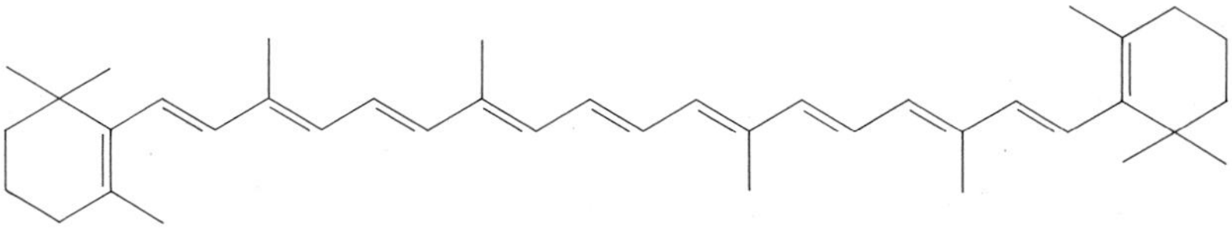

β-Carotene

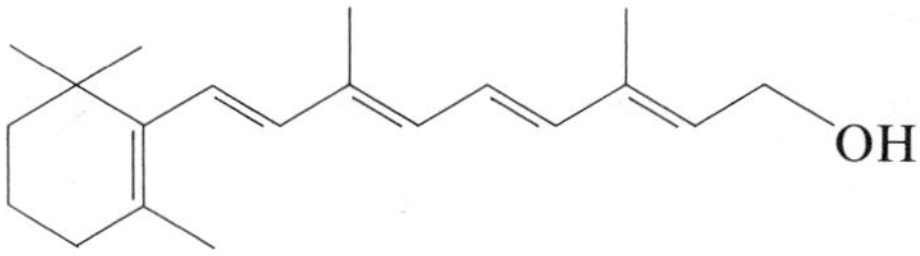

OH

Vitamin A

Figure 4-7
Carotenoids contain eight isoprene units. The beautiful colors of autumn leaves are
due to carotenoids. Organisms synthesize two vitamin A molecules by cleavage of a
β-carotene molecule.

Diagenetic degradation reactions acting on equal amounts of all four of these
substances in a reducing environment would produce more hydrocarbons
from the lipids than the other materials.

The quantity of lipids and other major constituents of typical fauna and
flora are shown in Table 4-2. Plants contain predominantly carbohydrates,
with the higher forms having lignin for strength. Animals are predominantly

Table 4-1
Average Chemical Composition of Natural Substances

	Elemental composition in weight percent				
	C	H	S	N	O
Carbohydrates	44	6			50
Lignin	63	5	0.1	0.3	31.6
Proteins	53	7	1	17	22
Lipids	76	12			12
Petroleum	85	13	1	0.5	0.5

Table 4-2
Composition of Living Matter

Substance	Weight percent of major constituents			
Plants	Proteins	Carbohydrates	Lipids	Lignin
Spruce wood	1	66	4	29
Oak leaves	6	52	5	37
Scots-pine needles	8	47	28	17
Phytoplankton	23	66	11	0
Diatoms	29	63	8	0
Lycopodium spores	8	42	50	0
Animals				
Zooplankton (mixed)	60	22	18	0
Copepods	65	25	10	0
Oysters	55	33	12	0
Higher invertebrates	70	20	10	0

Note: Dry, ash-free basis. There is great variability for different species of each organism. For example, Blumer et al. (1964) has reported *Calanus* copepods to contain 27 to 57 percent lipids. Lipid variability is partly due to nutrient availability and health of the organisms. If food supply is limited or there is crowding during growth, the organism increases its lipids. *Chlorella* grown in a favorable environment will contain 20 percent lipids, in an unfavorable environment, 60 percent lipids.

protein. Marine organisms such as corals and sponges also have mostly protein in their $CaCO_3$ matrix.

The lipid content of all forms of living matter is more than enough to account for the origin of oil. As mentioned in Chapter 2, less than 1 percent of the organic matter in sediments is required to form all the known petroleum. Lipids are more resistant to degradation in a reducing environment than are proteins and carbohydrates. Hydrocarbons are the most stable part of the lipids. Since these may be among the most important progenitors of early petroleum, they are discussed in more detail in the following section.

Hydrocarbons of Biogenic Origin: The Biological Markers

In the early work of API (American Petroleum Institute) Project 43 on the origin of petroleum, F. C. Whitmore and his associates isolated from kelp a whole series of hydrocarbons containing from 19 to 34 carbon atoms. Whitmore proposed that marine organisms alone could provide 60 million barrels of hydrocarbons a year, more than enough to account for that found in sediments (Whitmore 1943). It is now recognized that many of these hydrocarbons are cycled through the food chain, oxidized, and otherwise degraded. Only part of them is deposited and buried. From the standpoint of petroleum origin, it is important to determine the quantity of hydro-carbon surviving in the sediments. The types of structures synthesized by living organisms can give us considerable insight regarding the mechanisms and pathways of petroleum origin by tracing these biological markers from the living organisms to the sediments to the final petroleum accumulation.

Gerarde and Gerarde (1961–1962) have tabulated the occurrence of a large number of hydrocarbons, mostly in continental plants and animals. These include paraffins, cycloparaffins, olefins, and a few aromatics. In recent years, the biosynthesis of some of these hydrocarbons has been clarified through the use of ^{14}C-labeled progenitors. These studies have shown that there are several important pathways to hydrocarbon synthesis. Among the more significant hydrocarbon groups are the isoprenoids, the odd-numbered normal paraffins, and the iso- and anteiso-paraffins.

The Isoprenoid Hydrocarbons

The most important biosynthetic pathway that has existed since the beginning of life is the polymerization of the 5-carbon isoprene unit. The precursor of the phytol chain on chlorophyll and many of the lipids previously discussed, such as the sterols, the essential oils, and many plant and animal pigments, as well as some vitamins and hormones are all biosynthesized originally from isopentenyl pyrophosphate (Figure 4-8). The hydrocarbon portion of this unit polymerizes by displacement of the pyrophosphate group to give larger compounds in multiples of five carbon atoms.

The terpenes start with the C_{10} (mono-) and continue through the most common structures, as follows: C_{15} (sesqui-), C_{20} (di-), C_{30} (tri-), and C_{40} (tetraterpenes, including carotenoids). The occurrence of isoprene units is so general biologically that the "isoprene rule" was developed to assist in structure analysis. The rule states that a natural product containing some multiple of C_5 units probably has a structure divisible into isoprene units.

Some of the typical hydrocarbons formed by polymerization of the isopentyl unit are shown in Figure 4-8. The monoterpenes ($C_{10}H_{16}$) occur in the resins and oils of plants. Pinene is from conifers, limonene from lemon grass oil, and camphene from the camphor tree. Turpentine is largely a mixture of monoterpenes and their derivatives.

Sesquiterpenes (C_{15}) fall into two groups, depending on the naphthalene derivative formed on dehydrogenation: the codalene type, of which bisabolene is representative, and the eudalene type, such as selinene. Codalene dimers are in cotton seed, and selinene is in oil of celery. Some of the trimethylnaphthalenes in petroleum are probably derived from sesquiterpenes.

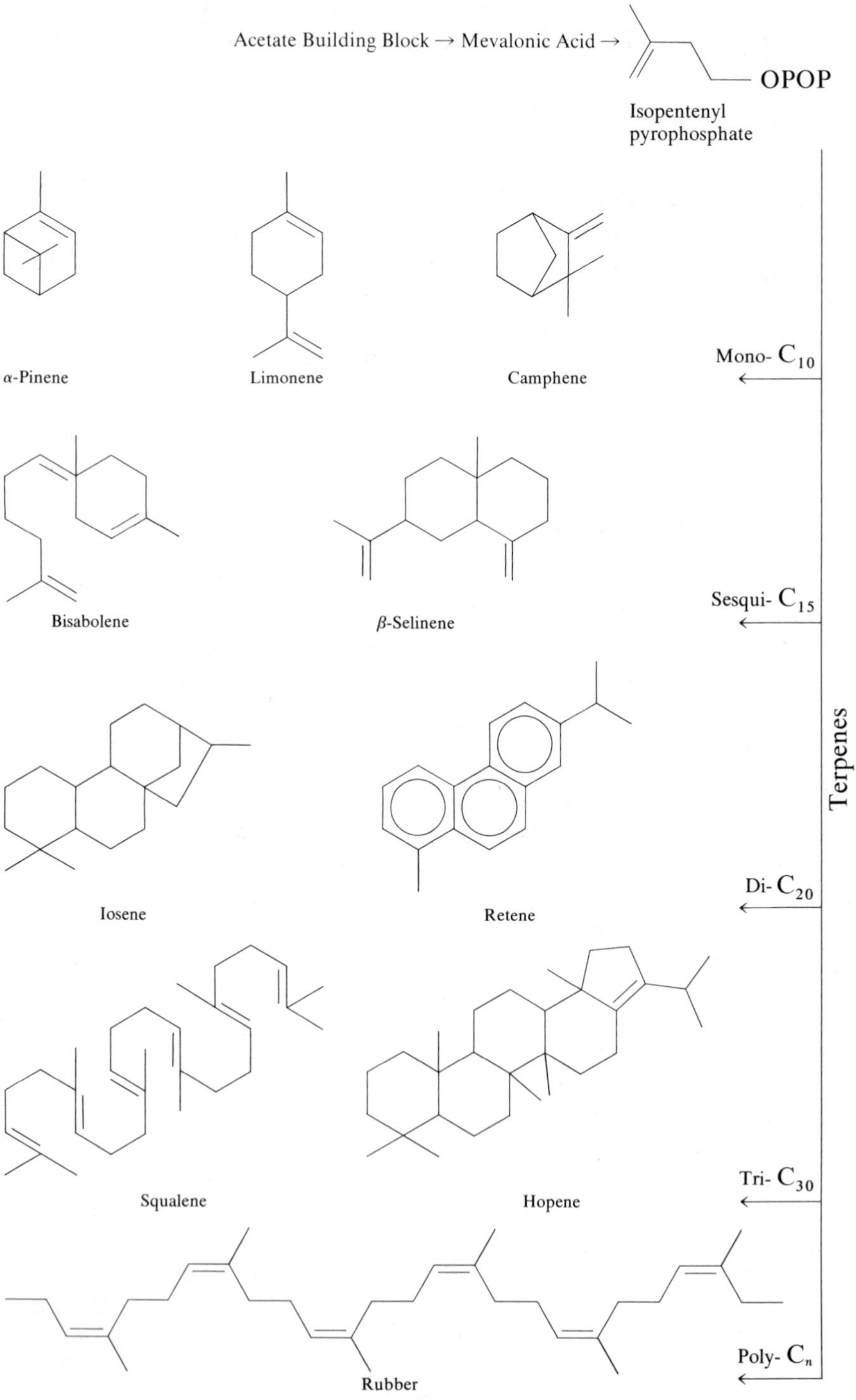

Acetate Building Block → Mevalonic Acid →

OPOP

Isopentenyl
pyrophosphate

α-Pinene

Limonene

Camphene

Mono- C_{10}

Bisabolene

β-Selinene

Sesqui- C_{15}

Iosene

Retene

Di- C_{20}

Squalene

Hopene

Tri- C_{30}

Rubber

Poly- C_n

Terpenes

Azulenes are blue, sesquiterpene hydrocarbons containing a conjugated highly reactive system of double bonds in two fused rings, shown as follows:

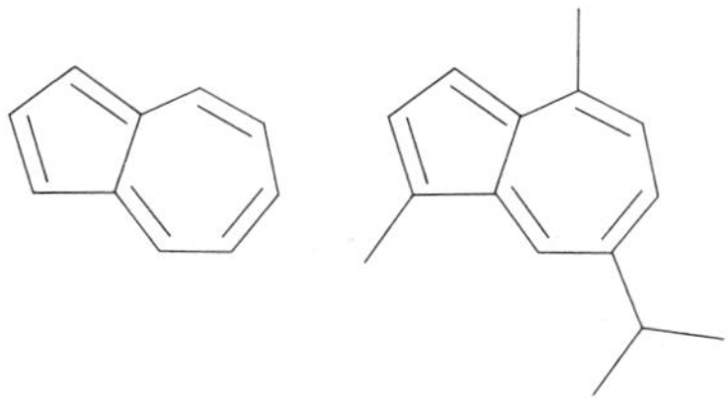

Azulene   Guaiazulene

Guaiazulene is the most abundant natural azulene from sources such as trees. Some sesquiterpene alcohols form substituted azulenes on fossilization. The azulenes were once involved in a fanciful hypothesis on the origin of oil. These terpenes were believed to cause the blue haze over certain forests, and when it rained the hydrocarbons were believed to return to earth to accumulate as oil. Terpenes do accumulate in the air over some plants. The dominant terpene in the blue haze is α-pinene. However, terpenes oxidize on prolonged exposure and are subject to bacterial attack. They may contribute to oil when buried unaltered in reducing sediments, as do other biogenetic hydrocarbons.

The diterpenes (C_{20}) are the most common isoprene derivatives in both living and fossil organisms. Most conifer resins are terpene acids, such as abietic acid (previously shown).

One of the best biological markers going from living organisms to sediments to crude oil is the diterpenoid phytol, which, as previously discussed, yields the hydrocarbons pristane and phytane. Blumer (1967; Blumer and Souza 1965; Blumer et al. 1963, 1971) has followed the movement of isoprenoid alkanes through the marine food chain. Algae at the beginning of the chain contains phytol as part of its chlorophyll, traces of pristane, and no phytane. The *Calanus* species of copepods (zooplankton) feeding on the algae contain 1 to 3 percent pristane in their body fat (fat content ranges from 27 to 57 percent). Mixed zooplankton from the Gulf of Maine were found to contain four C_{20} diolefins and three C_{19} mono-olefins, presumably derived from phytol. The mono-olefins were also found in the liver oils of various marine fishes and mammals that feed on the zooplankton. A more detailed study of the hydrocarbons in the digestive tract and liver of a basking shark showed the presence of five C_{19} olefins, four C_{20} diolefins, and

Figure 4-8
The synthesis of terpenes in nature from the five-carbon isopentenyl group. The tetra-C_{40} terpenes, carotene and lycopene, were shown in Figure 4-7. These are typical biological markers. Many of these structures, with slight modification, are recognized in shale extracts and crude oils.

pristane—all derived from phytol. Han and Calvin (1969) found pristane and phytane in both photosynthetic and nonphotosynthetic bacteria. Their results indicate that anaerobic, nonphotosynthetic bacteria may have an enzymatic system that can produce pristane directly in the absence of the phytol chain of chlorophyll.

Squalene ($C_{30}H_{50}$) is the most common triterpene (C_{30}). When Tsujimoto isolated it from shark liver oils in 1906, he supplied the first evidence that isoprenoids occur in the animal kingdom. Prior to this, isoprenoids had been thought to be exclusively plant products. Squalene occurs in most marine organisms, although only the higher fauna appear to accumulate large amounts of it. For example, copepod lipids contain 0.01 percent squalene, whereas the lipids of basking shark livers may contain more than 20 percent squalene (Blumer 1967).

The importance of squalene (Figure 4-8) to all living things has been clearly shown by a series of experiments that have established it as the biological precursor of polycyclic triterpenoids in both plants and animals, such as the sterols (Clayton 1965). Cholesterol ($C_{27}H_{45}OH$) from animal and plant tissues, sitosterols ($C_{29}H_{49}OH$) from higher plants (Figure 4-4), ergosterol ($C_{28}H_{43}OH$) from yeast, and fucosterol ($C_{29}H_{47}OH$) from algae are some of the naturally occurring sterols. Cholesterol, β-sitosterol, and fucos- terol are the most abundant free sterols in deep seawater (Gagosian 1976). Desouza and Nes (1968) have reported steroids in several species of algae. Their counterparts in sediments and crude oils are the steranes and related tetra- and pentacyclic triterpanes. Much of the optical activity of crude oil is caused by triterpane structures originally derived from squalene.

The dominant tetraterpenes (C_{40}) are carotenes, which have already been discussed. Carotenes are photochemically active precursors to the diterpene vitamin A. This vitamin is highly concentrated in the liver oils of fish. Carotenes and their reducing product carotane have both been identified in sediments.

The polyterpenes are represented in nature by rubbers, such as gutta- percha and caoutchouc, the latter containing about 1,000 isoprene units. The latex of rubber trees and the fossil rubber balls occasionally found in lignites are examples of these natural polyterpenes.

The presence of many terpenoid-derived structures as hydrocarbons in crude oil, such as pristane, phytane, p-cymene, cholestane, triterpanes, and carotanes, constitutes one of the many convincing arguments for the biological origin of petroleum.

Odd-Numbered Straight Chains

One of the more intriguing aspects of the biological synthesis of hydrocar- bons is the control over the unsaturated (double-bond) content and the length of the chain. Long-chain hydrocarbons are waxes. Short chains are liquids.

Olefin (unsaturated) hydrocarbons are liquids. Saturated long-chain alkanes are solids. Thus the alkane $C_{21}H_{44}$ is a solid (melting point $= 41°C$, or $106°F$) but the olefin $C_{21}H_{42}$ is a liquid (MP $= 3°C$, or $37°F$).

Marine organisms need liquids for internal lipids for food storage, insulation, and buoyancy in water. Land plants need solid waxes for external lipids to prevent water from leaving the organism and to minimize mechanical damage and inhibit fungal and insect attack on external surfaces. Consequently, marine organisms synthesize mostly liquid paraffins and olefins up to about C_{21}. Land plants synthesize hydrocarbon waxes to C_{37}. Waxes also contain esters, fatty acids, and alcohols. The waxes in crude oil come almost entirely from land plants.

The lipids both of marine and land plants contain from 1 to about 80 percent paraffin hydrocarbons. One of the most hydrocarbon-rich algae is *Botryococcus braunii,* which Maxwell et al. (1968) have reported can contain 76 percent of its dry weight as hydrocarbons. The percentage of hydrocarbons will vary with the availability of nutrients during growth. Boghead coals, such as the Scottish torbanite and coorongite, consist almost entirely of colonies and remains of *B. braunii.* At the other extreme is the Brazilian palm leaf (carnauba) wax, with only 1 percent hydrocarbon. Animal lipids contain 1 to 10 percent hydrocarbons, with insect lipids having up to 75 percent hydrocarbons. Approximate ranges of lipids and hydrocarbons in organisms are shown in Table 4-3.

A most significant discovery regarding biosynthesized normal paraffins was made by Chibnall and his associates in 1934. He found that plants synthesize almost exclusively paraffins with an odd number of carbon atoms in the chain. He identified odd-numbered paraffins from C_{25} to C_{37} in a variety of plants. Accompanying the paraffins were alcohols and acids with exclusively even-numbered carbon chains from C_{24} to C_{36}. Even-numbered paraffin chains and odd-numbered acids and alcohols were present, but in very small amounts (Waldron et al. 1961). Table 4-4 shows the data of Chibnall and Waldron and others on paraffins in land plant lipids. The most dominant

Table 4-3
Lipids and Hydrocarbons of Various Organisms

	Weight percent, dry basis	
	Lipids	Hydrocarbons
Land plant leaves	1–15	0.04–10
Coastal grasses	1–2	0.01–0.07
Phytoplankton	4–28	0.01–0.5
Zooplankton	3–57	0.10–1
Bacteria	6–35	0.01–3

Table 4-4
Straight-Chain Hydrocarbons in Land Plants

Source	Carbon atoms in chain																
	21	22	23	24	25	26	27	28	29	30	31	32	33	34	35	36	37
Apple skin							12	2	86								
Grape skin	1	1	6	4	17	4	20	3	22	2	15	1	2				
Rose petal	3		3		3	1	18	2	21	1	33	1	11				
Palm tree leaf (carnauba)	1			2	6	1	15	3	25	1	29		6				
Sempervivoideae tree leaf			1	2	4	5	12	7	22	6	31	1	4				
Sugar cane wax			1	1	7	5	56	3	13	2	4		2				
String bean wax						1	3	2	20	3	60	2	9				
Runner bean leaf							2	2	42	2	45	2	5				
Brussels sprout									88	2	10						
Turkish tobacco leaf					2	1	12	2	11	7	34	10	19				
Cactus leaf	1										2	1	15	6	65		11
Sunflower seed oil	1	2	3	3	3	3	12	3	37	3	25	1					
Barley				2	7	3	16	3	28	3	27	3	5	3			
Clover			6				7	1	23	7	38	3	8				
Cocksfoot grass							1	1	35	4	49		10				
Rye grass							7	1	40	5	39		8				
Oats				2	18	2	22	2	23	2	22	1	5	1			

Note: Values shown are percent of total straight-chain hydrocarbons. Values less than 1 are not shown. Up to 4 percent of green leaves and up to 15 percent of dried leaves is wax, of which 1 to 80 percent is hydrocarbon.

Source: Data from Waldron et al. 1961, Vandenburg and Wilder 1970, and a compilation by Clark 1966.

hydrocarbons are C_{27}, C_{29}, and C_{31}. About 90 percent of the paraffin fraction of apple skin, brussels sprouts, broccoli, and cabbage is a C_{29} alkane. Tree leaf and grass waxes have about equal amounts of C_{29} and C_{31} alkanes. The odd-carbon chain length strongly predominates over the entire range from C_{21} to C_{37}. The cactus leaf, which needs a hard wax of particularly high molecular weight to conserve water, has its dominant hydrocarbon at C_{35}.

Kaneda (1969) observed that the internal hydrocarbons isolated from spinach leaves showed the same C_{29} and C_{31} predominance as did the external hydrocarbons, but they also contained a series of hydrocarbons from C_{16} to C_{26} that had no odd/even predominance. However, since there were 15 times as much external as internal hydrocarbons, the latter would be of lesser significance in terms of preservation of the odd predominance.

Low-molecular-weight, odd-carbon alkanes with C_7, C_9, and C_{11} chain lengths are found in the oil of certain pine trees but are not common in other plants.

Marine plants, compared to land plants, synthesize smaller odd-carbon chain lengths of C_{15}, C_{17}, C_{19}, and C_{21}, as shown in Table 4-5. Furthermore, the marine plants contain a higher percentage of odd-numbered, straight-chain olefins having from one to six double bonds in the C_{15} to C_{21} range. The olefins, which are shown in italics in Table 4-5, are predominantly composed of C_{19} and C_{21} chain lengths. Blumer et al. (1970) have shown that several species of plankton algae contain only one hydrocarbon, an olefin with six double bonds called *3,6,9,12,15,18-heneicosahexaene* ($C_{21}H_{32}$).

A fresh- to brackish-water alga, *Botryococcus,* has been synthesizing large amounts of lipids since the Carboniferous. It has the expected odd-carbon-chain predominance of continental organisms. An analysis of the *n*-paraffins in *Botryococcus braunii* showed the following chain lengths: 56 percent C_{29}, 28 percent C_{31}, 11 percent C_{27}, 2 percent C_{33}, and 2 percent C_{17}. The *Botryococcus* is the principal organic constituent of many boghead coals.

Grasses growing in marine and brackish coastal waters show an odd-numbered-hydrocarbon predominance intermediate between the marine algae and the land plants. Dominant chain lengths appear to be C_{21}, C_{23}, and C_{25}. The *Ruppia*, which grows equally well in fresh and marine waters, peaks at C_{27}.

Although plants contribute most of the organic matter to sediments, it is of interest to compare hydrocarbon distributions in bacteria and higher organisms (Table 4-6). In general, the hydrocarbons of bacteria lipids show no odd-carbon predominance, although in several species the C_{17} chain length is dominant. Over half of the hydrocarbon in the marine bacteria *Vibro marinus* is a C_{17} olefin.

The hydrocarbon fractions of both marine and land plants previously discussed contain only traces of the isoprenoid hydrocarbons. In contrast, some bacteria contain more pristane, squalene, and related isoprenoid

Table 4-5
Straight-Chain Hydrocarbons in Marine Plants

Plant species	Carbon atoms in chain																			
	14	15	16	17	18	19	20	21	22	23	24	25	26	27	28	29	30	31	32	33
Rhizosolenia setigera								86												
Eutrepiella sp.				10				*4*												
Ascophyllum nodosum		98		19				77												
Corallina officinalis		14		76												1				
Sargassum	2	55	2	7	2	2	1	1	1	1	1	2	2	3	5	6	4	3		
Spongomorpha arcta						*17*		*82*												
Ectocarpus fasciculatus		7		1				*91*												
Pilayella littoralis								*98*												
Scytosiphon lomentaria		38		11 / *20*		1 / *23*					*1*									
Chorda tomentosa		31				5		*64*												
Laminaria digitata		64		2		17		*14*												
Ascophyllum nodosum		56 / *4*		1 / *29*		*2*	2	*6*												
Fucus vesiculosis		65		2				*16*												

Porphyra leucosticta	*15*		17	68							
Tribonema aequale	14	32 *54*									
Rhodymenia palmata	1	99									
Polysiphonia urceolata		96	3								
Coelastrum microsporum		100									
Scenedesmus quadricauda	1	26	7				43				
Tetrahedron sp.		30	40	3	20	6					
Anacystis nidulans	23	8	44 *20*	2							
Anacystis montana		12	9	8	15	4	38				

Brackish and Marine Coastal Plants (grasses)

Ruppia	16	4	3	8	1	22	2	51	2	13	1		
Diplanthera	12	2		16	1	20	1	13	1	8	1		
Syringodium	6	1	4	20	2	44	1	7	3	3	1	5	1 5
Halophila	14	4	23	1	20	2	36						
Thalassia	9	1	11	2	34	4	26	3	6	1	3		

Note: Values shown represent percent of total straight-chain hydrocarbons. Values less than 1 are not shown. Saturated paraffins are in regular type, olefins are in italics. Hydrocarbon concentrations in dry algae ranged from about 10 to 500 ppm with the average around 200 ppm. *Tribonema aequale* is the only freshwater alga on this list. *Ruppia* grows in both fresh and saline waters.

Source: Data from Blumer et al. 1971, Clark and Blumer 1967, Youngblood et al. 1971, Gelpi et al. 1970, and Attaway et al. 1970.

Table 4-6
Hydrocarbons in Bacteria and Higher Organisms

Source	Carbon atoms in straight-chain hydrocarbons										
	15	16	17	18	19	20	21	22	23	24	25
Nonphotosynthetic bacteria											
E. coli		2	6	28	12	10	6	6	8	7	6
P. shermanii	2	3	13	4	4	4	4	4	3	2	1
Clostridium acidurici	1	14	50	5	5	3	2	2	1	1	1
Desulfovibrio essex 6		2		3	16	34	26	9	3	1	1
Desulfovibrio Hildenborough		2		3	10	17	11	5	3	5	8
Photosynthetic bacteria											
Vibrio marinus	4	2	24	2							
	3		*56*								
Rhodopseudomonas spheroides	1	3	43	19	19						
Chlorofrium sulfurbacteria	2	1	50	1	1	1	2	3	4	7	11
Rhodospirillum rubrum			4		1						
Rhodomicrobium nanniclii								1	1	1	1
Higher organisms											
Zooplankton lipids[a]	11		7		*112*	79					
Beeswax									4	1	8
Cow manure										1	2

[a]Values are percent of total hydrocarbons, except for zooplankton, which is in parts per million (ppm). Percent or ppm of olefins is shown in italics. Hydrocarbons in bacteria range from 100 to 400 ppm dry weight of cells.

Source: Data from Han and Calvin 1969, Oro et al. 1967, Blumer 1967, and a compilation by Clark 1966.

hydrocarbons than straight-chain hydrocarbons. This same difference is noted in zooplankton lipids, which contain over 50 times as much pristane and squalene as straight-chain hydrocarbons.

Bacteria partially modify the organic matter in sediments, but they do not seriously alter the odd-chain-length preference of biosynthesized hydrocarbons. This characteristic, as will be shown later, is not only a significant biological marker but also a useful tool in characterizing source rocks of petroleum. Knoche and Ourisson (1967) showed the stability of this odd-carbon preference in comparing a fresh horsetail plant (*Equisetum brongniarti*) growing today with one fossilized in the Triassic (2×10^8 years old). In both

26	27	28	29	30	31	32	33	34	35	Isoprenoid hydrocarbons		
										Pristane	Phytane	Squalene and others
3	3		1									
										47	1	
										2	1	
13	11	5		1						1		
										10	1	
13	2											
												95
1												93
										10^4		140
1	30	1	17	1	19	2	16					
2	8	4	18	5	27	5	22	2	4			

the fresh and fossil horsetail plants, the dominant hydrocarbons were C_{23}, C_{25}, C_{27}, and C_{29}. Furthermore, the relative proportions of these hydrocarbons in the living and fossil plants were almost identical.

The biosynthesis of these odd-numbered hydrocarbon chains is believed to occur via even-numbered fatty acid chains. The acetate building block (Figure 4-8), acetyl coenzyme A, which leads to isopentenyl pyrophosphate and all terpenes, is also the starting material for fatty acids. A special enzyme, fatty acid synthetase, hooks together the two carbon acetyl units ($CH_3C\equiv O$) on an assembly-line basis to form fatty acids with even-numbered carbon chains. These are then decarboxylated (loss of CO_2) to form the odd-numbered hydrocarbon chains that are widespread in nature. The length of the chain is controlled by the type of fatty acid synthetase in the organism. The enzymes in marine organisms make shorter chains than those in continental organisms.

Iso- and Anteiso-Branched paraffins

Another characteristic of biosynthetic hydrocarbons is the predominance of the 2- and 3-methyl-substituted carbon chains called *iso-* and *anteisoalkanes,* as shown below.

2-Methyldecane (iso-C_{11})   3-Methylnonane (anteiso-C_{10})

Detailed studies of plant paraffins have failed to uncover the many thousands of structural isomers that are theoretically possible. For example, only three biological alkanes of the formula $C_{31}H_{64}$ are known with certainty. These are the normal alkane, the 2-methylalkane, and the 3-methylalkane, which are present in decreasing abundance as listed. Theoretically, more than a million alkanes of $C_{31}H_{64}$ are possible. The fact that biosynthetic isoparaffins consist predominantly of 2- and 3-methylalkanes is typical of the simple spectrum of hydrocarbons in living organisms compared to the wide spectrum of hydrocarbons in ancient sediments.

The quantity of iso- and anteiso-branched paraffins is less than the normal paraffins in land plants, and they have not been identified in many marine plants. An example of the relative distributions of normal, iso-, and anteisoparaffins in tobacco leaf is shown in Table 4-7. The odd-carbon preference of the normal paraffins in both Turkish and Bright tobacco is

Table 4-7
Total Paraffin Hydrocarbons in Tobacco Leaf

Source	HC type	Carbon atoms in molecule										
		25	26	27	28	29	30	31	32	33	34	Total
Turkish	*n*	1	1	9	2	8	6	23	7	13		70
	i					2		7		5		14
	a						5		9		2	16
												100
Bright	*n*	2	1	6	2	6	3	25	4	7		56
	i					3		14		6		23
	a						7		11		3	21
												100

Note: Values rounded off to whole numbers. Code: n = normal, i = iso, a = anteiso. Total hydrocarbons constituted 2,000 to 2,800 ppm of the dry tobacco. Hydrocarbons smaller than C_{25} represented less than 0.5 percent of this. A small amount of olefins (neophytadiene) were present.
Source: Data from Mold et al. 1963.

evident. In the corresponding branched paraffins, the iso- derivatives occur only in molecules of odd-numbered carbon atoms, and the anteiso- derivatives in even-numbered chains. The quantities of the 2- and 3-methyl derivatives are present in almost equivalent amounts. The sum of the two branched paraffins ranges from 30 to 44 percent of total paraffins in the example shown. Iso- and anteisoparaffins have also been found in rose petal, wool, and sugarcane waxes. They probably exist in the waxes of most land plants.

A few midbranched paraffins have been identified as major hydrocarbons of marine plants. Sever (1970) found 7-methyl- and 8-methylheptadecane in *Nostoc muscorum* (14 percent) and *Lyngbya lagerhaimii* (9 percent of total alkanes). She quotes others who have found these same hydrocarbons in plants. The remaining hydrocarbons in these plants were straight-chain paraffins, as follows: *Nostoc* 83 percent C_{17}, 1 percent C_{16}, 1 percent C_{15}; *Lyngbya* 85 percent C_{17}, 3 percent C_{16}, 1 percent C_{15}.

Aromatic Hydrocarbons

Most aromatic rings in nature exist in the lignin structure, with some occurring in the essential oils and pigments. Most of these contain oxygen atoms in the molecule, although a few, such as *p*-cymene and styrene, are hydrocarbons.

p-Cymene   Styrene

The compound *p*-cymene occurs in the oils of thyme, eucalyptus, and caraway. Styrene occurs in styrax balsam. The occurrence of free aromatic hydrocarbons is rare, possibly because most of the aromatics are toxic to living organisms. None of the common aromatics of crude oil—benzene, toluene, the xylenes, and naphthalene—are known to occur free in living things although they can be formed from natural products and do occur in sediments.

The high-molecular-weight polycyclic aromatic hydrocarbons (PAH) are found in trace amounts in many fauna and flora. Yellow and wilting parts of plants may contain up to 100 μg/kg or 0.1 ppm. Borneff et al. (1968) reported that the alga *Chlorella vulgaris* grown in a ^{14}C acetate medium was able to synthesize the carcinogen 3,4-benzpyrene and other PAHs. Incorporation of the ^{14}C indicated synthesis of 10 μg PAH/kg of dry plant. Others have reported the synthesis of trace amounts of polycyclic aromatic hydrocarbons by plants and various bacteria. Hase and Hites (1976), however, point out that bioaccumulation of PAH by various organisms may explain some of the reported syntheses. Such PAH compounds as 3,4-benzpyrene are common constituents of soil samples from nonindustrial areas (Blumer 1961). The source is believed to be forest fires.

DIAGENESIS OF ORGANIC MATTER

When organisms die, their organic matter undergoes a variety of reactions, some microbial, such as the formation of methane by anaerobes, and some purely physical or chemical, such as dehydration and oxidation. The combined attack of weathering and microbes converts much of the organic matter either to gases that escape into the atmosphere or soluble products that are carried off by groundwater. In high-energy oxygenated environments, additional material is consumed by benthic filter feeders and burrowing organisms in nearsurface sediments. Environments that preserve unusually large amounts of organic matter in the sediments are stagnant lakes and silled basins, where the bottom waters are strongly reducing, as in the Black Sea. In such areas, the organic content of the sediment frequently exceeds 15 percent. At the other extreme are the red clays of the oceanic abyssal plains, where slow rates of deposition, aerobic waters, and little contribution of organic matter results in sedimentary organic contents of less than 0.1 percent.

Neither of these extremes is typical of most petroleum-forming environments. Oil is formed from the organic matter deposited in the aerobic waters of sedimentary basins, where the waters occasionally may be locally anaerobic but the sediments are nearly always anaerobic below the first few centimeters. The more resistant organic matter, including humic material, resins, waxes, and lipids are preferentially preserved. The organic content of such sediments, which eventually become source beds of petroleum, generally ranges between about 0.5 and 5 percent, with a mean around 1.5 percent.

In recent years, detailed studies of the organic structures in sediments and petroleum have made it possible to explain some of the mechanisms and conditions of oil formation. The overall process can be summarized briefly as follows: *The simple molecular hydrocarbon spectrum of living organisms becomes the complex spectrum of petroleum through the diagenetic formation of a wider group of hydrocarbon derivatives from the original organic molecules and the addition of large quantities of hydrocarbons formed by thermal alteration of deeply buried organic matter. The largest quantity of petroleum hydrocarbons is formed from organic matter heated in the earth to temperatures between about 60 and 150°C (140 and 302°F).*

For example, chlorophyll is the only porphyrin structure synthesized in significant amounts by living organisms. As chlorophyll is subjected to diagenesis at increasingly greater depths, it loses the phytol chain, and the green chlorins become red porphyrins, which gradually form an increasingly wider spectrum of derivatives. Ancient sediments contain thousands of porphyrins derived from the chlorophyll molecule. The simple phytol chain from chlorophyll forms pristane, phytane, and related derivatives that are found in sediments and crude oil. The relatively simple hydrocarbon pattern of algae (three or four hydrocarbons) becomes increasingly complex with burial and diagenesis. Eventually, at the increasing temperatures of greater depths, thermal and catalytic cracking reactions begin to break down the organic matrix (kerogen) to yield hundreds of hydrocarbons that mix with the original simple biogenic mixture. *The most important factor in the origin of petroleum is the thermal history of the source rock.*

Diagenesis (Figure 4-9) is the process of biological, physical, and chemical alteration of the organic debris prior to a pronounced effect of temperature. It occurs with burial over a depth range where temperatures are too low for significant cracking of large molecules to occur. The organic matter deposited in sediments consists primarily of the biopolymers of living things: the carbohydrates, proteins, lipids, lignin, and subgroups such as chitin, waxes, resins, glycosides, pigments, fats, and essential oils. Some of this material is consumed by burrowing organisms, some may be complexed with the mineral matter, while some is attacked by microbes that use enzymes to degrade the biopolymers into the simple monomers from which they were originally formed. Some degraded biomonomers undergo no further reaction but others condense to form complex high-molecular-weight geopolymers that are the precursors of kerogen. During diagenesis, this complex mixture of geo- and biopolymers and monomers undergoes a whole series of low-temperature reactions that result in the formation of more hydrocarbon-like materials through the loss of oxygen, nitrogen, and sulfur. In this text, diagenesis is defined as covering the temperature range from the surface to about 50°C (122°F).

The process by which organic material is altered due to the effect of increasing temperature is called *catagenesis*. Temperatures in the earth increase from about 2 to 5 degrees C per 100 meters (1 to 3 degrees F per 100 feet) of depth. A linear increase in temperature causes a logarithmic increase in reaction rate for most reactions involving petroleum formation. Increasing temperatures also increase the solubility of some organic compounds in the sediment fluids and convert solids to liquids and liquids to gases, thereby increasing their ability to migrate. In this text, catagenesis is defined as ranging from about 50°C (122°F) to about 200°C (392°F). Higher temperatures are considered to represent *metamorphism,* during which all organic material is ultimately converted to methane and graphite.

It should be emphasized that the changes from diagenesis to catagenesis and from catagenesis to metamorphism are gradual, covering a temperature range rather than an isotherm.

ORGANIC MATTER AND PROCESSES   MATURATION RANGE

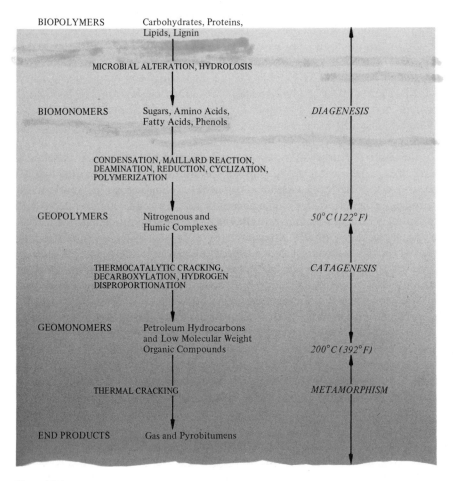

Figure 4-9
The transformation of organic matter in nature. Most petroleum is formed during catagenesis.

Quantity, Source, and Composition of Organic Matter in Sediments

The *quantity* of organic matter preserved in sediments varies enormously, depending on biological activity and the environment of deposition. In dry desert areas of continents, practically all organic matter is converted to carbon dioxide and water, except for the bones of mammals and some high-molecular-weight waxes and resins of plants. The marine counterparts are the biological deserts of the ocean, the abyssal plains where the small amount of life and slow rates of deposition form red clays and carbonate oozes with only traces of organic matter. Hundreds of analyses of cores from the abyssal regions have shown them to contain an average of only 0.05 percent organic carbon (Deep Sea Drilling Projects, Initial Reports, 1971–1978).

Next lowest in organic preservation are the high-energy parts of coastal areas, where productivity is adequate but where strong currents and the high oxygen content of the waters intensifies both biological and chemical degradation of the organic matter. Coarse sands rarely contain more than 0.5 percent organic carbon and usually closer to 0.2 percent. Preservation of organic matter generally increases from the high-energy to low-energy sediments (sands to muds). Inland seas, such as the Caspian, and silled basins, such as Lake Maracaibo, both show their highest organic carbon content associated with the fine-grained sediments and the lowest in the coarse sediment.

The increase in organic content with a decrease in sediment grain size was first recognized by Trask et al. (1932) in the Channel Islands region of California, where he found that clays with a median diameter of less than 5 μ had twice the organic content of silts with a diameter between 5 and 50 μ and four times that of fine sand whose median diameter was 50 to 250 μ. Gorskaya (1950) in a study of Recent clastic sediments reported the following organic matter contents in weight percent: sands 0.77, silts 1.2, and clay muds 1.8. Later, Emery (1960) showed that southern California shelf and beach sediments with a median grain diameter of over 100 μ had less than 0.2 percent organic matter compared with sediments of the offshore basins, which had particle sizes between 3 and 9 μ and organic contents in the range of 5 to 9 percent. Bordovskiy (1965) cites several examples in the Soviet Union of organic contents increasing with a decrease in sediment grain size. For example, organic carbon concentration in Bering Sea silts increases uniformly as particle size decreases. Bordovskiy also cites Strakhov and others as showing that the accumulation of organic matter in sediments is affected by the morphological features of a basin, such as width, depth, and bottom relief.

Low-energy coastal areas and inland sedimentary basins where fine-grained clay and carbonate muds are deposited generally contain 0.5 to 5 percent organic matter, in the range of most oil-forming sediments. Shallow inland seas, narrow seaways between continents, and restricted areas of quiet deposition are the typical environments for source beds of petroleum.

Even larger quantities of organic matter are preserved in areas where oxygen is eliminated and microbial activity suppressed. Sediments with organic contents exceeding 10 percent are found in stagnant, silled basins, such as the Norwegian fiords and the Black Sea, where hydrogen sulfide in the bottom waters eliminates all microorganisms except anaerobes, such as sulfate reducers. The lack of oxygen restricts decomposition to reducing processes and the poisonous effect of hydrogen sulfide kills all biota venturing into the area. Stagnant lacustrine environments, such as Lake Kivu in Africa, also accumulate large amounts of organic matter in their bottom sediments.

The densest accumulation of organic matter occurs in the coastal swamps of high-vegetation areas. When large amounts of vegetation are deposited in shallow, stagnant, fresh- to brackish-water swamps, the pH falls to the range of 3.5 to 4, at which level microbiological activity is so low that the rate of organic deposition exceeds microbiological decay and forms a peat bed, which is essentially pure organic matter. As long as a critical balance between drainage, sedimentation rate, and subsidence is maintained, the peat swamp may grow to considerable thickness. Muller (1964) estimated that it took 4,000 years to form 12 m (about 40 ft) of peat (equivalent to 1 m, or about 3.3 ft, of coal) in the tropical area of Borneo. Peat deposits cover very large areas when the swamps are continuously formed behind a regressive shoreline.

A more specialized type of organic accumulation occurs when the spores and resins of plant materials and the organic remains of algae and plankton accumulate in a swamp environment or the stagnant parts of lakes to form a watery ooze called *sapropel.* The term *gyttja* refers to any organic, rich sediment deposited in open waters, whereas sapropels are considered to form in waters low in or free of oxygen. Wind- or water-borne spores and the remains of surface algae are the major contributors to sapropel, which ultimately changes with depth of burial to cannel and boghead coals.

Open ocean areas generally contain less organic matter than inland seas and coastlines, as shown in Table 4-8. These represent the mean organic carbon contents for several hundred deep sea samples, except for the Cariaco

Table 4-8
Organic Carbon Contents of Ocean Sediments

Sediments	Organic carbon, weight percent
North Atlantic	0.1
South Atlantic	0.1
North Pacific, east	0.1
North Pacific, west	0.1
Aleutian Island Area	0.5
Panama Basin	0.4
Western Continental Margin (USA)	0.6
Gulf of Mexico	0.6
Arabian Sea	0.6
Cariaco Trench	0.6–5
Black Sea	0.5–15

Trench and Black Sea where ranges are shown. Coastal areas and upwelling areas produce over 90 percent of the life in the ocean, but production is less important than preservation in determining the organic content of ocean sediments. Demaison and Moore (1978) have pointed out that the critical factor related to sediment organic content is the development of anoxicity. Preservation is greatest where the oxygen minimum in the water column impinges on the coastline as along parts of Peru, Southwest Africa, Arabia, and India. In upwelling areas without anoxic conditions, as along Newfoundland and the Guiana Shelf, there is very little preservation.

Uspenskii (1956) estimated that 0.8 percent of the primary production of organic matter is fossilized in the sediment. He estimated this to be 1.04 percent for shelf areas, 0.37 percent on the continental slope, and 0.06 percent in the abyssal region of the ocean. Datsko (1959) estimated 0.6 percent preservation for the Caspian Sea, and Emery (1960) estimated 0.6 percent preservation of the organic matter produced on the continental margin of southern California (Table 4-9).

In summary, the wide range in organic content of recent sediments can be attributed to three factors: the rate of organic deposition versus mineral deposition, the availability of oxygen, and the level of biological activity.

The *source* of organic matter in sediments is a combination of marine and terrestrially derived organic compounds. Phytoplankton (plants) and zooplankton (animals) constitute over 95 percent of the life in the oceans. The main producers of organic matter among the phytoplankton are unicellular diatoms with a siliceous skeleton, found mainly in temperate and cold zones; peridineans with an algulose skeleton, found in warm waters; and coccolithophores, unicellular plants with a calcareous skeleton that are abundant in warm seas. The primary consumers of phytoplankton are small herbivorous zooplankton, such as copepods. They serve as food for the larger carnivorous zooplankton and for fish, the third step in the marine food chain.

Table 4-9
Organic Budget of Marine Basins

Organisms	Annual production (dry wt.), 10^3 MT per 1,000 km^2	
	Caspian Sea	Southern California Margin
Phytoplankton	460	560
Zooplankton	34	44
Benthos	41	19
Fish	2	1.3
Lost to sediment (0.6 percent of total)	3.2	3.8

Source: Data on Caspian Sea from Datsko 1959. Data on Southern California Margin from Emery 1960.

Added to this marine source is the often considerable contribution of land-derived organic matter. Wind-blown spores, pollen, other organic debris, and woody and recycled organic matter draining from continents by rivers, submarine discharges, and runoff can substantially alter the original marine matrix. Organic matter that falls to the seafloor is further worked over by benthic organisms and bacteria, which contribute their own products.

Land- and marine-derived organic matter can be recognized from an analysis of the normal paraffins in the sediment. As previously explained, land organisms synthesize predominantly C_{27}, C_{29}, and C_{31} normal paraffins, whereas marine organisms synthesize C_{15}, C_{17}, and C_{19}. Coastal sediments nearly always contain a mixture of marine- and continental-derived organic matter, as indicated by the presence of both these hydrocarbon ranges. The Baffin Bay and Harbor Island (U.S. Gulf Coast) sediments analyzed by Sever (1970) show a bimodal normal paraffin distribution, with the peaking at chain lengths synthesized by marine and continental organisms.

The carbon isotope composition of organic matter also tends to contain more ^{13}C in ocean sediments compared to river sediments (Hunt 1970). Deuser and Degens (1967) explain this as being caused by the greater utilization of ^{13}C-deficient CO_2 by organisms in the more acid rivers and lakes compared to the use of HCO_3^- in the alkaline ocean.

The *composition* of Recent sediment organic matter varies widely with both source and environment of deposition. Figure 4-10 shows the organic composition of Recent sediments and living things based on the major components of life, as previously shown in Table 4-2. The data in Figure 4-10 are from Trask et al. (1932), Shabarova (1955), and Degens et al. (1964). The biggest difference between living things and Recent sediment organic matter is the large decrease in carbohydrates and increase in lignin–humus complexes and nitrogenous compounds. As previously discussed, these changes are caused by microbial and chemical reactions that hydrolyze some of the organic matter to sugars and other simple molecules, which then polymerize to form the lignin–humus and nitrogenous compounds that are precursors of kerogen. Degens et al. (1964) have shown that the organic matter of carbonate muds, as in Florida Bay, is comparable to that in the carbonate shell matrix of marine organisms, being high in nitrogenous material and low in carbohydrates and humic material. Clastic muds, in contrast, have high contents of lignin–humus material, as in the San Diego Trough and Caspian Sea.

Humic substances, which constitute a substantial part of the organic matter in continental and nearshore marine sediments, are yellow to brown-colored, high-molecular-weight polymers that have no counterpart in the living organisms. It was once thought that humus was formed primarily from the degradation of lignin, but it is now recognized that humus arises from the degradation of nearly all plant components. Differences in humus arise from variations in the starting products. Thus the humus of marshes and swamps would tend to have more condensed aromatic structures derived from lignin,

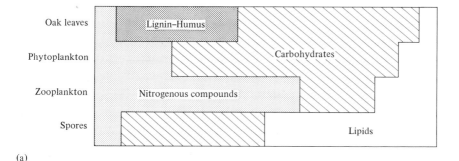

(a)

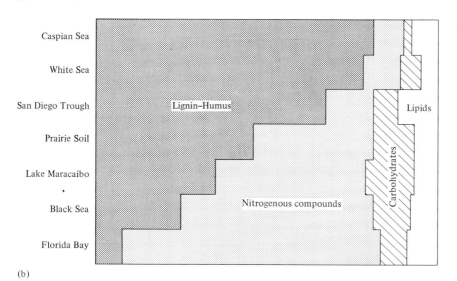

(b)

Figure 4-10
Composition of organic matter in (a) organisms and (b) sediments.

compared to the humus of offshore marine sediments, which form from marine plants, such as algae, that have no lignin. Bordovskiy (1965) recognized this difference in pointing out that humic acids in bottom sediments of the Bering Sea have an atomic hydrogen/carbon ratio of 1.4, compared to 0.6–1.0 for humic acids of soil. The lower ratios indicate a greater content of fused naphthenic and aromatic rings. For comparison, coal has an H/C ratio in the same range as humic acids of soils, whereas petroleum has a ratio of 1.85.

Kononova (1966) published an excellent review on soil organic matter, including methods for separating and characterizing humus fractions. The main fractions are the alkali-insoluble *humin*; the alkali-soluble, acid-insoluble *humic acid;* and *fulvic acid,* which is soluble in both alkali and acid.

In the marine environment, carbonate sediments have much less humus than clay sediments, since they contain less land-derived organic matter.

The bitumen and hydrocarbon content of Recent sediments varies in general with the carbon content, although there are significant differences, because of variations in the humic (low-hydrogen) or sapropelic (high-hydrogen) source of the organic matter. Geochemists have used the term *bitumen* to cover all substances extracted from a rock with oil solvents, such as benzene and chloroform. Since different solvents and extraction techniques will remove different quantities of bitumen, the bituminosity of a rock is not a reliable number for comparison, except in a relative manner. Quantitative determination of the hydrocarbons in bitumen extracts by liquid chromatography is a more useful number for comparison.

Table 4-10 shows the bitumens of several Recent sediments subdivided into paraffin and aromatic hydrocarbons and into nitrogen, sulfur, and oxygen compounds. The data are for hydrocarbons in the range C_{15}–C_{40}, because the analytical procedures tend to lose lower and higher compounds. These sediment samples varied from 0.5–13 percent organic carbon. The total hydrocarbon yields are within one order of magnitude of each other when computed on a uniform organic content.

The soluble NSO compounds of Recent sediments generally decrease with depth because of complexing into insoluble kerogen and because the NSO

Table 4-10

Quantities of C_{15+} Hydrocarbons and NSO Compounds in Recent Sediments ($\mu g/g$ sediment)

	Depth (m)	Paraffin hydrocarbons	Aromatic hydrocarbons	NSO compounds
Eastern Black Sea	1	3	31	649
Western Black Sea	1	1	35	635
Santa Cruz Basin, California	1	10	18	480
North Cooking Lake, Canada	0.1	22	51	662
Caricoa Trench, Venezuela	4	12	10	552
Caricoa Trench, Venezuela	85	8	12	173
Grande Isle, Gulf Mexico	1	19	6	285
Grande Isle, Gulf Mexico	33	56	14	152
DSDP Hole 280A, Tasmania	109	55	6	1,214
DSDP Hole 280A, Tasmania	203	133	17	1,275
DSDP Hole 280A, Tasmania	400	34	51	444
DSDP Hole 280A, Tasmania	443	18	40	436
DSDP Hole 280A, Tasmania	511	16	39	521
DSDP Hole 189, Bering Sea	84	20	36	352
DSDP Hole 232, Red Sea	138	8	22	239

Note: All yields adjusted to a uniform organic carbon content of $C_{org} = 1$ percent.

structures reduce to yield NH_3, H_2S, and H_2O respectively. When heavy metals are present, as in clays, sulfides are formed. In carbonate sediments, where metals are frequently absent, H_2S is formed. The decrease in NSO compounds at the three localities shown in depth—Cariaco Trench, Grande Isle, and Deep Sea Drilling Project (DSDP) Hole 280A—is clearly evident.

Hydrocarbons in the C_{15+} range in Recent marine sediments worldwide rarely exceed 100 $\mu g/g$ with the average close to 50 $\mu g/g$. The geometric mean of 30 Recent clastic sediments examined by Hunt (1961) was 50 $\mu g/g$ and, of carbonates studied by Gehman (1962), 75 $\mu g/g$. Ancient fine-grained sediments, by contrast, contain 100 $\mu g/g$ hydrocarbons in carbonates and 180 $\mu g/g$ in shales (Vassoevich et al. 1967). The difference is caused by the hydrocarbons generated during catagenesis. This is only part of the hydrocarbons generated, since some are lost to the coarse-grained sediments by compaction and migration processes. Hunt (1977a) estimated that 10 percent of the C_{15} to C_{40} hydrocarbons are formed during diagenesis, with 90 percent being formed during catagenesis.

Soils and Recent sediments also contain varying percentages of relatively inert organic matter that is in a high state of maturation, indicating it has passed through the thermal catagenesis stage. Part of this matter is recycled allochthonous carbon from the erosion of sedimentary rocks. Another part is the burned residue of forest and prairie fires. Blumer (1976) has shown that the distribution pattern of polycyclic aromatic compounds in soils and Recent sediments is characteristic of that derived from the pyrolysis of wood but differs from those derived from combustion in furnaces and engines and from crude oil. The polycyclic aromatics of Recent sediments are mostly unsubstituted aromatics, with some alkyl derivatives typical of those formed by pyrolysis at about 400 to 800°C (752 to 1472°F). The presence of unsubstituted polycyclic aromatics such as pyrene, coronene (Figure 3-13), chrysene, triphenylene, and perylene (shown below) in Recent sediments was first reported by Meinschein (1969), and has subsequently been observed by many investigators.

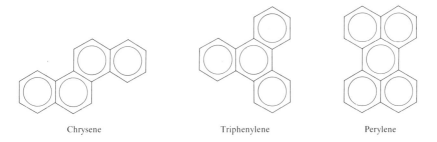

| Chrysene | Triphenylene | Perylene |

Some polycyclic aromatic hydrocarbons, such as perylene, appear to form during diagenesis. Perylene concentrations in the aromatic fraction of sediment extracts generally increase with depth at varying rates. Wakeham

(1977) estimated perylene to be forming at the rate of about 4 $\mu g/g$ sediments/1,000 years in Lake Washington sediments. Perylene is the dominant hydrocarbon at sediment depths beyond 1 m (3.3 ft) in many lakes. Aizenshtat (1973) concluded that perylene is formed from terrigenous organic matter, because it is frequently associated with land-derived compounds. Naphthalene quinones and dihydroxylperylene quinone, both of which are found in some organisms, have been postulated to be perylene precursors.

Enzymatic Microbiological Alteration

The first major chemical alterations of organic matter deposited on the seafloor are carried out by bacteria. The floors of oceans and the top few centimeters of the soil of continents are teeming with microorganisms that decompose and rework the organic matter. ZoBell (1964) states that the bacterial biomass of Recent marine sediments and soils ranges from 0 to 500 g/m^3 in the topmost layers. When plants and animals die, the dead tissues are attacked by microorganisms utilizing enzymes that are capable of digesting and oxidizing many of the tissue components. When the available oxygen is all used, conditions may become anaerobic in the immediate area of the dead material, and fermentative organisms may develop. All sediments contain mixtures of aerobes and anaerobes that can shift from the dormant to active state as the nutrients and environment changes. The catalytic power of microorganisms in bringing about chemical transformations is enormous. Their small size (0.2 to 2 μ in diameter) enables them to penetrate the interstices of sands and silts, and it also gives them a large surface-to-volume ratio compared with higher plants and animals. This permits a rapid exchange of substrates and waste products between the cells and their environment. In favorable environments, a single bacterium dividing every 20 minutes can increase its numbers over a thousand in a little over three hours.

The activity of microorganisms varies markedly with the environment. The symbol Eh represents a measure of the oxidizing or reducing intensity of a chemical system. It shows the ability of a system to be either positive or negative; that is, to accept or give up electrons relative to the standard hydrogen electrode. In sediments containing available oxygen, the Eh varies between 0 and +400 millivolts, whereas in reducing sediments containing hydrogen sulfide it varies from 0 to −400 millivolts. Emery (1960) found that the surface sediments in offshore California basins were usually positive. At sediment depths greater than about 2 meters (6.6 feet), the Eh values generally became zero, and at greater depths the values went as low as −300 millivolts. At zero Eh values, the dissolved oxygen disappears and hydrogen sulfide appears. Positive Eh values are generally found in clastic sediments of oxygenated environments, whereas negative Eh values exist in stagnant water bodies and quiet carbonate muds. Obviously, aerobes would flourish in the positive Eh environments and anaerobes in the negative Eh.

ZoBell (1964) has outlined the requirements of bacteria as free water, essential minerals in solution, utilizable nitrogen, fixed or free oxygen, a favorable temperature, and a source of energy. Bacteria have been found alive and active in temperatures as low as $-10°C$ ($14°F$) and as high as $105°C$ ($221°F$) under a pressure of 1,000 atm. Specialized bacteria will grow in pH ranges from 1 to 11, but they grow best in the range from 6.5 to 8. They can flourish in freshwater or in saturated salt solutions.

Despite the widespread occurrence of bacteria, their activity decreases rapidly with depth in fine-grained sediments. Surface aerobes and anaerobes are measured in the millions per cc whereas only a few hundred exist at depths greater than 3 m (10 ft). Lindblom and Lupton (1961) found that the aerobes in a carbonate mud from Florida Bay disappeared within the first meter (3.3 ft), whereas in a clay mud on the Orinoco Delta viable bacteria were reported to a depth of about 50 m (164 ft). They attributed the bacterial decrease in the carbonate mud to a strongly reducing environment, with hydrogen sulfide forming at about 1 m (3.3 ft). Beyond this depth, only anaerobes would survive. How deep anaerobes can survive has never been determined. Smith (1954) reported about 100 viable bacteria per gram of sediment at a sediment depth of about 140 m (459 ft) in the Gulf of Mexico. Methanobacillus are still capable of forming methane at 40 m (131 ft) sediment depth in the Cariaco Trench.

When microbes attack organic matter, they use enzymes to decompose the high-molecular-weight protein, carbohydrates, and lipids into simple molecules that they can assimilate. They resynthesize part of this into bacterial cell material and convert part of it to gas. The major reactions involve hydrolysis of cellulose to sugars, protein to amino acids, and fats to fatty acids, by saccharolytic, proteolytic, and lipolytic bacteria, respectively. In addition, there are a number of organisms that decompose lignin, mainly by oxidation and demethylation. Lignin decomposition is the slowest of the reactions, partly because the products do not serve as an energy source for the microorganisms, as do the amino acids, sugars, and glycerol from the other reactions.

The number and complexity of reactions that can be carried out by microorganisms is truly amazing. In lignin decomposition, microorganisms can cause cleavage of the side chains, decarboxylation, demethylation, demethoxylation, hydroxylation, and formation of quinones, which can then dimerize or polymerize to the humins.

Some microbes remove the nitrogen from amino acids, and others oxidize alcohols to ketones and acids. Others produce hydrogen, and still others utilize the hydrogen to reduce organic compounds. Some of the reactions form unstable organic compounds that can react chemically or be attacked by other bacteria. When reactions are completed, the principal free products under aerobic conditions are water, carbon dioxide, and sulfate, ammonium, and phosphate ions. Under anaerobic conditions, the products are carbon dioxide, methane, hydrogen, hydrogen sulfide, water, and ammonium and phosphate ions. Hydrogen sulfide can be produced both from sulfate ions and from the sulfur-containing amino acids. The reactions that remove nitrogen, sulfur, oxygen, and phosphorus either as gases or in solution would result in a concentration of the carbon and hydrogen, the principal elements of petroleum. There also is some evidence that both paraffinic and cyclic hydrocarbons in the C_{15+} range are synthesized in small amounts by microorganisms as part of their cell structure. Lijmbach (1975) has presented data indicating that the bodies of degrading microorganisms contribute substantially to the organic source material of petroleum. Although the importance of microbes in the origin of petroleum is still not fully understood, there is no doubt that they are a powerful influence in converting sedimentary organic matter to a more petroleum-like material.

Chemical Alteration

Small amounts of petroleum hydrocarbons are formed by chemical reactions during diagenesis. Several reactions can be duplicated in the laboratory under abiotic conditions. These are considered to be nonbiological reactions in sediments, although in the natural environment it is sometimes difficult to distinguish whether the reactions are catalyzed by enzymes or clay. There is evidence that all of the types of reactions shown in Figure 4-11 can occur during diagenesis, at temperatures up to 50–60°C (122–140°F).

The process of *hydrogen disproportionation* involves a series of reactions among the same molecules such that some of the molecules become depleted in hydrogen and others enriched in hydrogen. For example, Skrigan (1951) demonstrated that the principal component of turpentine, α-pinene, gradually changes during the fossilization of wood stumps to p-cymene and p-menthane. Molecules of α-pinene are converted to aromatic and naphthene hydrocarbons. Both of these compounds occur in crude oil. The conversion of abietic acid to retene and fichtelite (Figure 4-11a) was observed by Skrigan (1964) to occur in wood stumps sunk in peat for different lengths of time. Aromatization is probably the prime mover in these reactions.

Reactions involving *decarboxylation and reduction* (Figure 4-11b) convert unsaturated fatty acids to alkane hydrocarbons. Palmitoleic acid, which is abundant in marine fats, is converted to pentadecene by decarboxylation and to pentadecane by hydrogenation. Pentadecene has been identified in Recent sediments of the Cariaco Trench (Simoneit et al. 1973), and pentadecane is a

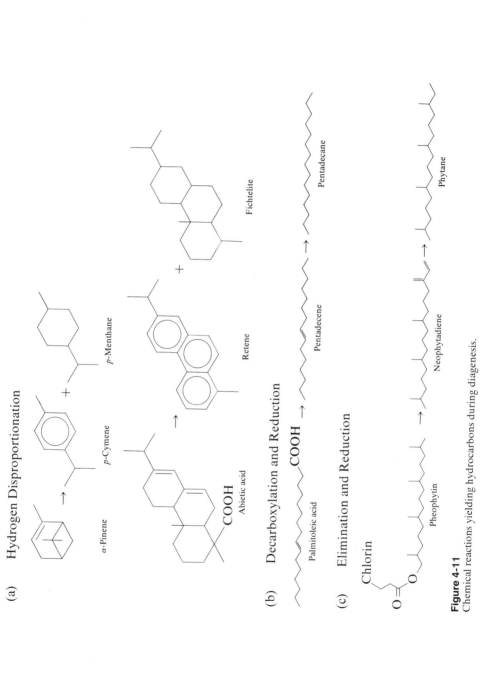

Figure 4-11
Chemical reactions yielding hydrocarbons during diagenesis.

Figure 4-11 (*continued*)

(d) Dimerization and Aromatization

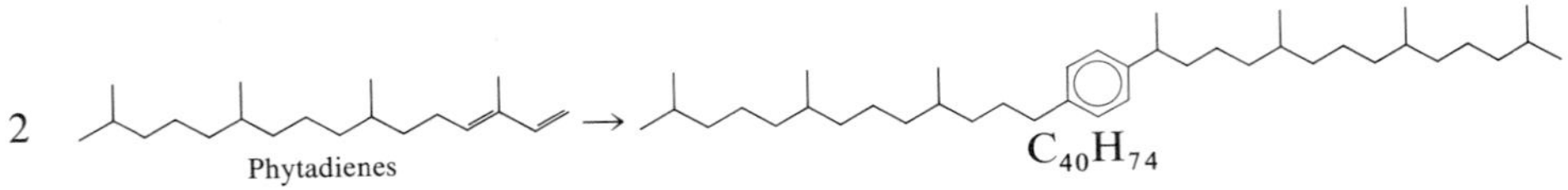
2
Phytadienes
$C_{40}H_{74}$

(e) Dehydration and Reduction

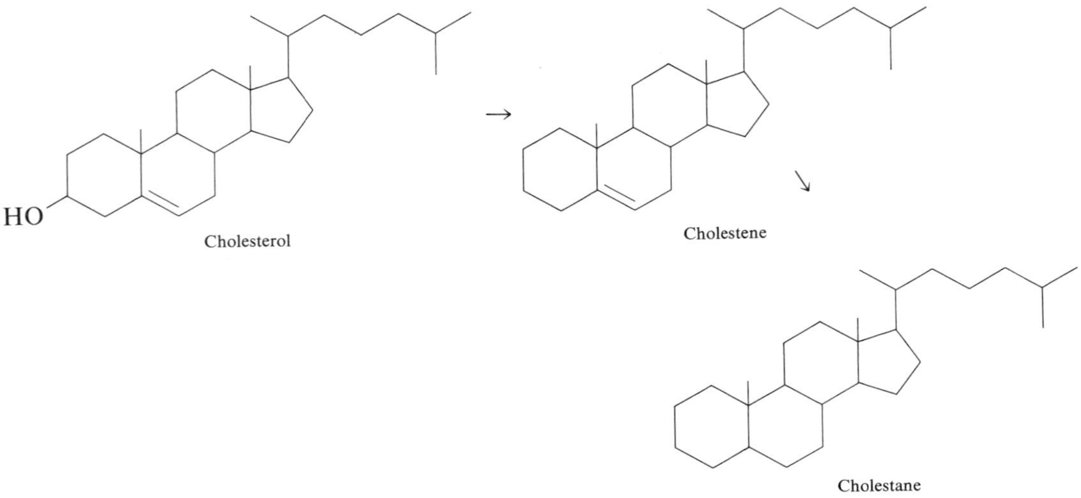
HO
Cholesterol
Cholestene
Cholestane

(f) **Deamination, Decarboxylation, and Reduction**

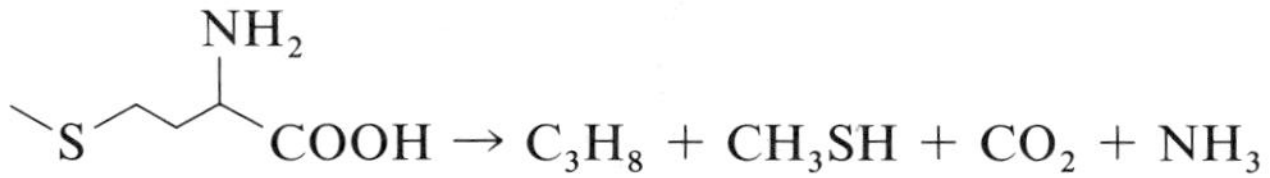

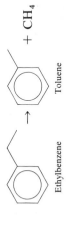

Methionine

$\rightarrow C_3H_8 + CH_3SH + CO_2 + NH_3$

Propane and
methylmercaptan

(g) **Deformylation**

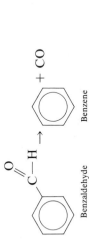

Benzaldehyde $\rightarrow$ Benzene $+ CO$

(h) **β-Carbon Dealkylation and Reduction**

Ethylbenzene $\rightarrow$ Toluene $+ CH_4$

common constituent of the paraffin fraction of Recent sediments and crude oil. Johns and Shimoyama (1972) concluded from laboratory experiments that the decarboxylation of unsaturated fatty acids occurs mainly in the diagenesis zone up to a temperature of 60°C (140°F).

The conversion of pheophytin to neophytadiene and to phytane (Figure 4-11c) involves elimination and reduction reactions. (Baker and Smith 1974). Two of the most prominant biological markers in petroleum are phytane and pristane, which are degradation products of the phytol chain on chlorophyll (Figure 4-6). Both phytane and pristane are found in Recent sediments, although the former is nearly always dominant. Simoneit et al. (1973) noted that in Cariaco Trench sediments the phytadienes were concentrated in the upper section and disappeared at a depth of 140 m (495 ft), presumably forming phytane.

The process of _dimerization_ involves the linking of two molecules to form a dimer. In Figure 4-11d, two phytadienes condense by a Diels–Alder reaction followed by hydrogen rearrangement and _aromatization._ Aromatization involves the formation of an aromatic ring, as shown in the case of p-cymene. Phytadiene dimers have been observed in several studies of Recent sediments. Dimerization reactions between olefins may form some of the naphthenic and aromatic hydrocarbons observed in Recent sediments.

The conversion of alcohols to olefins and to alkanes involves dehydration and reduction reactions. In Figure 4-11e, cholesterol, which is one of the most common sterols found in sediments, is converted to cholestene, which is then reduced to cholestane. Simoneit et al. (1973) found cholesterol, ergosterol, and β-sitosterol to occur at 6 m (20 ft) in the Cariaco Trench. At 67 m (220 ft), the olefins cholestene and stigmastene were found with the saturated tetracyclic naphthenes cholestane and stigmastane. Both of these hydrocarbons have been identified in petroleum. Minor amounts of the pentacyclic triterpanes $C_{30}H_{52}$ and $C_{31}H_{54}$, which are steroidal derivatives were found at 138 m (453 ft). The triterpanes are another major group of biological markers found in petroleum.

The removal of nitrogen and CO_2 involves deamination and decarboxylation reactions. In Figure 4-11f, methionine, an essential amino acid, is converted to propane through removal of sulfur as methyl mercaptan, oxygen as CO_2, and nitrogen as ammonia.

The reaction of _deformylation_ involves the loss of carbon monoxide, and _β-carbon dealkylation and reduction_ involve the loss of the carbon group that is in the beta position relative to the aromatic ring. In Figure 4-11g and 4-11h, the methane, benzene, and toluene formed in these reactions are all found both in Recent sediments and in petroleum.

It has been estimated that about 9 percent of the C_4–C_{40} hydrocarbons existing in all sedimentary rocks are formed in the diagenetic stage (Hunt 1977a). Most of this is in the C_{15+} range and comprises mainly simple paraffin, naphthene, and aromatic structures derived from living organisms,

Table 4-11

C_4–C_7 Hydrocarbons in Black Sea Cores,
DSDP Hole 380/380A, Leg 42B

	Hydrocarbon yield (ng/g dry sediment)	
	175 m Depth	356 m Depth
Isobutane	3.6	5.1
n-Butane	2.5	4.0
Isopentane	4.9	10.9
n-Pentane	0.7	2.6
2,2-Dimethylbutane	0.3	4.5
Cyclopentane	0.2	0.7
2,3-Dimethylbutane	0.2	3.7
2-Methylpentane	0.3	10.0
3-Methylpentane	0.5	11.7
n-Hexane	0.3	123
Methylcyclopentane	1.0	19.8
2,2-Dimethylpentane	1.7	0
Benzene	0.6	10.8
Cyclohexane	1.7	12.0
3,3-Dimethylpentane	0.4	0
2-Methylhexane	0.3	1.5
2,3-Dimethylpentane	1.5	3.2
3-Methylhexane	1.0	2.7
1-trans-3-Dimethylcyclopentane	0.5	1.8
1-trans-2-Dimethylcyclopentane	0.8	5.0
n-Heptane	0.2	3.2
Methylcyclohexane	3.2	8.3
Toluene	2.0	25.2
Total	28.4	270

Source: Some data from Hunt 1979.

with minor modifications. Trace amounts of light-gasoline-range hydrocarbons (C_4–C_7) also are present (Hunt 1975b), indicating that some of the reactions leading to light hydrocarbons are initiated in the diagenetic stage. Table 4-11 lists the light hydrocarbons that have been identified in Black Sea sediments. All of these hydrocarbons are present in crude oil. The total quantity of the C_4–C_7 hydrocarbons and the number of individual compounds increases with sediment depth in the Black Sea.

During diagenesis, the bulk of the organic matter undergoes changes that render it more inert to strong acids and bases. Both the alkali- and acid-soluble organic matter is higher in surface sediments than in ancient sedimentary rocks. A standard analysis for organic carbon in sediments uses dilute hydrochloric acid to remove carbonate carbon prior to combustion of organic carbon. As much as 50 percent of the organic carbon in surface

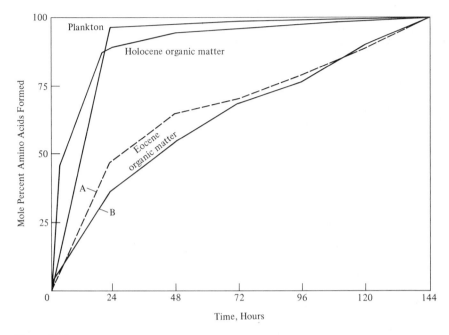

Figure 4-12
Curves showing the rate of decomposition to amino acids of the nitrogenous material
(proteins and peptides) of plankton, of Mediterranean Sea mud, and of two Eocene sediments (A = Green River Shale, Utah; B = marls of northern Spain), as observed by Connan
(1967). The unstable organic matter of Recent muds becomes complexed to minerals and
converted to the more stable kerogen during late diagenesis and early catagenesis.

sediments may dissolve in hydrochloric acid, compared to only a few percent
in sedimentary rocks. Consequently, the acid solution must be evaporated to
dryness rather than filtered, to avoid loss of soluble carbon in young
sediments.

Connan (1967) interpreted this difference between Recent and ancient
organic matter as partly due to the stability of bonding of proteinaceous
material to clay minerals and the organic matrix. He subjected plankton and
the organic matter of a Holocene sediment and two Eocene sediments to
prolonged extraction with 6N HCl. Figure 4-12 shows that for the plankton
and Holocene organic matter nearly all of the protein decomposition to
amino acids occurred in the first 24 hours. To reach the same stage of
decomposition of proteinaceous material in two Eocene carbonate rocks,
about 144 hours of refluxing was required. Clearly, the organic matter of the
Eocene sediments is much more stable and firmly complexed than the
Holocene material. Similar results were obtained by Huc et al. (1978) from
their study of the organic matter in a core from the Black Sea. The organic
matter at the surface was about 40 percent hydrolyzable, whereas at 600 m
(1,969 ft) only 20 percent was hydrolyzable.

Baedecker et al. (1977) found that only 44 percent of the organic matter in a surface sediment from Tanner Basin (off California) was insoluble in acids, bases, and organic solvents. Heating the sample at 65°C (149°F) for 30 days increased the insoluble fraction to 88 percent of the organic matter. This insoluble complex polymer formed during diagenesis is called the *kerogen* of lithified rocks. Decomposition of kerogen during catagenesis forms most of our oil and gas.

CATAGENESIS OF ORGANIC MATTER

The compaction of a sedimentary basin causes the organic matrix to be subjected to increasingly higher temperatures with greater depth of burial. These increasing temperatures cause the thermal degradation of kerogen and associated organic compounds to form petroleum-range hydrocarbons in a reducing environment. These hydrocarbon-forming reactions have been demonstrated many times in the laboratory and have been observed in the natural environment.

Some typical reactions are shown in Figure 4-13. The concept of low-temperature cracking reactions leading to petroleum was first studied in detail by Engler (1911–1912). He heated compounds such as oleic acid at low temperatures over clays to obtain hydrocarbons. More extensive studies were later carried out by the Russian physical chemist A. V. Frost (1945). For example, he heated cyclohexanone at low temperature in the presence of clay, obtaining a series of light liquid hydrocarbons, including cyclohexane, methylcyclopentane, and benzene. Treatment of octyl alcohol at 150°C (302°F) for 12 hours yielded saturated and unsaturated hydrocarbons containing from 8 to about 30 carbon atoms. Frost's experiments led him to conclude that plant and animal remains consisting of resins, acids, alcohols, and ketones were converted to a mixture of hydrocarbons through the catalytic action of minerals such as clays as the sediments were warmed to temperatures above 100°C (212°F). Frost's original hypothesis was essentially correct, although it has taken many years of research to define the mechanism and temperature range of oil formation more accurately than was possible in his time.

Bogomolov and Panina (1961) carried out the oleic acid reaction at temperatures as low as 150°C (302°F) and obtained a complete range of petroleum hydrocarbons. About 50 percent of the hydrocarbons were in the gasoline and kerosine range with 50 percent in the distillate oil fractions and residue. The hydrocarbon type distribution was comparable to crude oil in that the naphthenes were dominant with lesser amounts of paraffins and aromatics.

Petrov et al. (1969) examined the thermal decomposition products of oleic and stearic acids and their mixture in detail. They found that the distribution

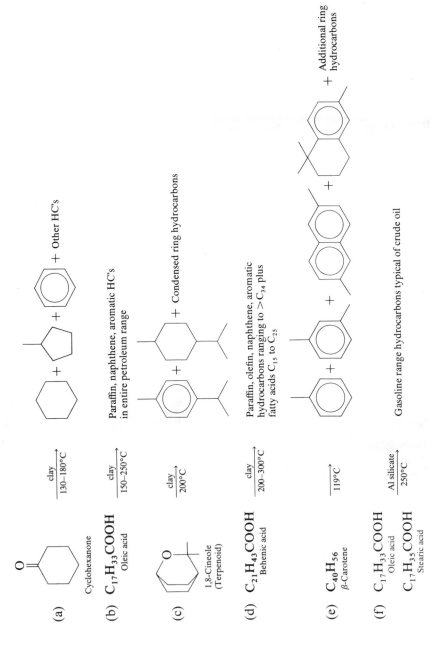

(a) Cyclohexanone $\xrightarrow[130-180°C]{clay}$ + + + Other HC's

(b) $C_{17}H_{33}COOH$ Oleic acid $\xrightarrow[150-250°C]{clay}$ Paraffin, naphthene, aromatic HC's in entire petroleum range

(c) 1,8-Cineole (Terpenoid) $\xrightarrow[200°C]{clay}$ + + Condensed ring hydrocarbons

(d) $C_{21}H_{43}COOH$ Behenic acid $\xrightarrow[200-300°C]{clay}$ Paraffin, olefin, naphthene, aromatic hydrocarbons ranging to >C_{34} plus fatty acids C_{15} to C_{25}

(e) $C_{40}H_{56}$ β-Carotene $\xrightarrow[]{119°C}$ + + + + Additional ring hydrocarbons

(f) $C_{17}H_{33}COOH$ Oleic acid $C_{17}H_{35}COOH$ Stearic acid $\xrightarrow[250°C]{Al\ silicate}$ Gasoline range hydrocarbons typical of crude oil

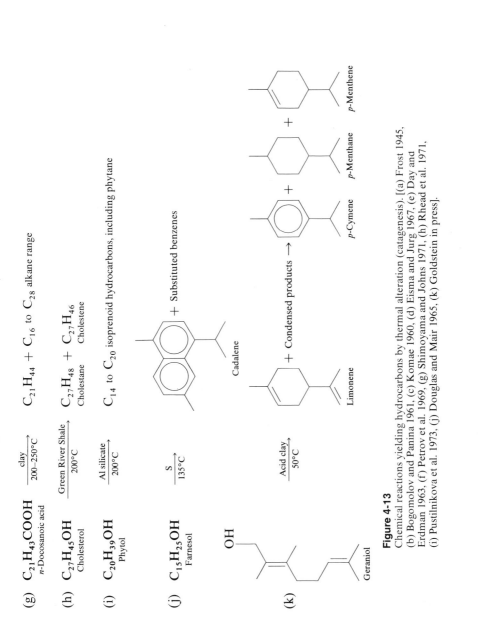

Figure 4-13
Chemical reactions yielding hydrocarbons by thermal alteration (catagenesis). [(a) Frost 1945, (b) Bogomolov and Panina 1961, (c) Komae 1960, (d) Eisma and Jurg 1967, (e) Day and Erdman 1963, (f) Petrov et al. 1969, (g) Shimoyama and Johns 1971, (h) Rhead et al. 1971, (i) Pustilnikova et al. 1973, (j) Douglas and Mair 1965, (k) Goldstein in press].

(g) $C_{21}H_{43}COOH$ $\xrightarrow[200-250°C]{clay}$ $C_{21}H_{44}$ + C_{16} to C_{28} alkane range
n-Docosanoic acid

(h) $C_{27}H_{45}OH$ $\xrightarrow[200°C]{Green River Shale}$ $C_{27}H_{48}$ + $C_{27}H_{46}$
Cholesterol   Cholestane   Cholestene

(i) $C_{20}H_{39}OH$ $\xrightarrow[200°C]{Al silicate}$ C_{14} to C_{20} isoprenoid hydrocarbons, including phytane
Phytol

(j) $C_{15}H_{25}OH$ $\xrightarrow[135°C]{S}$ Cadalene + Substituted benzenes
Farnesol

(k) Geraniol $\xrightarrow[50°C]{Acid clay}$ Limonene + Condensed products → *p*-Cymene + *p*-Menthane + *p*-Menthene

of heptane isomers, octane isomers, C_7 naphthenes, and C_8 naphthenes that formed were very similar to the distribution of these hydrocarbons in crude oil. For example, the heptane isomers obtained from the decomposition of stearic acid included 59 percent 2- and 3-methylhexane, 20 percent *n*-heptane, and 10 percent 2,3-dimethylpentane. The distribution of the corresponding heptane isomers in East Echabi gasoline is 61 percent, 19 percent, and 12 percent.

Other biological markers that have been heated to form hydrocarbons include carotene, the yellow pigment precursor to vitamin A; cholesterol; and phytol. Both carotenes and the structurally related xanthophylls, which are widely distributed in nature, can form pristane and phytane, as well as the other hydrocarbons shown in Figure 4-13.

The formation of single-ring and condensed-ring aromatic hydrocarbons by the low-temperature heating of β-carotene in benzene solution is of particular importance as a possible source of the aromatics in petroleum. Day and Erdman (1963) obtained toluene, *m*-xylene, 2,6-dimethylnaphthalene and ionene plus additional ring hydrocarbons in this noncatalytic experiment at 119°C (246°F).

Aromatic hydrocarbons are also formed by the low-temperature reaction of sulfur with organic compounds such as farnesol, a biological precursor of the sterols. Douglas and Mair (1965) reacted both farnesol and cholesterol with sulfur and obtained a range of aromatic hydrocarbons including 1-, 2-, and 3-ring condensed aromatics. Sulfur is known to react with many organic compounds, producing hydrogen sulfide as a by-product. Consequently, any sediments low in iron, such as carbonates, that have free sulfur may form sour crudes by low-temperature reactions.

Goldstein (in press) heated geraniol at 70°C (158°F) with an acid catalyst and monitored a series of reactions over a 350-hour period. Cyclization of the geraniol yielded limonene, which then underwent double-bond isomerization and condensation to yield a polymer. The product mixture gradually decomposed by oxidation and reduction (hydrogen disproportionation) to yield *p*-cymene, *p*-menthane, and several *p*-menthene isomers as the dominant products at the end of the experiment. This reaction pathway, demonstrated in the laboratory by Goldstein, may be what is happening in nature with α-pinene, as shown in Figure 4-11a. In sediments, the *p*-menthene would be reduced to *p*-menthane.

Mechanism of Cracking Reactions

The formation of light hydrocarbons such as gas and gasoline from kerogen and other high-molecular-weight organic compounds requires the breaking of carbon to carbon bonds. Each bond contains a pair of electrons, so the bond may be broken in two ways. The electron pair may be split with each carbon atom retaining one electron. This is called *homolytic cleavage* or *free-radical reaction*. Or a reaction may occur in which both electrons are transferred from one atom to another (*heterolytic cleavage*). The molecular fragment that loses the electron pair has a positive charge and is called a *carbonium ion*. The fragment that gains two electrons is called a *carbanion*. These terms refer only to organic compounds and not to acids, bases, or other chemicals that may donate or accept electrons.

For example, when a hydrocarbon such as *n*-hexane is heated above 500°C (932°F) without a catalyst, a carbon–carbon bond cracks to yield free radicals. These free radicals are extremely reactive and will immediately react further. The following example illustrates that a paraffin hydrocarbon cracks to yield a paraffin plus an olefin. The intermediate steps are much more complicated than shown and several free-radical reactions may occur to form a series of paraffins and olefins. The chemical bonds are shown as lines except for the electron pair involved in the reaction.

$$
\begin{array}{c}
\text{H}\ \text{H}\ \text{H}\ \ \text{H}\ \text{H}\ \ \text{H} \\
|\ \ \ |\ \ \ |\ \ \ \ |\ \ |\ \ \ \ | \\
\text{H}-\text{C}-\text{C}-\text{C}-\text{C}:\text{C}-\text{C}-\text{H} \xrightarrow{500°\text{C}} \\
|\ \ \ |\ \ \ |\ \ \ \ |\ \ |\ \ \ \ | \\
\text{H}\ \text{H}\ \text{H}\ \ \text{H}\ \text{H}\ \ \text{H} \\
\textit{n}\text{-Hexane}
\end{array}
$$

$$
\begin{array}{c}
\text{H}\ \text{H}\ \text{H}\ \text{H}\ \ \ \ \ \ \text{H}\ \text{H} \\
|\ \ \ |\ \ \ |\ \ \ | \ \ \ \ \ \ \ |\ \ \ | \\
\text{H}-\text{C}-\text{C}-\text{C}-\text{C}\cdot \ +\ \cdot\text{C}-\text{C}-\text{H} \\
|\ \ \ |\ \ \ |\ \ \ |\ \ \ \ \ \ \ \ |\ \ \ | \\
\text{H}\ \text{H}\ \text{H}\ \text{H}\ \ \ \ \ \ \text{H}\ \text{H} \\
\text{Intermediate free radicals}
\end{array}
$$

$$\downarrow$$

$$
\begin{array}{c}
\text{H}\ \text{H}\ \text{H}\ \text{H}\ \ \ \ \ \ \text{H}\ \text{H} \\
|\ \ \ |\ \ \ |\ \ \ |\ \ \ \ \ \ \ |\ \ \ | \\
\text{H}-\text{C}-\text{C}-\text{C}-\text{C}-\text{H} \ +\ \text{C}=\text{C} \ +\ \begin{array}{l}\text{Other}\\ \text{products}\end{array} \\
|\ \ \ |\ \ \ |\ \ \ |\ \ \ \ \ \ \ |\ \ \ | \\
\text{H}\ \text{H}\ \text{H}\ \text{H}\ \ \ \ \ \ \text{H}\ \text{H} \\
\textit{n}\text{-Butane} \ \ \ \ \ \ \ \ \ \ \ \text{Ethylene}
\end{array}
$$

Free-radical reactions tend to produce straight-chain hydrocarbons from straight-chain precursors, because there is usually no rearrangement of the carbon skeleton in this process. In refinery operations, this process is called *thermal cracking*, since the reaction rates are increased primarily by raising the temperature.

The second cracking mechanism frequently involves a Lewis acid catalyst to form a carbonium ion. A catalyst is a substance that accelerates a chemical reaction but is not consumed by it. A Lewis acid is an electron pair acceptor. In the preceding example, the *n*-hexane would form a 6-carbon carbonium ion by transferring a hydride ion, H^-, to an acid. The unstable carbonium ion would undergo decomposition and rearrangement through a series of reactions to ultimately yield isobutane and ethylene among the products. Isobutane would form instead of *n*-butane, because acid catalysts tend to cause rearrangement as shown in the following figure.

$$A^+ + H:\underset{\substack{|\\CH_2\\|\\CH_3}}{\overset{\substack{CH_3\\|}}{CH}} \rightleftharpoons A:H + {}^+\underset{\substack{|\\CH_2\\|\\CH_3}}{\overset{\substack{CH_3\\|}}{CH}}$$

Lewis acid

        n-Butane   Secondary carbonium ion

↓

$$A^+ + H:\underset{\substack{|\\CH_3}}{\overset{\substack{CH_3\\|}}{C}}-CH_3 \leftarrow A:H + {}^+\underset{\substack{|\\CH_3}}{\overset{\substack{CH_3\\|}}{C}}-CH_3$$

Lewis acid

        Isobutane   Tertiary carbonium ion

Carbonium ion reactions such as this in the presence of acid catalysts are very complex and yield a variety of products, including straight-chain paraffins (Galwey 1972). Branched hydrocarbons are the dominant products, however, because the stability of the carbonium ion decreases in the order tertiary→secondary→primary.

This process is called *catalytic cracking* in refinery operations because the reaction rate is increased by the presence of the catalyst. Catalytic cracking is used to produce large quantities of branched alkanes, since they have higher octane numbers than do the straight-chain alkanes, as discussed in Chapter 3.

When Brooks (1948) proposed that surface active minerals such as clays are required to form light petroleum oils, he assumed that such formation was a carbonium ion mechanism comparable to catalytic cracking in a refinery. Brooks felt that limestone reservoirs tended to contain heavy oils with few light constituents because the carbonates contained no surface active minerals to give a Lewis-acid-type catalytic effect.

Eisma and Jurg (1967) carried out some carefully controlled experiments involving the thermal decomposition of behenic acid ($C_{21}H_{43}COOH$) over clay, with and without water, in the temperature range from 200 to 300°C (392 to 572°F). Alkanes and fatty acids with carbon chains both shorter and longer than behenic acid were formed, along with olefins, naphthenes, and some aromatics. The ratio of branched-chain alkanes to straight-chain normal alkanes, such as isobutane to normal butane and isopentane to normal pentane, ranged from 9 to 38 in the experiments without water and from 0.17 to 1 in the experiments with water. This and other reaction products led Eisma and Jurg to conclude that the dry catalytic cracking involved a carbonium ion mechanism, whereas the wet cracking involved a free-radical mechanism.

Almon and Johns (1977) investigated in considerable detail the kinetics and mechanism of behenic acid decarboxylation under conditions simulating burial diagenesis. They not only substantiated the conclusions of Eisma and Jurg but also studied the effects of several smectite mineral catalyst parameters on the rates of decarboxylation. They found that the branching ratio of the hydrocarbons formed (branched alkanes to n-alkanes) was 0.1 in the presence of water, suggesting mainly a free-radical mechanism, while in the anhydrous system it was about 4.5, indicating a carbonium ion mechanism. The smectite in the first reaction contained about 7.5 percent absorbed water. The mechanism of this reaction was studied further by using a free-radical initiator (H_2O_2) that increased the reaction rate and by using electron paramagnetic resonance (EPR) to determine the nature of the free-radical intermediate. The EPR spectrum showed that an organic free radical formed only after the system was heated sufficiently to induce decarboxylation. They also showed, by coating the crystal edges of the smectite catalyst, that both the interlayer sites and crystal edge sites participate in the decarboxylation reaction. The reaction was more efficient with the expanding smectite-type clays than with kaolinite, probably due to the greater surface area of the former.

It has been known for many years that mica-type layer silicates, such as smectite, are capable of exchanging their cations for organic compounds. Weiss (1963) reported that about 9,000 organic derivatives of smectite have been prepared. This includes the common building blocks of living organisms—the sugars, amino acids, fatty acids, and alcohols. Aromatic hydrocarbons are vigorously absorbed with the aromatic rings arranged parallel to the silicate layers. Weiss observed that gentle heating of these organic clay complexes in the absence of oxygen yielded mixtures of aliphatic, naphthenic, and aromatic hydrocarbons comparable to petroleum. I have prepared a number of clay organic complexes by substituting the sodium ion in a sodium smectite with amines and amino acids. Gentle heating causes decarboxylation and deamination to yield a variety of hydrocarbons as shown in Tables 4-12 and 4-13. Lysine and phenylalanine are two of the nine amino acids essential for the growth of animals. All three of the amino acids shown are found in Recent sediments. Hydrocarbons were detected at temperatures as low as 100°C (212°F), and the yields increased logarithmically, with a linear increase in temperature. Lysine, which contains one more CH_2 group than ornithine, yielded about five times more total hydrocarbon. Naphthenes were not formed until the temperatures were raised to about 200°C (392°F). Also, isobutane and isopentane only appeared at the higher temperatures. The formation of aromatic hydrocarbons from an amino acid containing an aromatic ring, as shown in Table 4-13, is important in explaining the origin of some aromatic structures in crude oil. The favored reaction is cracking of the carbon side chain at the position beta relative to the aromatic ring, to yield toluene.

Phenylalanine smectite complex   Benzene   Toluene   Ethylbenzene

Metaxylene   Paraxylene   Orthoxylene

Complete removal of the chain would yield benzene. Simple decarboxylation and deamination would yield ethylbenzene, which then could isomerize to the xylenes. Since Almon and Johns' (1977) studies showed that at least half of the decarboxylation reactions occur within the crystal lattice, it is likely that smectite-type clays play an important role in the origin of petro-

Table 4-12
Low-Temperature Formation of Hydrocarbons
from Amino Acid – Clay Complexes

	Yield (nanograms/gm of complex)	
Hydrocarbon formed	NH_2 $H_2NCH_2CH_2CH_2CHCOOH$ Ornithine	NH_2 $H_2NCH_2CH_2CH_2CH_2CHCOOH$ Lysine
Methane, ethane	19	190
Propane	0	13
n-Butane	128	447
n-Pentane	3	16
2,3-Dimethylbutane	25	241
n-Hexane	8	14
Methylcyclopentane	3	11
Cyclohexane	0	5
3,3-Dimethylpentane	5	33
n-Heptane	30	49

Note: Smectite clay absorbed about 3 percent by weight amino acid. Hydrocarbons formed by cracking of complex when heated at 200°C (392°F) for 20 minutes in helium atmosphere.

Table 4-13
Formation of Aromatic Hydrocarbons from Phenylalanine–
Smectite Complex at Various Temperatures

Hydrocarbon formed	Yield in μmoles/kg of complex at T (°C)			
	100°	150°	200°	250°
Benzene	0.04	0.15	24	519
Toluene	0.40	5.2	95	1,320
Ethylbenzene	0	0	6.5	217
Meta- and paraxylene	0	0	0.24	5.9
Orthoxylene	0	0	0.02	1.4

leum hydrocarbons through the low-temperature catalytic decomposition of organic clay complexes in the fine-grained source rock.

The rates of chemical reactions are profoundly influenced by temperature. An increase in temperature of 10°C can double or triple the reaction rate. A doubling of the rate means that at a sediment depth equivalent to a temperature of 120°C (248°F), the reaction rate is approximately 1,000 times the rate near the surface, where the temperature is 20°C (68°F). The relationship between temperature and reaction rate is expressed in the Arrhenius equation as follows:

$$k = Ae^{-(E_a/RT)}$$

where k = the reaction rate constant related to change in concentration of parent substance with change in time
   A = frequency factor
   E = activation energy
   R = gas constant
   T = temperature in degrees Kelvin (°K = 273 + °C)
The A factor is a constant representing the frequency with which molecules collide in proper orientation to enable a reaction to occur. The activation energy (E_a) is the amount of energy that must be absorbed by a molecule or molecular complex to break the bonds and form new products.

Thermal cracking rates are primarily a function of temperature and the concentration of the reactants. Catalytic cracking rates are a function of temperature, reactant concentrations, and the activity and concentration of the catalyst.

Excess water is known to lower the activity of many catalysts. Critics of the importance of catalysis in petroleum origin have argued that the presence of water and the alkalinity of marine deposits prevent acid-catalyzed reactions from occurring in most sedimentary basins. Goldstein (in press) has made a significant study showing that this argument is erroneous. Acid-catalyzed reactions do occur in marine sediments, and the presence of water slows, but does not stop them.

The activity of mineral catalysts in general is caused by the presence of Lewis acid sites such as ions Al^{3+} and Fe^{3+} and the acid character of interlayer water in the mineral. The activity of different catalysts can be compared on a relative basis by using them to catalyze a specific reaction. Goldstein (in press) used the decomposition of t-butylacetate to isobutylene and acetic acid to measure the relative catalytic activity of sediments and compare them with known cracking catalysts, such as zeolites. Specific catalytic activity was measured as the number of cubic centimeters of isobutylene formed per minute per gram of catalyst used. Initial activity was measured, since the rate decreases with decreasing concentration of t-butyl-acetate. Specific activity for 16 shales varied between 0.1 and 84. Excluding the two highest and lowest activities gave a mean activity of 2.4 for 12 shales. In comparison, two limestones showed an activity of 0.2. A third limestone, which may have contained clay, had an activity of 5, which was greater than 13 of the 16 shales tested. The age of the sediments ranged from Cretaceous to Devonian. Several of the sediments also were shown to catalyze the condensation and hydrogen disproportionation reactions shown in Figure 4-13k. The specific activity of cuttings from a well varied between about 0.1 and 5 through a depth interval from 5,000 to 10,000 ft (1,524 m to 3,048 m).

Part of the wide variation in activity is due to the nature of the cations present. Goldstein exchanged the cations on a bentonite clay and found that the activity of the Al^{3+} clay was three orders of magnitude greater than that of the Na^+ clay. It is possible that the presence of small amounts of iron or aluminum compounds in carbonates may contribute to their catalytic activity along with clays. Anhydrous $CaCO_3$ showed no catalytic activity. However, in the natural environment $CaCO_3$ would contain many impurities that could act as catalysts.

The presence of water slowed down but did not terminate the reactions. A linear decrease in the rate of t-butylacetate decomposition occurs in going from 5 percent water to 40 percent water on a silica alumina catalyst, to a level of about half that at 5 percent water (Figure 4-14). Since most oil-forming sediments contain less than 40 percent water, this does not appear to be a serious factor in limiting catalytic activity.

There is no direct way to determine the rate of catalytic cracking occurring under geological conditions, since the activity and concentration of the catalysts in nature are unknown. It is possible to compare catalytic and thermal cracking rates in the laboratory and to extrapolate these over geologic time in order to determine the feasibility of such reactions in nature. Goldstein (in press) used published data plus his own analyses to show that catalytic cracking is probably the dominant process in petroleum generation up to a subsurface temperature of at least 125°C (257°F). At higher tem-

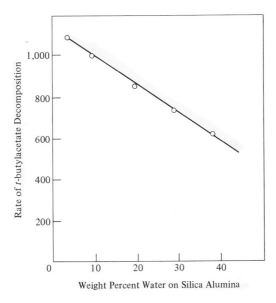

Figure 4-14
Effect of water on catalyst activity. [Goldstein in press]

peratures, thermal cracking becomes increasingly important. Figure 4-15 shows the residence time versus temperature on a log–log plot for the catalytic and thermal cracking of *n*-hexadecane. Data previously published by Greensfelder et al. (1949) involved thermal cracking over quartz chips and catalytic cracking over a low-activity silica alumina catalyst. Goldstein cracked *n*-hexadecane with a high-activity catalyst for comparison. The activation energies of both the high- and low-activity catalysts were about 35 kcal/mole, which is evident from the similar slopes in Figure 4-15. The slope of the line is related to the activation energy. Higher activation energies will have a steeper slope and lower energies the opposite. The activation energy for the thermal reaction is 65 kcal/mole (steeper slope). Although both catalysts have the same activation energy, the reaction proceeds at a faster rate with the high-activity catalyst because the geometry of the catalyst and the concentration of active sites is more favorable than in the low-activity catalyst.

The extrapolation of the laboratory data from high temperatures to low temperatures indicates that at 100°C (212°F) the high-activity catalyst requires only a few months for the cracking of *n*-hexadecane, whereas the low-activity catalyst requires about 1,000 years, and the thermal reaction requires a time period greater than the age of the earth.

The approximate time and temperature for petroleum formation also is shown in Figure 4-15, and the data suggests that catalytic reactions form

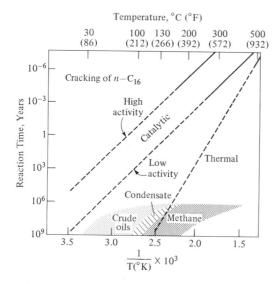

Figure 4-15
Residence time (reaction period) required for thermal and acid-catalyzed cracking of *n*-hexadecane as a function of temperature. The shaded area indicates the age and temperature range of oil- and gas-producing sediments. [Goldstein in press]

much of the low-temperature petroleum, while thermal reactions are probably important in the high-temperature formation of condensate and gas. Because temperature is important in catalytic as well as in thermal reactions, the former are often referred to as *thermocatalytic* in the literature.

Although the catalytic activity of most natural sediments is lower than the two examples shown in Figure 4-15, it is probable that the reactions generating hydrocarbons in nature occur at lower activation energies (lower slope) than the cracking of *n*-hexadecane shown in Figure 4-15. Sokolov's data (1960) on the formation of hydrocarbon gases from low-temperature heating of marine oozes indicates an activation energy of 12 kcal/mole. Galwey's (1972) experiments show an activation energy of 24 kcal/mole for the formation of benzene from peat and 27 kcal/mole for the formation of hexane from peat. Galwey also mixed *n*-dodecanol with a crushed sandstone containing shale lenses and produced several hydrocarbons on heating, the activation energy for hexene production being 23 kcal/mole.

Activation energies are normally calculated for pure compound reactions; however, a pseudo energy of activation can be calculated for an overall reaction, such as the formation of many hydrocarbon components from kerogen. Tissot (1969) calculated a pseudo energy of activation for the generation of various petroleum compounds from kerogen as follows: 20 kcal/mole for hydrocarbons, 15 kcal/mole for hydrocarbons plus resins plus asphaltenes, and 14 kcal/mole for resins plus asphaltenes. Karweil (1969) calculated activation energies between 8.4 and 30 kcal/mole for natural carbonization. Connan (1974) calculated an activation energy of 13.8

kcal/mole and a frequency factor (A) of 10^6 sec^{-1} for the threshold of intense oil generation; that is, the temperature at which large quantities of oil are formed from the kerogen.

Jüntgen and Klein (1975) studied the kinetics of methane formation from coal and suggested that a group of several simultaneous reactions could show a lower activation energy than any one of the reactions by itself. They showed an example of eight reactions all with the same frequency factor (10^{15} min^{-1}) but with activation energies varying between 48 and 62 kcal/mole. These reactions combined to yield an apparent activation energy of 20 kcal/mole and a frequency factor of 10^4 min^{-1}. This may be one explanation for the frequently calculated low activation energies in nature. There probably also are many individual reactions of low activation energies whose kinetics have not been studied because of the complexity of the system. The presence of a wide distribution of light hydrocarbons in shallow young sediments (Table 4-11) indicates that low-energy reaction pathways may be more common in nature than has been previously believed.

Activation energies calculated from laboratory experiments and extrapolated into geological time periods are rough approximations at best. Snowdon (in press) has pointed out that the difference between high values observed in the laboratory compared to the low values in nature also could be caused by such factors as nonlinearity of the Arrhenius equation, changes in reaction mechanisms, partially altered reactants, diffusion effects in the reaction kinetics, and changes in the reversibility of the reaction in response to temperature changes.

Time and Temperature in the Origin of Petroleum

There are limits to the time and temperature (depth) at which petroleum is formed in economic quantities. An awareness of the reasoning behind these limits is important in assisting the operating personnel in making exploration decisions. The concept of activation energies (E_a) has been stressed because it is important in understanding the interrelationships between time and temperature.

Temperature can be traded for time, depending on E_a and providing the Arrhenius equation holds. For example, if the formation of hydrocarbons from kerogen required an activation energy of 18 kcal, then the reaction rate will approximately double for each 10°C (18°F) rise in temperature. This means that the quantity of oil formed in 25 million years at 110°C (230°F) would take 50 million years to form at 100°C (212°F) or 100 million years at 90°C (194°F). Actually, there are so many variables affecting yield, such as the nature of the individual kerogen components, each of which could be involved in many simultaneous reactions with different activation energies, that the example is hypothetical. But it does show the importance of increasing temperatures in generating petroleum.

If two reactions have the same A factor and one has a lower activation energy, it will be faster at any temperature than the reaction with a high activation energy. As the temperature increases, however, the rate of the slower reaction with the high E_a will increase more rapidly. For example, the conversion of amorphous kerogen to oil by certain reactions involving a low activation energy might start at 20°C (68°F), but the rate would only increase slowly with increasing temperature. In contrast, a high E_a reaction such as the formation of methane from a coaly kerogen might not be observable at low temperatures, but at high temperatures the rate could triple for each 10°C (18°F) rise in temperature. It could be the dominant reaction above 150°C (302°F).

The rate of hydrocarbon generation from kerogen seems to double on the average for each 10°C (18°F) rise in temperature, but kerogen is so heterogeneous that this is only an approximation. Each source rock forms oil and gas at different rates with increasing temperature, depending on its content of the various types of kerogen and the possible catalytic activity of its mineral constituents.

Larskaya and Zhabrev (1964) were the first geochemists to demonstrate that the generation of hydrocarbons from the kerogen of shales increases logarithmically with depth (increasing temperature). They found that the bitumen content (hydrocarbons plus soluble nonhydrocarbons) of shales in the western Ciscaspian region changed very little in the temperature range from 20 to 50°C (68 to 122°F), after which it increased markedly. In Jurassic, Cretaceous, Paleocene, and Miocene deposits, the most intensive hydrocarbon generation started at 60°C (140°F) and 300 atmospheres pressure equivalent to sediment depths of 1,200 to 1,500 m (3,937 to 4,921 ft). Hydrocarbon yields in the fine-grained shales increased by factors ranging from 3 to 7 in the deep sediments compared to the shallow. Later, Philippi (1965), Louis and Tissot (1967), Albrecht (1970), and others observed the same increase in hydrocarbon generation with depth in other sedimentary basins.

Vassoevich (1969) described this intensive generation zone as representing the "principal phase of oil formation." Connan (1974) called the depth at which a significant increase in hydrocarbons occurred the "threshold of intense oil generation." He developed a way of using these depth–yield data in a practical manner. He constructed a time–temperature Arrhenius plot to define the threshold of intense generation for several sedimentary basins. In Figure 4-16, this threshold is shown as a depth of 1,370 m (4,495 ft) equivalent to a temperature of 65°C (149°F) in the Douala Basin. At this depth, there is a logarithmic increase in oil generation, indicating that the rates of kerogen

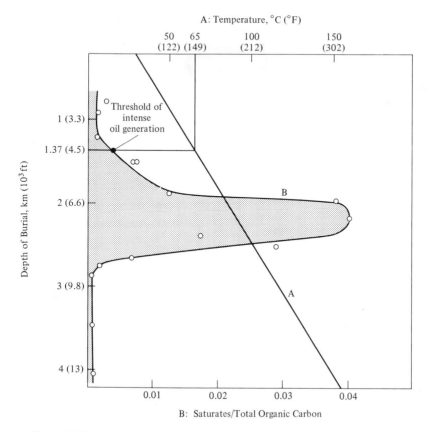

Figure 4-16
Generation of hydrocarbons with depth, Douala Basin, Cameroon. Line A, temperature, °C; line B, saturates per total organic carbon. [Albrecht 1970]

degradation reactions are increasing rapidly. In Figure 4-16, line A represents the increase in temperature with depth, while line B represents the quantity of saturated hydrocarbons in relation to the total organic carbon. It should be emphasized that these hydrocarbon yields are for the fine-grained source rocks and are not determined from any reservoir rock.

Connan (1974) obtained samples of fine-grained sediments from many sedimentary basins in order to analyze for the same type of data, on a worldwide basis, as was shown by Albrecht in the Douala Basin. He also utilized data published by others to obtain a time–temperature–depth relationship for the threshold of intense generation.

Prior to the threshold of intense oil generation, there is very little degradation of the kerogen to form hydrocarbons. Consequently, Connan (1974) assumed that up to this time the ratio of original kerogen to degraded kerogen was approximately constant. This enabled him to quantify the time–temperature relation from the Arrhenius equation as follows:

$$ln\ t = \frac{E}{RT} - A$$

where T = absolute temperature in degrees Kelvin
   t = time
   A = frequency constant

This equation means that a linear relationship exists between the logarithm of time and the reciprocal of the observed temperature. Connan plotted these data for different basins on semilog paper, a modification of which is shown in Figure 4-17. The central line in the main oil zone is the general relationship between time and temperature for the threshold of intense generation in the sedimentary basins shown. The basins shown are only those where the measured temperature does not differ very much from the maximum reached in the past. Connan excluded examples of basins in which orogenic activity, erosion, volcanism, or hydrothermal influence may have affected the temperature threshold of intense oil generation. Also, the temperatures shown are the formation temperatures measured by logging instruments. These are usually 10–15°C (18–27°F) lower than the true formation temperatures. Such a diagram can only be constructed from data on petroleum source beds. As previously mentioned, data on reservoired oils are subject to the whims of migration and may bear no genetic relation to the adjacent fine-grained rock.

Some geochemists and palynologists use the terms *immature* and *mature* to indicate the initiation of oil formation. Connan considers the threshold of intense oil generation to be this transition point between the immature and mature sediments.

The slope of the line in Figure 4-17 is the activation energy, E_a. Connan (1974) calculated this to be in the range of 11–14 kcal/mole, depending on whether he obtained the regression-line coefficients from his own data or the published data. These activation energies are in the same range as those previously reported by Tissot and by others for the generation of hydrocarbons from sediment organic matter. Connan also observed that the threshold temperature tended to be higher in carbonate rocks than in shales. This is consistent with the idea that the activation energy for kerogen decomposition in carbonates would be higher than in clays because of the greater catalytic effect of the latter.

The E_a values calculated from this diagram are actually slightly lower than the true E_a values for the oil-forming process. The reason is that a more exact calculation of E_a takes into account that in sedimentary basins the temperature rises in roughly linear fashion with time from the surface temperature of about 15°C (60°F) to the observed temperature at depth (assuming a uniform burial rate and no change in geothermal gradient). An increment calculation made every 5°C gives an E_a value of 15.5 kcal/mole, instead of the 14

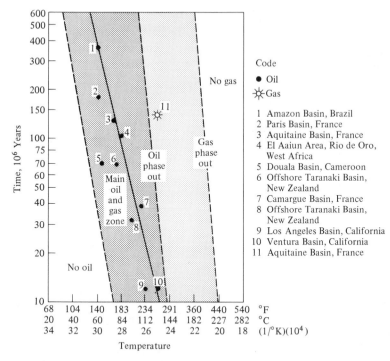

Figure 4-17
Time–temperature of petroleum genesis. This figure can be used to evaluate drilling prospects. In basins where the geothermal gradient is low and deposition rates are high (lower left of figure), the sediments will be immature and nonprospective. Basins where geothermal gradients are high and deposition rates are low (upper right of figure) are equally nonprospective, because no more oil is being generated and what was generated may have been destroyed. The prospective areas for oil are within well-defined time–temperature limits, as shown by the dark shading. In using this figure, temperatures must be maximum temperatures to which the sediments have been exposed, not necessarily present-day temperatures. [Connan 1974]

kcal/mole from Connan's diagram. The difference is small, because most of the reaction takes place in the last 15°C of the final temperature. Hood et al. (1975) recognized this when they developed a simplified method of predicting the organic level of maturation of a sediment from its maximum temperature and effective heating time. The latter was defined as the time during which a specific rock has been within 15°C of its maximum temperature.

Figure 4-17 only applies to relatively continuously subsiding basins. Obviously, if a Devonian source rock is buried to a temperature of 30°C

(86°F) for 380 million years and then subsides rapidly to a temperature of 100°C (212°F) for only 10 million years, the heating time would not correlate with the age of the formation. Tissot, Deroo, and Espitalié (1975) have discussed the interpretation of time–temperature data for basins that have periodically subsided and uplifted.

The dotted lines in Figure 4-17 at a slight angle to Connan's time–temperature line are based on vitrinite reflectance values, as discussed in Chapter 7. They indicate the initiation and termination of the oil generation zone. Very few data are available on the phasing out of the gas zone, but studies of kerogen from many basins indicate that there is a maximum depth for gas generation that is well within the range of current drilling capability.

Exploration geologists can use Figure 4-17 in a general way to predict the probability of oil or gas in sedimentary basins. Figure 4-17 says that oil is not generated in young, cold sediments because temperatures are not high enough to initiate the threshold of intense oil generation. Also, no hydrocarbons are being generated in old, hot basins because any generating capability has long since been destroyed. Consequently, the places to prospect are the young, hot basins or the old, cold basins, within the limits of Figure 4-17. Obviously, subsurface data are needed to determine if the old, cold basins have always been relatively cold and how long the young, hot basins have been hot. The thermal history of a basin is more important than the present-day temperature.

Field Observations of Petroleum Generation

The origin of petroleum hydrocarbons with increasing depth and temperature has been documented in many sedimentary basins. Methane is formed biogenically at the surface and accumulates in sediments along with hydrocarbons derived from the living organisms and a complex mixture of nitrogen, sulfur, and oxygen compounds representing microbiologically degraded organic matter. Some generation of hydrocarbons occurs in the first thousand feet of burial from reactions requiring very low activation energies. The quantities are small, and it is not until the burial temperature rises to the threshold of intense generation that large quantities of hydrocarbon are formed. This threshold varies somewhat in different sedimentary basins, depending on the ease of decomposition of the organic matter and on geological factors such as rate of sedimentation. In a typical basin, intense oil generation might start around 50°C (122°F), peak at about 90°C (194°F) and terminate at about 175°C (347°F). This example is only intended to show that the origin of petroleum occurs within a finite temperature range of the earth's crust. In any specific sedimentary basin, hydrocarbon yield data must be obtained on cuttings through a sufficient depth interval to define these limits. Some examples reported in the literature are briefly discussed as follows.

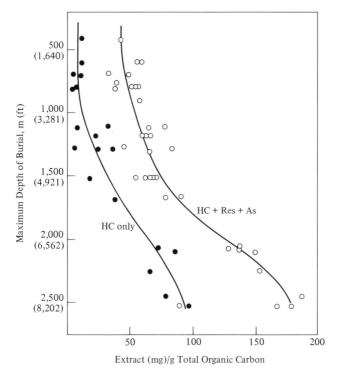

Figure 4-18
Increase in the C_{15+} hydrocarbons (HC only) and total chloroform
extract (hydrocarbons plus resins plus asphalts) relative to total
organic carbon as a function of depth in the lower Toarcian shales,
Paris Basin. [Tissot et al. 1971]

The most detailed study, which involved the origin of individual hydrocarbon groups, was made by Tissot et al. (1971) in the Paris Basin. Figure 4-18
shows the increase in both heavy hydrocarbons (C_{15+}) and heavy hydrocarbons plus asphaltic constituents with depth in the range of 1,200–1,400
m (3,937–4,593 ft) equivalent to a present-day subsurface temperature
of 60°C (140°F). Durand and Espitalié (1972) obtained similar curves for the
C_1–C_{15} hydrocarbons in the Toarcian shales (Figure 4-19). It is interesting
to note here that the threshold for the paraffin and naphthene hydrocarbons
is about 1,300 m (4,265 ft), compared to about 2,000 m (6,562 ft) for the
aromatic hydrocarbons. The activation energies required for the reactions
forming aromatics are higher than those for the reactions forming the saturated hydrocarbons, or the catalytic activities are different.
    The threshold of intense oil generation in the Maikop shales of the West
Pre-Caucasus, USSR, is at about 2,000 m (6,562 ft), equivalent to a subsurface temperature of 75°C (167°F), according to Kartsev et al. (1971, Figure

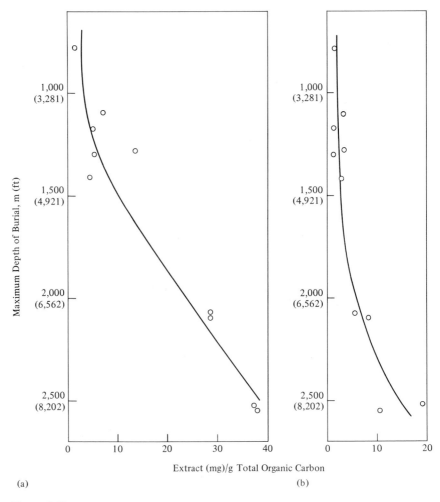

Figure 4-19
Increase in (a) the C_1–C_{15} paraffin plus naphthene hydrocarbons and (b) the C_6–C_{15} aromatic hydrocarbons with depth in the lower Toarcian shales, Paris Basin. [Durand and Espitalié 1972]

4-20). Russian geochemists consider this threshold as the advent of the principal phase of oil formation.

Philippi's (1965) work in the Los Angeles Basin showed the threshold to occur at a present-day temperature of about 120°C, or 248°F (Figure 4-21). A similar study of the Ventura Basin showed the threshold to start about 10°C higher, in the range of 130°C (266°F). The data in Figures 4-20 and 4-21 are based on the yield of C_{15+} hydrocarbons relative to organic carbon or to the total rock.

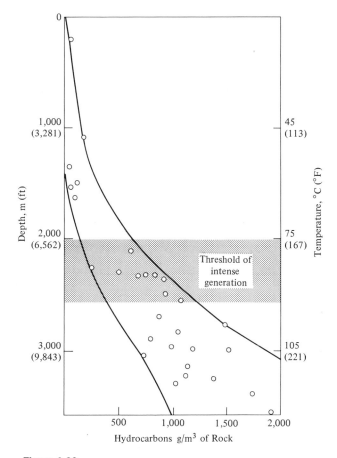

Figure 4-20
Increase in the C_{15+} hydrocarbons with depth in the Maikop shales of the West Pre-Caucasus, USSR. [Kartsev et al. 1971]

In the Aquitaine Basin, intense oil generation is initiated at a temperature of about 72°C (162°F), equivalent to a subsurface depth of 2,500 m, or 8,202 ft (Le Tran 1972). The generation of oil peaks at about 100°C, or 212°F (Figure 4-22). Intense gas generation begins much deeper, below 4,100 m (13,451 ft), equivalent to a temperature greater than 120°C (248°F).

LaPlante (1974) used a different method of analysis for estimating the threshold in Gulf Coast samples. He analyzed the kerogen with depth for carbon, hydrogen, and oxygen and determined from this the depth at which an increasing percentage of the kerogen was converted to hydrocarbons. Figure 4-23 shows the initiation of intense generation to occur at about 13,000 ft (3,962 m) equivalent to a temperature of 91°C (196°F).

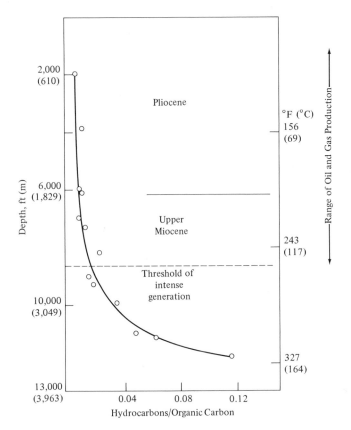

Figure 4-21
Increase in the C_{15+} hydrocarbons relative to total organic carbon with depth in the fine-grained source rocks of the Los Angeles Basin. [Philippi 1965]

The same type of curve showing the depth range of oil generation can be determined on samples that have been buried deeply and then uplifted to the surface. If weathering has not altered the organic matter, the imprint of increased heavy hydrocarbon generation will still be observable in the rock samples. An example of this is the study of C_{15+} hydrocarbons and the soluble nitrogen, sulfur, and oxygen compounds in the Phosphoria Formation of Wyoming. A study by Claypool et al. (1978) was conducted largely on unweathered outcrop samples that at one time had been buried very deeply. Through a reconstruction of the geological data, Claypool was able to show the peak zone of hydrocarbon generation (Figure 4-24), which was also accompanied by a peak generation of asphaltic by-products. During the

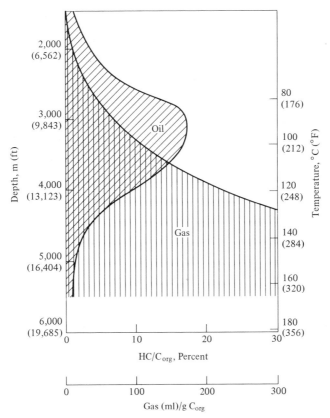

Figure 4-22
Increase and decrease in C_{15+} hydrocarbons with depth in the fine-grained carbonates of the Aquitaine Basin compared to generation of C_1–C_3 hydrocarbons. This shows the peaking of oil generation at about 90°C (194°F) prior to the beginning of intense gas generation. [Le Tran 1972]

intense generation, the steranes and triterpanes present in the shallow samples disappeared, probably due to their conversion to condensed-ring aromatic compounds. This is an example of the fact that the imprint of the biological compounds tends to be modified or destroyed during the period of intense oil generation. Also of interest is the fact that the ratio of hydrogen to carbon in the kerogen decreased markedly through the oil generation zone, from a value of 1.26 to 0.6. This occurs because the kerogen is providing the hydrogen to form the oil.

All of these examples shown represent the generation of hydrocarbons within fine-grained sediments. Clearly, the origin of petroleum occurs over a

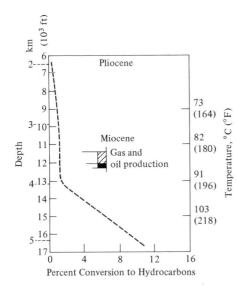

Figure 4-23
Generation of hydrocarbons from kerogen with depth in West Delta area, Gulf Coast. [LaPlante 1974]

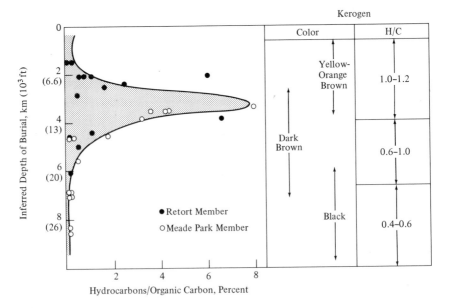

Figure 4-24
Increase and decrease in C_{15+} hydrocarbons relative to total organic carbon and the change in kerogen composition versus inferred maximum depth of burial in Permian Phosphoria Formation, western United States. Maximum conversion of kerogen to hydrocarbons occurs in the 2.5–4.5 km (8,202–14,764 ft) depth range. [Claypool et al. 1978]

finite temperature range that can be observed in the natural environment and can be simulated by heating experiments in the laboratory. In exploring for oil and gas in wildcat areas, it is important to know whether or not the sediments have passed through this generation range and, if so, at what depths generation was initiated, peaked, and terminated. Such data alone will not pinpoint the location of economic petroleum accumulations, since the latter are affected by many other geological factors. However, such data does bracket the depth ranges in which the hydrocarbon source beds occur, and it does indicate the most likely subsurface zones in which to prospect for oil and gas. There is no point in drilling a hole to 20,000 ft (6,096 m) if the kerogen is too immature to generate hydrocarbons or if it is so depleted in hydrogen that the generating capability is gone. The limited depth–temperature zone of hydrocarbon generation is a natural phenomenon resulting from the thermodynamic instability of the organic matrix in the rocks. Although we do not understand all aspects of the origin process, enough of it is clear to enable subsurface geochemical analyses to be effectively used in making exploration decisions.

METAMORPHISM

Geochemists define mineralogical changes that can be reasonably attributed to the action of heat and pressure at depth as metamorphic. The low-temperature end of the metamorphic scale is generally considered to be at about 200–300°C (392–572°F). For example, kaolinite is converted to muscovite in this range. This also appears to be the range in which the last small amount of methane is formed from the organic matter that is gradually converted to an anthracitic and graphitic material. As previously mentioned, the high-energy reactions initiated in this range go to completion rapidly, leaving a black carbon residue in the rocks. The end products of all sedimentary organic matter are methane and graphite, and it is in the metamorphic stage that the last act is played in the thermal alteration process. At sediment temperatures greater than 250°C (482°F), no more hydrocarbons are generated in significant amounts from organic matter. At such temperatures the H/C ratio of the kerogen is generally less than 0.3 typical of the kerogen of a phyllite. The 30,000 ft (9,144 m) well drilled to the Cambro-Ordovician Arbuckle Formation in the Anadarko Basin of Oklahoma contained kerogen with an H/C ratio of about 0.25. This is below the ratio of 0.3 that is empirically considered to be the minimum for generating methane.

The whole process of the origin of petroleum may be summarized as shown in Figure 4-25. Some hydrocarbons are formed from living organisms during their life processes and undergo little change before becoming part of petroleum. Other hydrocarbons form from bacterial residues and from early

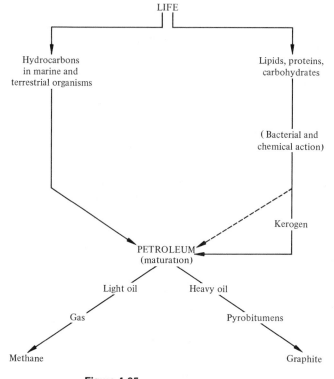

Figure 4-25
Origin and maturation of petroleum.

diagenetic reactions of lipids, proteins, and carbohydrates. Most petroleum hydrocarbons, probably 80–95 percent, form from the thermal transformation of kerogen with temperature and time. This process is referred to as *maturation,* or *thermal alteration,* during the catagenetic stage. After petroleum forms, it undergoes continuing maturation in both source rock and reservoir, eventually forming mainly methane plus pyrobitumens and graphite.

DISTRIBUTION OF HYDROCARBONS
IN SEDIMENTARY ROCKS

The thermal generation of hydrocarbons from the organic matter of sedimentary rocks has resulted in the distribution of oil and gas within specific subsurface temperature intervals of the earth's crust. Migration processes have not significantly altered this distribution since more than 80 percent of the hydrocarbons in most fine-grained sediments are autochthonous (Hunt

1977a). Table 4-14 contains an approximation of the mass of carbon present as hydrocarbons and asphaltic compounds (nonhydrocarbons) in the three stages of organic maturation (Hunt 1977a). Hydrocarbons are divided into methane (C_1), ethane plus propane, butanes through tetradecanes, and pentadecanes through tetracontanes (C_{40}). In petroleum terminology, the first two groups represent gas; the C_4–C_{14} range is gasoline, kerosine, and light gas oil; and the C_{15}–C_{40} range is gas oil through lubricating oil. The residuum above C_{40} is mostly asphaltic compounds (nonhydrocarbons).

Figure 4-26 shows these data in relative concentrations in the rocks during diagenesis, catagenesis, and metamorphism. The areas under each curve represent the relative quantities of hydrocarbons, in terms of carbon, that are present in each stage. The temperature at which the concentrations peak is indicated as about 150–160°C (302–320°F) for C_1–C_3 (predominantly methane), 110–120°C (230–248°F) for C_4–C_{14}, and 100°C (212°F) for C_{15}–C_{40}.

Approximately 7 percent of the hydrocarbon gases, 9 percent of the liquid hydrocarbons, and 40 percent of the asphaltic compounds are present in the diagenetic stage. The catagenetic stage contains 82 percent of the gases, 91 percent of the liquid hydrocarbons, and 60 percent of the asphaltic compounds. The metamorphic stage contains about 11 percent of the gases and traces of soluble hydrocarbons and asphaltic compounds. Clearly, the largest quantities of hydrocarbons and asphaltics are present in the catagenetic stage.

Table 4-14
Estimated Quantities of Hydrocarbons and Soluble Asphaltic Compounds Present in Nonreservoir Sediments of Earth's Crust

Maturation facies	Immature	Mature[a]	Metamorphosed	
Stage of organic transformation	Diagenesis	Catagenesis[b]	Metamorphism	
Temperature range	0–50°C (32–122°F)	50–200°C (122–392°F)	>200°C (>392°F)	
	Mass present as carbon in 10^{18} grams			Total
Hydrocarbons				
C_1	2	20	3	25
$C_2 + C_3$	0.001	2	0	2
C_4–C_{14}	1	37	0	38
C_{15}–C_{40}	20	179	1	200
Asphaltic compounds	110	164	1	275
Total	133	402	5	540

[a]The facies concept of Evans and Staplin (1971) defines about 175°C as the end of the mature (oil and condensate) stage. Their metamorphosed facies is the beginning of dry gas. Catagenesis in this text extends to 200°C, near the end of dry gas.

[b]Includes the third brown coal substage of protocatagenesis, mesocatagenesis, and the first lean coking coal substage of apocatagenesis as defined by Vassoevich et al. (1967).

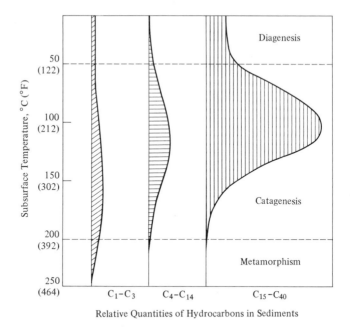

Figure 4-26
Relative quantities of hydrocarbons in fine-grained, nonreservoir rocks.
Areas under curves are proportional to masses as carbon. [Hunt 1977a]

In fact, most basin studies show that the most hydrocarbons are found at subsurface temperatures between 60 and 150°C (140 and 302°F).

It should be emphasized that these curves represent the total carbon as hydrocarbons in nonreservoir rocks. The picture in reservoir rocks would be considerably different due to the greater mobility of the lighter hydrocarbons. Hedberg (personal communication, 1977) has pointed out that many reservoir oils in the principal oil-generating stage have gas/oil ratios of several hundred to a thousand, and this gas is largely methane. If Figure 4-26 were drawn for reservoir accumulations, it would show the C_1–C_3 curve area to be one-third to one-half the size of the other two curves combined, based on the data of Vassoevich et al. (1967).

The hydrocarbons present in the diagenetic stage to 50°C (122°F) are mostly of high molecular weight and are associated with a high content of asphaltic compounds. Any accumulation of these early oils would be heavy and asphaltic until buried to higher temperatures. A possible example of an early asphaltic oil is the Rozel Point asphalt, which seeps from Recent sediments into the Great Salt Lake of Utah. Accumulations of such early oils are exceptional. Most shallow asphaltic accumulations are old oils that are weathered and microbially degraded.

The hydrocarbon distributions in Figure 4-26 indicate that nearly all hydrocarbon generation, except for some methane, is completed by the end of the catagenetic stage. This means that any deep drilling in strata where temperatures are much in excess of 200°C (about 400°F) must rely on the hope that previously formed hydrocarbons have not diffused out or been destroyed by high-temperature reactions.

The estimate for methane in the metamorphosed facies is based on very limited data and will change as more information from deep wells becomes available. This range should contain methane from the maturation of both oil and kerogen. However, petroleum geochemists, such as Bayliss (personal communication, 1976), who have studied large numbers of samples from many sedimentary basins observe that methane tends to be absent, or present only in traces, in fine-grained rocks at temperatures much above 200°C (about 400°F). Either most of the conversion to methane has occurred be-before this, or methane formed in high-temperature rocks tends to diffuse out or be destroyed. Elemental sulfur is known to react with methane at high temperatures to form hydrogen sulfide. Also, Barker and Kemp (1977) have postulated that methane may react with water to form carbon dioxide and hydrogen in high-temperature strata.

SUMMARY

1. Petroleum originates from a combination of the hydrocarbons formed directly by living organisms and deposited with sediments plus the hydrocarbons formed from the thermal alteration after burial of much of the originally nonhydrocarbon organic matter of these organisms. About 9 percent of the total hydrocarbons in sediments are formed by the first process, 91 percent by the second.

2. The quantity of hydrocarbons formed by inorganic processes is negligible from an exploration standpoint.

3. Biological markers are molecules whose carbon structures can be traced from living organisms to Recent sediments to lithified rocks and to crude oil. Typical biological markers are the porphyrins (chlorophyll), the terpenoids (pristane, phytane, steranes, hopanes), the odd-numbered n-paraffin carbon chains, and the iso- and anteiso-branched carbon chains.

4. The organic matrix (kerogen) of rocks contains structures derived from carbohydrates, proteins, and lipids, all of which can be thermally altered to form hydrocarbons. The lipid structures are closest to hydrocarbons in composition and are the most likely progenitors of petroleum.

5. Living organisms typically contain a very simple hydrocarbon spectrum, with only a few hydrocarbons dominant in contrast to crude oil, which has a complex spectrum of hundreds of hydrocarbons.

6. The odd-carbon-chain *n*-paraffins and olefins are synthesized by marine plants primarily in the C_{15} to C_{21} range and by land plants primarily in the C_{27} to C_{35} range. Brackish-water plants synthesize in the intermediate range C_{19} to C_{27}.

7. Biosynthesis and early diagenesis forms odd-chain-length *n*-paraffins and 2- and 3-methyl branched paraffins, naphthenes (from terpenoids and steroids), and traces of aromatic hydrocarbons (from aromatization of naphthenes and terpenoids, and from lignin degradation).

8. The three major stages of petroleum formation within the sediments of the earth are *diagenesis,* the biological, physical, and chemical alteration of organic matter prior to a pronounced effect of temperature; *catagenesis,* the thermal alteration of organic matter; and *metamorphism,* high-temperature thermal alteration. The approximate temperature range for these stages are diagenesis, to 50–60°C (122–140°F); catagenesis, to 175–200°C (347–392°F); and metamorphism, above 200°C (392°F). The largest quantity of petroleum hydrocarbons is formed from organic matter heated in the earth to temperatures between 60 and 150°C (140 and 302°F).

9. The most important factor in the origin of petroleum is the thermal history of the source rock.

10. Petroleum hydrocarbons are cracked from the kerogen-mineral complex by mechanisms that require an apparent activation energy of 10 to 20 kcal/mole to break the bonds. Catalytic cracking appears to be the dominant process in petroleum generation in the subsurface temperature range up to about 125°C (257°F). Thermal cracking becomes increasingly important at higher temperatures.

11. In sedimentary basins, the threshold of intense oil generation occurs in the subsurface temperature range of 50 to 130°C (122 to 266°F). The precise temperature depends on the composition of the kerogen, the catalytic activity of the mineral matrix, the chemical environment of the source beds, and the rate of deposition. The threshold in any drill hole can be determined from geochemical analyses of the well cuttings.

12. The time–temperature history of a basin should be evaluated to determine if it is prospective. Oil can form in old, cold basins as well as in young, hot basins. It cannot be generated however, in young, cold basins except in trace amounts, and it is usually destroyed in old, hot basins, assuming that subsidence has been continuous.

13. The distribution of oil and gas in sedimentary rocks of the earth's crust is as follows: About 7 percent of the hydrocarbon gases, 9 percent of the liquids, and 40 percent of the asphaltic nonhydrocarbons are present in the diagenetic stage. The catagenetic stage contains 82 percent of the gas and 91 percent of the liquid hydrocarbons, plus 60 percent of the asphaltic nonhydrocarbons. The metamorphic stage contains about 11 percent of the gases and traces of higher hydrocarbons.

14. Only small amounts of methane have been found to date in the fine-grained rocks heated to temperatures much above 200°C (about 400°F). This suggests 200–250°C (400–480°F) is the limit to the methane-generating capability of source rocks. This can adversely affect exploration in the range over 30,000 ft (9,144 m) if there are appreciable methane losses, through physical or chemical processes, from deeply buried reservoirs.

SUPPLEMENTARY READING

American Association of Petroleum Geologists (AAPG). 1950–1969. *Origin of petroleum.* Reprint Series Nos. 1 and 9 (selected papers reprinted from *Bulletin of the American Association of Petroleum Geologists*). Tulsa, Okla.: American Association of Petroleum Geologists. 402 pp.

Breger, I. A. 1963. *Organic geochemistry.* New York: Pergamon Press. 658 pp.

Davis, J. B. 1967. *Petroleum microbiology.* New York: Elsevier. 604 pp.

Dott, R. H., and M. J. Reynolds. 1969. *Sourcebook for petroleum geology,* Tulsa, Okla.: American Association of Petroleum Geologists. 471 pp.

Eglinton, G., and M. T. J. Murphy. 1969. *Organic geochemistry.* New York: Springer-Verlag. 828 pp.

Manskaya, S. M., and T. V. Drozdova. 1968. *Geochemistry of organic substances.* New York: Pergamon Press. 347 pp.

Vassoevich, N. B. 1971. The source of Petroleum: A biogenic carboniferous substance. *Priroda,* no. 3, 58–69.

5

How Gas Forms

Natural gas is the gaseous phase of petroleum. Typically, a reservoir gas contains 70–100 percent methane, 1–10 percent ethane, and lower percentages of higher hydrocarbons through the hexanes and traces up through nonanes (C_9H_{20}). Percentages of nonhydrocarbon constituents, such as carbon dioxide, nitrogen, and hydrogen sulfide, may vary from very low to 100 percent. Up to 8 percent helium and 15 percent hydrogen have been found in some reservoir gas. Natural gas is classified in the field as dry gas or as wet gas if it has less than 0.1 or more than 0.3 gallons of condensible liquids per 1,000 ft³. The terms *sweet* and *sour* refer to gases that are low and high, respectively, in hydrogen sulfide. Reservoir gas may occur in the free gaseous state, or as liquefied gas, or as gas dissolved in oil or water. Vassoevich et al. (1967) estimated that reservoirs worldwide contain two or three times as much oil as gas.

Most of the gas in the subsurface, however, is disseminated in the nonreservoir rock and dissolved in formation waters. Estimates of the ratio of dispersed and dissolved gas to reservoir gas in sedimentary basins range from 10–200 to 1 (Vassoevich et al. 1967; Zor'kin et al. 1974). Nonreservoir rocks worldwide contain about ten times more oil than gas (Hunt, 1977a). It follows that reservoirs contain a higher percentage of the gas generated by the organic matter of fine-grained rocks than of the oil generated, unless it is assumed that more than half of all the gas in reservoirs is from the cracking of reservoir oil.

This chapter discusses the origin of natural gas and some characteristics of its accumulation. More details on gas occurrence are presented in the three succeeding chapters.

SOURCES OF NATURAL GASES

Both the hydrocarbons and nonhydrocarbons in natural gas have multiple sources. Table 5-1 lists the sources and their probable order of importance. The dominant source for the major gases—methane, CO_2, N_2, and H_2S—is the thermal degradation of organic constituents of sedimentary rocks. Large volumes of CO_2 also are formed from the high-temperature, high-pressure dissolution of carbonate rocks. The maturation of coal beds, as well as that of disseminated coal in sediments, generates CH_4, CO_2, and NH_3; the last is oxidized to N_2 in some environments (Getz 1977). Reactions in deep, high-temperature reservoirs gradually convert oil to gases such as CH_4 and H_2S. Helium and some nitrogen are believed to diffuse from igneous and metamorphic rocks (Muller et al. 1973). A small fraction of the nitrogen may be attributed to trapped air (Zartman et al. 1961).

The order in Table 5-1 is an average for gas generation from source rocks worldwide. The source of gas for any specific accumulation will not necessarily be in this order. Thus, the CO_2 in the Delaware–Val Verde Basin of the western United States is believed to come from the thermal decomposition of $CaCO_3$ because of igneous intrusions. Much of the methane in deep, high-temperature reservoirs may originate from the thermal transformation of reservoired oil to gas. The geologic history of an accumulation must be thoroughly understood to determine which possible source is most important. The origin of the principal components of natural gas is discussed in terms of diagenesis and catagenesis.

Table 5-1
Sources of Natural Gas Constituents

Source	Probable order of importance for each source					
	CH_4	CO_2	N_2	H_2S	He	H_2
Thermal degradation of organic matter in sediments	1	1	1	1		1
Microbial degradation of organic matter in sediments	2	4				
Maturation of coal beds	4	3	2			
Thermal alteration of reservoired oil and asphalt	3			2		
Diffusion from igneous and metamorphic rocks			3		1	
Thermal dissolution of carbonate rocks		2				
Trapped air			4			

DIAGENESIS

Microbial activity produces the gases CO_2, H_2, H_2S, CH_4, NH_3, and N_2 in surface sediments. Methane survives as the major component of free gases in shallow sediments, along with small amounts of CO_2 and N_2. Most of the CO_2 exists as HCO_3^- in solution; the H_2 is utilized to reduce S, N, and O compounds; the H_2S reacts with heavy metals or is oxidized to S; and part of the N_2 is reduced to NH_3, which is adsorbed by clays as NH_4^+ or stays in solution. Hammond (1974) observed that gas pockets in sediment cores from the Cariaco Trench contained about 90 percent CH_4, 8 percent CO_2, and the remainder N_2 and C_2H_6. This is comparable to the composition of the biogenic gas produced for several years in Japan (Marsden and Kawai 1965).

GASES FORMED IN MARINE SEDIMENTS

	Reaction	Product
Aerobic zone	$CH_2O + O_2 \rightarrow CO_2 + H_2O$	Carbon dioxide
	$2H_2S + O_2 \rightarrow S + 2H_2O$	
Anaerobic zone	$CH_2O + H_2O \rightarrow CO_2 + 2H_2$	Hydrogen
Sulfate reduction	$4H_2 + SO_4^{2-} \rightarrow S^{2-} + 4H_2O$	Hydrogen sulfide
Carbon dioxide reduction	$4H_2 + CO_2 \rightarrow CH_4 + 2H_2O$	Methane
Nitrogen fixation	$3H_2 + N_2 \rightarrow 2NH_3$	Ammonia
Nitrate reduction	$10H_2 + 4HNO_3 \rightarrow 2N_2 + 12H_2O$	Nitrogen

The reactions producing gases in surface sediments are summarized here. A reducing condition with available hydrogen is required before the anaerobic reactions can be initiated. In marine sediments, sulfate-reducing bacteria are dominant at the top of the anaerobic zone until all available sulfate is reduced. The S^{2-} (as HS^-) combines with heavy metals such as iron (FeS), or in the absence of heavy metals it is released as H_2S. The H_2S readily dissolves in groundwater, and on contact with the aerobic zone it is converted to free sulfur particles by sulfur oxidizing bacteria. For more details on the nature and sequence of aerobic and anaerobic reactions in marine sediments, see Claypool (1974).

The state of oxidation or reduction of a sediment, as discussed in Chapter 4, is generally expressed as Eh. ZoBell (1946b) found Eh values to range from $+0.35$ to -0.50 volt over a pH range from 6.4 to 9.5 in 1,000 samples of ocean bottom sediments. Reducing conditions increased with depth, the fine-grained sediments being more reducing than the coarser sediments. Oxygen practically disappears a few meters into most sediments except in well-aerated sands. Under anaerobic conditions, there are enteric bacteria that ferment organic matter to produce hydrogen and CO_2. ZoBell (1947) observed hydrogen production at Eh values ranging from -0.05 to -0.45 volt and pH ranges from 5.5 to 9.8.

Although hydrogen is produced under widespread conditions, it is almost immediately utilized for the reduction of nitrogen, sulfur, and oxygen compounds. If sulfate is absent as in freshwater environments, methane is generated by carbon dioxide reduction. Claypool (1974) noted that if sulfate is present as in a marine environment methane is normally not formed until all of the sulfate is reduced. Apparently the methanobacillus that generate methane cannot compete with sulfate reducers for available hydrogen. If methane production has been initiated, however, it does not appear to be terminated by a later influx of sulfate waters. Emery and Hoggan (1958) noted that methane production in the Santa Barbara Basin started at a depth of about 100 cm. Nakai (1962) observed methane production in the first 15 cm depth in two Japanese lakes. Whelan (1975) found that methane production began at a sediment depth of about 70 cm in the Mississippi Delta after sulfate had decreased below an average concentration of 3 millimole/liter. Ammonia is formed from reduction of N_2 and deamination of proteins. In most sediments, nitrogen disappears with depth and ammonia increases. Many recent sediments also contain large amounts of free sulfur, which forms from oxidation of the released H_2S by sulfur-oxidizing bacteria.

Bacteria can form methane from methanol and from other organic compounds with oxygen-containing groups. Traces of ethane and propane have been produced with CH_4 from microbial cultures. Olefins such as ethylene are formed by fungi and by the ripening of fruits. Davis and Squires (1954) were able to form ethane, ethylene, propane, and propylene in the microbial fermentation of cellulose. Juranek (1958a) also observed ethane, propane, and butane production from cellulose fermentation. Although the amounts formed were small, such formation may explain the consistent observation of traces of higher gaseous hydrocarbons in analyses of Recent sediments.

Weber and Turkel'taub (1960) carried out extensive studies on the formation of gaseous hydrocarbons in Recent marine sediments in the USSR. They hermetically sealed mud samples from the Caspian, Azov, and Black seas in glass jars for periods ranging from several days to 9.5 years. Analysis of the free gas phase showed the presence of ethane-through-pentane hydrocarbons in yields up to 0.07 volume percent of the gas phase. They also produced these hydrocarbons from the decomposition of marine vegetation such as *Zostera*.

Malyshek et al. (1962) conducted experiments on the fermentation of muds under more controlled conditions approaching the natural environment. Gaseous products were removed at various stages of formation. They noted that methane was the only hydrocarbon produced in sediments where fermentation was relatively slow. Tests were conducted over periods of several months to 2.5 years. In sediments of more rapid fermentation, only methane was formed in the early months, after which small amounts of propane, butanes, and pentanes began to appear. Malyshek concluded that the higher hydrocarbons form at later stages in the decomposition of the organic matter by mechanisms that may or may not involve microbes. It was also noted that the type of organic matter and the reducing conditions were more important in producing higher hydrocarbons than was the organic content of the sediments.

Analysis of the ethane-through-pentane hydrocarbons in a 1,000 m (3,281 ft) hole drilled in the Black Sea (Hunt and Whelan 1978a) showed a systematic increase in the heavier hydrocarbons with depth (Table 5-2). Although the concentration of ethane increased a hundredfold to 1,000 m, it still represented only 0.1 percent of the methane at that depth. Claypool (1976) determined ethane/methane ratios on gas samples from deep sea drilling cores taken in various ocean basins. All samples showed an increase in ethane with depth, but the rate of increase varied considerably depending on the geothermal gradient, the age of the samples, and the quantity and nature of the organic matter. For example, at 200 m (656 ft) in the cool Aleutian Trench the C_2/C_1 ratio was 10^{-5}. In the southern Red Sea, where the geothermal gradient is $7°C/100$ m, the C_2/C_1 ratio at 200 m was 10^{-3}. As yet there is no clear evidence whether these higher hydrocarbons in sediments are originating from low-temperature abiotic or biotic reactions although it is believed to be the former (Claypool 1974).

In all of these observations, it is clear that although ethane-through-pentane hydrocarbons are formed during diagenesis, the quantities are negligible in terms of any exploration value. Methane is the only hydrocarbon that will form significant accumulations during diagenesis. Even the formation of methane is highly variable. It will only form where reducing

Table 5-2
Ratio of Higher Hydrocarbons to Methane in Black Sea Sediments, DSDP Leg 42B, Hole 380

Hydrocarbon ratio (volume percent)	At 50 m (164 ft)	At 1,000 m (3,281 ft)
C_2H_6/CH_4	10^{-5}	10^{-3}
C_3H_8/CH_4	0.5×10^{-5}	10^{-4}
C_4H_{10}/CH_4	10^{-6}	0.5×10^{-4}
C_5H_{12}/CH_4	0.5×10^{-6}	10^{-5}

conditions are sufficiently strong to remove both dissolved oxygen and sulfate from the pore water. Claypool and Kaplan (1974) have estimated that under these conditions biogenic methane production appears to be self-limiting and results in dissolved methane concentrations on the order of 10 to 20 millimole/liter in the interstitial water of muds. Bacterial methane production may continue to depths of several hundred meters in some fine-grained sediments.

Biogenic Gas Accumulations

The major gas production in Japan from the mid-1920s to the 1960s was from Pliocene-Pleistocene biogenic methane dissolved in water at near saturation. The biogenic origin of this gas was first established by Nakai (1962) and Sugisaki (1964). A detailed description of this type of deposit has been published by Marsden and Kawai (1965). The gas consists of 90–97 percent CH_4, 1–8 percent CO_2, 0.5–3 percent N_2, and less than 0.01 percent C_2H_6. It is produced from brines located in highly permeable sands or gravels in the synclines of marine or lagoonal sedimentary basins. Meteoric water upstructure confines the brines to the downstructure permeable beds. Pay intervals range from 10 to 40 m (33 to 131 ft) with porosities 30 percent and permeabilities from several to 50 darcys. The high permeabilities are required for economic production since gas-to-water ratios vary from 0.5 in the shallow zones to 2.7 in the deep zones, and reservoir pressures are generally low. Producing depths range up to 2,000 m (6,562 ft), but most production is from less than 1,000 m (3,281 ft).

Gas/water ratios as shown in Figure 5-1, are very close to the maximum solubility of methane at reservoir temperatures and pressures. The curve represents the change in the theoretical gas-to-water ratio with depth, and the circles represent samples from the southern Kanto gas-producing region (Sugisaki 1964). Production at about 1,000 m (3,281 ft) is 2 m^3/kiloliter, equivalent to 11 ft^3/bbl. Salinity of the brine increases from 2,000 mg/liter at 150 m (492 ft) to 34,000 mg/liter at 1,500 m (4,921 ft). Sulfate is absent, as would be expected in order for methane to be produced.

An example of unusual biogenic methane is that in Lake Kivu, an African Rift lake that contains about 50 km^3 of methane in its deep water. Carbon isotope studies (Deuser et al. 1973) indicate that the methane is formed by bacterial reduction of CO_2 from inorganic sources such as local volcanic emanations. The methane reaches a maximum concentration of 0.37 m^3/kl (2 ft^3/bbl).

Coleman (1976) obtained radiocarbon dates on glacial drift gas from 22 wells in the Illinois Basin and found them all to be younger than 40,000 years. The methane source was bacterial decomposition of organic matter in the sediments.

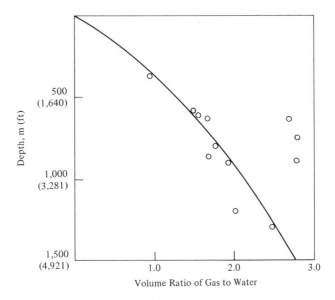

Figure 5-1
Curve showing variation of theoretical gas/water ratio with depth.
Circles represent samples from the southern Kanto gas-producing
region. [Sugisaki 1964]

Methane Hydrates

Gas hydrates are crystalline compounds in which the ice lattice of water
expands to form cages that contain the gas molecules. The hydrates are solids
resembling wet snow in appearance, and they form both above and below the
freezing point of water under specific pressure–temperature conditions. The
water molecules form two kinds of unit cell structures (for details, see Hitchon
1974b). The smaller unit structure contains 46 water molecules, which will
hold up to 8 methane molecules. Gases such as CH_4, H_2S, CO_2, and C_2H_6 will
fit into this structure. The larger unit cell contains 136 molecules of water.
Gases such as propane and isobutane will fit in it. These are the only gaseous
hydrocarbons that will form hydrates. The pentanes and n-butane molecules
are too large. The relationships between the size of the gas molecule and the
ratio of gas to water molecules in the hydrate are shown in Figure 5-2. A
methane hydrate in which all of the ice stages are completely filled with
methane will contain about 172 m³ of methane at STP per m³ of hydrate.
Trebin et al. (1966) determined that the hydrate will form with as little as
one-third of the ice cages filled or only about 60 m³ of methane at STP. A
completely filled hydrate would have a formula of $CH_4 \cdot 5.75H_2O$. Measured
hydrates generally show six or seven H_2O molecules per CH_4 molecule.

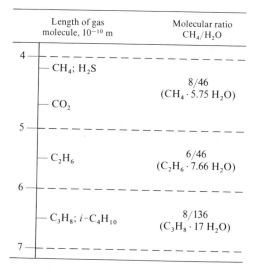

Length of gas molecule, 10^{-10} m	Molecular ratio CH_4/H_2O
CH$_4$; H$_2$S	8/46 (CH$_4 \cdot 5.75$ H$_2$O)
CO$_2$	
C$_2$H$_6$	6/46 (C$_2$H$_6 \cdot 7.66$ H$_2$O)
C$_3$H$_8$; i-C$_4$H$_{10}$	8/136 (C$_3$H$_8 \cdot 17$ H$_2$O)

Figure 5-2
Relationship between hydrate structure and gas/water molecular ratio in hydrate.

The pressure-temperature diagram for methane and 0.6-gravity gas is shown in Figure 5-3 (Katz et al. 1959). Hydrates are formed by increasing pressures and are decomposed by increasing temperatures. Assuming a normal pressure gradient, a methane hydrate will form under about 900 ft (274 m) of permafrost, providing there is sufficient methane available. On the continental margins, where ocean bottom temperatures are around 36°F (2°C), a methane hydrate can form in the surface sediments under about 1,100 ft (335 m) of water. The other gases shown in Figure 5-2—ethane, propane, isobutane, CO_2, and H_2S—all tend to lower the pressure requirements when forming mixed hydrates with methane. The 0.6-gravity gas at 40°F (4.4°C) forms the hydrate under 250 psia* compared to 600 psia required to form the methane hydrate. The salinity of water lowers the temperature at which the hydrate forms, the effect being greater at high temperatures than at low temperatures. Figure 5-3 shows a dotted line for the methane seawater hydrate limit.

Since the pressure required to form gas hydrates increases logarithmically with a linear increase in temperatures it is apparent that the hydrates in most sedimentary basins will decompose in the temperature range 70–80°F (21–27°C), because the pressures are inadequate to preserve them. This puts a depth limit of about 5,000 ft (1,524 m) on hydrate formation.

*Pounds per square inch absolute.

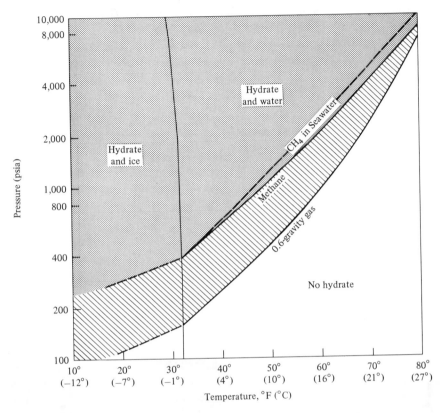

Figure 5-3
Pressure–temperature diagram for gas hydrates. [Data from Katz et al. 1959]

Katz (1971) developed a temperature–depth curve to predict the depths at which gas hydrates will occur (Figure 5-4). This shows the curve for methane and 0.6-gravity gas assuming a hydrostatic head of 0.435 psi per foot of depth and considering salinity effect as negligible. In order to determine the hydrate zone, it is necessary to know the geothermal gradient and the depth of the permafrost when drilling in permafrost territory, or the sea bottom temperature when drilling offshore. Assuming a constant geothermal gradient, a line can be drawn, as shown on Figure 5-4 for the Cape Simpson area, using Katz's data. The methane hydrate zone here would be only a few hundred feet thick, but the 0.6 gas hydrate would be about 1,900 ft (579 m) thick. Obviously, areas of high geothermal gradients will form thinner hydrate zones than areas of low geothermal gradients. The maximum thickness of the potential hydrate zone can be more accurately estimated if subsurface temperatures are available. In Figure 5-4, the temperature data from Prudhoe Bay (Holder et

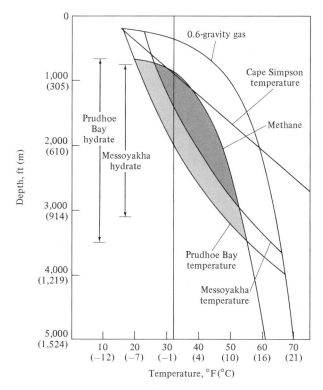

Figure 5-4
Depth–temperature curve for predicting the depth and thickness
of gas hydrates. Geothermal gradient is lower in permafrost than
in deeper section at Messoyakha and Prudhoe Bay. [Data from
Holder et al. 1976; Makogon et al. 1971]

al. 1976) and from the Messoyakah field (Makogon et al. 1971) are plotted.
The permafrost layer at Prudhoe Bay extends to about 2,000 ft (610 m). The
potential-hydrate depth interval overlaps the permafrost extending from
about 700 to 3,500 ft (213 to 1,067 m). For a 0.6-gravity gas, it would extend
on down to 4,000 ft (1,219 m). At Messoyakah, the hydrate interval has been
measured from about 1,148 to 2,854 ft (350 to 870 m). If an area with as
much as 3,000 ft (914 m) of permafrost were found, it would move the
geothermal gradient curves down but they would still intersect the gas
hydrate curves at depths shallower than 5,000 ft (1,524 m).

Hydrates form initially from biogenic gas in near-surface unconsolidated
muds containing 40–70 percent water. Once formed, the hydrates immobilize
the pore water. Any methane migrating into this zone may be converted to
hydrate assuming saturation in the required temperature–pressure range.

Further deposition of methane muds will build up the hydrate until the bottom of the crystallized mass reaches the decomposition temperature and is converted to a gelatinous ooze. Hedberg (1974) has suggested that decomposition at the base of hydrates could cause mud diapirs, mud volcanoes, and other phenomena of overpressuring. The Russians have observed pressures of 100,000 psi from the decomposition of methane hydrates in closed systems.

The cooling of the crust during ice ages could create thick sections of permafrost and gas hydrates, which would later be decomposed during warm intervals. Makogon et al. (1972) point out that the permafrost is now receding in both West and East Siberia. It was much more extensive in the past and, therefore, they expect large accumulations of gas from the decomposition of hydrates.

Hydrates can be recognized in the subsurface by seismic data and variations in drilling rates, but sampling with a pressure core barrel is required for proof of a hydrate. Stoll et al. (1971) based their suggestion that there were gas hydrates in the deep ocean on the fact that seismic reflectors tended to follow surface contours, as would a hydrate, rather than to follow the bedding planes. Stoll formed hydrates in the laboratory and found that the seismic velocity increased from 1.85 km/sec to 2.69 km/sec as the solid crystalline hydrate was formed. Drillers on the Deep Sea Drilling Project (DSDP) have noted a marked decrease in drilling rates from less than 1 minute per meter to 5 or 6 minutes per meter when crystalline hydrates were encountered.

The bottom of a hydrate interval is usually well defined on seismic records due to the sharp drop in velocities from around 3 km/sec to 0.5 and even 0.2 km/sec. Sometimes a bright spot shows up, indicating free gas to be below the hydrate.

Hydrate sections are known to thin up to 50 percent over salt diapirs, because of the heat conductivity of the salt.

Hydrates that are formed in pressure chambers in the laboratory are generally decomposed by allowing the temperature to increase and by bleeding off gas periodically to keep the pressure from rising. This same technique can be used to positively identify hydrates in the subsurface with a pressure core barrel. Pressure core barrels, which have been used on an experimental basis in recent years, involve taking a subsurface core at the in situ pressure. When the core barrel is brought to the surface, a gauge is attached to measure the internal pressure, and a small amount of gas is bled off as the barrel warms up. If only free gas is present, the pressure decline will be approximately as shown in Figure 5-5. If a gas hydrate is present and the gas is released at the same rate at which free gas is formed by hydrate decomposition, the pressure curve will follow the dotted line at the top of Figure 5-5. There will be no change in pressure until the hydrate is completely decomposed, after which there will be a normal pressure decline.

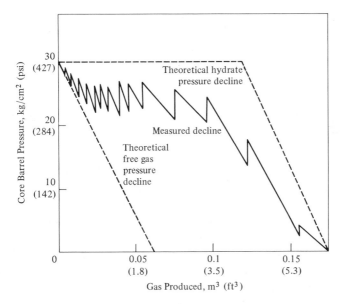

Figure 5-5
Technique for identifying hydrates in pressure core barrels.

With intermittent bleeding of the gas, a zigzag line will result, as shown. Release of a small amount of gas will cause a pressure drop. Closing the valve will allow the pressure to return to the original 30 kg/cm² (427 psi) if one waits sufficiently long. Normally, warming will be controlled gradually to avoid too rapid a decomposition. Consequently, additional gas will be released before the pressure increases to 30 kg/cm². The increase in pressure toward the theoretical hydrate pressure each time the valve is closed confirms the presence of the hydrate.

Gas hydrates have excited the interest of many operating personnel, because a hydrate reservoir can hold six times as much gas as free gas held in the same space. However, this enrichment factor decreases with depth, because, as previously mentioned, free gas compresses considerably with depth. The ratio of gas in a hydrate to free gas in a reservoir of the same size decreases from 6 at about 900 ft (274 m) to 2 at 2,500 ft (762 m) and 1.25 at 4,000 ft (1,219 m). This is assuming a normal hydrostatic gradient and a geothermal gradient of 1.5°F/100 ft.

When the Russians first found extensive gas hydrate deposits in Siberia, they estimated the reserves at around 15 × 10¹² m³ (530 × 10¹² ft³). Unfortunately, it has been difficult to figure out how to produce the hydrate gas economically. Both methanol and salt solutions have been used to decompose the hydrates, with marginal results. Hydrates can represent a hazard in that melting of the hydrate around the well bore would result in conversion of a crystalline solid to a gelatinous mass that could cause the casing to collapse. It is necessary to use small, well-insulated casings through the whole length of the hydrate zone and to maintain adequate blowout prevention equipment even on production wells. Also, the driller must be prepared for possible high gas pressures at the bottom of the hydrate zone. The technology for the drilling and optimum recovery of gas from gas hydrates is so little developed that thus far they are more of a hazard than a benefit.

Bily and Dick (1974) report the experience of Imperial Oil Ltd. in drilling through gas hydrates in the Mackenzie Delta. The hydrate zones can be recognized by various logging techniques. Total mud gas readings are significantly high, the sonic log shows increased velocity, the dual-induction laterolog shows high resistivity, and the SP (self-potential) curve shows little deflection opposite hydrates compared to free gas or water-bearing zones.

Drill-stem test pressure readings in one hydrate zone showed a straight-line buildup in reservoir pressure during each shut-in period (comparable to Figure 5-5). Bily and Dick (1974) concluded that hydrated gas is unlikely to be produced at significant rates with conventional completion and production techniques, given the heat requirements for hydrate decomposition and the amount of water released. They feel that hydrate-bearing zones could be drilled safely by using cool mud and controlled penetration rates. Also, they advise casing off the hydrates before drilling deeper, to minimize mud gasification.

Hydrate formation may make some reservoir oils heavy by removing the gaseous fraction. Holder et al. (1976) have shown that crude oil can be denuded of lighter constituents when contacted with water at low temperature and high pressure. They suggest that some shallow Alaskan oil reserves have been denuded of the methane through isobutane hydrocarbons by hydrate formation.

CATAGENESIS

The largest quantity of natural gas constituents is generated during the catagenesis stage. Increasing temperatures over geologic time result in the gradual removal of hydrogen, nitrogen, oxygen, and sulfur from the organic matter as hydrocarbons, ammonia, carbon dioxide, water, and hydrogen sulfide. Eventually the fine-grained rock contains only a graphitic form of carbon comparable to that in very old metamorphosed rocks. The rates of formation of the natural gas constituents determine to some extent the depths at which they are found and their areal distribution within the basin. The latter is also affected by the large differences in the ability of the different

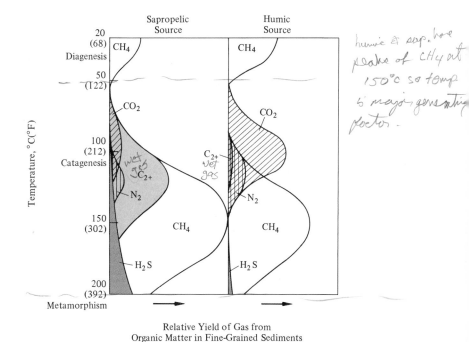

Handwritten margin note: humic & sap. have peaks of CH₄ at 150°C so temp is major generating factor.

Relative Yield of Gas from
Organic Matter in Fine-Grained Sediments

Figure 5-6
Generation of gases with depth. C_{2+} represents hydrocarbons
heavier than CH_4 in gas phase. N_2 is generated initially as NH_3.

gaseous constituents to diffuse and migrate through the sedimentary column. Figure 5-6 is a schematic illustration of the temperature ranges in which the various gases are formed, assuming a normal geothermal gradient of about 3°C/100 m (1.7°F/100 ft). Under a very high geothermal gradient, these gases would form at a higher temperature over a shorter time, and vice versa. The curves show in a very general way the relative volumes of the different gases from sapropelic (amorphous and herbaceous) organic matter and from humic (woody and coaly) organic matter when both are deposited in an aquatic environment (marine or nonmarine). The biogenic methane formed during diagenesis is estimated to represent about 10 percent of the total methane in the temperature range to 200°C (392°F), as discussed in Chapter 4. The quantity varies considerably with depositional environment. Most biogenic methane accumulations are in sand–shale sequences high in continental organic matter rather than in carbonate–evaporite sequences high in marine or lacustrine source material (Rice 1975; Polivanova 1977). Polivanova attributes this difference to a suppression of methanogenic bacteria activity in evaporite deposition.

Figure 5-6 does not show the CO_2, N_2, and H_2S formed during diagenesis, because very little of it survives early microbial and chemical reactions. Some large sulfur deposits are due to oxidation of diagenetic H_2S.

During catagenesis, the amorphous, sapropelic organic matter generates about the same amount of hydrocarbon gas as does humic source material. However, part of the previously formed oil that is in sapropelic source beds also is converted to hydrocarbon gas (Harwood 1977). This results in an overall yield of gas from sapropelic source material from 1.5 to 2 times the gas yield from humic material, depending on the amount of oil remaining and its ease of conversion to gas.

Humic source material contributes most of the organically derived CO_2 in sediments. The Cretaceous Mannville Formation of the Western Canada Basin contains humic organic matter with a high CO_2 yield. Other source relationships shown in Figure 5-6 are discussed in the following section.

Nitrogen

A large percentage of the nitrogen in proteinaceous organic matter is converted to ammonia (NH_3), which dissolves in pore waters and is adsorbed by clays in the diagenetic stage (Bordovskiy 1965). Hare (1973) noted that the ratio of ammonia to total hydrolyzed amino acids increased from 2 to 1 in near-surface sediments of the Cariaco Trench to 10 to 1 at a sediment depth of 110 m (361 ft). In addition, ammonia is released during the catagenetic stage, from the thermocatalytic cracking of organic matter. Part of this dissolved and adsorbed ammonia is believed to be converted to nitrogen during burial through contact with heavy metal oxides or meteoric waters carrying oxygen. High-nitrogen gases are often found associated with red beds. Guseva and Fayngersh (1973) pointed out the presence of gases with over 90 percent nitrogen in red beds of the central European and Chu-Sarysuy oil–gas basins. Nitrogen gases in both basins occur mainly in red beds of Carboniferous, Permian, or Triassic strata overlain by thick salt beds. The authors suggested that the nitrogen may be formed by the reaction of ferric oxide with nitrogen-bearing organic compounds.

In the North Sea off the coast of Germany, high-nitrogen gases are also found in red beds of the Rotliegendes Formation under a thick cover of Zechstein salt. Lutz et al. (1975) attribute its origin to maturation of underlying Carboniferous coal measures and the thermal alteration of dispersed organic nitrogen compounds in shales.

The high-nitrogen gases in the North Sea are believed by Getz (1977) to indicate the presence of much larger accumulations of methane entrapped nearby. He states that methane and ammonia are emitted in a volume ratio of about 80 to 15 during the maturation and degasification of coal. He believes

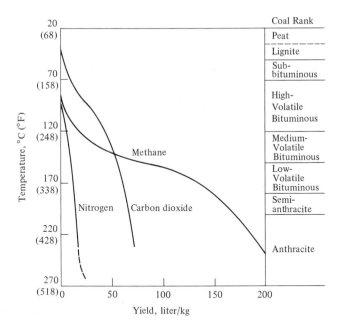

Figure 5-7
Calculated curves of gases generated from coal during coalification.
[Data from Karweil 1969]

that oxidation of ammonia by ferric iron would yield nitrogen, which would migrate further than methane because of its smaller molecular size.

Studies on the origin of natural gas from the maturation of coaly sediments indicate that nitrogen begins to be emitted (as NH_3) near the end of the high-volatile bituminous coal stage. Figure 5-7 shows the estimated loss of nitrogen, carbon dioxide, and methane based on coalification studies by Karweil (1969), Jüntgen and Klein (1975), and others quoted in their papers. The ratio of methane to carbon dioxide to nitrogen given off in going from peat to semianthracite is approximately 10 to 4 to 1. The ratio is probably similar for continental-derived organic matter deposited in a marine basin.

Some nitrogen may come from the high-temperature decomposition of relatively stable heterocyclic organic nitrogen compounds during the metamorphism of sedimentary rocks. Klein and Jüntgen (1972) heated various ranks of coal and measured the N_2 given off. They found that the activation energies required for release of nitrogen increased from 39 kcal/mole for high-volatile bituminous coal to 50 kcal/mole for anthracite. As was discussed in Chapter 4, an increase in the activation energy required means that the atomic bonds in the more mature coal are more stable and require higher temperatures to crack. An interesting finding of Klein and Jüntgen's studies was that the release of nitrogen occurred in two peaks, the first prior to the release of H_2S (which starts at about 130–150°C, or 266–302°F) and the second at higher temperatures. The interpretation was that the less stable paraffin and cycloparaffin nitrogen compounds would break up early, whereas very high temperatures were required to break the nitrogen bonds in heterocyclic nitrogen—carbon rings. An evaluation of their data relative to Figures 5-6 and 5-7 would mean that the less stable nitrogen compounds would be given off during catagenesis, whereas the more stable compounds would not be released until the metamorphism of the sediments at temperatures above 200°C (392°F).

The possibility of nitrogen coming from the degassing of basement rocks is suggested by observations that gases occluded in igneous rocks are frequently high in nitrogen and that the nitrogen content of some gas accumulations increases toward basement (Beebe and Curtis 1968). Petersil'ye et al. (1970) found 24 to 40 percent N_2 and 0.6 to 3.7 percent helium in methane gas issuing from holes drilled in ultramafic rocks of the Kola Peninsula. The frequently observed correlation of nitrogen and helium in natural gas has indicated a deep-seated source for some nitrogen.

Argon is an inert, easily measurable constituent of the atmosphere. Consequently, the ratio of nitrogen to argon in subsurface gases can be used to estimate how much of the nitrogen came from air trapped in sedimentary rocks. Most analyses have shown the N_2/A ratio is much higher in the subsurface than in the atmosphere, indicating that very little N_2 has an atmospheric origin. Zartman et al. (1961) carried out detailed isotopic studies on natural gases and concluded that only a small fraction of the nitrogen in reservoir gas accumulations can be attributed to the incorporation of air.

Carbon Dioxide

The somewhat erratic distribution of carbon dioxide (CO_2) in reservoirs is caused by a variety of factors, such as multiple sources, high solubility in formation fluids, and high reactivity. The solubility of CO_2 in formation waters increases with increasing pressure. At one atmosphere and 20°C (68°F), about one volume of CO_2 dissolves in one volume of water. At a pressure of 300 atmospheres and a temperature of 100°C, or 212°F (equivalent to about 10,000 ft of burial), 30 volumes of CO_2 (at STP) will dissolve in one volume of water.

A major source of CO_2 in the subsurface is from the thermocatalytic decomposition of oxygen containing groups in organic matter to yield CO_2 and H_2O. Decarboxylation of COOH groups occurs mainly in the diagenetic stage but much of this CO_2 is either utilized by microbes or dissolves as HCO_3^-. The CO_2 peaks in Figure 5-6 are mostly due to the decomposition of carbonyl (C=O), methoxyl (—OCH$_3$), phenolic hydroxyl (—OH) and other oxygen groups in the catagenetic stage. Continental derived humic materials contribute most of the organic source CO$_2$. Up to 75 liters of CO$_2$ per kilogram of coal (2,700 ft^3/ton) is produced during the maturation of coal from the lignite into the anthracite stage (Karweil 1969).

LEACHING OF CaCO$_3$

The dissolution of CaCO$_3$ during burial may be represented by the equation:

$$CaCO_3 \rightleftharpoons Ca^{2+} + CO_3^{2-}$$

The activity product of $(a_{Ca}{}^{2+})$ $(a_{CO_3}{}^{2-})$ is a constant at equilibrium under any given pressure–temperature conditions. When CO$_3^{2-}$ enters the pore water, it reacts with H$^+$ to form HCO$_3^-$. This reduces the activity of CO$_3^{2-}$, thereby allowing more CaCO$_3$ to dissolve.

High pressures at great depths increase the partial pressure of CO$_2$, thereby increasing the H$^+$ through the following reaction:

$$CO_2 + H_2O \rightleftharpoons H^+ + HCO_3^-$$

This allows more CO$_3^{2-}$ to react and more CaCO$_3$ to dissolve. The process ends when CaCO$_3$ reaches saturation under the existing pressure–temperature–pCO$_2$ conditions. If CaCO$_3$ does not reach saturation as with some meteoric waters, leaching of CaCO$_3$ will continue creating huge caverns in the carbonate rock.

Some of the CO$_2$ in deeply buried, high-temperature limestone reservoirs is believed to have an inorganic source from the thermal decomposition of carbonates. Early experiments in heating pure calcium carbonate indicated that temperatures higher than those encountered in sedimentary basins were required to produce appreciable amounts of CO$_2$. Germann and Ayres (1942), however, demonstrated that impure limestones with traces of alumina, magnesium, iron, manganese and silica yielded 8 to 17 times as much CO$_2$ as pure calcite at temperatures as low as 98°C (208°F). These experiments were run in the presence of water. They interpreted from their data that the thermal decomposition of impure calcium carbonate can account for much of the CO$_2$ observed in natural gas accumulations.

Kissin and Pakhomov (1967) demonstrated that water-wet, dolomitized limestones evolve CO$_2$ at temperatures as low as 75°C. The yields at 75°C (167°F), 150°C (302°F), and 243°C (469°F) were 0.013 liter/kg, 0.4 liter/kg, and 3.7 liter/kg respectively. Later experiments by Pakhomov and Kissin (1968) showed that CO_2 was formed by the high-temperature leaching of a whole series of rock types containing carbonate minerals. The presence of magnesium accelerated the reaction. The authors concluded that CO_2 generation was most intense where rocks or subsurface waters of marine origin were involved. From field examples, they concluded that the largest increases in CO_2 due to $CaCO_3$ decomposition occurred when subsurface temperatures approached 150°C (302°F).

High temperatures can result from either deep burial or the presence of igneous dikes. Miller (1938) calculated that if a 9.75-m (32-ft) igneous dike intruded a carbonate rock at a temperature of 1,300°C (2,372°F), it would heat the carbonate rock to a temperature of 100°C (212°F) about 7.9 m (26 ft) from the edge of the dike within a year. This was based on heat conduction only and did not allow for heat transmitted by evolving vapors and liquids.

The possible relationship between CO_2 generation and high-temperature igneous intrusions was discussed by Holmquest (1965) for the Delaware–Val Verde basins of New Mexico and Texas. Figure 5-8 shows the change in CO_2 concentration of gases from the Ellenburger Dolomite of Cambro-Ordovician age. Ellenburger gases in the regions closest to the igneous intrusion of the Diablo platform and Marathon uplift contain the highest percentages of CO_2. The CO_2 concentrations decrease away from the Tertiary igneous activity toward the Central Basin platform, where CO_2 concentrations are zero. There is a vertical upward decrease in CO_2 from the Ellenburger as well as a horizontal decrease away from the intrusion. Table 5-3 shows high CO_2 values next to basement for some of the fields in Figure 5-8, with much lower values further up in the section (Farmer 1965).

The higher temperatures encountered in the deeper parts of sedimentary basins are adequate to form CO_2 from carbonates without the additional heat of igneous activity. Colombo et al. (1969) studied the composition of gases in a multilayer gas field of southern Italy composed of alternating sand–shale layers of Pliocene age. The substratum consists of carbonates. The CO_2 in gases from a series of producing strata extending over a depth interval of

Table 5-3
Increase in CO_2 of Natural Gases with Depth Approaching Basement in Delaware–Val Verde Basins, West Texas

	Percent CO_2 in natural gas at			
	Elsinore	Toyah	Brown–Bassett	Puckett
Pennsylvanian			0.2	
Devonian	45			3
Silurian		7	3	
Cambro-Ordovician	100	98	50	40

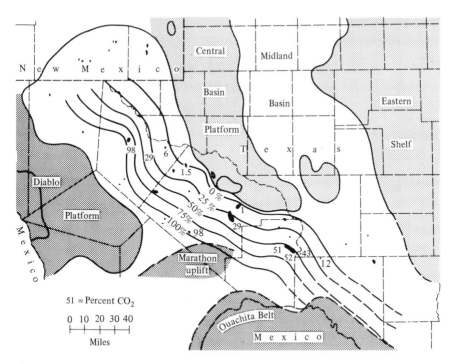

Figure 5-8
Contours of percent CO_2 in gases from Ellenburger Formation in the Delaware–Val Verde basins of New Mexico and Texas. [Holmquest 1965]

about 1,500 m (4,921 ft) averages less than 1 percent. A gas sand below this, partly lying on the unconformity on top of the limestone, contains from 51 to 98 percent CO_2.

Hydrogen Sulfide

Hydrogen sulfide (H_2S) is the most deadly gas produced in large quantities in nature. As little as 0.1 percent by volume in air causes respiratory paralysis and sudden death from asphyxiation. Although the nose can detect only a few ppm in air, the gas deadens the sense of smell so that workers are overcome before they realize it. The gas is very corrosive and causes hydrogen embrittlement in both drill pipe and in pipelines. Mixed with H_2S, CO_2 promotes the embrittlement. This requires the use of high-carbon steels in transporting or drilling for high H_2S gas.

Large amounts of H_2S are formed from the bacterial reduction of sulfates in near-surface sediments. The gas is so soluble and reactive that practically all of it is converted either to free sulfur, metallic sulfides, or organic sulfur compounds in sediments. The solubility of H_2S at 20°C (68°F) and 1 atmosphere pressure is 2.6 volumes of gas per volume of water, which is more than twice the solubility of CO_2. The huge sulfur deposits associated with salt domes in the U.S. Gulf Coast are the result of biogenic H_2S being oxidized and trapped as sulfur (Thode et al. 1954). The amount of biogenic H_2S that becomes a constituent of natural gas, however, is negligible compared to H_2S from other sources.

The reservoir H_2S comes from the thermal alteration of organic sulfur compounds in both the source beds and the reservoirs plus the thermocatalytic reduction of sulfate in formation waters when in contact with reservoired hydrocarbons (Le Tran 1972; Orr 1974). The generation of H_2S from these various sources is not significant until the sediments are buried to temperatures in the range of 100°C (212°F). Hydrogen sulfide is the last of the natural gas constituents to be formed and is often found with methane in the deepest

Table 5-4
Natural Gas Deposits with High H_2S Content

Region	Reservoir Age	Lithology	Depth (m)	Percent H_2S
Lacq, France	Upper Jurassic and Late Cretaceous	Dolomite and limestone	3,100–4,500	15
Pont d'As-Meillon, France	Upper Jurassic	Dolomite	4,300–5,000	6
Weser-Ems, West Germany	Permian (Zechstein)	Dolomite	3,800	10
Asmari–Bandar Shahpur, Iran	Jurassic	Limestone	3,600–4,800	26
Ural–Volga, USSR	Late Carboniferous	Limestone	1,500–2,000	6
Irkutsk, USSR	Late Cambrian	Dolomite	2,540	42
Alberta, Canada	Mississippian	Limestone	3,506	13
	Devonian	Limestone	3,800	87
South Texas, USA	Late Cretaceous (Edwards)	Limestone	3,354	8
	Upper Jurassic (Smackover)	Limestone	5,793–6,098	98
East Texas, USA	Upper Jurassic (Smackover)	Limestone	3,683–3,757	14
Mississippi, USA	Upper Jurassic (Smackover)	Limestone	5,793–6,098	78
Wyoming, USA	Permian (Embar)	Limestone	3,049	42

parts of basins. The largest quantities of H_2S are in carbonate reservoirs. Table 5-4 summarizes the data for gas deposits with more than 5 percent H_2S (Le Tran 1972). Several factors cause more H_2S to be in carbonate reservoirs than in sands: (1) the sapropelic kerogen of carbonate source beds generates more H_2S than the humic kerogen of shales, (2) carbonate minerals catalyze the reaction of sulfur with hydrocarbons to form H_2S, and (3) the ratio of iron in shale to iron in carbonates is $12/1$ and that of iron in sand to iron in carbonates is $3/1$. This means that from 3 to 12 times more H_2S can be changed to iron sulfide in sands and shales than in carbonates.

Hydrogen sulfide is common in Paleozoic limestones in areas such as Ohio, Indiana, Illinois, Michigan, and western Kentucky. Although Kansas and Oklahoma have practically no H_2S in younger formations, there are large amounts in the Cambro-Ordovician carbonates of both states.

Hydrogen sulfide will occur in sands if meteoric sulfate containing waters contact a deep hydrocarbon reservoir. Orr (1974) has pointed out that sulfate reduction by hydrocarbons can produce large quantities of H_2S during petroleum maturation. Sheet sands outcropping at the surface or in contact with anhydrite beds could provide a source of sulfate.

Le Tran et al. (1974) demonstrated that most H_2S in the subsurface originates from the thermal decomposition of organic matter in fine-grained rocks. Le Tran analyzed the adsorbed hydrogen sulfide and methane in sediment samples with depth in three wells of the Aquitaine Basin. He also analyzed for the C_{15+} liquid hydrocarbons in order to determine the zones of heavy hydrocarbon generation. Results are shown in Figure 5-9 for the Jurassic Mano Formation. Depth is plotted linearly with the gas yield in ml/g of carbon plotted logarithmically. The location of the C_{15+} hydrocarbon yield peak and approximate present-day temperatures also are indicated. Similar results were obtained in a study of the Lower Cretaceous Barremian-Neocomian Formation.

The data show that the thermal cracking of organic sulfur structures in the dispersed organic matter to form H_2S starts later than the generation of methane. The largest quantities of both gases are formed at temperatures greater than 120°C (248°F). This temperature is minimal, since the geological evidence indicates that paleotemperatures were higher. The oil generation peak is at about 90°C (194°F) in Figure 5-9. The peak in methane generation is in the range of 140°C (284°F), while the H_2S generation does not peak at the maximum depth analyzed, equivalent to almost 170°C (338°F).

Le Tran (1972) also analyzed for adsorbed H_2S in cores and cuttings of wells from other basins. He found that the increase in H_2S always paralleled the increase in CH_4 in the deepest part of the basin. Also, the formation of H_2S was always related to the dispersed organic matter. Where there was little or no organic matter in the fine-grained rocks, there was no H_2S, and zones

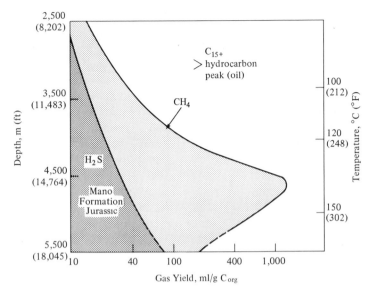

Figure 5-9
Generation of CH_4 and H_2S from thermal decomposition of organic matter in carbonate source rock of Aquitaine Basin, France. Peaking of oil generation (C_{15+} hydrocarbons) occurs prior to peaking of CH_4 generation, which occurs prior to peaking of H_2S generation. [Data from Le Tran et al. 1974]

high in organic matter were high in H_2S, other things being equal. The yields of H_2S always increased with depth of burial, with the largest increase occurring after the peaking of oil generation.

Chapter 8 contains a more detailed discussion of the formation of H_2S from reactions in the reservoir.

Trace Gases

Helium and argon are the noble gases, unreactive in the natural environment, so they last unchanged through geologic time. The helium atom has a diameter about one-half that of a methane molecule. It is so small that it readily migrates through the geologic column. Helium is produced continuously by the disintegration of radioactive elements in Precambrian and Phanerozoic granite rocks and minerals. A gram of uranium will generate up to 1.2×10^{-7} ml of helium in a year. The quantities of helium in a particular reservoir are usually related to the distance from a granitic basement and the impermeability of the reservoir cap rock.

Hydrogen is a common constituent of many well gases and subsurface waters, but it is rarely reported in the literature because it is not included in conventional gas analyses. Zinger (1962) found up to 43 percent hydrogen in the gas dissolved in oil field waters of Paleozoic rocks of the lower Volga region of the USSR. Samples were taken in new wells directly after perforation of the casing and after the water attained constant chemical composition, so the hydrogen could not be attributed to acid reactions with the casing. Zinger's paper also reports other Russian data on hydrogen in natural gases and in coal gases. In western Siberia, Nechayeva (1968) found hydrogen in about 15 percent of all gas samples analyzed. About 60 percent of the samples with hydrogen had under 1 percent. Peak hydrogen concentrations were 0.9 percent for gas fields, 6 percent for gas condensate fields, and 11 percent for oil fields. Highest concentration were in gases from organic-rich Jurassic sediments suggesting that the hydrogen was coming from thermal decomposition of the organic matter. Analyses of both well gases and sediments from the North American continent have shown hydrogen to be relatively common, with up to 15 percent in some natural gases.

Hydrogen is so mobile and reactive that it cannot be permanently retained in a geological trap. Its presence indicates that it is either being actively generated from reactions in the reservoir or adjacent source beds or is diffusing up from deeper sources.

Methane and Heavier Hydrocarbon Gases

About 82 percent of the methane and practically all of the heavier hydrocarbon gases are formed in the catagenetic stage (see Chapter 4). Ethane, propane, and the butanes are formed in the temperature range from about 70 to 150°C (158 to 302°F), with peak generation around 120°C (248°F). Methane forms in a higher range with peak generation around 150°C (302°F) (Figure 5-6). The quantity of gaseous hydrocarbons formed varies considerably with the organic source material. The woody, humic, continental organic matter (right side of Figures 5-6 and 5-10) generates methane with traces of higher hydrocarbons. The amorphous, sapropelic, marine material (left side of Figures 5-6 and 5-10) generates most of the C_2H_6, C_3H_8, and C_4H_{10} hydrocarbons plus more methane than the humic organic matter. Methane begins to form as early as oil, but the bulk of it comes after oil generation from both kerogen and heavier hydrocarbons.

Figure 5-10 is a schematic illustration of the basic difference in sapropelic source (oil-generating) and humic source (gas-generating) organic matter. The oil-generating organic matter has many long chains and small groups of rings that can break up to form the liquid fractions of petroleum. As the chains break off, they strip hydrogen from the remaining structure, which

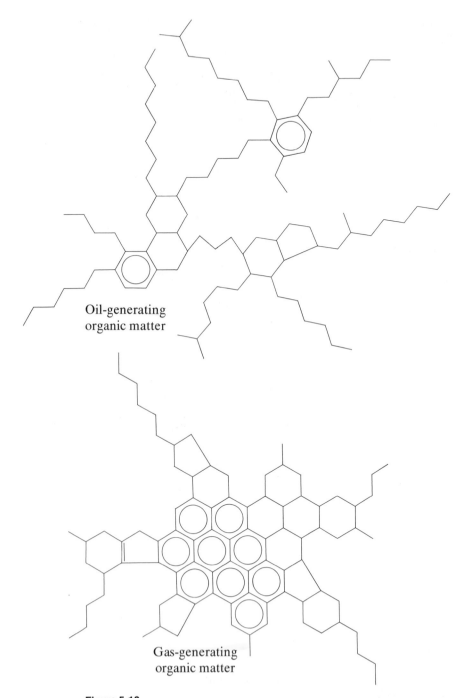

Oil-generating
organic matter

Gas-generating
organic matter

Figure 5-10
Schematic structures of oil-generating and gas-generating organic matter.

condenses to form the compact ring system of the gas-generating organic matter. The major generation of gas occurs after the organic matter has formed oil (see Chapter 7 for more details).

The humic gas-generating material has only a few side chains plus single methyl groups (shown by one line) and a large number of condensed rings. The gas-generating material can form a trace amount of ethane, propane, and butane, but its major product will be methane, leaving behind a residue that becomes graphitic.

This means that a gas-prone area is either an area that has always contained only gas-generating organic matter or an area in which oil was formed and destroyed, leaving organic matter capable of forming only gas. There are various geochemical tests that will indicate the present capability of a fine-grained rock to generate only gas or oil plus gas. For example, one technique is simply to heat the organic matter and determine the ratio of ethane or a heavier hydrocarbon to methane. As shown in Table 5-5, the ratio of ethane to methane in an oil source material is considerably higher than that in a gas source material. Ratios of higher hydrocarbons such as C_5-C_{10}/C_1 show even larger differences. Other techniques for distinguishing between gas and oil source rocks are discussed in Chapters 7 and 10.

The terms *marine source* and *continental source* have sometimes been used to describe oil-generating and gas-generating organic material, respectively. As a generalization, it is true that most of the humic material is formed on the continents and sapropelic material in the marine environment. However, the wax fraction of crude oil is entirely from a continental source, and there is marine humus that presumably yields only gas. Consequently, it is better to classify organic matter types in terms of their chemistry rather than their source.

Table 5-5
Ratio of Ethane to Methane Obtained from Heating
Organic Matter

Gas Source	Ethane/Methane
Atoka Shale kerogen, Carboniferous, Oklahoma	1.6[a]
Davis Shale kerogen, Cambrian, Missouri	1.0
Bituminous coal, Pennsylvania	0.3
Gunflint Chert kerogen, Precambrian, Canada	0.05

Oil Source	
Viking Shale kerogen, Cretaceous, Canada	3.5[a]
Green River Shale kerogen, Eocene, Utah	3.3
Monterey Shale kerogen, Miocene, California	3.2

[a]C_2-C_4/C_1
Source: From Hunt 1963b.

Coals yield dry methane on heating. When the heavier gases, ethane through butanes, are formed from coal, it is in the lower temperature ranges. In the Munsterland I bore hole in Europe, Teichmüller and Teichmüller (1968) reported that traces of pentanes were in the coal gas only to a depth of about 2,100 m, 85°C (6,890 ft, 185°F), and butanes only to a depth of 2,600 m, 95°C (8,530 ft, 203°F). Propane represented more than 10 percent of the residual gas at 2,000 m (6,562 ft) but less than 1 percent at 3,000 m (9,843 ft).

Kozlov (1960) analyzed several coals from the Donets Basin and found ethane, propane, and butanes plus higher hydrocarbons present in most samples, although the yields were generally less than 1 percent. In some, ethane reached concentrations of 0.8 percent, with the higher hydrocarbons representing 0.1 to 0.2 percent. Kim (1973) found ethane and higher hydrocarbons in a variety of coals in both the eastern and western United States. Eastern coals had up to 1.5 percent ethane, with 0.01 percent higher hydrocarbons. Although these concentrations are very small, their presence does show that some structures in coals are capable of forming the heavier gaseous hydrocarbons.

CARBON ISOTOPIC COMPOSITION OF METHANE

When methane is formed biogenically, as shown at the top of Figure 5-6, it has a different carbon isotopic composition from that of methane formed abiotically during catagenesis. Figure 2-3 showed the ^{13}C value of various carbon reservoirs compared to the standard Peedee belemnite. The lightest petroleum on that scale had a $\delta^{13}C$ value of -32 ‰. When methanogenic bacteria form methane by the reduction of CO_2, they consume mainly the lighter $^{12}CO_2$ rather than the heavier $^{13}CO_2$. Depending on the ^{13}C value of the carbon source, the bacteria can create methane with a $\delta^{13}C$ of -88 ‰. Normally, biogenic methane ranges in $\delta^{13}C$ values from about -55 ‰ to -75 ‰. In contrast, methane formed by thermal cracking has been shown by Sackett et al. (1966) to have $\delta^{13}C$ values in the range of -5 ‰ to -30 ‰ relative to the parent material. This means that a kerogen with a $\delta^{13}C$ of -24 ‰ would yield methane in the range of -29 to -54 ‰ by thermal cracking. The more negative values were obtained at lower cracking temperatures equivalent to shallower depths. Later experiments by Frank et al. (1974) indicated that most methane is formed by cracking processes after rock temperatures reach about 130°C (266°F). This agrees with the results in the natural situations discussed in this chapter.

Sackett (1978) also observed that catalytic cracking of a hydrocarbon by a carbonium ion mechanism showed almost no carbon isotope fractionation. He suggested that this might explain the occasional natural occurrence of anomalously heavy methane.

The range in the carbon isotope composition of methane in natural gas accumulations versus age is shown in Figure 5-11. The Quaternary and Pliocene gases, which are largely biogenic, are in the expected range of light methane, with a low ^{13}C content. With increasing age, there is an irregular progression toward a higher ^{13}C content in older sediments. This change is only evident until the Cretaceous, after which all formations show considerable overlap. The increase in ^{13}C through the Tertiary would be expected, since the youngest sediments would be subjected to the lowest cracking temperatures and therefore have the highest ^{12}C content (lowest ^{13}C). The ^{13}C increase could also be caused by the effect of diluting biogenically formed methane with some thermally formed methane. The Quaternary data are mostly from Japanese gas fields (Nakai 1960), and the Pliocene data are mainly from Italian fields (Columbo et al. 1969). Variations in older sediments are probably caused by a small population sample and isotope exchange effects. North German gas accumulations associated with Upper Carboniferous (Pennsylvanian) coals have somewhat higher ^{13}C contents than does methane from kerogen in petroleum areas. The data gathered by Stahl (1974) on coal gas show that the highest ^{13}C content is in gases associated with anthracites, while the lowest is in gases associated with high-volatile bituminous coals.

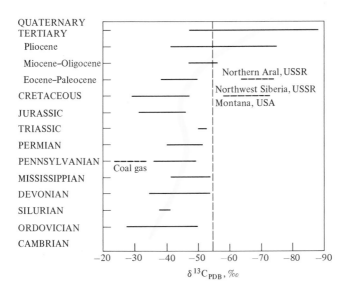

Figure 5-11
Carbon isotope composition of gas (predominantly methane) accumulations.

The difference in the $\delta^{13}C$ value of biotic and abiotic methane should be maintained during millions of years of burial, providing gases from the two sources do not mix and providing the carbon in methane does not undergo isotopic exchange with other carbon compounds, such as CO_2. Two examples where biogenic methane accumulations have survived in ancient sediments in the USSR are shown in Figure 5-11. The Akkulkovo–Bazay gas accumulation in the North Aral region is in sandy clays of Late Eocene age at a depth of 320 to 350 m, or 1,050 to 1,148 ft (Avrov and Galimov 1968). Analyses of gases from ten wells have $\delta^{13}C$ values ranging between -64 and -72 ‰. This is clearly within the range of microbiologically formed methane. Geological study of the area indicated that the origin of the methane must have occurred after deposition of the Lower Oligocene Chagan cap rock, which suggests that biological generation of methane can take place at depths of 300–350 m (984–1,148 ft). Theoretically, methane production will continue as long as the bacteria survive and the organic matter contains oxygen compounds and nutrients. Reserves in this accumulation are in the tens of billions of cubic meters.

An example of biogenic methane accumulations in Cretaceous formations are the Northwest Siberian fields studied by Yermakov et al. (1970). Gases (mostly methane) in nine fields ranging in age from Aptian through Cenomanian had $\delta^{13}C$ values from -58 to -68 ‰. This area was a lowland during the Cretaceous, with the watershed occupied by peat bogs, swamps, and lakes where large amounts of continental plant material accumulated. The organic carbon content of the Cretaceous rocks ranges between 3 and 6 percent, which is high compared to most petroliferous basins. Gases produced in these same fields from deeper Jurassic horizons have $\delta^{13}C$ values ranging from -39 to -46 ‰, which is typical of abiotically formed methane. Also, gases from the same Cretaceous formations in gas provinces outside of the Northwest Siberian plain area show $\delta^{13}C$ values in the range of -30 to -46 ‰. The combination of these data is very good evidence that the Cretaceous gases of the Northwest Siberian plain have a biogenic origin that is distinct from the abiotic origin of other methane accumulations in the region. Examples of Cretaceous biogenic methane in the United States are cited by Rice (1975) to depths of 1,400 m (4,700 ft). These range in $\delta^{13}C$ from -63 to -72 ‰. Rice feels that large resources of biogenic gas are yet to be discovered.

DEPTH AND BASIN POSITION OF NATURAL GASES

Our discussion of Figure 5-6 leads to the conclusion that dry biogenic methane would be found in the shallow sediments of a basin, methane plus heavier gases (wet gas) in the deeper catagenetic zone, followed by dry gas in the deepest and oldest sediments. This hydrocarbon depth distribution

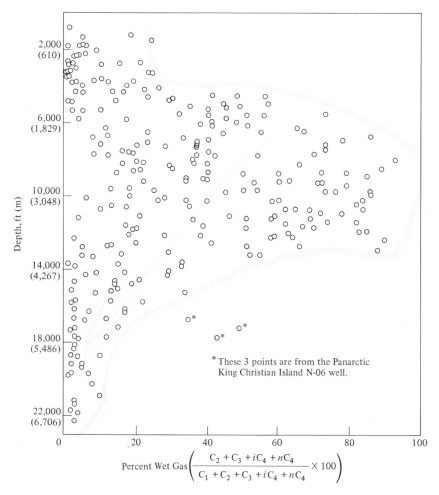

$$\text{Percent Wet Gas}\left(\frac{C_2 + C_3 + iC_4 + nC_4}{C_1 + C_2 + C_3 + iC_4 + nC_4} \times 100\right)$$

*These 3 points are from the Panarctic King Christian Island N-06 well.

Figure 5-12

Average percent wet gas in cuttings from 500-ft (152-m) intervals versus maximum depth of burial for 14 Sverdrup Basin wells. [Snowdon and Roy 1975]

pattern of dry gas followed by wet gas and followed again by dry gas is repeated in many sedimentary basins throughout the world. The heavier gaseous hydrocarbons are formed through the same temperature–depth interval as the liquid hydrocarbons, which is below the biogenic methane but above the maximum generation of abiotic methane.

An example of this is shown in Figure 5-12, which is a plot of the average percent wet gas in shale cuttings from 500-ft (152-m) intervals for 14 wells in the Sverdrup Basin, Canada (Snowdon and Roy 1975). Wet gas is low to about

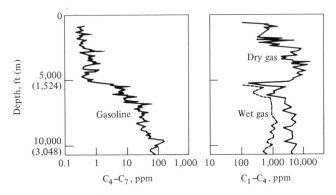

Figure 5-13
Cuttings gas analysis of well in Beaufort Basin, Northwest Territory, Canada. Methane and ethane plus gas yields obtained from the laboratory analysis of cuttings canned at the well site. Only methane was obtained from cuttings to a depth of 5,000 ft (1,524 m) through the diagenetic zone, while wet gas was obtained in the catagenetic zone to total drilling depth. [Evans and Staplin 1971]

4,500 ft (1,372 m), where it increases markedly. The wet gas zone continues to about 14,000 ft (4,267 m), where it begins to drop off to lower values. The gas becomes drier with increasing depth.

Vertical migration in some areas will distort this pattern but the general relationship is the same everywhere. In the Beaufort Basin, Northwest Territory, Canada, Evans and Staplin (1971) observed dry gas in well cuttings of fine-grained sediments to a depth (Figure 5-13) of about 5,000 ft (1,524 m) and wet gas from this depth to about 10,000 ft (3,048 m). On the eastern updip side of the Western Canada Basin, they also observed this transition from dry gas to wet gas with depth. In the deeper and older Devonian sediments of northern Alberta they observed the transition from wet gas to dry gas at a depth equivalent to a paleotemperature of about 320°F (160°C). From their associated studies on the color of kerogen particles, they developed an organic facies concept for the Western Canada Basin that is illustrated graphically in Figure 5-14. Added to this is the distribution of nonhydrocarbon gases as observed in Canada and in many other sedimentary basins. The immature facies (equivalent to the diagenetic stage in this text) contains kerogen with light-yellow spore and pollen particles, indicating that it has never been thermally altered. The dominant gases are biogenic methane and nitrogen. The deeper, mature facies (equivalent to the catagenetic stage in this text) contain spore and pollen ranging from deep amber to red-brown in color, indicating that the increasing temperatures have begun to cook the kerogen. Wet gas and oil is in the upper part of this range, with condensate in the lower part and with carbon dioxide throughout. In the deepest metamorphosed

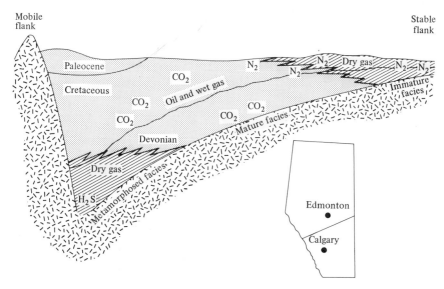

Figure 5-14
Organic facies changes in Western Canada Basin. [Evans and Staplin 1971]

stage, near the end of catagenesis, the spore and pollen in the kerogen are dark brown to black, indicating that the cooking process has eliminated most of the heavier hydrocarbons. The dominant gases in this zone are methane and hydrogen sulfide. Evans and Staplin (1971) estimated that the wet gas and liquid hydrocarbons in the Western Canada Basin were formed mainly in the temperature range between 60 and 170°C (140 and 338°F). This is almost the same range discussed in Chapter 4 for petroleum formation on a worldwide basis. It is the zone referred to by Vassoevich et al. (1969) as the principal stage of petroleum formation. In all these facies, methane is generally the dominant reservoir gas, but the relationships show where the other gases are apt to be found with methane.

The depth concept illustrated in Figures 5-6 and 5-14 relates to a vertical increase in temperature with depth. Temperature may increase horizontally, because of different geothermal gradients within the same basin. In the Rainbow area of Alberta, Canada, the Middle Devonian carbonates in the east are rich in wet gas, but moving west we find a change to dry gas (Evans and Staplin 1971). Middle Devonian temperatures increase horizontally at the same depth westward more than 10°C (50°F) from the wet gas to the dry gas area. In the Dnieper–Donets depression of the USSR, there are oil and gas pools to the northwest, but only gas and condensate to the southeast within the same Paleozoic horizons (Kravets 1974). The depression is a graben formed along a system of faults in the Precambrian crystalline basement and filled by 2.5 to 11 km (8,200 to 36,090 ft) of Devonian through Cenozoic sediments. Subsurface temperature measurements at the northwest edge of the oil-producing area show a geothermal gradient of 2°C/100 m (1.1°F/100 ft), whereas at the southeast edge of the gas-producing area the gradient is 3°C/100 m (1.7°F/100 ft). Clearly, the changes in facies in Figure 5-14, such as wet gas to dry gas, are caused by increasing temperature, not depth. When the temperature increases horizontally in a formation, it has the same effect as a temperature increase vertically.

Among the nonhydrocarbon gases, nitrogen is found in highest concentrations on the stable shelf of basins, as shown in Figure 5-14. This distribution is partly related to diagenetic ammonia being oxidized in near-surface sediments and partly to the ease with which nitrogen migrates. Among natural gas constituents, nitrogen is the smallest molecule next to helium. The approximate molecular diameters of the various gases in meters are as follows: He, 2.6×10^{-10}; N_2, 3×10^{-10}; CH_4 and H_2S, 4.1×10^{-10}; CO_2, 4.7×10^{-10}; C_2H_6, 5.5×10^{-10}; C_3H_8, 6.5×10^{-10}. Nitrogen generated at any depths should be able to migrate out of a basin more easily than the other major gases.

Zor'kin and Stadnik (1970) studied the composition of gases in pore waters of Mesozoic and Paleozoic sediments of the Caspian depression. They found that the dissolved gases were primarily nitrogen in the eastern, shallower part of the depression and along its margin. Down the section, toward the central, deeper part, the concentration of methane and its homologs increased. Zor'kin et al. (1972) stated that subsurface waters of the marginal areas of the Baltic Basin, the Mid-Russian Basin, and the Volga–Ural oil and gas province all are high in nitrogen. Devonian formation waters on the south flank of the Baltic shield are fresh to brackish and contain nitrogen-rich gas. Going basinward, the methane content increases, and the nitrogen content decreases. On the Russian platform, the gases become richer in hydrocarbons and poorer in nitrogen going down toward the deeper parts of the L'vov, the Baltic, and the Moscow basins and the Yarensk trough.

The change in gas facies from nitrogen to carbon dioxide to hydrogen sulfide with depth is shown in the cross section of the Western Canada Basin in Figure 5-14. An aerial view of the nonhydrocarbon gas distribution relative to the "hot lines" of this basin is shown in Figure 5-15. The hot lines were defined by Evans and Staplin (1971) as the transition zone from the wet gas to dry gas facies. This zone represents, in a general way, the westerly limit of oil exploration for each individual strata. The limit is caused by the deepening of the basin and increased temperatures going westward. Line A is the westerly limit of wet gas and oil for the Mississippian; line B for the Upper Devonian Wabamun; C, Upper Devonian Woodbend–Winterburn; and D, Middle Devonian Elk Point. Dry gas will be west of these hot lines, while wet gas and

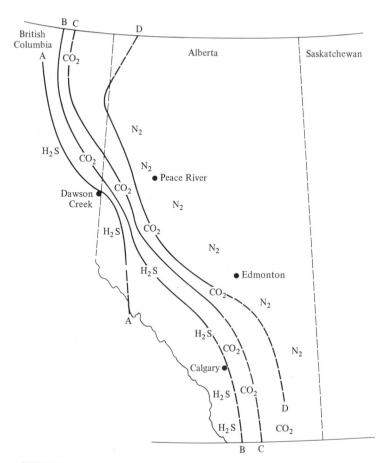

Figure 5-15
Distribution of N_2, CO_2, and H_2S relative to the hot lines of the Western
Canada Basin. The change from oil to gas going west occurs at line D in the
Devonian Elk Point Formation and at line A in the Mississippian. [Evans
and Staplin 1971]

oil will be to the east. The whole area east of line A is prospective for
Mississippian oil, but only the area east of line D is prospective for Devonian
Elk Point oil. Areas of gas accumulations with high N_2, CO_2, and H_2S
concentrations are based on Hitchon's (1963) data. The nonhydrocarbon
gases tend to follow these maturity relationships in a general way, as shown in
Figure 5-14. Since H_2S forms last among natural gas components, it is found
only in the deepest, highest temperature regimes, generally west of the hot
lines in each formation of Figure 5-15. Thus, H_2S can be anywhere west of
line D in the Devonian Elk Point Formation but it would normally only be

west of line A in the Mississippian Formation. Carbon dioxide, which is concentrated in the wet gas facies, would tend to be east of line A in the Mississippian and east of line D in the Devonian Elk Point Formation. To some extent, of course, the gas distributions are affected by migration within the formations away from their source area. The profile does show the general relationship of nitrogen on the shallow, stable flank to the east, with CO_2 in the deeper transition zone and H_2S in the deep, high-temperature sediments on the mobile flank to the west. The overlying Cretaceous formations would follow the same general trend, although they would be too shallow to contain any H_2S.

The Lower Cretaceous Mannville Shale of Alberta contains a land-derived humic kerogen that is a major generator of CO_2. McIver (1967) calculated that a cubic meter (35 cubic feet) of this shale would generate 7 cubic meters (247 cubic feet) of CO_2 at STP during thermal alteration through a 2,000-m (6,562-ft) depth interval. Carbon dioxide is common in Mannville reservoir gases paralleling the hot-line zones from Calgary to British Columbia.

The older Appalachian Basin in the eastern United States shows the same trend in hydrocarbon and nonhydrocarbon gases illustrated in Figure 5-14. In the deep, mobile side of the basin along the Allegheny front, there is dry methane. Moving northwest into the central basin, there is wet gas; continuing onto the shelf area extending from New York through Ohio into Kentucky, the gas is dry again. Although nitrogen is a minor problem in the Appalachian area, the highest nitrogen contents are on the stable, shallow shelf of Ontario, Ohio, and Kentucky.

SUMMARY

1. Methane is formed during diagenesis from the microbiological alteration of organic matter and during catagenesis and metamorphism by the thermal degradation of organic matter, coal, and oil. Sapropelic organic matter generates up to twice the amount of methane formed by humic material, because the former yields methane from both organic matter and previously formed oil, whereas the latter yields it only from organic matter. Some methane forms along with the earliest oil, but peak methane generation is after oil in the range of 150°C (302°F).

2. The heavier hydrocarbon gases, ethane, propane, and butane, are mostly formed after the biogenic methane and before the peak thermal generation of methane. This difference in the time of formation of the hydrocarbons causes a vertical distribution in many sedimentary basins of dry gas on the stable shelf in the diagenetic zone, wet gas in the deeper catagenetic zone, followed by dry gas in the deepest part of the basin.

3. Large quantities of biogenic methane are trapped in the form of hydrates under permafrost areas and offshore areas in low-temperature, high-pressure regimes. New areas for prospecting for gas reserves are indicated beneath the hydrates where the permafrost is retreating and where there are adequate sediments in the deep ocean to drill through the hydrates.

4. Most nitrogen originates from the thermal degradation of organic matter, although some appears to be diffusing from igneous and metamorphic rocks, and a small amount may be trapped air. Nitrogen is often found in reservoirs on the shallow, stable shelves of sedimentary basins. High-nitrogen gases are often associated with red beds.

5. Much carbon dioxide originates from the thermal decomposition of organic compounds. In carbonate formations, appreciable quantities of CO_2 result from the dissolution of carbonate rocks. Some CO_2 also comes from the thermal degradation of carbonates caused by the heat generated by igneous intrusions. In sedimentary basins, the high-CO_2 gases tend to be in the intermediate depth intervals between the shallow nitrogen gas and the deeper H_2S gas. High-CO_2 gases also are present in carbonate sequences near igneous intrusions.

6. Hydrogen sulfide is formed from the high-temperature thermal alteration of organic matter in source and reservoir rocks and from the reduction of sulfate and sulfur by hydrocarbons in the reservoir. A large amount of H_2S is formed by the microbiological reduction of sulfate near the surface, but most of this H_2S is oxidized to sulfur or forms metallic sulfides. High-H_2S gases associated with dry methane occur most frequently in deeply buried carbonate reservoirs. The high H_2S concentration in carbonates is primarily caused by a lack of sufficient heavy metals, such as iron, to react with H_2S.

7. The typical pattern of gas composition in many sedimentary basins in the world is nitrogen plus dry gas on the shallow, stable shelf; CO_2 plus wet gas in the intermediate depth intervals associated with oil production; and H_2S plus dry gas in the deepest part of the basin against the mobile flank.

SUPPLEMENTARY READING

Beebe, B. W., and B. F. Curtis. 1968. *Natural gases of North America.* American Association of Petroleum Geologists Memoir 9, vols. 1 and 2. Tulsa, Okla.: American Association of Petroleum Geologists. 2493 pp.

Kaplan, I. R. 1974. *Natural gases in marine sediments.* New York: Plenum Press. 324 pp.

6

Migration and Accumulation

The concept that oil originates in fine-grained nonreservoir muds and during compaction moves out with the pore fluids into coarse-grained sediments was enunciated in 1909 by Munn as the hydraulic theory for the migration and accumulation of oil. In the intervening 70 years, we have learned a great deal about fluid movement in the subsurface. Numerous hypotheses have been proposed as to exactly how petroleum is concentrated in reservoir sands. Nevertheless, the least clearly understood phenomenon in the overall process of origin, migration, and accumulation is the mechanism of migration. The problem is complex, because it may involve several simultaneous mechanisms operating at different rates for different hydrocarbon ranges in different types of sediment, some operating independently of the water.

Geologists have long recognized the importance of understanding the migration process. In 1933, V. C. Illing summarized these feelings in stating, "No theory of oil origin is secure until we know the limits of oil migration. No intelligent study of oil accumulation can be made until we know the features which influence oil migration and its storage in bulk." In the same paper, Illing defined primary and secondary migration and outlined some of the possible mechanisms. The process of *primary migration* is the movement of oil and gas out of the source rocks into the permeable reservoir rocks, while *secondary migration* is the movement of the fluids within the permeable rocks that eventually leads to the segregation of the oil and gas in certain parts of these rocks.

In this classic paper, Illing asked four questions that are pertinent to our discussion today: "How does the oil move from the source rock to the reservoir rock? How is it trapped in the reservoir rock? How does the water–oil mixture of the source rock, in which water predominates, become a rich oil in the reservoir? How does the oil separate from the water in the process of migration?" The concept of the oil being trapped in the reservoir by the capillary barrier of the water-wet, fine-grained cap rock was beginning to be accepted in Illing's day. Today we have many more concepts in varying stages of acceptance, such as the semipermeable membrane effect of clays, the ratio of free to structured water in fine-grained rocks, the generation of hydrocarbons, and differential compaction as causes of abnormal pressures. The subject of this chapter is how these old and new concepts are clarifying our still-incomplete understanding of the migration and accumulation of petroleum.

WATER

Water (H_2O) is a liquid at standard temperature and pressure (STP). In contrast, other major elements in the life cycle—carbon, hydrogen, nitrogen, and sulfur—form gaseous hydrides, namely CH_4, H_2, NH_3, and H_2S. Why is water a liquid? The reason is that the hydrogen of a water molecule has such a powerful affinity for oxygen that it forms temporary hydrogen bonds with the oxygens of neighboring water molecules. This causes the water to be a liquid polymer at STP. Its boiling point is about 160°C (320°F) higher than it should be, compared to the hydrides of the other adjacent chemical elements in the periodic table.

The polymerization of water molecules causes liquid water to have many anomalous physical properties compared to other substances. For example, it has the highest heat capacity of all solids and liquids at STP. It conducts heat more readily than any other liquid. Its surface tension, dielectric constant, and latent heat of evaporation are the highest of all liquids. In general, water dissolves more substances and in greater quantities than any other liquid. Its maximum density occurs at temperatures above the freezing point. These anomalous properties of water compared to other liquids are important in chemical, biological, and geological processes in nature. For example, the high heat capacity results in large heat transfer by water movement, which prevents extreme ranges in temperature. Our uniform body temperature is maintained through the heat transfer of fluids that are mostly water.

Water is mostly monomeric in the gaseous form, as steam. In the solid form, as ice, the water molecules are arranged in a tetrahedral structure as shown in Figure 6-1. When ice melts, only about 15 percent of the hydrogen bonds are broken. Most of the water remains structured as tetrahedra. For example, the dielectric constant, which is a property highly dependent on the spatial disposition of atoms and charge sites, changes very little. It is about 74 in ice and 88 in water at 0°C (Dorsey 1940). Other physical properties, such as the x-ray radial distribution curve for water and the density of water and ice, indicate that the tetrahedral arrangement still exists in the water phase. The

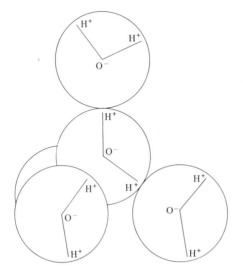

Figure 6-1
Tetrahedral structure of water
molecules forming a polymer.

continuity and stability of this structured water is a matter of debate, but it is
clear that water prefers to form its own internal structures to the exclusion of
foreign materials. For example, when water is slowly frozen any entrained
electrolytes are surrounded by a water structure and excluded from the main
ice lattice formation.

Structured Water

The affinity of hydrogen for oxygen also expresses itself in the structuring of
water next to mineral surfaces. The hydrogen of water bonds to the oxygen of
the mineral surface. Pseudomorphism of water molecules on smectite or illite
can occur because the basal oxygen network of the silica tetrahedra is similar
in size to the tetrahedra structure of water. When exposed to water, smectite
swells because the free energy of the pseudomorphic water is less than that of
the free water in the external solution. Hence the free water enters the
interlayer region to achieve equilibrium. This water next to mineral surfaces
has been referred to as *structured, abnormal, perturbed, adsorbed, anomalous,
ordered,* and *fixed.* In this text, it will be referred to as *structured* or *ordered*
water, in contrast to the *bulk* or *free* water away from mineral surfaces.

Structured water adjacent to mineral surfaces has many properties, such as
the partial specific volume, the coefficient of thermal expansion, and the

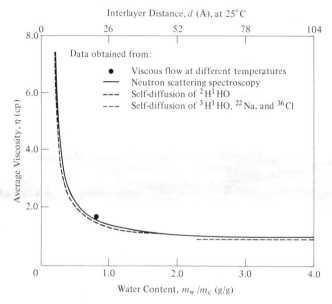

Interlayer Distance, d (Å), at 25°C

Figure 6-2
Change in viscosity of water approaching a mineral surface. High viscosity is caused by structuring of water. cp = centipoise; d = distance from flat smectite surface, in Angstroms; η = viscosity of water, in centipoise; m_w/m_c = ratio of mass of water to mass of clay, in grams/gram. [Reproduced from Low 1976—*Soil Science Society of America Journal*, Volume 40, No. 4, p. 502, 1976, by permission of the Soil Science Society of America.]

viscosity, that differ from bulk water. The marked increase in viscosity of water within 1 nm (10 Å) of the mineral surface is evident in Figure 6-2. Water next to clay surfaces is less dense than bulk water (Anderson and Low 1958). The water has no capacity to dissolve materials. Ions are squeezed out, except for those cations on specific positions. Out to 0.6 nm, the water is highly structured. Some ordering occurs out to several nm. If salt is added to the water or pressure is applied to the system, the structure breaks down in stages. Overburden pressure is not adequate to destroy the last two or three layers of water. At room temperature, 60,000 psi is required to remove the last two layers. Increasing temperatures reduce the ordering of water on mineral surfaces, but the last two or three layers will probably remain at temperatures approaching 200°C (392°F). In sodium and potassium smectites, all water is structured to some extent out to several nm. Calcium and magnesium clay

lattices seldom show ordering beyond 1 or 2 nm. The change from a sodium-
to a calcium-clay lattice with depth would reduce the ratio of structured water
to bulk water.

Organic matter causes structuring of water. Methane hydrates (clathrates)
were discussed in Chapter 5. Water cages exist around many organic
molecules, such as protein. These water structures are referred to as *flickering
clusters* because of their short half-life, in the range of 10^{-10} seconds. As one
structure decays, others are formed, so there is a continuum of structured
water molecules swimming in free water.

In smectitic shales, the ratio of structured water to free water is greater than
in kaolinitic shales. This means that a greater pressure is required to cause
water to flow through a smectitic shale compared to a kaolinitic shale. Flow
through the ordered-water state does not occur until a threshold gradient is
reached. The water has a yield point and then begins to flow, shearing in
non-Newtonian flow. The water peels away from the center. As the gradient is
reduced, the water reforms into a structure. This presence of ordered water in
smectitic shales is one reason why these shales fail to compact under
overburden pressure at the same rate as adjacent sediments that contain very
little smectite. This differential response to overburden pressure may be one
of the causes of abnormal pressure zones in the subsurface. More details of
the properties of structured water have been published by Drost-Hansen
(1969).

Formation Waters

Water occupies all of the pore spaces in sedimentary rocks below the water
table except those containing oil or gas. The water contained in the small
spaces between the mineral grains is called *interstitial* or *pore water*.
Interstitial waters contain ions in solution in varying concentrations depending on their source and environment of diagenesis. As Table 6-1 shows, the
bulk of the ions are contributed by seawater. River, lake, and rain waters act
primarily to dilute the seawater.

During the first few hundred meters of burial, several changes occur in the
chemical composition of the pore waters. Analyses of samples from 200 drill
holes in the Atlantic, Pacific, and Indian oceans have shown the following. In
clay muds, about 70 percent of the samples are depleted in sulfate from 10 to
100 percent. Magnesium is depleted 20 to 90 percent in about a third of the
samples. Both sulfate and magnesium depletion occur least in the carbonates,
but they do occur. Calcium does not change very much until most of the
bicarbonate is used up. After this, the calcium is enriched gradually up to
from 2 to 6 times its concentration in seawater in about half the samples.

Table 6-1
Chemical Composition of Various Waters, $\permil$

	River water	Sea water	Mud pore water[a]		Sand formation water[b]	
			9.5 m (31 ft)	335 m (1,099 ft)	1,570 m (5,151 ft)	1,814 m (5,951 ft)
Cations						
Na^+	0.006	10.8	10.5	7.8	53.9	57.0
K^+	0.002	0.4	0.4	0.3	—	—
Mg^{2+}	0.004	1.3	1.3	0.4	2.1	2.2
Ca^{2+}	0.015	0.4	0.4	2.7	15.0	18.0
Anions						
Cl^-	0.008	19.4	19.6	23.4	115.9	126.0
SO_4^{2-}	0.011	2.7	2.8	2.8	0.1	0.07
HCO_3^-	0.059	1.4	0.1	0.05	0.05	0.06
Total	0.105	36.4	35.1	37.4	187	203

[a]Interstitial water in deep sea drilling carbonate mud samples obtained in Hole 292 east of the Philippines. Sediment depth below sea floor shown in meters and feet. Sodium by difference (White 1975).

[b]Pennsylvanian sands of the Tonkawa and Morrow formations in Texas and Oklahoma (Dickey and Soto 1974). Samples were not analyzed for potassium.

There seems to be no difference in calcium enrichment in clays and carbonates, providing bicarbonate has been reduced. About a third of all the samples studied have shown practically no changes in any of the ions discussed (Sayles and Manheim 1975).

The change in the ratio of calcium to magnesium in many samples appears to occur quite early. In seawater, magnesium exceeds calcium by about 3 to 1. In the pore water analyses shown in Table 6-1, this ratio is maintained in the first 9 m (29 ft) of burial but it is sharply reversed by the time the sediments reach 335 m (1,099 ft), where calcium exceeds magnesium by about 7 to 1. DSDP Hole 293, which also drilled through carbonate muds in the Philippine Basin, contained pore waters with a magnesium/calcium ratio of 2/1 at 95 m (312 ft) depth. This ratio was reversed with calcium exceeding magnesium 3 to 1 at a depth of 445 m (1,460 ft).

As sediments are buried deeper, the sand waters are found to contain markedly increasing quantities of salts, whereas salinities in the shales are comparable to seawater. Table 6-1 shows the composition of two formation waters at depths of over 1,500 m (4,921 ft) in the oil fields of Texas and Oklahoma. This increase with depth can result in formation water salinities ten times greater than seawater. In contrast, Dickey and Baharlou (1973), Schmidt (1973), and Hedberg (1967) all reported shale salinities to be one-third to one-half that of the adjacent sands. A typical analysis showed the

Bartlesville Shale of Oklahoma to contain 41,900 mg/liter total dissolved solids,* whereas the associated sand contained 167,000 mg/liter. The cause of this increase is not fully understood. The two most accepted hypotheses involve salt concentration caused by shales acting as imperfect semipermeable membranes or by diffusion of salts from deeply buried evaporites. Manheim and Horn (1968) have presented evidence for salt diffusion in sediments of the continental margin of the eastern United States. They also point out the practical difficulties in applying a membrane filtration concept to the Atlantic continental margin in particular and to geologic environments in general. Studies of Deep Sea Drilling Project cores have shown that salt concentrations increase in pore waters in areas having evaporite deposits at depth. Pore waters resembling oil field brines in salt content occur in the Red Sea, Mediterranean, West African continental margin, Brazilian continental margin, and Timor Sea. In the USSR, many brine formations are spatially related to solid evaporite formations.

However, salt diffusion cannot explain increases in formation water salinities where no salt beds are known to exist. Also, the change from a low to a high to a low salinity in going upward from a shale to a lenticular sand to an overlying shale cannot be explained by salt diffusion.

White (1965) and Collins (1975) have summarized experiments relating to the membrane filtration concept. Hanshaw and Coplen (1973) and Kharaka and Smalley (1976), among others, have demonstrated that natural materials behave as semipermeable membranes at the temperatures and pressures encountered in the subsurface. There is an opinion that shales extrude salts from their larger pore openings during the early stages of compaction, when the membrane is only partially effective. The large pores contain most of the ions, while the small pores are filled with fresh structured water. As compaction proceeds, all the large pores lose their salts to the sands, until finally only small pores remain and the extruded water begins to freshen. Meanwhile, water escaping through updip reservoir cap rocks is filtered of the salt, which gradually accumulates in the sands.

Fossil reservoir fluids that have not been in contact with the atmosphere since deposition are called *connate waters*. As shown in Table 6-1, connate waters are dominantly chloride waters, with almost no bicarbonate or sulfate and a calcium/magnesium ratio in the range of 5 or 10 to 1. The Russian geochemist V. A. Sulin (1946) classified these as chloride–calcium waters and

---

*True pore water values are probably lower, because salinities in shales should not be as high as those of seawater. Theoretically salt should be expelled from the interlayer region as the clay layers come together. Therefore, even though the concentration of exchangeable ions remains high in the interlayer region, the concentration of free salt (as indicated by the anions) should be low. The high electrical conductivity of a shale is caused by the surface conductance of the exchangeable cations. A high conductivity does not mean that the concentration of free salt must be high.

recognized them as typical of deeply buried residual water left in a compacting system.

Although the chloride–calcium brines are considered the norm for connate waters, wide variations in connate waters result from the complexities of geological and geochemical processes in sediments. For example, Sulin also recognized chloride–magnesium waters, which are characteristic of evaporite sequences. Connate waters rich in sulfate have been found in the Permian Hugoton gas field and in the Pennsylvanian Morrow sands in the Oklahoma Panhandle area (Dickey and Soto 1974). Also, in overpressured zones the retardation of compaction results in the formation fluids receiving fewer ions from the shales and consequently being less concentrated than waters in normally pressured zones (Dickey et al. 1972).

The increase in the total dissolved solids of formation fluids with depth ranges from 50 to 300 mg/liter/m (15,000 to 100,000 ppm/1,000 ft) in many areas. The rate of increase in total solids for various areas is shown in Figure 6-3 (Dickey 1969). This change in concentration with depth can be maintained even though the sediments may be later uplifted and eroded. When this occurs, the exposed sediment will have a seep of concentrated chloride–calcium brine. These compositions can be changed by the infiltration of fresh-water from the surface. Such mixing has occurred on a wide scale in formation waters of the Rocky Mountains of the United States.

If a sedimentary section is deeply buried and compacted and later uplifted and eroded, groundwater enters the formation through permeable outcrops. Freshwater entering in this manner is called *meteoric water*. Such water is relatively fresh, containing less than 10,000 ppm dissolved total solids, most of which is sodium. Meteoric water is high in bicarbonate and sulfate, with almost no calcium and magnesium. Meteoric water often contains both microbes and oxygen. Consequently, it can react both biologically and chemically with any oil accumulation that it encounters. The degradation of crude oils by meteoric waters is discussed in Chapter 8.

Sulin (1946) classified meteoric waters as sulfate–sodium or bicarbonate–sodium, depending on which of the anions is dominant. The composition of meteoric waters is highly variable. Many of its ions are picked up during migration. For example, meteoric waters enter elevated Paleozoic outcrops in the Powder River Basin of Wyoming and move toward the center of the basin across strata containing substantial amounts of gypsum and anhydrite. This causes high sulfate concentrations in these waters.

There are approximately 260 giant oil and gas fields in the world, defined as containing 500×10^6 or more barrels of recoverable oil or a minimum of 3.5×10^{12} cubic feet of recoverable gas (Halbouty et al. 1970). About half of these fields contain waters that have salinities of less than 40,000 ppm. About a fourth contain salinities greater than 150,000 ppm, while the remaining fourth is in the intermediate range. The mean salinity in sandstone reservoirs

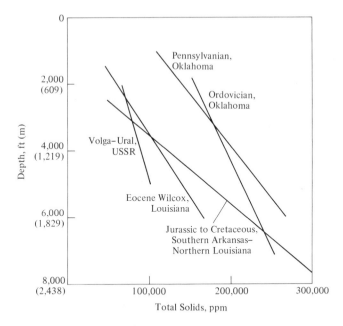

Figure 6-3
Change in salinity of reservoir waters with depth. [Dickey 1969]

is about 27,000 ppm, compared to around 90,000 ppm for carbonate reservoirs. This is because artesian water can flush continuously permeable sand reservoirs more easily than it can flush the intermittently permeable carbonate reservoirs. Salinities in Paleozoic sediments are about four times higher than in Cenozoic sediments. This is probably caused by the former reservoirs undergoing a longer period of ion filtration. Studies of the composition of subsurface waters represent a rich field for research on fluid migration. Almost every physical and chemical process in the subsurface has left its imprint on the ion composition of the waters. Some of these processes, such as the downward migration of sulfate-rich artesian waters or the upward diffusion of chloride from buried evaporites, are processes we can understand and interpret. More complicated is the occasional presence of high-sulfate or high-bicarbonate connate waters in formations where there is no evidence of contact with artesian flows. These may be ancient meteoric waters, trapped in uplifted previously compacted sediments and subsequently reburied. Unfortunately, there is little hard evidence to explain these anomalous situations. Other interesting phenomena include the enrichment of lithium in oil-field waters up to 2,000 times its concentration in seawater (Collins 1975). Hopefully, future generations of students will unravel some of these interesting stories that formation waters have to tell.

PRIMARY MIGRATION

The primary cause of the movement of fluids from a source sediment to a reservoir rock is compaction. Reservoir beds show little compaction in comparison with source beds. The process of compaction begins immediately after sedimentation occurs. The expressed fluids pass through the compacting sediments, away from the position of greater potential, usually upward during the early stages of compaction and later outward toward adjacent, less compressible reservoir beds with an outlet to the surface. Sands compact by rearrangement of the grains shortly after burial. Further compaction by deformation of the coarse grains is minor until very great depths are reached. Clays undergo particle rearrangement over a long period of time but the strongly adsorbed water tends to prevent grain contact. They do undergo plastic deformation. Their compaction is further complicated by processes such as expulsion of water from between clay layers.

The more permeable silt and sand bodies within compacting muds are the main channels of fluid migration. The permeability of sands is very high, compared with that of fine-grained clays; therefore, their pressure gradients are negligible, and their pressure corresponds with the pressure of the immediately surrounding clay layer. This means that fluid movements tend to converge toward coarse-grained silts and sands from surrounding fine-grained muds, except at the points of low potential, which act as exits for the fluids. Usually, but not always, this is the highest point in the formation. As compaction proceeds, there is a continuous movement of fluids from the more compacted fine-grained rocks to the less compacted coarse-grained rocks, the fluid movements being guided by the permeability of the sediments. Compaction reduces the permeability of clay muds very early, of carbonate muds somewhat less, and of silts and sands the least.

The compaction of carbonate muds is more complicated than that of clays, because the former frequently undergo early lithification, chemical alteration, solution, and recrystallization. Most of the discussion in this section relates to the migration of fluids in sand–shale sequences. A section on carbonate migration is included at the end.

The fabric of clays controls their porosity and permeability. Scanning electron micrographs (Borst and Keller 1969) have shown most clay minerals to consist of packets of flat particles. Kaolinite consists of small pseudo-hexagonal books and plates in varying orientations. Illite consists of ir-

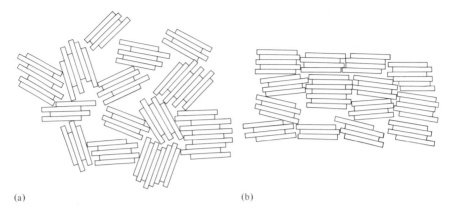

(a)   (b)

Figure 6-4
Stacking of clay particle packets based on scanning electron micrographs. (a) Massive shale,
poor packet orientation. (b) Fissile shale, high orientation.

regularly shaped packets of plates and flakes, some with curled edges.
Smectite consists of highly crumpled and curled flakes in sheets of varying
thickness and lateral extent. There is very little morphological symmetry.
Some smectites consist of extremely thin films or sheets that are diffused and
wisplike, with irregular, fingerlike extensions. A typical bentonite, with 85
percent smectite, 10 percent quartz, and 5 percent feldspar, was found to have
crystallite particles 20 to 30 nm in length and 1 nm in thickness (Borst and
Shell 1971). The crystallite particles are stacked like overlapping packets of
flakes, as shown in Figure 6-4. Spaces between groups of layers range from
about 6 to 16 nm.

The orientation of clay particles, which is important in directing fluid
movements, is not strictly a matter of compaction. Odom (1967) noted that
the orientation of clay particles varied from essentially parallel to nearly
random within a few inches stratigraphically in compacted sediments.
Although it is generally accepted that the compaction of a clay sediment
produces an increase in the parallelism of the particles, it is also true that their
orientation can vary considerably, depending on the conditions of deposition
and the composition of the sediment.

Massive shales in which smectite is the major clay mineral component show
poor particle orientation to considerable depths (Figure 6-4a). Clays with
more than 10 percent carbonate minerals commonly show poor clay particle
orientation. Large amounts of silt and other minerals also tend to reduce
parallel orientation.

Sediments with laminated or fissile structure have a high degree of
orientation, as shown in Figure 6-4b. When laid down in bands or varves,

organic material imparts a fissile structure to the shale, with planes occurring parallel to the bands of enriched organic matter.

Momper (1978) believes that the organic matter in most argillaceous rocks is not dispersed throughout the mineral matrix but is concentrated in layers along bedding surfaces, partings, and laminations. In cross-section view, it is compressed, but in plain view it appears as fluffy patches and blobs with dimensions that can exceed 100 μm. Some bedding surfaces appear to be almost blanketed with organic matter.

Compaction—Equilibrium and Disequilibrium

The direction and extent of fluid migration depends on the fluid potential gradient and the permeability. The *hydrostatic gradient* is the pressure increase with depth of a liquid in contact with the surface. The gradient for freshwater is about 9.8 kPa/m (kilopascals per meter) or 0.433 psi/ft. The *lithostatic gradient* is the total pressure increase caused by rock grains and water. It averages 24.4 kPa/m or about 1.08 psi/ft. If a hole were drilled through a highly permeable sand, extending from the surface to 20,000 ft (6,096 m), the pressure at all depths would be hydrostatic following along line A, shown in Figure 6-5 for 10.4 kPa/m (0.46 psi/ft). This is the gradient for most reservoirs. An increase in salinity with depth will cause the pressure gradient to increase 0.98 kPa/m (0.0043 psi/ft) for each increase of 0.01 g/cc in fluid density (Levorsen 1967, p. 685). Any departure from hydrostatic pressure is an *abnormal pressure.* This includes *overpressures* generally ranging above 12 kPa/m (0.53 psi/ft) and *underpressures* below 9.8 kPa/m (0.43 psi/ft).

Compaction requires not only the application of a load but also the expulsion of pore water. It has long been recognized that good drainage is required for shales to lose their fluids and follow compaction equilibrium with depth (Hedberg 1926, 1936; Athy 1930). If the water cannot escape fast enough, because of the low hydraulic conductivity of the rock, then all or part of the applied load will be carried by the water, and compaction will be retarded. This condition of compaction disequilibrium results in (1) the fluid pressure being greater than hydrostatic and (2) the porosity of the shales being greater than they would be at the same depth under compaction equilibrium.

Variations in the degree of compaction of shales at various distances from permeable beds have been observed in the field. Teslenko and Korotkov (1966) compared the Maikop clays from the West Kuban marginal down-warp. They observed that clays of the same age, occurring in the same depth intervals, and virtually in the same tectonic environment are compacted at different rates, depending to a considerable extent on the number and

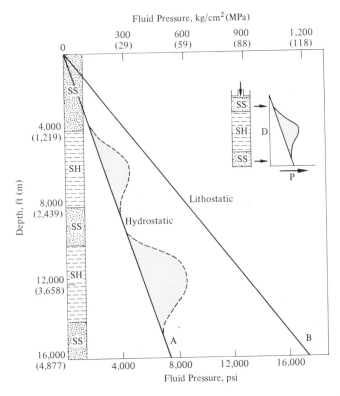

Figure 6-5
Increase in fluid pressure with depth. Line A is hydrostatic gradient for most petroleum basins, 0.46 psi/ft (10.4 kPa/m). Line B is lithostatic gradient, 1.08 psi/ft (24.4 kPa/m). Dotted line shows fluid pressures in sand–shale sequence to the left. Diagram in upper right shows fluid pressure change when pressure is applied to layers of sand and clay mud. Center of the shale is a pressure barrier to fluid movement. Fluids are escaping upward and downward to sand beds.

thickness of highly permeable rocks within the section. Thin clays interbedded with permeable rocks had much lower porosities at the same depth than thick clays that were a long distance from a permeable rock.

An example of their observation is shown in Figure 6-6a. In this figure, the shaded area represents the range of porosities for various sedimentary basins that have been published (see Figure 6-7). Line A represents the average porosity change with depth for Maikop shale core samples that were deposited with rapidly alternating, highly permeable sandstone beds. Line B represents the median porosity for shale cores at a distance of 10–50 m (33–164 ft) from the nearest reservoir rock. Line C represents shale samples that are over 500 m (1,640 ft) from the nearest permeable beds. The bottom samples for lines B and C at a depth of almost 4.5 km (14,764 ft) have a

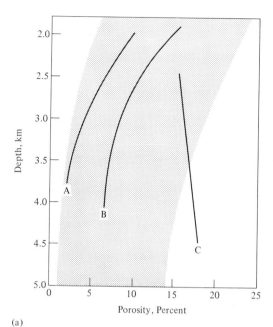

(a)

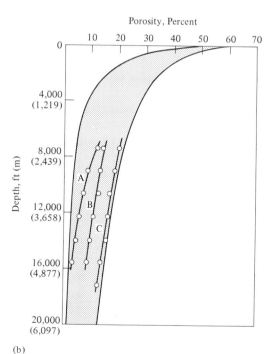

(b)

Figure 6-6
Range in porosities of shales
caused by differing abilities to
release water during compaction.
Interbedded sands and shales
release water faster—Line A in (a)
and (b)—than thick shales—Line
C in (a) and (b). Thick shales are
in compaction disequilibrium
(abnormally pressured). (a)
Maikop shales, USSR; A, alternating sands and shales; B,
shales 10–50 m (33–164 ft) from
sands; C, shales over 500 m
(1,640 ft) from sands. [Teslenko
and Korotkov 1966] (b) Tertiary
shales, South Caspian, USSR; A,
shales 20–50 m (66–164 ft) thick;
B, shales 100–300 m (328–984 ft)
thick; C, shales 500–1,000 m
(1,640–3,280 ft) thick. [Durmish'yan
1973]

porosity difference of about 10 percent. The increase in porosity for line C is probably caused by overpressured shales in the deeper sections.

Teslenko and Korotkov (1966) noted that as the overburden pressure increased the fluids were squeezed out first from shale members in direct contact with the over- and underlying sandy beds. These beds had to be thick and highly permeable sandstones. Isolated thin and poorly permeable beds and small members of sandstones and siltstones had no appreciable effect on the degree of shale compaction. Shale cores remote from the reservoir rocks required much higher overburden pressures to squeeze out their waters. The authors concluded that highly permeable and areally extensive beds of sandy rocks with surface connections such as unconformities, large rifts, and outcrops were required as drainage channels for the sedimentation waters in order for the shales to compact under equilibrium conditions.

In effect, Teslenko and Korotkov (1966) had observed on a large scale what is shown schematically on a small scale in the upper right of Figure 6-5, where sand, clay mud, and sand are shown in alternate layers in a container. If pressure is applied on the top of these beds and fluids are allowed to escape from the top and bottom sand, differential compaction will occur within the clay mud. The edges of the mud near the sands will compact more readily than the center of the mud.

This means that fluid pressures will be higher in the center. This differential fluid pressure can become much greater than hydrostatic as shown in the example to the left of Figure 6-5. A column of sands and thick shales is shown to 16,000 ft (4,877 m). A schematic indication of the variability in pressure for this sedimentary column is shown as a dotted line between lines A and B. This dotted line assumes a uniform rate of compaction for the section and continuity of sand permeability to the surface. Under these conditions, all three sands would be normally pressured (equilibrium), with the shales overpressured (disequilibrium) to varying degrees depending on their level of hydraulic conductivity. Time is also a factor. If, in this example, the pressure were increased very slowly over millions of years, as in a very slowly compacting basin, fluids would have more time to escape, so pressures could be only slightly above equilibrium.

Magara (1971) developed equations calculating the depth in the subsurface at which shale permeability would drop below the level required to prevent abnormally high fluid pressures. In the Gulf Coast, this depth was calculated to be about 7,000 ft (2,130 m). Magara noted that the occurrence of overpressures in sedimentary basins would increase with increases in the sedimentation rate and the total thickness of shale.

Smith (1971b) described the compaction of shales with a mathematical model. He concluded that thick shales may develop pore water pressures much greater than normal and that these pressures may persist for tens or hundreds of millions of years.

Chapman (1972) pointed out that the concept of compaction disequilibrium solves the problem of explaining the timing difference between oil generation and fluid migration. Some geologists have contended that if most of the oil is generated after sediment temperatures reach 50°C (122°F), it is too late to participate in the major period of fluid migration. Shale porosities under compaction equilibrium at depths of 2,500 ft (762 m), equivalent to about 50°C, are around 15 percent compared to 30–35 percent under disequilibrium. This represents a large difference in the water available for primary migration. The disparity is even greater at 8,000 ft (2,439 m), where shales in disequilibrium may have more than five times the porosity of shales in equilibrium.

Disequilibrium conditions may be the rule rather than the exception. Chapman believes that most shales have been in disequilibrium to some extent since burial to a shallow depth. The difference between the volume of fluids available for migration at all depths can be seen in Figure 6-7. The curve to the far left represents a compacting basin with normal hydrostatic pressure throughout. The curve to the right represents a basin where most of the shales are abnormally pressured to some extent. Practically all of the porosity–depth curves that have been published in the literature over the last several decades fall between these two curves. Porosity–depth curves were originally determined from the measurement of porosities and densities of well samples. More recently, sonic and gamma–gamma logs (formation density logs) are used to determine the density or porosity of rocks. Since all types of samples under varying equilibrium conditions are used to plot these curves, the published data points do not show small differences in porosity between samples with different drainage conditions.

Hedberg's (1926) porosity curve and Athy's (1930) density and porosity curves, which were based largely on Paleozoic shales from Kansas and Oklahoma, represent stability attained over a long time span and follow the compaction equilibrium curve in Figure 6-7. Dickinson's (1953) porosity curve is based on the young, rapidly deposited Gulf Coast sediments, which contain high percentages of smectite. Dickinson's data follow the disequilibrium curve in Figure 6-7. Hedberg's (1936) porosity curve for the Tertiary shales of a well section in Venezuela containing numerous interbedded sands that had allowed the escape of water in step with compaction and Vassoevich's (1960) general curve are intermediate between these extremes.

Durmish'yan (1973) studied the porosity of shales in the productive series of Middle Pliocene and Miocene-Paleogene sediments of the northwest flank of the South Caspian depression. He came to the same conclusion as had Teslenko and Korotkov (1966), namely, that at the same depth the porosity of shales in alternating sand–shale sequences is much less than in thick clay deposits. Durmish'yan's data are plotted in Figure 6-6b within the porosity limits of Figure 6-7. Line A represents normal compaction of interbedded

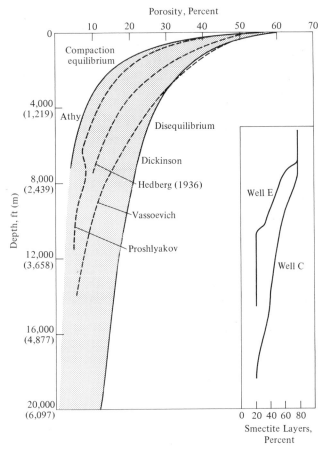

Figure 6-7
Change in porosity of shales with depth. Athy (1930) curve is from
Paleozoic Oklahoma shales at compaction equilibrium. Dickinson
(1953) curve is from young Gulf Coast shales under compaction
disequilibrium. Other curves indicate intermediate stages of compaction. Curves to right show smectite–illite change with depth (Perry
and Hower 1972). Proshlyakov (1960) porosity curve suggests clay
dewatering at about 8,000 ft (2,439 m).

sands and shales of Central Apsheron, where formation pressures are
essentially hydrostatic. Shale thicknesses range from 20–50 m (82–164 ft),
with thick sands between. Line B represents compaction disequilibrium in
shales 100–300 m (328–984 ft) thick in Southwest Apsheron. Sands are silty
and thinner. Line C indicates strongly retarded compaction with abnormal
fluid pressures close to lithostatic in shales 500–1,000 m (1,640–3,281 ft) thick
from the dome and flanks of structures in the Baku Archipelago. Durmish'yan emphasized the importance of retarded compaction in providing
fluids for primary migration of oil and gas formed during catagenesis.

Shales at equivalent depths in varying states of disequilibrium depending on drainage patterns are shown by Durmish'yan (1973) and by Teslenko and Korotkov (1966). The volume of water available for migration of even late-stage hydrocarbon generation at about 4,200 m (14,000 ft) is still equivalent to several percent porosity, as seen from Figure 6-7.

Clay Mineral Dehydration

The principal clay minerals in sediments are smectite, illite, and kaolinite. Smectite and illite are the most common clays in marine shales, although kaolinite is dominant in some. Smectite is composed of the smallest clay particles, with a surface diameter of 0.01 to 0.1 μ. Clay-particle surface areas for smectite, illite, and kaolinite, including the interlayer region, are about 800, 90, and 15 m^2/g respectively (Warner 1964). Smectite, due to its large internal surface area, adsorbs water in much greater amounts than the other minerals. The external surface area of smectite is only about 100 m^2/g.

Rock porosity is defined as the ratio of the interstitial volume to the total rock volume. The decrease in porosity shown in Figure 6-7 is caused by the loss of this water in the interstitial pore space. Rock porosity does not include the interlayer water. Such water constitutes part of the rock mineral volume. When smectite is converted to illite, this water is released from between the layers and becomes part of the pore volume. The water between the smectite layers has a lower density than the water in the larger pores. Consequently, when it is released in the formation of illite there is an overall loss in volume, and no abnormally high pressures are produced (Anderson and Low 1958). This has been verified by other investigators (Neglia, personal communication, 1977), so statements in the literature that clay mineral dehydration causes abnormal pressures are erroneous.

The importance of clay mineral dehydration to primary oil migration is in providing additional pore water for migration during the peak interval of hydrocarbon generation for those mechanisms involving water as the transporting agent. From Burst's (1969) data, it is estimated that the volume of water from dehydration can be 5–10 percent of the bulk volume of the sediment.

This phase change does not occur spontaneously under all conditions. Khitarov and Pugin (1966) noted that the conversion of smectite to illite is primarily a function of subsurface temperature. From combined laboratory and field observations, they found that the conversion occurred at various depths, depending on the geothermal gradient. In areas of high geothermal gradients, the conversion occurred at more shallow depths than in areas of low geothermal gradients.

In general, most of the dehydration occurs at 80–120°C (176–248°F) although Heling (1971) has observed it to start at 50–60°C (122–140°F) in areas with very high geothermal gradients from 7–10°C/100 m (3.8–5.5°F/100ft). These temperatures are within the oil generation range, as discussed in Chapter 4. The dehydration is also controlled by the chemistry of the system, since it involves a replacement of sodium by potassium in the interlayer region and a replacement of silicon by aluminum in tetrahedral sites. The reaction depends on pressure, with smectite more stable at lower pressures.

Perry and Hower (1972) followed the conversion of smectite to illite in shales from five Gulf Coast wells in Pleistocene to Eocene strata. Their results for two of the wells are shown to the right of Figure 6-7. Well E, near Galveston, Texas, with a geothermal gradient of 1.7°F/100 ft (3.1°C/100 m) showed the dehydration in two steps, occurring at a depth of about 7,000 ft (2,134 m) and 10,000 ft (3,048 m). Well C, off coastal Louisiana, with a geothermal gradient of 1.3°F/100 ft (2.3°C/100 m) showed the dehydration occurring over a longer interval from about 9,000–18,000 ft (2,743–5,486 m). In both cases, the dehydration stopped when there was 20 percent smectite in the resulting mixed-layer illite–smectite clay mineral.

Although clay mineral dehydration does provide more pore water for primary migration, it is not a requirement for the accumulation of commercial petroleum fields. Some producing basins, such as the Williston of the United States and Canada, contain no smectite in their source rocks (Dow 1974).

MECHANISM OF PRIMARY MIGRATION

Primary migration assumes that a small part of the hydrocarbons in the organic matter–mineral–water complex be released and transported through the fine pore openings of the source bed into the coarse-grained reservoir rock. Young and McIver's (1977) absorption studies on source sediment–crude oil pairs support the concept that hydrocarbon components of petroleum do migrate from nonreservoir to reservoir rock. Furthermore, a distribution equilibrium of components in the source and reservoir oils is reached in some old, deep source–reservoir pairs but not in relatively young, shallow accumulations. Migration and accumulation appear to be continuous processes occurring over long periods of time. Composition trends with depth indicate that paraffinic and naphthenic hydrocarbons migrate later than the aromatics. Young and McIver (1977) also concluded that the generation of petroleum components causes primary migration.

The migration mechanisms transport hydrocarbons through fine-grained sediments with permeabilities ranging from 10^{-3} to 10^{-11} md (millidarcys).

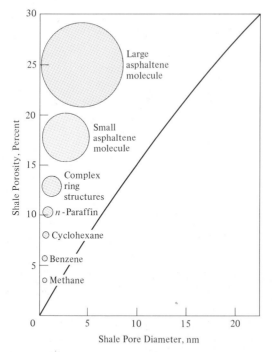

Figure 6-8
Approximate decrease in shale pore diameters with
porosity decrease. Hydrocarbon molecules in crude oil
are generally smaller than 2 nm. An asphaltene molecule
is 5–10 nm. High-molecular-weight, biogenic, nitrogen-,
sulfur-, and oxygen-containing bitumen aggregates
migrate prior to major oil generation, or they are
trapped and remain with the shale. [Welte 1972]

The permeability of shales varies with the quantity and type of clay minerals
and with the degree of compaction. Shale pore diameters also vary with these
factors. The approximate relationship between porosity and shale pore
diameters is shown in Figure 6-8. These data are based on measurements and
extrapolations by several investigators with the results summarized by Welte
(1972). They represent an approximate average of a range of pore diameters
within each porosity zone. Momper (1978) believes that in a typical compact
shale of low porosity, 70 percent or more of the pores are small (less than 3
nm) while the remainder can extend up to several hundred nm. Effective
diameters of some petroleum constituents and water are listed in Table 6-2.
The former also are shown in Figure 6-8.

The data indicate that even complex ring structures can migrate in shales of
5 to 10 percent porosity that extend to over 20,000 ft (6,096 m) depending on

Table 6-2
Approximate Effective Molecular
Diameters of Components of
Reservoir Gases and Liquids

Molecule	Effective diameter, nm
Water	∿0.3
Methane	0.38
Benzene	0.47
n-Alkanes	0.48
Cyclohexane	0.54
Complex ring structure	1–3
Asphaltene molecules	5–10

Source: From Welte 1972.

the degree of compaction. Asphaltene molecules would have difficulty migrating, but it is probable that many of them are formed in the reservoir rock. This does not consider the effect of structured water, which may occupy 1 nm or more of the available pore space. Also, organic matter may occupy considerable space. Consequently, any generated oil and gas could create high internal pressures in a restricted environment. This could cause the formation of microfractures if lithostatic pressures are approached inside the shale pores.

One important constraint on any migration mechanism is the ratio of oil to water during primary migration. The data in Figures 4-16 through 4-24 indicate that the principal phase of oil generation occurs through a depth interval of 2 to 3 km. The maximum porosity loss in 3 km of generation starting at 1,000 m (3,281 ft) and ending at 4,000 m (13,123 ft) would be about 20 percent from Figure 6-7 and the minimum about 5 percent. Consequently, a sediment layer 1 m thick and 100 km² in area (10×10^9 m³ in volume) would give up from 0.5 to 2×10^9 m³ of water during most of the oil generation period.

A typical source rock containing lenticular sands with modest reservoir accumulations is the Frontier Formation in the Powder River Basin of Wyoming. It contains 300 ppm hydrocarbon by weight (600 ppm by volume) in the oil generation range.

If our hypothetical 10×10^9 m³ of sediment also generated 600 ppm by volume of hydrocarbon during peak generation, it would probably release 15 percent of this (Hunt 1977b). This calculates to

$$\frac{600\,(0.15)}{10^6}\,(10 \times 10^9 \text{ m}^3) = 900 \times 10^3 \text{ m}^3$$

of hydrocarbon. Consequently, the minimum ratio of oil to water during primary migration would be

$$\frac{900 \times 10^3 \text{ m}^3}{2 \times 10^9 \text{ m}^3} = 450 \text{ ppm by volume}$$

The maximum ratio would be 1,800 ppm.

Giant oil fields require either very large gathering areas, or rich source rocks, or both. Source rocks with 3,000 ppm hydrocarbon would have hydrocarbon/water ratios in the 2,000–9,000 ppm range during primary migration.

These are the general conditions under which primary migration occurs. Several mechanisms have been proposed to move the hydrocarbons with or without the water as a carrier (for a comprehensive summary, see Cordell 1972).

Migration as a Colloid (Micelle)

Oil cannot move through shales in colloidal (micellar) solution because the shale pore diameters are too small and the soap requirement is too high. The reported mean size of neutral micelles is 500 nm and of ionic micelles 6.4 nm (Baker 1967). This compares with shale pore diameters in the range of 5 to 10 nm. Theoretically ionic micelles could penetrate some of the pores. However, laboratory experiments have shown that the semipermeable membrane effect of shales retards ions that are smaller than ionic micelles, so it is inconceivable that the latter could migrate freely. Experiments with flowing soap solutions through silty clays have shown they are retained by sediments with permeabilities less than 10^{-3} md. This includes practically all shales. Both Kennedy (1963) and Zhuze et al. (1971) found that the quantity of soap required to form micelles ranged from 600 to 6,000 ppm depending on the water temperature, whereas formation waters contain only 2–30 ppm of possible organic solubilizers. Trofimuk et al. (1977) stated that the total organic matter in formation waters of most of the USSR averages 0.01 g/liter, whereas several g/liter would be required for micellar solution. Another problem is that colloidal solubility maximizes at 70°C (158°F). It is drastically reduced at lower and higher temperatures (Trofimuk et al. 1977).

Migration Along Organic Matter

Migration of oil along a continuous organic network (candlewick hypothesis) is an old idea that recently has been reevaluated by McAuliffe (1978). Such flow would be independent of any water movement. Scanning electron micrographs of organic matter taken by McAuliffe after removal of the mineral matter, show a three-dimensional interconnected network. These networks were observed in shales with only 1 to 6 percent organic matter. The network is probably most complete in two dimensions along bedding planes (as observed by Momper 1978) with only occasional connections between planes. Differential pressures causing oil flow would result from such factors as compaction, oil and gas generation, and thermal expansion of hydrocarbons. McAuliffe (1978) estimated that the level of oil saturation required in the organic matter in order for oil flow to occur would be from 2.5 to 10 percent. This is based on the ratio of hydrocarbons to organic matter for many known source rocks.

An important question in reference to this mechanism is how much organic matter is required to maintain a sufficient network for appreciable oil flow? Some geochemists feel that this mechanism would not work in the Tertiary of the Gulf Coast, because of its low organic content. It is claimed that only the Cretaceous would qualify as the source of all Gulf Coast oils. Such a concept does not agree with numerous source–reservoir studies in that area including the age-dating of Gulf Coast oils by Young et al. (1977). They calculated the ages of both the gasoline and C_{15+} fraction of about 70 offshore Louisiana oils. The oldest age was 62×10^6 years, and all but six oils were younger than 20×10^6 years.

The network migration hypothesis may be important with some types of source rocks, but more research is needed to determine its feasibility.

Migration in Solution

Molecular solution is probably a transport mechanism for some of the methane and possibly a few of the light hydrocarbons in sediments. Although methane is relatively insoluble in water at STP (25 ppm or 0.2 ft³/bbl), its solubility increases rapidly with increasing pressures, as shown in Figure 6-9. At about 8,000 ft (2,438 m) the solubility of methane is 100 times greater than at the surface, and at 20,000 ft (6,096 m) it is about 300 times greater. At the latter depth, it is 4 times more soluble than benzene, which is the most soluble of the liquid hydrocarbons. This rapidly increasing solubility with increasing pressure and temperature means that compacting fluids will readily pick up methane from source beds and release it during any upward migration to regimes of lower temperature and pressure. The effect would be even more pronounced in methane released from abnormal pressure zones, where the solubility increases faster than in normally pressured sediments.

Ethane is about 30 percent more soluble than methane in water at STP, but their solubilities are similar around 500 psia, and at high pressures ethane is much less soluble. Consequently, the solubility of natural gases will be to the

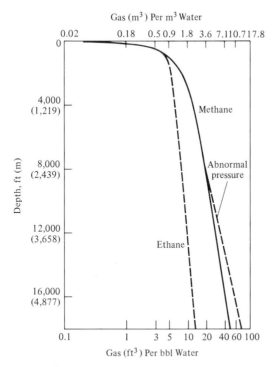

Gas (m³) Per m³ Water

Figure 6-9
The solubility of methane and ethane in distilled water with depth assuming a pressure gradient of 0.46 psi/ft and a geothermal gradient of 1.5°F/100 ft (2.7°C/100 m). Salinities around 35 ‰ would reduce solubilities 10–20 percent, depending on temperature. Abnormal pressure line calculated at 0.7 psi/ft and a geothermal gradient of 2°F/100 ft (3.65°C/100 m) starting at 7,000 ft (2,134 m). To convert ft³/bbl to ppm, multiply by 127. To convert ft³/bbl to mol fraction methane, divide by 7,370. [Data from Culberson and McKetta 1951]

left of the solid line in Figure 6-9, depending on the percentage of hydrocarbons heavier than methane.

The solubilities of heavier gaseous and liquid hydrocarbons are shown in Figure 6-10. McAuliffe (1966) measured the solubilities of 65 hydrocarbons in water at room temperature. He found that for each homologous series of hydrocarbons the logarithm of the solubility in water is a linear function of the hydrocarbon molar volume. In other words, the small molecules dissolve more readily than the large molecules. McAuliffe also found that forming a hydrocarbon ring from a chain increased the water solubility, and increasing the unsaturation of the ring further increased the solubility. This means that the increasing order of solubility for the major hydrocarbon groups is

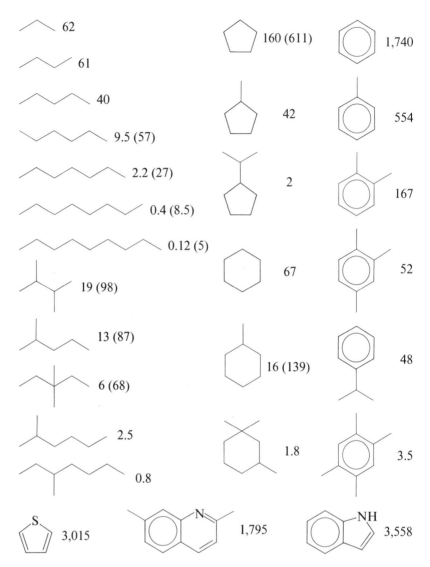

Figure 6-10
The solubility of hydrocarbons and nonhydrocarbons in pure water. [McAuliffe 1966; Price 1976] Solubility at STP on right side of structure. Solubility at 137°C (279°F) in parentheses.

paraffins < naphthenes < aromatics. The mono-olefins have solubilities comparable to the naphthenes (cycloparaffins). Di- and triolefins have proportionately greater solubilities.

Zhuze et al. (1971) quotes work of earlier Russian scientists showing that benzene, toluene, and methylcyclohexane solubilities all increase with temperature. Thus the solubility of methylcyclohexane more than doubled between room temperature and 115°C (239°F). They also found that water saturated with gas reduced the solubility of liquid hydrocarbons, depending on the hydrocarbon, the gas, and the temperature–pressure conditions. The solubility of a mixture of hydrocarbons was found to be about 50 percent lower than the sum of the solubilities of the individual hydrocarbons, indicating that hydrocarbons tend to displace each other in solution.

Price (1976) made a detailed study of the effect of temperature on the solubilities of hydrocarbons and crude oil fractions. He found that the aqueous solubility of hydrocarbons increases gradually to around 100°C (212°F), where a more drastic increase occurs because of a change in the solution mechanism. Price (1973) postulated that the change in slope of the solubility line could be caused by the break-up of aggregates of molecules to a true molecular solution at the higher temperatures. It also is possible that clustered water structures, which are prevalent at lower temperatures, would be broken up at higher temperatures, resulting in more unbonded water available to dissolve the hydrocarbons.

Price's (1973) data for the solubility of some representative crude oils are shown in Figure 6-11. The effect of temperature is clearly evident. Solubilities

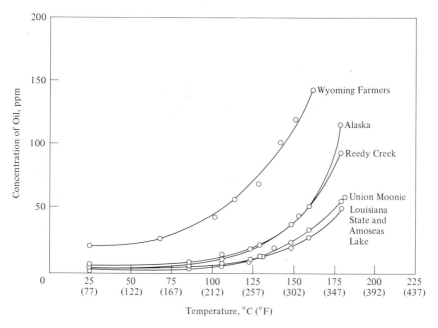

Figure 6-11
Solubilities in water of two whole oils (Wyoming Farmers and Louisiana State) and four topped oils (Amoseas Lake, Reedy Creek, Alaska, and Union Moonie) as a function of temperature at systems' pressure. [Price 1976] Topping temperature is 200°C (392°F).

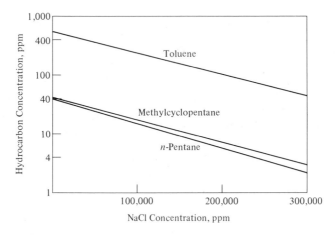

Figure 6-12
Solubility of toluene, methylcyclopentane, and *n*-pentane in aqueous
solutions at 25°C (77°F), 1 atm, as a function of NaCl concentration
on a semilog plot. [Price 1976]

of the topped oils increase from 0.52–2.5 ppm at 25°C (77°F) to levels of
50–125 ppm at 180°C (356°F). The topped oils would be equivalent to heavy
crudes, because their specific gravities would be greater than 0.9. The
solubilities of several individual hydrocarbons at 137°C (279°F) from Price's
(1976) paper are shown in parentheses in Figure 6-10.

Simonenko (1974) observed the same marked increase in the solubility of
oil components in the 80–150°C (176–302°F) range. He pointed out that
water at temperatures above 100°C (212°F) loses its polar properties and
becomes a low-polar solvent (lower dielectric constant). This enables water to
dissolve a larger proportion of nonpolar compounds such as hydrocarbons.

A solution mechanism for primary migration requires that some of the
hydrocarbons come out of solution in the reservoir rock. Exosolution could
be caused by the following changes: increase in salinity, decrease in pressure
and temperature, partitioning with an oil–gas phase, and increase in gas
saturation. Decrease in solubility of an aromatic, naphthene, and paraffin
with increasing salt concentration at 25°C (77°F) is shown in Figure 6-12.
There is a ten- to fifteenfold decrease in solubilities over the salt range shown.
These data suggest that connate waters would cause exosolution of hydrocarbons more readily than meteoric waters. Zhuze et al. (1971) measured toluene
solubilities in connate waters of the chloride–calcium type and meteoric
waters of the bicarbonate–sodium type. Solubilities in connate compared to
meteoric waters were 35 percent lower at temperatures below about 135°C
(275°F), but at higher temperatures the solubilities in sodium–bicarbonate
waters were lower. In other words, meteoric waters at high temperatures

could cause exosolution of hydrocarbons even more readily than connate waters based on Zhuze's data.

The decrease from the high intrapore pressure of shales to the normal pressure of sands would cause exosolution of gases, which could carry heavier hydrocarbons by entrainment. The decrease in temperature as an exosolution mechanism would be most effective in fluids migrating upward vertically through faults or fractures or updip to reservoir sands deposited along the rims of basins.

Exosolution is not necessarily completed at the entry point of the fluids into the sand. Hydrocarbons remaining in solution during movement through sands could be further depleted on contacting an oil–gas accumulation at an updip capillary barrier. The hydrocarbons would tend to partition into the oil phase in preference to the water phase.

Zhuze et al. (1971) found that hydrocarbons such as benzene, toluene, and methylcyclohexane are less soluble in water saturated with gases such as nitrogen, helium, carbon dioxide, and methane than in pure water. This effect was noted over the temperature range from about 50 to 150°C (122 to 302°F). This means that water saturated with almost any subsurface gas would cause more exosolution of hydrocarbons than would waters relatively free of gas.

The most serious problem for primary migration in solution is that hydrocarbon solubilities even at 100°C (212°F) are much lower than what appears to be required in water migrating during peak oil generation, except for methane, the aromatics, and a few of the low-molecular-weight saturates. If hydrocarbon solubilities measured in the laboratory are comparable to those in the pores of fine-grained rocks, then the migration of most hydrocarbons will have to be explained by mechanisms other than solution.

Gas-Phase Migration

Compressed gas can dissolve increasing amounts of heavy liquid hydrocarbons as pressure and temperature increase. Sokolov et al. (1963) and Sokolov and Mironov (1962) have demonstrated that subsurface gases will dissolve large amounts of liquid hydrocarbons under temperature–pressure conditions corresponding to depth ranges of 6,000–10,000 ft (1,829–3,048 m). Rzasa and Katz (1950) studied the critical pressure for binary systems with methane and found that methane–decane mixtures are single phase at pressures above 5,400 psia. Neglia (1979) has reported that the Malossa field in Italy is condensate with heavy liquid components dissolved in the gas phase. This field produces from a Triassic dolomite at a depth of about 20,000 ft (6,096 m). Reservoir pressure is 1,050 atmospheres and temperature is 307°F (153°C). The data of Rzasa and Katz (1950) indicate that hydrocarbons through C_{18} would be dissolved in the gas phase under these conditions.

Neglia (1979) believes that gases generated in source rocks migrate vertically through microfractures dissolving oil from adjacent pores. Molecular distillation of oil occurs because its vapor pressure in the liquid phase is higher than the vapor pressure in the gaseous phase. Eventually, however, the migrating gases reach a level where the reduced pressure and temperature results in retrograde condensation with formation of an oil phase.

Nearly all limestones, dolomites, and siliceous rocks have some fracture porosity and permeability. In recent years, the concept of microfractures has gained favor as a means of developing sufficient permeability for oil- and gas-phase migration. These microscopically visible fractures are believed to be caused by the high pore pressures resulting from the generation of oil and gas in the restricted pore systems of fine-grained rocks (Snarskii 1970). Momper (1978) cites several geochemical papers of the last 10 years that comment on the role of microfractures in primary migration.

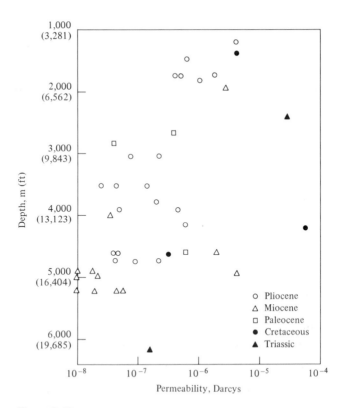

Figure 6-13
Semilog plot of permeability of shales versus depth. [Neglia 1979]

Figure 6-13 shows data published by Neglia (1979) indicating that microfractures may be forming in the oil-generating depth range. This plot of shale permeability with depth shows a decrease in permeability to about 3,500 m (11,483 ft), after which there is an increase. Assuming a geothermal gradient of 2.7°C/100 m (1.5°F/100 ft), this increase starts around a temperature of 114°C (237°F), which is within the oil-generating range for Tertiary sediments (Figure 4-17). The increase in permeability is attributed to the development of microfractures.

Rumeau and Sourisse (1973) concluded that the hydrocarbons in the carbonate reservoirs of the S.W. Aquitaine Basin probably entered in the gas phase because fluid flow from compaction ceased very early.

Gas-phase migration along vertical faults and through permeable Tertiary sandstones may explain the presence of aromatic rich condensates in the Pleistocene of the Gulf Coast. When gas is passed through oil, it tends to concentrate more of the light aromatics in the gas phase. Aromatic condensates also are found in the Western Canada Basin (Hitchon and Gawlak 1972) and in East Turkmen, USSR (Gavrilov and Dragunskaya 1963).

Gas-phase migration cannot readily account for giant oil accumulations such as those of the Middle East unless it is assumed that huge volumes of gas have been lost. Also, it does not account for oil accumulations at an early stage of generation as in shallow pools of the Paris Basin where large volumes of gas have not been formed. Nevertheless, it may be the most reasonable explanation for some accumulations of the Gulf Coast, the Niger Delta, the Mackenzie Delta, and the North Sea, where microfractures could direct the gases into major vertical fault systems. With upward migration, the reduction in pressure and temperature would cause retrograde condensation of the oil in reservoir-type rocks along the fault zone.

Oil-Phase Migration

The ratio of bulk pore water to structured pore water in shales decreases with decreasing porosity. Since an oil particle would be restricted to the bulk water phase, it is conceivable that at some depth the oil in the smaller pores would occupy enough of the bulk space to be expelled by capillary forces assisted by the fluid potential gradient into a coarser-grained (lower capillary pressure) zone of the shale. By moving into successively larger openings, it would eventually reach the reservoir rock. Hunt (1973) and Dickey (1975) both have suggested that structured water may limit the free pore space sufficiently to permit oil-phase migration. This mechanism would migrate all the components of crude oil, but, as Dickey suggests, considerable research is needed to verify it. Structured water would occupy more pore space in sodium than in calcium smectites, but the latter are more common in the subsurface. If calcium smectites and illites show ordering to 2 nm, then the pores would have 100 percent structured water at about 7 percent porosity (4 nm pore diameter) based on Figure 6-8. The problem is complicated by the fact that increasing temperatures and pressures reduce the extent of structuring. Also, structuring would not occur where organic matter covers the mineral surface.

Structuring of water is not needed to explain oil-phase migration in very organic-rich rocks. Such rocks may be oil wet, with organic matter and oil occupying much of the pore space and blanketing much of the mineral surfaces particularly along bedding planes. The lacustrine shales of the Altamount-Bluebell field in the Uinta Basin of Utah are a typical example, where water is probably the discontinuous phase (Meissner 1978a). Other examples of organic-rich rocks that are partially oil wet include the Bakken, Exshaw, and Nordegg formations of the Western Canada Basin and the La Luna Formation of Colombia and Venezuela. Dickey (1975) has suggested that the saturation at which oil will flow as a continuous phase may be less than 10 percent in shales that are partially oil wet.

The minimum organic content required for completely oil-wet pores was estimated by Byramjee (1967) from empirical observations to be about 30 percent. Under such conditions, oil would migrate out of shales, because of compaction in a manner analogous to the expulsion of water from water-wet rocks. The hydrocarbons would ooze from oil-wet shales into adjacent water-wet reservoir rocks, where they would accumulate by buoyancy.

Byramjee (1967) also noted that oil-wet shales are readily distinguished from typical water-wet shales by their abnormally high resistivities. More recently, Meissner (1978b) has used these high resistivities to identify the oil generation zone in organic-rich rocks such as the Bakken of the Williston Basin (see the box "Overpressures in Organic-Rich Rocks," pp. 238–239).

Barker (1974) has suggested that, if oil droplets block the pore openings in a source rock, then the water trapped behind the oil will expand as the temperature rises, forcing the oil through the pores. The increased pressure caused by the thermal expansion of water has been termed *aquathermal pressuring* by Barker (1972). He estimated that a depth increase of only 3 m (10 ft) will generate sufficient water pressure to force oil through constrictions. Such a situation would be limited to small pores where all exits are blocked by oil particles.

Cartmill (1976) has pointed out that conventional reservoir mechanics cannot explain the absence of residual oil stains at locations of reservoir entry from the source rock. He believes that under conditions of high temperature and pressure there may be discrete phases of variable oil–water mixtures with such low interfacial tensions that simultaneous flow will occur through the fine capillaries of the shale.

Momper (1978) believes that oil-phase migration on a significant scale does not commence until the organic matter has generated about 850 ppm (15 bbl/acre-ft) of total extractable bitumen (hydrocarbons plus asphaltic com-

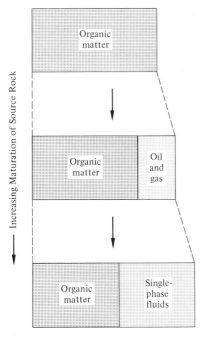

Increasing Maturation of Source Rock

Figure 6-14
Increase in net volume of organic matter plus
generated fluids with increasing maturation of
source rock.

pounds). He estimates that at peak oil generation the conversion of organic
matter to liquids and gases can cause a net volume increase of up to 25 percent
over the original organic volume (Figure 6-14). In the restricted pore space of
a fine-grained source rock, this creates a pressure build-up causing microfractures, or reopening existing fractures, with expulsion of the oil. After the
oil is expelled, the fractures close until the pressure builds up from
subsequent generation. This results in the pulsed expulsion of oil until the
generating system runs down. Momper (1978) also points out that any carbon
dioxide formed during oil generation would dissolve in the oil, lowering its
viscosity and improving its mobility. Carbon dioxide is very soluble in oil at
pressures of more than 1,000 psi (6.9 MPa, or megapascals) and temperatures
above 90°F (32°C).

The discussion as to whether hydrocarbons migrate in an oil or gas phase
becomes academic in very deep formations where pressures exceed the
critical point. This is the point at which the densities, viscosities and surface
tensions of oil and gas are so similar that they are considered to be in a single
phase. Katz et al. (1959, p. 465) consider most fields containing gas and oil at
pressures above 4,000 psi (28 MPa) and temperatures above 200°F (93°C)
to be single-phase fluids. In typical oil basins, this would correspond to depths
greater than 8,000 ft (2,439 m) at hydrostatic pressure (Figure 6-5). Pressures
would be much higher than hydrostatic within restricted shale pores as oil and
gas are generated. This suggests that primary migration as a single-phase fluid
may be the norm in all deep wells and may even occur in some formations as

shallow as 5,000 ft (1,524 m). The reduction in pressure on entering a reservoir rock would cause retrograde condensation; that is, formation of an oil and gas phase. If the reservoir rock also were overpressured, secondary migration might occur as a single-phase fluid.

Primary Migration by Diffusion

Hydrocarbon molecules, or particles representing small aggregates of molecules, possess energy with respect to their position. Since they are capable of motion in the subsurface, they will move from regions of higher to lower potential energy. Eventually they will come to rest in positions where their potential is at a minimum; that is, where they are surrounded by regions of higher potential energy. Consider, for example, the basis for the concept of diffusion induced by structured water. Cages of ordered water (clathrates) exist around organic molecules dissolved in water. These cages require energy to break them down. Energy also is required to break down the water structures existing on clay minerals. Figure 6-15 illustrates this relationship. The water cages around the hydrocarbon complexes are incompatible with the water structures next to the clay surfaces. Movement of a hydrocarbon–water cage to the left would tend to disrupt the clay-induced water structure, whereas movement to the right would avoid it. Although this figure shows a sharp boundary between the fixed water and the free water, there is actually a diffuse boundary, so that some ordering continues toward the hydrocarbon cages. The favored position, or region of lowest potential energy, is to the right, where the pore openings are much larger and where there is minimum disruption of the clay-induced water structures. This means that a potential energy gradient exists from left to right that tends to drive the hydrocarbons in the water cages in this direction. The important point is that this movement can occur without any movement of the clay particles or the bulk water. In other words, there is no requirement for compaction or fluid motion, so basically it is a diffusion mechanism.

Thermodynamically, the potential energy of the hydrocarbon complex in the surrounding water is expressed in terms of its activity, or escaping tendency. High activity coefficients indicate a strong tendency to escape, whereas low activities indicate a weak tendency. Activity coefficients for hydrocarbons in various rocks have been measured (Antonov 1953), and they generally show higher activity in clays than in sands, so that the hydrocarbons have a tendency to escape from the clays to the sands. The activity of a hydrocarbon entity is related to its solubility in the liquid phase and its tendency to be adsorbed by the solid phase. A hydrocarbon that is comfortable in a liquid, or near a mineral surface, has a low activity, whereas a hydrocarbon repelled by the liquid and solid phases has a high activity. This relationship is shown schematically in Figure 6-16. The numbers in this

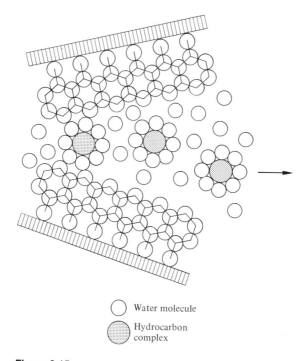

Figure 6-15
Migration induced by structured water. A decrease in potential energy from left to right is caused by the fact that energy is required to disrupt the clay-induced water structures. Hydrocarbons in water cages will travel to the right down this gradient toward larger pore openings without any movement of clay particles or liquid water.

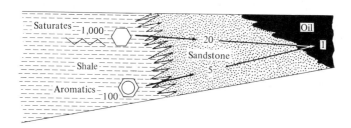

Figure 6-16
Thermodynamic activity, or escaping tendency, of saturated (paraffin, naphthene) and aromatic hydrocarbons. Hypothetical numbers indicate relative escaping tendencies.

illustration are hypothetical, but they demonstrate the principle of relative activities of different hydrocarbons in different environments. Hydrocarbons are most comfortable (lowest activity) in the oil phase designated as 1. In the shale, they have very high activities. They are repelled by the structured water of clay surfaces. If it were an oil-wet shale, they would be attracted and have low activity. The saturated hydrocarbons, paraffins, and naphthenes, are repelled more than are aromatics. The latter are polar and have a slight affinity for water and minerals, so their escaping tendency is less than for the saturates. Activities are much lower in the sand than the shale.

Activities also are related to particle size. As an oil drop becomes smaller, the pressure within the drop increases, relative to the pressure in the surrounding medium. An increase in pressure increases the activity of each hydrocarbon species within the drop. Therefore, any reduction in drop diameter with compaction will increase the tendency for all of the hydrocarbon species within it to escape into the reservoir.

The difference in the escaping tendencies of saturates and aromatics would imply that oil would fractionate in going from shale to sand with more saturates entering the sand and more aromatics remaining in the shale. This change has been observed in a number of comparisons between source rock oil and reservoir oil (Hunt 1961; Baker 1962).

Hydrocarbons would have less desire to migrate (lower escaping tendencies) in partially oil-wet, organic-rich oil shales. Also, there would be more bulk water (less structured water) in pores of partially oil-wet shales.

The escaping tendencies of hydrocarbons from source to reservoir rocks does not indicate the quantity of material that can be moved. Calculations using Fick's law of diffusion show that migration paths would be short, on the order of tens to hundreds of feet, to account for a commercial oil accumulation by diffusion in a reasonable time scale such as 10 million years. As most fine-grained rocks continue to compact through the oil generation interval, any diffusion movement would be enhanced by the fluid movement. Thus, in Figure 6-15, although diffusion can occur without the two mineral plates being pressed together, they probably are squeezed in the natural situation, so that fluid movement to the right would augment the diffusion process. Obviously, two or three migration mechanisms acting in parallel would increase the possible distance of migration.

Primary migration probably involves a combination of different mechanisms operating in varying degrees of intensity depending on the quantity and type of organic matter in the source rock. Rocks with less than 1 percent organic matter, of which an appreciable part is gas generating, would migrate hydrocarbons by diffusion, in solution, or in a gas phase until they reach a fracture or fault system where they could migrate as gas by buoyancy. Source rocks of the offshore U.S. Gulf Coast are in this category. Rocks with more than 5 percent organic matter, most of which is oil generating, would migrate hydrocarbons as an oil or gas phase. The Bakken Shale of the Williston Basin and the basal Green River–Wasatch of the Uinta Basin are examples of this. Source rocks with intermediate organic contents of mixed oil- and gas-generating capability probably involve different migration mechanisms for different types and sizes of hydrocarbon molecules.

There is a current tendency among petroleum geochemists to minimize the role of water in the migration process. While it is true that water movement is not required for most of the proposed mechanisms, there is no doubt that water can enhance many of these processes. Also, as soon as the hydrocarbons encounter a siltstone, sand, or fracture–fault system with permeabilities several orders of magnitude greater than the fine-grained source rock, water becomes the dominant pore fluid except in organic-rich fractured source rocks. As water has been moving out of the sediments all through the compaction history of a basin, it has established the pathways and directions of migration. These exert appreciable control over the movement of hydrocarbon fluids.

DIRECTION, PATHWAYS, AND DISTANCE OF PRIMARY MIGRATION

Fluids move in the direction of decreasing free energy.* Two major factors affecting this quantity are the pressure and the potential energy. Fluids move to the lowest potential energy only when this factor predominates. If it always predominated, fluids could only move downward in the earth's crust because the potential energy decreases downward in the earth's gravitational field. More often, the pressure predominates.

In shales with normal compaction, the pressure decreases upward, and this is the usual direction of fluid flow. In compaction disequilibrium, however, pressure barriers may be created within the shale that prevent the vertical migration of water. The concept of pressure barriers caused by compaction disequilibrium is important in understanding the drainage patterns carrying hydrocarbons from source to reservoir.

The first published observation of pressure barriers at shallow depth was made in three bore holes drilled to depths up to 58 m (190 ft) in clay muds of the Orinoco Delta (Kidwell and Hunt 1958). In Figure 6-17, the hydrostatic pressure in each well is shown as a vertical thin line, with the excess pressure above hydrostatic shown as a heavy line. Starting at the surface and moving down in the well at the far left, there is an increase in pressure above

---

*Some authors use the term *potential* instead of *free energy*. However, in water movement through a shale where the salts are restricted in their movement, there is a component to the driving force other than pressure and gravity. Only the free energy encompasses all components.

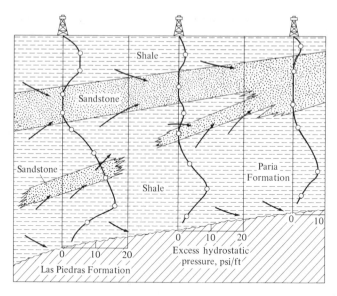

Figure 6-17
Directions of fluid flow during compaction of the Paria Formation in
the Orinoco Delta. Straight vertical lines under wells represent normal
hydrostatic pressure. Curved vertical lines are excess hydrostatic
pressure in atmospheres. [Kidwell and Hunt 1958]

hydrostatic approaching the center of the first shale. Below this is a sand
whose updip end outcrops at the surface. There is normal hydrostatic
pressure in the sand. Proceeding deeper into the thick lower shale of the Paria
Formation, there is an increase in excess hydrostatic pressure, which
continues to a peak in the lower part of the shale. The presence of the small
stratigraphic sand trap has no effect on these excess pressures, because the
sand is completely surrounded by the high-pressure shale. This excess
pressure peaks in the shale below the sand and then drops off toward
hydrostatic as the unconformity is approached. In the center well starting at
the surface, there also is an increase in pressure, which returns to normal in
the sand. This time, the second sand is connected to the outcrop, so pressure
returns to normal. Again, in the underlying shale, pressure builds up to a
peak and falls back approaching the unconformity. The well on the right
shows the same pattern of excess pressures building up in the shales and returning to normal in the sands and toward the unconformity. The centers
of the shales, where the pressures reach a peak, represent a barrier to fluid
movement. Fluids will move into the long, continuous sand both from the
shales above and the shales below, since these are the directions of lower free
energy. This means that the drainage area for shales into the top continuous

sand is very large. Waters are being squeezed from both shales during compaction as far to the left as the sand is highly permeable. Consequently, if a capillary barrier of fine-grained sediment were later deposited on the outcrop of this sand, a petroleum accumulation could result.

The small stratigraphic trap, in contrast, has a limited drainage area, namely, only the clay muds below and to the left of the sand. On the other hand, the unconformity also would cover a large drainage area since fluids for some distance to the left would be migrating down from the overlying shale and moving horizontally along the unconformity to the right. The unconformity represents a more permeable bed for fluid flow than the clay muds. Capping of the unconformity could form an updip petroleum accumulation.

Unconformities can drain fluids from enormous areas. Jones (1978) has suggested that the Athabasca heavy oil deposits of the Western Canada Basin resulted from fluids carrying bitumens eastward along the basal Cretaceous unconformity all through Tertiary time. Compaction fluids are believed to have been augmented by high volumes of meteoric water that penetrated to the unconformity along the mountain front.

The same principles can be applied to migration in other types of structures. Figures 6-18 and 6-19 show migration into an anticline and into stratigraphic traps. In Figure 6-18, the shale overlying the sand is thick on the flanks and thin at the top of the structure. A thin shale will reach compaction equilibrium much quicker than a thick shale, as previously discussed. This will cause the thick shale to have a pressure greater than hydrostatic (see dotted line on the flanks), which diminishes to normal on the thin shale at the crest. Since the sand in this example shales out on both sides, the lowest pressure and direction of fluid movement will be vertically upward at the crest. If the sand on the right did not shale out but encountered another anticline and moved further updip, then the fluids would all move through the sand to the right.

In Figure 6-19, fluids are shown migrating updip to form two stratigraphic oil accumulations. The trap on the left would form a small accumulation of

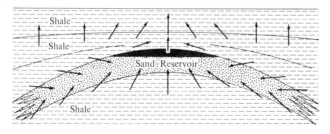

Figure 6-18
Directions of fluid migration into anticline.

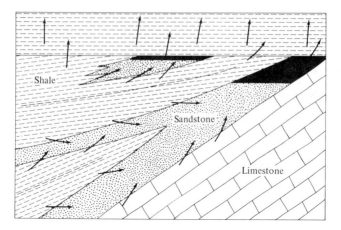

Figure 6-19
Migration into stratigraphic traps.

petroleum because the drainage area for fluids moving through the trap is restricted to a short section of shale immediately below. The trap to the right could have a large accumulation because fluids are migrating downdip from the shale on the left into the underlying sand as well as out of the second shale below. Fluids may also move out of the carbonate rock, assuming it is still compactable. In all of these examples, the water continues to move through the coarse-grained–fine-grained capillary barriers that cover the trap, while the petroleum and some salt ions are held back. The capillary barrier effect is discussed more fully in the section on secondary migration.

An interbedded sand–shale sequence (Figure 6-20) represents alternating permeable and compactible beds, which is the ideal situation for maximum drainage of shale pore fluids. Migration in the sands is lateral, since the shale beds are a barrier to vertical migration.

Fluid movement into a pinnacle reef is shown in Figure 6-21. The highly permeable reef acts as a fluid aquifer, directing pore waters from the compacting shales upward through the reef. The limey Ireton shale thins out over the reef so there is no pressure barrier. Davis (1972) noted velocity anomalies directly above the Leduc reefs, which he interpreted as indicating vertical permeability zoning and vertical brine movement.

These examples demonstrate the general concepts regarding the directions of fluid flow. Chapman (1973) discusses the mechanical aspects of the pressure barrier concept in more detail, and Evans et al. (1975) show practical applications of it, using Mackenzie Delta wells as examples. In an actual well, some indication of fluid movements can be obtained from well logs. Magara (1968) used conductivity logs to calculate shale porosities in Miocene

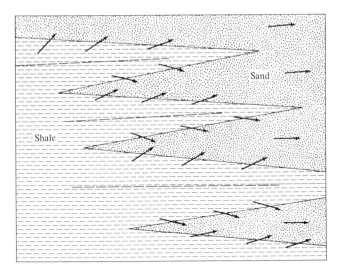

Figure 6-20
Migration from an interbedded shale–sand sequence.

sediments of the Nagaoka Plain of Japan. High porosities represented uncompacted shales with pressure barriers. Magara concluded that several of the hydrocarbon accumulations resulted from the downward migration of fluids from overlying source beds.

Figure 6-22 shows formation fluid pressures estimated from electric logs (Schmidt 1973) in a well of the Manchester field of Louisiana. There are no significant pressure barriers until about 7,000 ft (2,134 m), so the direction of fluid movement in the upper section would be mainly upward, except where differences in permeability would cause some horizontal migration.

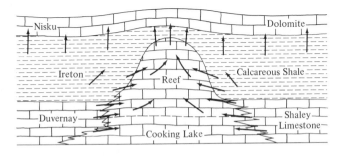

Figure 6-21
Migration into a pinnacle reef. This would be typical of the reefs in the Leduc area, Western Canada.

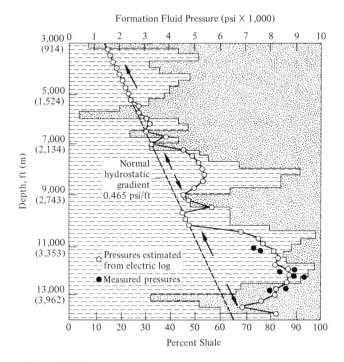

Figure 6-22
Directions of fluid movement in Manchester field well, Louisiana.
[Schmidt 1973]

The high-pressure barrier at about 8,000 ft (2,438 m) would cause fluid movement upward and downward into the overlying and underlying sands from the peak pressure horizon. Likewise, in the second high-pressure zone, fluid movement would be both upward and downward from the peak pressures. The direction of primary migration is important in interpreting and predicting the horizons in which oil will accumulate. Using the concept just described, Magara (1973) showed that most Mesozoic oil and gas pools in northwestern Alberta and northeastern British Columbia are concentrated where the largest volume of fluids is expelled downward from a pressure barrier in the Cretaceous shales. Magara used sonic and formation density logs plus examination of cores to determine porosity–depth trends. A typical cross section is shown in Figure 6-23. This section extends about 150 miles (241 km) across the northern corner of British Columbia and Alberta. The Cretaceous shales show varying intensities of compaction disequilibrium down to a depth equivalent to about 500 ft (152 m) below sea level. Below this depth, the porosities decrease as the shales approach equilibrium at the Triassic unconformity. The dotted line in Figure 6-23 shows the boundary within the Cretaceous shale between upward and downward fluid migration. Oil and gas pools in the Cretaceous, Jurassic, and Triassic formations are concentrated in the British Columbia area, where the most fluid is involved in downward migration. Figure 6-24 shows the thickness of the shale below the dotted line in Figure 6-23. Most of the oil and gas pools are found in the part of the sedimentary sequence where the thickness of the downward migration zone is greater than 500 ft (152 m). This area would involve the largest volume of downward-migrating fluids. The underlying carbonates also may have contributed some hydrocarbons to the oil and gas accumulations, providing they were able to release fluids after deposition of the overlying Cretaceous shales. In this example the overlying shales acted as both source rock and cap rock, since the pressure barrier inhibits vertical migration.

These examples show that the direction and distance of primary migration is controlled by the nearest highly permeable drainage beds, which are the pathways for fluid flow out of the basin. Such pathways can be continuous sands, unconformities, fracture–fault systems—any beds that are highly permeable compared to shales. The basic concept is that primary migration in low-permeability rocks is controlled by the pathways of secondary migration in high-permeability rocks established during compaction.

A simplified model of fluid expulsion directions and volumes involving both primary and secondary migration during basin compaction has been proposed by Magara (1976a). The basic equation he derived relates the direction of flow to the vertical and horizontal permeability and the sediments' rate of thickness change with distance for the study area, as follows.

$$\frac{q_H}{q_V} = \frac{k_H}{k_V}\left(\frac{l - h}{x}\right)$$

where q_H = volume of horizontal fluid movement
q_V = volume of vertical fluid movement
k_H = horizontal permeability
k_V = vertical permeability
x = horizontal distance
$l - h$ = change in thickness within distance x

For example, for the Cretaceous formations in the Western Canada Basin (as shown in Figure 6-25),

$$\frac{q_H}{q_V} = \frac{k_H}{k_V}\frac{9,500 - 1,000}{(400)(5,280)} = \frac{k_H}{k_V(250)}$$

This means that if the vertical and horizontal permeabilities throughout the Cretaceous were about the same, as if it were all a uniform plastic shale, the

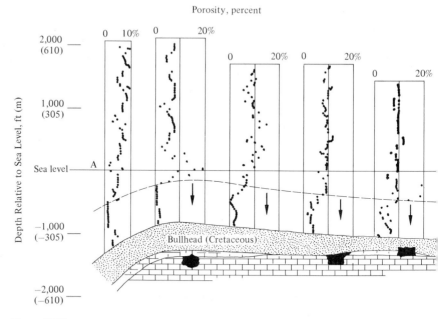

Figure 6-23

Cross section showing relation of shale porosity distribution to oil and gas accumulations in Mesozoic formations of northeastern British Columbia, Canada. Dotted line is boundary

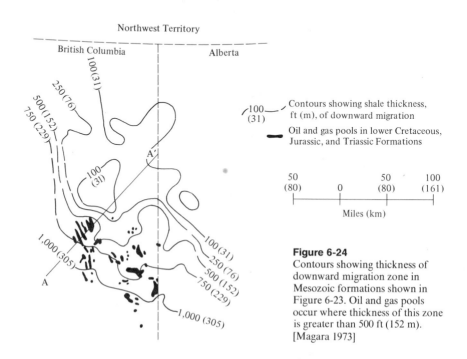

Figure 6-24

Contours showing thickness of downward migration zone in Mesozoic formations shown in Figure 6-23. Oil and gas pools occur where thickness of this zone is greater than 500 ft (152 m). [Magara 1973]

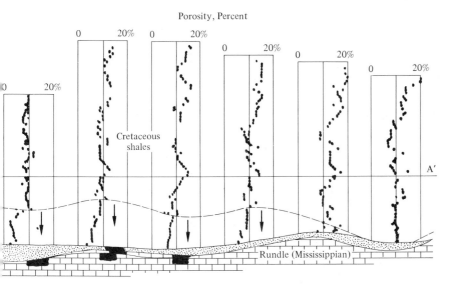

Porosity, Percent

between upward and downward fluid migration. Petroleum accumulations shown as solid blocks. Location of line A–A' is in Figure 6-24. [Magara 1973]

major fluid movement would be vertically upward. Actually, the Cretaceous represents interbedded sands and shales with the sands oriented horizontally updip to the outcrops in the east. Under these conditions, secondary migration in the sands controls the horizontal flow. Assuming a permeability greater than 50 md for the sands and less than 0.05 md for the shales and substituting for k_H and k_V in the equation gives

$$\frac{q_H}{q_V} = \frac{50}{0.05}\left(\frac{1}{250}\right) = 4$$

This means that more fluid would be moving horizontally than vertically.

In most sedimentary basins, the value of $l - h/x$ for young deposits would range from about 1/20 to 1/200. In the Tertiary of the Gulf Coast, it is about 1/40. The ratio becomes smaller in older sediments, being around 1/400 for the Devonian of western Canada. Since this ratio is small, the ratio of horizontal permeability k_H to vertical permeability k_V must be large if there is to be more horizontal fluid movement than vertical. Magara's (1976a) model showed that interbedded sand–shale sequences expelled most of their water horizontally, whereas continuous shale sequences expelled most of their water vertically. The volume of water moving horizontally through the

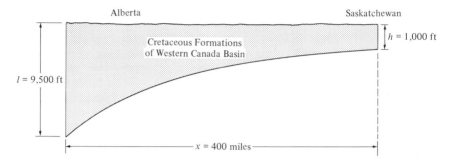

Figure 6-25
Diagram showing how thickness of Cretaceous formations in the Western Canada Basin
changes from 9,500 ft to 1,000 ft (2,896 to 305 m) in 400 miles (644 km).

interbedded zone increased as the horizontal distance of migration increased.

In the Gulf Coast, Magara (1976a) proposed a combined vertical and horizontal migration model as shown in Figure 6-26. The lower Tertiary in the Gulf Coast is an overpressured shale, while the upper section consists of interbedded sands and shales (Figure 6-26b). Fluids would move upward in the compacting shale and then be channeled horizontally along the interbedded sands. Cumulative water volume migrating horizontally would be considerably greater than that migrating vertically, as shown in Figure 6-26a. Magara noted that the water volume plot resulting from his migration model

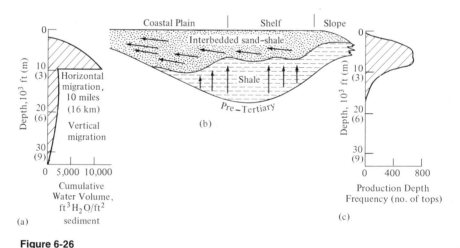

Figure 6-26
(a) Cumulative water-loss volumes from Gulf Coast shales for a sediment thickness of 33,000 ft (10 km) and horizontal migration distance of 10 miles (16 km). (b) Cross section of Texas part of Gulf of Mexico with arrows showing directions of fluid movement. [Magara 1976a] (c) Production depth frequency in Gulf Coast area showing tops for 5,368 production levels. [Burst 1969]

resembled the oil production frequency plot for the Gulf Coast shown in Figure 6-26c (Burst 1969). He suggested that the volume of water movement could be the controlling factor in hydrocarbon occurrence rather than the clay dehydration range as postulated by Burst. More importantly, the distribution in Figure 6-26c is caused by the greater frequency of reservoir sands above the overpressured shale.

Horizontal Migration

In many interbedded sand–shale sequences, vertical migration has been minor. The evidence is that crude oils within a vertical interval of a few thousand feet are chemically different yet are similar to oils from the same horizons several miles away. Bass (1963) cites several examples in northwestern Colorado and northeastern Utah (Figure 6-27). For example, the Ashley

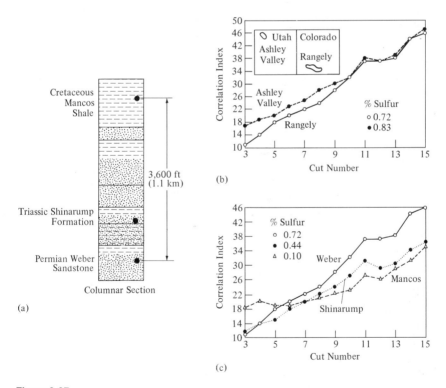

Figure 6-27
(a) Generalized columnar section of rocks showing Weber, Shinarump, and Mancos oil reservoirs as solid circles. (b) Comparison of Weber Sandstone oils from Ashley Valley and Rangely oil fields. (c) Comparison of Weber, Shinarump, and Mancos oils from Rangely oil field. [Bass 1963]

Valley and Rangely oil fields, in Utah and Colorado respectively, are 27 miles (43.5 km) apart. Crude oils from the Weber Sandstone in these two fields are similar based on the correlation index (CI) analysis and sulfur content (see Glossary for explanation of CI). Correlation of crude oils by CI is most accurate in the high boiling ranges, cut numbers 9 through 15. In the Rangely oil field, there are three producing horizons—the Weber Sandstone, the Shinarump Sand, and the Mancos Shale—all within a vertical distance of about 3,600 ft (1.1 km). These oils are quite different, based on CI and sulfur content, as shown in Figure 6-27. Bass concluded that vertical migration of hydrocarbons could not be occurring through this 3,600-ft section, or the oils in all three formations would be similar. This relatively short section of interbedded sands and shales covers a time period from Permian to Cretaceous. Under such conditions, there would be maximum expulsion of water and hydrocarbons from the shales, so that each thin shale source bed would expel its oil into the immediately adjacent sands. Although no source bed analyses were made, it is possible that the Weber oil originated in Permian or older sediments, the Shinarump oil in Jurassic–Triassic sediments, and the Mancos oil in Cretaceous sediments. Bass attributed the differences in these oils to differences in the source beds. This would result in a similar oil over a broad area horizontally but a limited zone vertically.

One of the best examples of limited vertical fluid movement and extensive lateral movement is in the Greater Oficina fields of Eastern Venezuela (Hedberg et al. 1947). According to Hedberg (personal communication),

> Here you have some 50 producing reservoir sands in vertical sequence, with strong evidence that each sand received its oil from the locally adjacent shales above and below the reservoir and that it then migrated laterally along the reservoir sand until trapped against a fault which formed the common updip barrier for all of the sands.

There are differences in the composition of the oils and in the salinities of the waters associated with different sands under one structure. This would be compatible with the retention of fluids within stratigraphic boundaries.

Vertical Migration

Fissuring and fracturing associated with faulting induces secondary permeability and porosity, which favor vertical fluid migration. Faults may act as conduits or seals, depending on a variety of factors (Eremenko and Michailov 1974). The normal stress across a fault and the nature of the fault plane and of the strata cut by the fault are among the critical factors. Tensional faults in hard, brittle rocks, such as limestones and dolomites, are frequently channels for vertical migration. Faults in clay or shale sections, however, commonly smear clay gouge across the faces of the abutting sands and thus seal them.

Link (1952) cites several cases of vertical migration of hydrocarbons along faults. Seeps associated with the Rothwell field of Ontario, Canada, leak from a fault interrupting a series of flat beds from Ordovician to Devonian in age. The Norman Wells field in Canada is a leaking reef limestone in which oil comes up along the fault zones on the edge of the reef. The faulting is caused by draping over the reef itself and is a classic example of the loss of oil from a stratigraphic trap through extension faults without folding.

Figure 6-28 shows a rather common type of seepage along a fault with sediments on one side and igneous rocks on the other. Fluid movements along the fault carry hydrocarbons from the adjacent sediments, resulting in a seep at the granite outcrop. Since the top section in this example consists of evaporites, which are essentially impermeable to hydrocarbons, the fluid migration must be occurring along the fault.

The Russian literature identifies many examples of bitumens along fault planes and of small accumulations associated with the updip end of fault planes. For example, Kudryatseva et al. (1974) cite migration along a fault zone in the Savan hot springs region on the eastern margin of the Golygino Basin in southwestern Kamchatka. The fault, which is more than 2 km (6,562 ft) long, has 100–120 m (328–394 ft) of visible throw. The fault zone is up to 20 m (66 ft) wide, and the rock throughout it is saturated with methane. Gas is liberated through two large vents at the rate of about

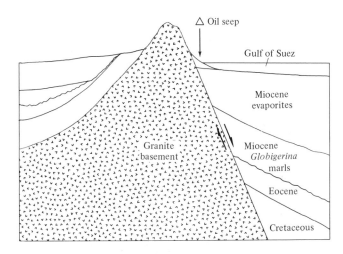

Figure 6-28
Migration of fluids along normal fault at Gebel Zeit, Egypt, causes a hydrocarbon seep at granite outcrop. [Link 1952]

30 m³/day (1,059 ft³/day). There are also several smaller vents in the fault zone. Core holes drilled on the site fill with water containing small amounts of kerosine. The kerosine consists almost entirely of condensed naphthenic or aromatic hydrocarbons.

On the other hand, Weeks (1958) cites several examples where faults act as seals, rather than conduits for migration such as the Velasquez field of the Magdalena Valley, Columbia. He notes that there are no seepages associated with the lower shelf and hinge belt of the Eastern Venezuela Basin, which has a maze of faults serving as barriers to trap oil in a large number of fields.

Levorsen (1967, p. 260) cites many examples of reservoir traps caused by tightly sealed normal, reverse, and thrust-type faults. Dickey and Hunt (1972) point out that hundreds of normal faults in the LaBrea–Parinas field of Peru act as seals, cutting the field into separate reservoirs.

It is clear that both leaking and sealing faults exist, and no general statement can be made regarding vertical migration along faults, simply because there is such a wide variation in the level and maintenance of continuous permeability.

Fracturing and fissuring is more common in brittle rocks such as carbonates and calcareous shales. Radchenko et al. (1951) cite many examples of smears and seeps of liquid hydrocarbons and impregnations of solid asphalts along bedding planes of fissures and faults in carbonate rocks, silicified sands, and some metamorphosed rocks. Goldberg (1973) found several types of bitumens in a dense network of nearly vertical fractures occurring through the argillaceous carbonate Ordovician cover of Cambrian oil reservoirs in the Baltic region. The highest concentration was in the lower 20–25 m (66–82 ft) of the Ordovician section. The vertical fractures ranged in thickness from less than 1 mm (0.04 in) to 3 mm (0.12 in). The paleopermeability caused by the joints and fractures was determined to be from 85 to 110 md before the filling of the fractures with bitumens. It was higher in some mineral-filled fractures. This situation is analogous to the vertical bitumen-filled joints in the dolomitic shales of the Uinta Basin reported by Hunt et al. (1954). In the latter study, the vein thicknesses ranged from about 1 to 50 mm (0.04 to 2 in) over a vertical distance of about 150 m (492 ft).

Teslenko and Korotkov (1966) in their study of the Maikop clays from the West Kuban downwarp noted that the calcareous clays at a depth of about 3,200 m (10,499 ft) showed a network of vertical microfractures filled with calcite. Migration of bitumens along the microfractures was evident from luminescence tests. Adjacent layers of noncarbonate clays did not contain any visible microfractures.

Jointing in brittle rock masses is recognized as a fundamental and widespread process regardless of the depth of burial. The general lack of lateral displacement indicates that many joints are natural tension fractures. Tension fractures at great depths have been discounted in the past on the basis that absolute tension is impossible very deep in the earth's crust.

However, these calculations did not take into consideration the role of fluid pressure in the tension fracturing process. Secor (1965) demonstrated from theoretical considerations that tension fractures can develop at increasingly greater depth in the earth as the ratio of fluid pressure to overburden pressure approaches 1. Secor also noted that previously formed fractures can be opened up by fluid pressures at depths equivalent to current drilling ranges.

Faults or fractures also may be opened up by the thermal expansion of water (Snarskii 1964; Barker 1972). Aquathermal pressuring, as defined by Barker is the increase in pressure of confined systems due to fluid expansion. If a confined fluid were already close to overburden pressure because of undercompaction, the addition of aquathermal pressuring might be enough to open existing faults, which would close again after release of the fluid pressure. The exact effect of aquathermal pressuring is difficult to estimate, because a temperature increase would decrease the viscosity of water, making it easier for pore fluids to escape before an appreciable increment of pressure.

Fluid movements along vertical fractures have been observed in the Gulf Coast. Forgotson (1969) reports a Louisiana well in which a fault became a passageway for upward movement of high-pressure fluids from a sandstone around 13,000 ft (3,962 m). The fluids fractured the formations above them on the downthrown side of the fault until the containing pressure of the formation and the diminishing pressure of the invading fluids were equalized.

Kushnareva (1971) investigated a fault extending from carbonate rocks of Upper Devonian to Upper Permian in the West Soplyas area at the southeast end of the Pechora–Kozhva arch. Oil accumulations occur wherever there are good reservoirs throughout this interval. In the fine-grained rock, regardless of composition, oil occurs in discordant microjoints, stylolites, and hairline cracks. Secondary accumulations of calcite, sulfides, and fluorite were also noted. Kushnareva concluded that the clear evidence of oil movement along discordant joints suggested a wide-front, upward vertical migration of fluid. The fault was the main channel for percolation of hydrothermal fluids and then of the oil.

Some geologists claim that microscopic examinations of cores from deep wells show microfractures to exist through virtually all shales and limestones with low tensile strength. If such a network of fractures developed in deeply buried, overpressured shales, it would enable more fluids and consequently more hydrocarbons to be released from the shales in a short geological time.

Evaporites

Evaporites represent the tightest barrier to hydrocarbon migration existing in sedimentary basins. For many years, L. G. Weeks (1958, 1961) emphasized the importance of evaporites as cap rocks for oil accumulations. He pointed out that there are many cycles of deposition involving organic-rich carbonate marls or muds that end with evaporites. The evaporites act as an excellent seal, which effectively traps most of the hydrocarbons generated during preevaporite sedimentation. Evaporites overlie many of the huge oil fields of the Middle East.

The ability of evaporites to prevent vertical migration of hydrocarbons is caused by the capability of salt to undergo plastic flow at elevated temperatures and the small size of the crystal lattice of sodium chloride. The distance between NaC1 lattice units is 2.8×10^{-10} m whereas the smallest hydrocarbon, methane, has a molecular diameter of about 4×10^{-10} m. The only way that hydrocarbons can migrate vertically through salt even by diffusion is if the salt contains impurities that render it brittle and capable of fracturing under tectonic stress. Antonov et al. (1958) examined four specimens of rock salts ranging from Devonian to Permian in age and found that they did have a mosaic of very fine fractures that permitted some vertical diffusion of methane. As a rule, however, evaporites may be considered to be the strongest mineralogical barriers to vertical migration of hydrocarbons.

Continental Facies

As a generalization, continental facies represent the opposite end of the permeability scale to evaporites. They are the sieve of the subsurface. Continental beds generally contain high percentages of sands and silts, which have much greater permeability than the smectitic shales of marine facies. Bailey (1947) described the nonmarine red bed facies of the Sespe Formation of the Ventura Basin of California as a conduit for vertical migration of hydrocarbons from deeper Eocene beds.

Surface geochemical surveys on the shelf off the Gulf Coast have been useful in delineating regional areas of gas and oil accumulation because of the high permeability of the top section of continental sandstone facies. Geologists who are having subsurface samples analyzed for hydrocarbons as part of an exploration program should also include red bed sequences. Although red beds are not source rocks, they frequently contain light hydrocarbons that give evidence of migration pathways through the red beds.

Distance of Migration

Primary migration in fine-grained sediments with permeabilities from 10^{-3} to 10^{-11} md will only go as far as necessary to reach a permeable bed. This may be millimeters, inches, tens or hundreds of feet, or meters, but not kilometers or miles. It would be unusual to move very far in any direction in a geological column without encountering permeabilities appreciably larger than 10^{-3} md.

Within more permeable sequences, hydrocarbons can and probably do move very long distances. The Athabaska oil sands must be the result of a fluid drainage area extending over hundreds of miles. The Pleistocene production of the Gulf Coast, except for biogenic methane, almost certainly had to migrate from older and deeper beds or from the deeply buried Pleistocene bathyal facies beneath the continental slope. This could entail a migration distance of 50–100 miles (80–161 km). The fact that hydrocarbons can migrate with fluids long distances through permeable beds means that the geochemists must be cautious in condemning a section because it is too shallow and too cool to generate hydrocarbons. If this nongenerating section is connected to a more deeply buried source rock through continuous sandstones, unconformities, fracture–fault systems, or continental facies, it can contain commercial petroleum accumulations.

ABNORMAL PRESSURES

Retarded Migration

Abnormal overpressures ($>$12 kPa/m, or $>$0.53 psi/ft) are created and maintained by the inability of pore fluids to migrate within a reasonable geologic time period when subjected to stresses causing increased fluid pressure. Several types of stresses can increase fluid pressure such as (1) rapid loading, (2) thermal expansion of fluids, (3) compression by tectonic forces, and (4) generation of oil and gas from organic matter in the rock matrix.

The first has been discussed as compaction disequilibrium in young rapidly deposited sediments, and the second as aquathermal pressuring. The third type of stress has been described by Berry (1969) as the source of abnormal pressures in the California coast ranges, where two giant granitic blocks are squeezing the thick wet clays of the Franciscan and Great Valley geosynclinal sediments. An example of the fourth stress is gas and oil generation, which Hedberg (1974) and Meissner (1978a,b) each describe as a cause of overpressures. Also, Tissot and Pelet (1971) demonstrated in a laboratory experiment that the generation of N_2 gas from organic matter in a confined shale increased the pore fluid pressure from 43 to 53 megapascals (6,258 to 7,680 psi) before the development of microfissures caused release of the gas.

There are probably other forms of stress in addition to the four listed above, but all of them point to the inability of the fluids to migrate as the cause of abnormal pressure. Some processes, such as clay dewatering, have been shown not to be a cause of abnormal pressure, but sometimes a result of it, caused by the increase of temperature.

OVERPRESSURES IN ORGANIC-RICH ROCKS

The discussion relating to fluid pressures in typical water-wet rocks does not necessarily apply to organic-rich rocks that may be largely oil wet. A sediment containing 8 percent organic carbon by weight would have approximately $(8 \times 1.22 \times 2) = 20$ percent organic matter (OM) by volume, where 1.22 is the conversion factor for C_{org} to OM and 2 the difference in OM and mineral densities. At depths greater than about 4,000 ft (1,219 m) the organic matter would occupy the major part of the intergranular pore space and the rock would be oil wet. In very fine-grained rocks, the matrix permeabilities would be so low that the generation of oil

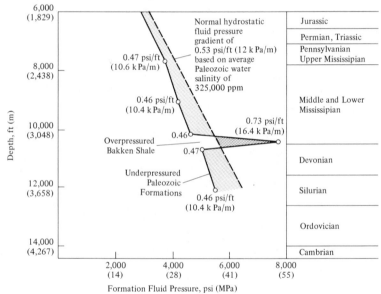

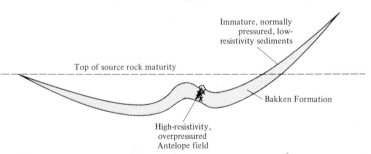

Relation of Fluid Pressures to Oil Generation in Antelope Field, Williston Basin, North Dakota. [Meissner 1978b]

OVERPRESSURES IN ORGANIC-RICH ROCKS (continued)

and gas from the kerogen could create very high internal fluid pressures. An example of overpressures caused by the generation of oil and gas is the Antelope field, McKenzie County, North Dakota (Meissner 1978b). This field produces oil from fractured shales and siltstones of the Bakken Formation. The Bakken is an organic-rich (up to 10 percent C_{org}) source rock that Dow (1974) estimates generated 10 billion barrels of oil in the Williston Basin. The 70 ft (21 m) Bakken interval at Antelope was over-pressured at about 7,600 psi (52 MPa) compared to underpressures of 4,700 and 5,100 psi (32 and 35 MPa) above and below this interval as shown. To the east, where the Bakken is above the threshold of intense oil generation (about 165°F, or 74°C), the formation is normally pressured. Meissner (1978b) noted that nearly infinite resistivities occur in the oil generation zone of the Bakken, due to the pores being filled with oil and organic matter. He proposed using resistivity logs to recognize the oil maturity zone in rocks. Overpressures in the Altamont–Bluebell field (Uinta Basin, Utah) also are caused by the generation of petroleum in fine-grained, organic-rich rocks.

Abnormal pressures are found throughout the world (Fertl 1972; Rehm 1972). Figure 6-29 shows the typical range of abnormal overpressures in producing areas of Europe. Of particular interest is the Paleogene formation in the Tadzhik depression (USSR) where the pressure is 11 MPa (1,595 psi) at a depth of 480 m (1,575 ft) (Kalomazov and Vakhitov 1975). This is about twice the hydrostatic pressure. The oil in these high-pressure sands is lighter than those in normal-pressure regions of the Tadzhik petroleum basin. In the Beshtentyak field, the Paleogene shales at a subsurface depth between 480 and 2,787 m (1,575 and 2,787 ft) have a mean formation pressure 1.8 times hydrostatic. At about 2,700 m (8,858 ft), the fluid pressure was reported to be 2.3 times hydrostatic, which is about equivalent to the lithostatic pressure.

Abnormal pressures due to the causes listed above may approach, but usually will not exceed, lithostatic pressure in normal sedimentary sequences (Hubbert and Rubey 1959; Chapman 1973, p. 61). When fluid pressures reach lithostatic, the overburden is supported not on rock, but on water. Since water is the continuous phase throughout a basin, it will adjust by leakage, even from so-called isolated or constant-volume reservoirs. There are no true constant-volume reservoirs in the subsurface except possibly a limestone tightly cemented on all sides, and even this would probably fracture near overburden pressure. Even tight nonconducting faults would be forced open briefly if fluid pressures in adjacent reservoirs approached lithostatic. (A good discussion of abnormal pressures is given in Chapman 1973, pp. 48–80.)

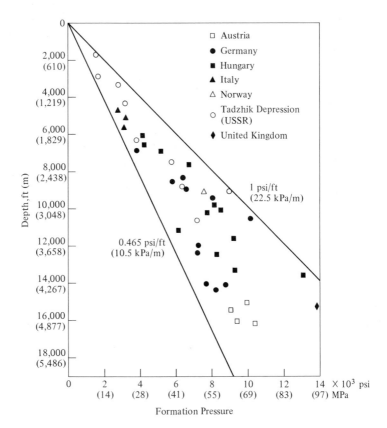

Figure 6-29
Abnormal formation pressures in European basins. [Fertl 1972; Kalomazov and Vakhitov 1975]

Abnormal overpressures in oil wells are generally noticed at depths greater than 4,000 ft (1,219 m), although they have been encountered as shallow as 1,500 ft (460 m). They are most common in young, rapidly deposited sediments although they are found in rocks of practically all ages.

Abnormal pressures are nearly always accompanied by an increase in the porosity and a decrease in the pore water salinity of the shale and by an increase in temperature. Electrical resistivities decrease, conductivities increase, acoustic transit time increases, shale bulk density decreases, and the pulsed neutron log signal decreases (Fertl and Timko 1972).

Chemical changes include an increase in the soluble organic matter at the top of the overpressured zone, and an increase in sulfate and bicarbonate ion concentration in shale pore water. Sulfate concentration may double or triple in shale pore waters when entering a high-pressure zone. Both sulfate and bicarbonate exceed chloride in the shale pore waters of high-pressure zones (Weaver and Beck 1969; Schmidt 1973).

Actually, there has been no physical increase in porosity or change in salinity, even though it appears this way. The situation is that a mass of shale has moved from a shallow depth to a deep horizon with only a negligible loss of water, whereas the shales above it have been undergoing normal compaction. Moving this abnormally pressured shale mass up in the section, to the point where its porosity matches that of the normal shale porosity decline curve, will indicate the depth at which overpressuring began, assuming no clay dewatering. The latter process will freshen the water and increase the porosity, even though it has no effect on the pressure.

An important phenomenon of many overpressured zones from the standpoint of the origin and migration of petroleum is the increase in temperature. Well flowline temperature gradients in areas such as the Gulf Coast, the North Sea, and the South China Sea have been observed to increase appreciably prior to and/or when entering a high-pressure zone, as much as 18.2°C/100 m (10°F/100 ft), according to Fertl (1976). The close correlation between increased temperatures and increased pressures has been observed throughout the world. Figure 6-30 shows the increase in temperature with pressure in 60 Gulf Coast reservoirs. All of the reservoirs at temperatures below 93°C (200°F) have pore pressures lower than 13.5 kPa/m (0.6 psi/ft). Above 93°C (200°F), particularly in the range from about 112 to 140°C (235 to 285°F), most reservoirs are overpressured.

A similar relationship is observed in the oil fields of Hungary near Szeged, where the highest geothermal gradients are associated with the highest

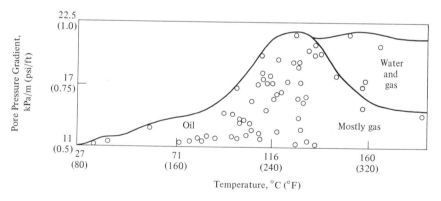

Figure 6-30
Relationship of pressure and temperature in 60 Gulf Coast reservoirs. The highest abnormal pressure gradients, over 0.85 psi/ft, are in rocks at temperatures between 112 and 140°C (235 and 285°F). Primarily gas reservoirs occur at the higher temperatures (greater depths), as would be expected, given the late generation of catagenic methane. [Timko and Fertl 1972]

abnormal pressures (Alliquander 1973). These high geothermal gradients are due to the lower thermal conductivity of overpressured shales. Lewis and Rose (1970) developed the concept that overpressured shales with very low thermal conductivities act as heat insulators, causing abnormal temperature gradients to exist across them.

The observed increase in well flowline temperatures in high-pressure zones has been attributed by some geologists to the disturbance of the geothermal gradient by changes in the drilling mud operation. Drilling slower, with less circulation through smaller casing will show higher temperature readings than fast drilling in the larger hole above the pressure zone. This is not the entire explanation, however, because temperature logs in cased wells that have been idle for two years or more show sharp differences in geothermal gradients, because of differences in thermal conductivities (Gretener 1976).

Dogleg Geothermal Gradients

Heat moves from the center of the earth outward through the sedimentary crust into the ocean or atmosphere, where it is lost as radiant energy. The steady-state, one-dimensional heat flow through a medium is computed as a product of the temperature gradient times the thermal conductivity of the material in which the temperature gradient is measured. The equation is

$$Q = \lambda \frac{\partial T}{\partial Z}$$

where Q = heat flow in cal/cm² sec

λ = thermal conductivity of the substance in cal/cm sec °C

$\frac{\partial T}{\partial Z}$ = geothermal gradient in °C/cm, where Z is positive downward

Although heat flow, Q, varies considerably in different areas, it is usually constant for a specific locality in a basin. In the Western Canada Basin, $Q = 1.46 \times 10^{-6}$ cal/cm² sec, which is typical of many oil-producing basins. At a well location, Q would be essentially constant from top to bottom, assuming no disturbing effects such as a nearby salt dome. Since Q is constant, a decrease in thermal conductivity in a sedimentary section results in an increase in the geothermal gradient, and vice versa.

The thermal conductivity, λ, of rock components varies considerably as shown in Figure 6-31 (Zierfuss 1969; Gretener 1976). In clastic sequences, clays have lower conductivities than sands, so, as a generalization, an increase in the sand/shale ratio increases the conductivity and decreases the geothermal gradient. In nonclastics, carbonates have lower conductivities than evaporites, so evaporite seals show a lower geothermal gradient. These

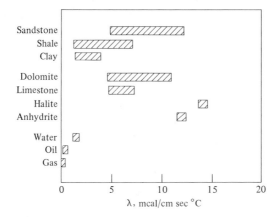

Figure 6-31
Thermal conductivity of rocks and pore fluids at 20°C.
[Data from Zierfuss 1969; Gretener 1976]

differences in the thermal conductivities of lithologic sections cause dogleg geothermal gradients; that is, the gradient will show an abrupt angle at the lithologic change.

Pore fillers (oil, gas, and water) have much lower conductivities than the rock minerals and correspondingly higher geothermal gradients. This means that an increase in porosity will reduce the conductivity (raise the geothermal gradient) of most rocks.

Overpressured shales have high porosity, and their overpressuring is commonly caused by a lack of permeable silt and sand bodies as channels of migration. *The combination of high porosity and lack of sands results in a geothermal gradient in the overpressured shale, usually higher than that of normally pressured sections above or below.* Consequently, overpressured shales frequently show a dogleg in the geothermal gradient.

An example of this dogleg effect is shown in Figure 6-32 for the Texas Gulf Coast (Jones and Wallace 1974). This same dogleg in bottom-hole temperatures, accompanied by a decrease in salinity and an increase in porosity, has been observed in overpressured zones in the North Sea, the South China Sea, and many other places. In addition, doglegs in paleogeothermal gradients have been recognized through vitrinite reflectance measurements (see Chapter 7). Since a geothermal gradient leaves a permanent record on the organic matter of sedimentary rocks, the vitrinite reflectance technique can recognize past changes in geothermal gradients and can recognize corresponding past abnormally pressured zones that may be normal today. A reverse dogleg in the geothermal gradient occurs when an over-

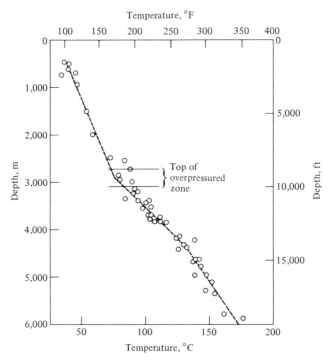

Figure 6-32
Bottom hole temperature data for nine wells in Cameron County,
Texas. Dogleg geothermal gradient with the change in temperature
increase occurring at the top of the abnormally pressured zone.
Geothermal gradients in a six-county area of the Texas Gulf Coast
range from 2.5 to 10°C/100 m (1.4 to 5.5°F/100 ft), according to Jones
and Wallace (1974).

pressured shale is underlain by more normally pressured sediments. Matviyenko (1975) reports that in West Siberia the abnormally pressured
Paleogene and Upper Cretaceous thick shales in the depth interval between
400 and 1,100 m (1,312 and 3,609 ft) have geothermal gradients as high as
3.6–5.3°C/100 m (2–3°F/100 ft), with a thermal conductivity of $3.2 \cdot 10^{-3}$
cal/cm sec °C. Below these thick shales is a more normally pressured section
composed of alternating sands and clays of Mesozoic age extending from
about 1,100 m to basement. These interbedded sediments have geothermal
gradients of 2.8–3.8°C/100 m (1.6–2.1°F/100 ft) and thermal conductivities
of $4.2–5.3 \cdot 10^{-3}$ cal/cm sec°C. Clearly, this decrease in the geothermal
gradient is due to a higher thermal conductivity resulting from an increase in
the sand/shale ratio and a decrease in the porosity of the interbedded shales.
From these data, the heat flow, Q, is about $1.4 \cdot 10^{-6}$ cal/cm² sec.

The increased temperature of overpressured shales results in a faster generation of oil and gas from kerogen, an early dewatering of clays, a greater solubility of hydrocarbons for fluid transmission, and a greater solubility of silica. High-temperature pods represent an ideal situation for the maximum generation of hydrocarbons, and if the high temperatures also result in a fracturing of the shale there can be both generation and accumulation within the source bed without the aid of significant amounts of water for migration. This is the situation in the Altamont field of the Uinta Basin of Utah, where oil is produced from fracture-enhanced permeability in the black shale source rocks of the Upper Wasatch (Baker and Lucas 1972). Parts of the producing section have a geothermal gradient of 7.2°C/100 m (4°F/100 ft), a temperature of 111°C (232°F), and pressure gradients up to 18 kPa/m (0.8 psi/ft). Clay dewatering has no effect on the porosity, since the shale is composed of fine-grained silica and carbonates with less than 5 percent clay.

Mud Volcanoes

When a highly porous, undercompacted mud, with a low density of 2g/cc, is buried under a normally compacted rock of greater density, a mechanically unstable system is created. The undercompacted mud may contain large amounts of gas, causing greater instability. Eventually, buoyancy causes the mud to flow upward along a vertical zone of weakness until the system stabilizes. Hobson and Tiratsoo (1975, p. 102) describe a model experiment that illustrates the cause of mud volcanoes. They placed corn syrup above crude oil in a closed glass container, with a rubber diaphragm on the bottom. The slightest pressure on the diaphragm disturbed the unstable system, causing a dome of the lighter oil to well up through the denser syrup.

Sokolov et al. (1969) have described violent eruptions of mud volcanoes in the South Caspian Basin, which have released hundreds of millions of cubic meters of gas. Over the past 155 years, the eruptions have occurred with some regularity, indicating periodic buildup and release of gas from solution within the clay mud. There are more than 200 mud volcanoes in Azerbaijan, and it is estimated that they have given off 10^{11} metric tons (Mg) of gas in the last million years. The gas in mud volcanoes is nearly always methane, although Al'bov (1971) has reported carbon dioxide mud volcanoes in the USSR.

Hedberg (1974) points out that most mud volcanoes are related to lines of fracture, faulting, or sharp folding. The source of mud volcanoes can often be traced to a thick subsurface layer, or diapir, of overpressured, plastic, undercompacted clay mud. Hedberg cites a large number of mud volcano areas in the world, generally in Cenozoic or Late Mesozoic sediments. The source or roots of the mud volcano can be very deep, greater than 20,000 ft (6,096 m).

Shaulov (1973) compared the Maikop clays from different parts of the West Kuban Downwarp. In the southeast, the clays are thin and interbedded with sands, so diapirism and mud volcanism are absent. In the northwest, the Maikop clays contain no sands or silts, and porosity is still 17 percent in the 3,000 to 4,500 m (9,843 to 14,764 ft) depth range. Also, the clays here contain a large amount of organic matter. The absence of sands and silts results in the clays being very plastic. Mud volcanoes are common in this northwest area, and the solid debris contains fragments of Eocene, Paleocene, and Cretaceous rocks. It was concluded that the roots of several of the mud volcanoes extended far below the Maikop, reaching the Lower Cretaceous.

Mud volcanoes move not only hydrocarbons but also heat. Sukharev et al. (1970) determined the heat flow in wells drilled near a fossil mud volcano on the Apsheron Peninsula. They found heat flow to be greatest at the crest of the structure. The vent of the mud volcano acted as a channel, along which the transport of subsurface heat was more intense.

Not all mud volcanoes originate from deep-seated muds. Biogenic methane mixed with plastic muds can create gas-charged sediment cones originating at depths from a few hundred to a few thousand feet. These gas-charged cones are common in the Gulf Coast and have been described in detail by Sieck (1973). There is seismic evidence that similar gas-charged cones are present in other areas such as Alaska, South America, Africa, and the North Sea.

TIME OF MIGRATION

Biogenic methane begins migrating in the earliest stages of compaction. Evidence of methane migration in Pliocene sandstones at sediment depths of less than 400 m (1,312 ft) have been observed by Whelan (1979). Because most hydrocarbons are not formed until sediment temperatures reach about 50°C (122°F), the maximum volume of primary migration would not occur until that time. After that, the time of migration would depend on the migration mechanism, the quantity of hydrocarbons available, and the molecular weight range of the hydrocarbons. Gaseous hydrocarbons will migrate from fine-grained source rocks, as they are generated at all depths. Migration, both in solution and as an oil phase, would occur mainly during periods of compaction and fluid movement. Uplifting of sediments would terminate primary migration except for diffusion. If uplifted sediments are buried beyond their previous depth, a second release of hydrocarbons could occur, assuming that enough additional hydrocarbons are available for migration. This is the "second-squirt" hypothesis, and oils of the Magdalena Valley of Columbia are believed to have originated in this manner (Dickey, personal communication).

PRIMARY MIGRATION IN CARBONATES

Primary migration of petroleum in argillaceous carbonates can be via compacting waters, but this is not the case in pure carbonates, because they lithify mainly by recrystallization and cementation. Micrites—calcareous muds of particle size under 4μ—have primary porosities of 50 to 70 percent. In some deep sea sediments, these porosities show no significant change under sediment loads over 1,000 ft (305 m) in thickness. In contrast, many carbonates deposited in inland seas and on continental margins undergo porosity reduction down to 1 to 2 percent within the first 10 to 100 ft (3 to 31 m) of burial, owing to carbonate cementation. Whether cementation occurs or not is a complex problem, depending on the physical and chemical environment.

Cool, humid climates with acid soils (high CO_2, low pH) keep Ca^{2+} in solution, but warm, dry climates (low CO_2, high pH) tend to precipitate calcite. A high pH can result from biological activity and from inorganic processes, such as the presence of feldspars in the paths of moving brackish waters, which lead to a concentration of the OH^- ion, raising the pH (Chilingar et al. 1967). The equation relating $CaCO_3$ precipitation and dissolution is as follows:

Cool, polar   *Warm, tropical*

$$H_2O + CO_2 \rightleftharpoons H_2CO_3 \rightleftharpoons HCO_3^- + H^+ \rightleftharpoons CO_3^{2-} + 2H^+$$

pH = 5, acid,   pH = 9, alkaline,
reduced activity of   increased activity of CO_3^{2-}
CO_3^{2-} so more $CaCO_3$ so $CaCO_3$ precipitates
dissolves

Friedman (1975) has discussed the ups and downs of limestone cementation. He points out that pH levels of 9 to 10.5 may be reached on a microscale in the marine environment, causing precipitation of carbonate cement. In a Red Sea reef, the porosity of a coral was almost entirely eliminated by cementation within 60 cm of the live coral at the surface. Friedman also cites evidence for early carbonate cementation in other areas, such as the Persian Gulf and parts of the Bahama Bank. Both Friedman (1975) and Sayles and Manheim (1975) emphasize the importance of microbial activity in causing cementation. Terrigenous, organic-rich sediments deposited along continental margins undergo sulfate reduction and deamination leading to removal of SO_4^{2-}, enrichment of HCO_3^-, and formation of NH_3, all of which results in more alkaline pore waters. The microbial reaction is

$$2CH_2O + SO_4^{2-} \rightleftharpoons H_2S + 2HCO_3^-$$

where CH_2O is the source of organic carbon. The hydrogen sulfide combines with heavy metals, such as iron, to form mineral sulfides. The bicarbonate ion precipitates calcium out of solution as $CaCO_3$ cement.

Evidence that compaction is not significant in ancient pure carbonate sediments is shown by pellets and other carbonate particles in their original shapes with no indication of being crushed. Yet recrystallization and cementation do appear to redistribute and concentrate the organic matter in some sediments. Polonskaya et al. (1974) noted organic matter filling both large pores and passages between dolomite grains and recrystallized carbonates representing various stages of diagenesis. Subsequent microfracturing along grain boundaries would make larger amounts of organic matter available for migration, because it is driven out of the mineral phase during recrystallization. Light hydrocarbons generated by this organic matter could diffuse in formation fluids to regions of high permeability. Chalk beds are an exception to many carbonates, in that they do compact to some extent with increasing overburden (Polak 1954; Scholle 1975). Polak studied chalk compaction in rocks of the Ural–Emba region, and Scholle studied it in the North Sea. Chalk ooze has porosities around 80 percent. After burial to 350–400 m (1,148–1,312 ft) porosities are in the range of 38 to 48 percent and permeabilities from 4 to 13 md. In the North Sea at 1,500 to 2,000 m (4,900 to 6,600 ft), porosities had decreased to 15–30 percent with permeabilities of 0.1–1 md. This compaction is not simple water loss caused by compressive stress as with clays but also involves dissolution of unstable carbonate crystals at grain-to-grain contact and local reprecipitation within foraminiferal chambers or as overgrowths on coccolith plates (Scholle 1975). Further burial to 2,700–3,300 m (9,000–11,000 ft) reduced porosities to 2–25 percent and permeabilities to 0–0.5 md. In some instances, porosity and permeability were enhanced or retained by fracturing or early entry of oil into the pore spaces.

Because fluids are continually being lost during chalk diagenesis, any of the previously discussed primary migration mechanisms could be operable.

Stylolites that occur between blocks of limestones and dolomites are indicators of fluid loss during compaction. Dunnington (1967a) estimated that 2,000–3,000 ft (610–914 m) of overburden is required to initiate pressure solution of carbonates to form stylolites. The stylolites, which occur as toothlike partings commonly perpendicular to the bedding, are filled with the insoluble mineral residue and organic matter left from dissolution of the lithified carbonate. Many stylolytic limestones have lost 20–35 percent of their rock volume since lithification, with losses as high as 40 percent being estimated. Reprecipitation of the dissolved carbonate updip in a lower pressure regime could cause cementation of shallower rocks. Hydrocarbons generated in the limestone prior to pressure solution could be carried off with the migrating fluids.

The above refers mainly to clay-free carbonates, whereas many carbonates contain some clay mineral. Zankl (1969) observed that the purity of the

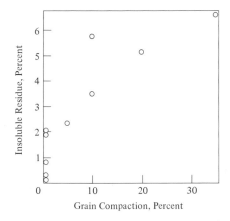

Figure 6-33
Relationship between insoluble residue of
lime muds and their ability to compact.
Lime muds with less than 2 percent insoluble residue do not compact. Compaction
increases with increasing residue, particularly if it is clay minerals. [Zankl 1969]

lime mud determines whether carbonates lithify early or compact. Lime
muds containing less than 2 percent acid-insoluble residue tend to undergo
cementation and recrystallization. These are the examples quoted by Friedman. Lime muds with more than 2 percent insoluble residue, particularly clay
minerals, undergo compaction, the extent of compaction increasing with the
increase in clay mineral content. Zankl's results are shown in Figure 6-33,
where the insoluble residue is plotted against grain compaction for a whole
series of limestones. Except for one data point, there is almost a straight-line
increase in grain compaction with insoluble residue. Zankl also noted that
limestones containing more than 2 percent clay minerals are predominantly
composed of micritic grains.

McCrossan (1961) determined porosity–depth curves for Upper Devonian
shales of the Western Canada Basin containing different amounts of calcium
carbonate. These curves, after correcting for the removal of overburden, are
shown in Figure 6-34. Also shown in this figure is the range of porosities for
different depths from Figure 6-7. It is clear from McCrossan's data that
carbonates containing 10 percent or more of clay material undergo fluid
losses through compaction, which would enable any of the discussed primary
migration mechanisms to operate. Migration in the relatively pure carbonates, except for chalks, would be limited to a diffusion mechanism or to
gas-phase migration.

Abnormal Pressures in Carbonates

Carbonate cementation can be so tight that there is essentially no fluid
transmission. Under such conditions, deep burial of the carbonates in
high-temperature regimes will result in abnormal fluid pressures caused by
the generation of gases from organic matter and the thermal expansion of

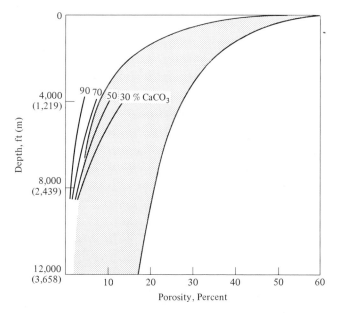

Figure 6-34
Porosity–depth curves for carbonates of varying insoluble residue
contents from Western Canada Basin. The less calcareous rocks have
a higher initial porosity and a faster decrease in porosity with depth
than do more calcareous rocks. Rocks with less than 70 percent $CaCO_3$
have compaction curves comparable to clays. [Data plotted from
McCrossan 1961, with 457 m (1,500 ft) of overburden added based
on Magara 1976b]

water. Parker (1973) gives examples of fluid pressure gradients as high as 24
kPa/m (1.06 psi/ft) in the deep Smackover Formation of Mississippi.
Overpressures are erratic, with isolated high-pressure blocks being completely surrounded by lower pressure regimes. Methane, CO_2, and H_2S are
the principal gases in the Smackover at about 20,000 ft (6,096 m). Reservoir
temperature at this depth is 360°F (182°C). All of these gases can result from
high-temperature reactions, as previously discussed in Chapter 5. In addition
to generation of CH_4 from organic matter, Parker presented evidence that
preexisting oil accumulations in the Smackover could have been converted
to methane and solid asphalts through hydrogen disproportionation reactions. These reactions, whose rates increase at greater temperatures, cause
reservoir oils to decompose to lighter hydrocarbons, the hydrogen being
obtained from polymerization and aromatization of the heavy asphaltic
molecules. Eventually this process leads to methane and solid asphalt, both
of which were observed in the Smackover sandstones.

Erratic, unpredictable, abnormal pressure zones can characterize any deeply buried carbonates where cementation has terminated fluid movements. Such formations may hold their overpressures indefinitely, unless microfractures or faults enable the pressures to be dissipated.

SECONDARY MIGRATION AND ACCUMULATION

Secondary migration is considered to be the movement of fluids within reservoir rocks leading to oil and gas segregation. Tertiary migration is movement of oil and gas after formation of a recognizable accumulation. When hydrocarbons pass from a fine-grained to a coarse-grained rock, they encounter lower pressures and higher salinities. If the fluids are moving updip through continuous sands out of a compacting basin, lower temperatures also would be encountered. All of these factors would cause exosolution of part of the gases and liquid hydrocarbons. The precipitated hydrocarbons could form as highly dispersed colloidal-size particles or as globules. Secondary migration through reservoir rock could continue in solution, as a dispersion, or as an oil or gas phase until a permeability barrier initiates coagulation and accumulation.

Buoyancy will cause hydrocarbon particles to rise to the top of the reservoir rock, providing the particles are small enough to pass through the throats and pores of the reservoir rock. Most hydrocarbon particles passing from fine-grained shales to coarse-grained sands are small enough to move freely through the sands. Only after they begin coagulating into large particles in the reservoir rock would their movement be more restricted.

There is some evidence that dispersed droplets of oil can travel long distances through sandstones. Cartmill and Dickey (1970) demonstrated that oil droplets $0.5-1.5\ \mu$ in diameter in a concentration of $20-40$ ppm oil in water passed freely through a sand with a permeability of 53 darcys. Oil particles this size would carry a negative charge, so they would repel each other rather than coagulate, and they conceivably could migrate long distances in the dispersed state.

There is field evidence that hydrocarbons can travel long distances as a dispersion. Vyshemirskii and Yamkovaya (1970) examined 55 samples of nonproductive sands in the West Siberian oil and gas province. Each stratigraphically correlative sample group included samples from the top and sides of structures. They found that the structurally high samples contained 0.08 percent hydrocarbon, whereas those downstructure contained half as much. These were migrated hydrocarbons, according to the authors, but the migration could not possibly have been in a continuous phase at that concentration.

Although the rates of fluid migration during compaction are very low, they are sufficient to carry dispersed oil long distances through permeable beds in reasonable geologic time periods. Fluids moving 20 ft (6 m)/100 years would migrate 37 miles (60 km) in a million years.

Agglomerations of hydrocarbons that grow large enough to form a continuous phase also can migrate long distances through permeable reservoir beds. Levorsen (1967, p. 543) cites the migration of oil accumulated in Pennsylvanian sands of Oklahoma over distances of 75 miles (121 km) because of a reversal in dip after accumulation. Movements of oil pools over such distances could not leave behind 5–10 percent residual oil saturation over the entire migration path without exhausting the accumulation en route. Hobson and Tiratsoo (1975, p. 72) suggest that the slow rates of movement would preclude the disturbances that cause disruption of stringers of oil at the waist or necks in pore throats, thus minimizing the formation of isolated oil globules that account for residual oil saturation.

Direction and Pathways of Migration

The geochemical correlation of oils in reservoirs and oils extracted from source rocks has shown that vertical and horizontal migration through permeable sands, fracture zones, and unconformities occurs over much greater distances than previously had been believed. Dow (1974) and Williams (1974) identified the source beds for three types of oils in the Williston Basin of the United States and Canada. Figure 6-35 shows the structural section with the presumed migration path of Bakken oil along vertical fracture zones and the Madison–Jurassic unconformity into updip Madison reservoirs. The oil is found in reservoirs extending nearly 100 miles (161 km) beyond the limits of effective Bakken source rocks. Vertical migration occurred in fracture zones located primarily on the Nesson anticline. Oil from Bakken source rocks probably filled Mission Canyon and Charles reservoirs to closure on the anticline and then spilled northward up the axis of the south-plunging structure. The Charles evaporites acted as a seal, and oil migrated beneath them updip until trapped by porosity pinchouts within the Madison or at the Madison–Jurassic unconformity. Jurassic red shales seal the unconformity accumulations. Dow estimated that about 10 billion barrels of oil was expelled by the Bakken Shale, a third of which is in commercial reservoirs. The rest is either dispersed in noncommercial accumulations or in undiscovered pools.

In the Middle East, hydrocarbons previously accumulated in Middle and Lower Cretaceous reservoirs of Iran and Iraq, have migrated vertically through fracture zones into overlying Tertiary (Asmari) reservoirs. This conclusion was made on the basis of field evidence (Dunnington 1967b) and geochemical analyses of the oil. Extensive vertical migration of oil in Iraq was originally proposed by Dunnington (1958), based on evidence of pressure communica-

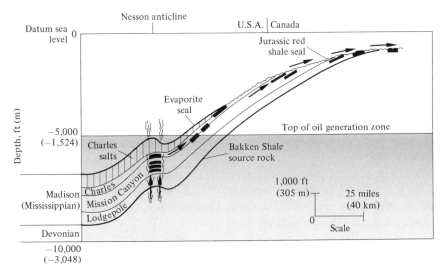

Figure 6-35
Vertical and lateral migration paths of oil generated in Bakken Shale and reservoired in updip Madison traps. Lateral migration occurs along unconformity and in porous carbonates. Vertical migration occurs along fracture zones of Nesson anticline. Oil accumulations shown as black bars. [Dow 1974]

tion and fluid flow between Eocene and Cretaceous reservoirs at the Ain Zalah field. Geochemical analyses of several Tertiary and Cretaceous oils showed them to be virtually indistinguishable except for small differences in sulfur content, gravity, and asphaltenes. Also, heavy oil staining was found in fracture zones extending hundreds of feet below oil–water contacts. In his later paper, Dunnington pointed out that pressure communication exists between several Asmari and Cretaceous reservoirs, which are highly fractured and therefore extremely permeable.

Many of the oils in overlying Asmari pools are lighter in gravity than those in the underlying Cretaceous reservoirs. The reason is that the Cretaceous oils were probably gravity-segregated in their original accumulation. Fracturing would release the gases and light oils first, resulting in a higher-gravity oil accumulating above a lower-gravity oil (°API). Although the vertical migration changed the distribution of molecular sizes in the oils, resulting in smaller molecules with less sulfur in the Tertiary reservoirs, it did not change the hydrocarbon types within a specific boiling range. Both the Tertiary and Cretaceous oils have a high paraffin content in the naphtha–gas oil fractions.

Vertical migration through fractures in the Middle East fields have occurred mainly in Iraq and Iran on the mobile side of the basin. There is some evidence for migration along faults on the stable shelf side, but the volumes of migrating oil are small.

In the discussion published with Dunnington's paper (1967b), Falcon pointed out that a deep test was drilled below the Masjid-i-Suleiman Asmari reservoir in the basinal facies of the Jurassic and Lower Cretaceous where fractures during folding would be unlikely to form. The test well was subsequently cemented and abandoned, but the cement job failed. The man-made fracture so produced was able to repressure the huge Asmari reservoir at a rate of about 69 kPa/month (10 psi/month). This represented an enormous transfer of gas from a deep, high-pressure zone to a shallow, low-pressure zone (depleted by production) over a very short geological time scale.

The age of eight Miocene reservoir oil samples in Iran was calculated by Young et al. (1977) from hydrocarbon ratios that change with time. More details of this are given in Chapter 8. They found six of the oils to have an average calculated age of 104 million years, with two being older than 120 million years. This substantiates Dunnington's observation that Tertiary oils of the Middle East migrated from Cretaceous or Jurassic reservoirs.

Accumulation: Barriers to Petroleum Migration

Any reduction in the permeability of rock during migration can result in accumulation. The concept of a capillary barrier was demonstrated long ago in experiments by Illing (1933). He placed alternate layers of coarse and fine sands in two tubes, one being wet with oil and the other wet with water. When an oil–water mixture was passed through the coarse–fine layers of water-wet sands, the oil was segregated in the coarse layers and the water in the fine layers. When the oil–water mixture was passed through the oil-wet coarse–fine sand layers, the water segregated in the coarse sand and the oil in the fine sand. This clearly indicated the effect of a capillary barrier. A small opening wet with water rejects oil, whereas a small opening wet with oil rejects water. Because the majority of rocks in the subsurface are water wet, water can pass uninhibited from a sand to a shale, but oil is trapped by the capillary barrier of the fine, water-wet pore openings.

The pressures developed by capillary barriers are enormous. Hubbert (1953) calculated that the capillary displacement pressure in a shale with a particle diameter of 10^{-4} mm would be about 40 atmospheres compared to less than 0.1 atmosphere for a sand. This means that a globule of oil attempting to enter a shale from a sand would be rejected from the shale by a pressure on the order of tens of atmospheres. The capillary pressure in a siltstone can be several times greater than that in the sand. Almost any change from a coarse to a fine particle size will result in some accumulation of oil.

The experiments of Illing and others were for oil-phase migration, but Cartmill and Dickey (1970) have demonstrated that even dispersed oil can be trapped by a reduction in pore dimensions. In their experiments, a layer of glass beads 37–88 μ in diameter was placed downstream from a layer of beads 200 μ in diameter. When a dispersion of 20–40 ppm oil in water was passed through the two beds, about 80 percent of the oil was screened out at the coarse–fine interface. This occurred even though the diameter of the oil globules was less than 2 μ, considerably smaller than the pore openings of the fine-grained layer. They suggested that electrostatic charges may have caused some coalescence of oil at the coarse–fine interface. This reduced the entry permeability and caused further filtering out of oil particles. Consequently, it appears that any size oil particles except those in true solution will be filtered out by a coarse–fine mineral interface that is water wet. As soon as an oil or gas phase is formed, even hydrocarbons in solution will be partially retained, because they would tend to partition into the hydrocarbon phase.

Capillary barriers at coarse–fine-grained sediment interfaces represent the most common type of barrier to petroleum migration. Other types of barriers include such things as secondary mineralization and asphalt accumulation.

Oil accumulations in carbonate reservoirs can be retained by the postaccumulation formation of stylolites. Dunnington (1967a) cites examples of Middle East reservoirs where the late formation of stylolites in the porous and permeable parts of the reservoir has sealed off any further petroleum migration.

Asphalt can act as a barrier to oil and gas movement. Levorsen (1967, p. 336) cites several examples of fields where surface outcrops of asphalt prevent the escape of downdip accumulations of oil and gas. The asphalt forms from the weathering and bacterial degradation of seepages. Most notable are the Bolivar coastal fields of Maracaibo, Venezuela, where asphalt and heavy degraded oil is updip and lighter oil downdip in the same reservoirs (Dickey and Hunt 1972). Subsurface deasphalting of oil or degradation by meteoric waters may cause asphalt barriers at oil–water contacts. Such asphalt barriers have been reported from Hawkins field (Texas), Prudhoe Bay (Alaska), Frannie (Wyoming), and Ghawar (Saudi Arabia). They may cause some tilted oil–water contacts.

The difficulty in pinpointing oil accumulations is to a large extent caused by the many, sometimes conflicting factors entering into secondary migration. Drilling on structures was initiated by the fact that buoyancy causes oil and gas to seek the highest permeable part of the reservoir. If cementation or pinchout of the reservoir rock has partially sealed the top of the structure, the oil may occur downdip. Capillary forces direct oil into the coarsest-grained sediments first and into successively finer-grained sediments later. Permeability barriers in the reservoir will channel the oil in a somewhat random distribution. Oil accumulations in carbonate rocks are often erratic, because part of the original void spaces may have undergone cementation after lithification. In large sand bodies, barriers formed by thin layers of fine silt or dense shale may hold the oil at various levels. When crustal movements of the earth occur, oil pools will be shifted away from the place in which they originally accumulated. Faults sometimes cut through reservoirs, either allowing part of the oil and gas to escape or creating a tighter barrier than previously existed. Vertical fracturing may allow the same source rocks to feed oil into a series of overlying reservoirs. Uplift and erosion can bring oil accumulations near the surface where the lighter hydrocarbons evaporate, leaving a low-°API-gravity oil in the reservoir. Wherever differential fluid potentials exist and permeable openings provide a path, oil and gas will move.

SUMMARY

1. Water exists as a polymer of H_2O and forms structures next to mineral surfaces because the hydrogen of the H_2O bonds with the oxygen of silicates and with other water molecules.

2. Permeable subsurface formations contain connate waters usually high in chloride. Total solids content of connate waters in many areas increases 50–300 mg/liter/m (15,000–100,000 ppm/1,000 ft) with depth. This increase has been attributed to shales acting as semipermeable membranes and to salt diffusion from evaporites. About half the giant oil and gas fields in the world contain waters with salinities less than 40,000 ppm.

3. Highly permeable formations with subaerial outcrops and hydrodynamic flow contain meteoric waters usually with less than 10,000 ppm total solids high in sodium, bicarbonate, and sometimes sulfate.

4. Shales compact at different rates depending on the number and thickness of interbedded highly permeable rocks. Thin shales interbedded with sands are in compaction equilibrium, and their pore waters are at hydrostatic pressure. Thick shales are often in compaction disequilibrium, and the pore water pressures range between hydrostatic and lithostatic. At 8,000 ft (2,439 m), shale porosities range from about 5–20 percent and at 16,000 ft (4,877 m) from about 2–14 percent because of the variation in compaction disequilibrium.

5. The conversion of smectite to illite (clay dewatering) is controlled by the temperature and chemistry of the system. Clay dewatering can increase the porosity of shales 5–10 percent but does not cause abnormal pressures. It provides additional water for primary migration, but it is not required for hydrocarbon accumulation.

6. Primary migration is the movement of oil and gas out of the fine-grained source rocks into the coarse-grained reservoir rocks. Four

transport mechanisms are believed to carry most of the hydrocarbons: diffusion, molecular solution, and oil-phase and gas-phase migration.

7. Fluids move in the direction of decreasing free energy. Factors affecting this are the potential energy, which decreases downward, and the pressure, which usually decreases upward in normally compacted sediments. The centers of thick, overpressured shales act as fluid pressure barriers directing fluid movements downward and upward away from the center.

8. Evaporites are the tightest sedimentary barriers to hydrocarbon migration in the subsurface. Continental facies such as red beds are among the most open. Red beds sometimes show evidence of hydrocarbons migrating through them.

9. Abnormal fluid overpressures are created and maintained by the inability of pore fluids to migrate within a reasonable geologic time period when subjected to stresses causing increased fluid pressure, such as rapid loading, thermal expansion of fluids, compression by tectonic forces, and generation of gas and oil from organic matter in the rock matrix.

10. Geothermal gradients are frequently higher (thermal conductivities lower) in overpressured shales than in nearby normally pressured formations, because the former have a higher porosity and lower silt and sand content.

11. Primary migration in carbonates with 10 percent or more clay minerals and in some chalks can occur by any of the four mechanisms listed in item 6, because these carbonates undergo compaction. Migration mechanisms in pure carbonates that lithify by recrystallization and cementation would be limited to diffusion and gas-phase migration.

12. The distance of primary migration in fine-grained sediments is usually short—only as far as needed to reach a permeable bed. Secondary migration in continuous sands, fractures, faults, unconformities, or permeable carbonates will proceed until a permeability barrier, a seal or cap rock, is encountered. This could be a few feet or meters, or hundreds of miles or kilometers. Migration of hundreds of miles is more common than previously believed.

13. Fissuring and fracturing associated with faulting induces secondary permeability and porosity in brittle rocks, which favors vertical fluid migration. Faults may act as conduits or seals at different times in their history, depending on the stress across the fault and on the nature of the fault plane and of the strata cut by the fault.

14. Major avenues of fluid migration during basin compaction are sandstones and unconformities. Other pathways include permeable fracture–fault systems, biohermal reefs, and piercement salt structures.

15. Buoyancy causes hydrocarbons to rise to the top of reservoir rock, and capillary plus electrostatic forces screen hydrocarbons from water at coarse–fine sediment interfaces, causing petroleum accumulation.

SUPPLEMENTARY READING

American Association of Petroleum Geologists (AAPG). 1974. *Abnormal subsurface pressure.* Reprint Series No. 11, selected papers reprinted from the *Bulletin of the American Association of Petroleum Geologists.* Tulsa, Okla.: AAPG. 205 p.

Collins, A. G. 1975. *Geochemistry of oilfield waters.* New York: Elsevier. 496 p.

Cordell, R. J. 1972. Depths of oil origin and primary migration: a review and critique, *Bull. Amer. Assoc. Petrol. Geol. 56* (10), 2029–2067.

Magara, K. In press. *Compaction and fluid migration.* New York: Elsevier.

Weeks, L. G. 1961. Origin, migration and occurrence of petroleum. In G. B. Moody (ed.), *Petroleum Exploration Handbook.* New York: McGraw-Hill, pp. 5–1 to 5–50.

Young, A., and J. E. Galley. 1965. *Fluids in subsurface environments.* Memoir 4. Tulsa, Okla.: American Association of Petroleum Geologists. 414 p.

III

HABITAT

7

The Source Rock

What is a petroleum source rock? This question has intrigued geologists since oil and gas were first discovered. Newberry (1860) described the oil production of the Berea sand near Mecca, Ohio, as being formed by the low-temperature heating of organic matter in the Hamilton bituminous shale. Later the "Ohio black shale" became established as the source of oil and gas in Ohio and Kentucky. The early geologists believed that Kentucky's best prospects for oil fields were where the Cumberland Sandstone was immediately overlain by the black shale.

Decades later, Snider (1934, p. 51) summarized the general opinion of petroleum geologists in stating, "There seems to be a very nearly universal agreement that these organic materials are buried principally in argillaceous mud and to a less extent in calcareous muds and marls and in sandy muds. Coarse sands and gravels and very pure calcareous deposits, are generally without any notable content of organic material. Consequently, shales and bituminous limestones consolidated from muds and marls are generally regarded as source rocks for petroleum and natural gas." Snider went on to point out that some shales and carbonates with porous and permeable sections within them can act as both source and reservoir.

In their search for oil, petroleum geologists continued to emphasize the location of a proper structure, the presence of suitable reservoir permeability and porosity, and some type of cap rock. When all of these conditions were present but no oil or gas was found, it was generally attributed to the lack of source rocks. This conclusion was based on intuitive reasoning rather than on any hard data on the characteristics of the source rock, because these characteristics were only vaguely understood.

A petroleum *source rock* may be defined as a fine-grained sediment that in its natural setting has generated and released enough hydrocarbons to form a commercial accumulation of oil or gas. Definitions that do not include migration and accumulation are too general, because in a sense practically all fine-grained sediments form some hydrocarbons. Techniques for recognizing source rocks are usually based on case histories of the type, quantity, and maturation level of organic matter in rocks associated with production. Also, crude oil–source rock oil correlations have helped establish that a particular rock has yielded oil to a reservoir.

In 1931, the American Petroleum Institute and the U.S. Geological Survey supported a monumental study by Parker D. Trask to determine diagnostic criteria for recognizing source beds of petroleum. Trask examined 32,000 well samples and 3,000 outcrop samples from all over the United States over the next ten years. Unfortunately, analytical techniques in those days were very crude. Trask determined such properties as carbon content; reduction number, which is the cc of 0.4 normal chromic acid reduced by 100 mg of sediments; nitrogen content; color; and volatility. The only property that showed a consistent difference with respect to the distance of the source bed from a reservoir was the ratio of nitrogen to the reduction number. It was low in sediments near oil zones and high in sediments far from oil zones (Trask and Patnode 1942). Trask did extract bitumens from source rocks with boiling carbon tetrachloride. The yields averaged 550 ppm. This bitumen undoubtedly contained liquid hydrocarbons comparable to crude oil, but unfortunately Trask did not have modern techniques of chromatographic separation and analysis that would have enabled him to recognize petroleum in the source rocks.

The first proof of liquid crude oil in source rocks with a composition comparable to reservoir oil was published by Hunt and Jamieson (1956). They pulverized rocks to a very small particle size and extracted them with a combination of an oil-miscible organic solvent and a water-miscible organic solvent—usually benzene and methanol. The extracts were separated by column chromatography and fractionated on a microstill in order to make a direct comparison with oils from reservoirs. This study showed that liquid hydrocarbons comparable to reservoir oils were indigenous to practically all shales and carbonate rocks. Also, since the volume of nonreservoir rock far exceeded that of the reservoirs, it was apparent that enormous quantities of oil were still locked up in the source beds. In an 800-square-mile area (2,072 square kilometers) of the Powder River Basin, it was estimated that the Frontier shales still contained 3 billion barrels of crude oil. The extraction techniques developed by Hunt and Jamieson in the early 1950s were later used by Smith (1954) to identify the presence of liquid hydrocarbons in Recent sediments.

This early work led to the issuance of a patent on the benzene–methanol–acetone technique for extracting hydrocarbons and identifying source rocks (Hunt and Meinert 1954). In the patent, good source rocks were identified as

Table 7-1
Source Rock Quality as Defined by Extractable Hydrocarbon Content

Hunt and Meinert 1954	Philippi 1957	Baker 1972	Hydrocarbon content, ppm
	Excellent		>5,000
	Very good	Less common, better	1,500–5,000
Good	Good		500–1,500
	Fair	Common, adequate	150–500
Fair	Poor		50–150
Poor	Very poor	Inadequate	0–50

containing 3–30, fair source rocks 1–3, and poor source rocks 0–1 barrels of hydrocarbon per acre-foot of rock.

Meanwhile, Philippi at Shell had been obtaining liquid hydrocarbons from source rocks by extracting with 2,3-dimethylbutane followed by diethylether. In 1957, he published analyses of shales from several areas and also defined source rock quality on the basis of hydrocarbon content. These data, along with a later evaluation by Baker (1972) based on his study of the Cherokee Formation, are shown in Table 7-1. These quality definitions are empirical, and their main difference is at the low and high end of the scale. Hunt and Meinert and Baker considered 50 to 150 ppm hydrocarbon as adequate in a source rock, whereas Philippi considered this level to be associated with marginally commercial accumulations.

These early studies defined source rocks on the basis of the quantity of hydrocarbons in them, whereas it is now recognized that many other factors, such as the stage of maturation, are more critical than the actual hydrocarbon content. The kinds of questions being answered today by sophisticated analyses of fine-grained sediments are as follows. Does the rock have the right kind of organic matter to generate oil or gas? How sensitive is the organic matter to thermal degradation? How long and at what temperature has the organic matter been heated to produce hydrocarbons? Is the rock capable of releasing hydrocarbons after they are formed? As answers to these questions accumulate in sedimentary basins throughout the world, the recognition of petroleum source rocks becomes less empirical and more based on fundamental chemical and physical parameters.

QUANTITY OF ORGANIC MATTER AND HYDROCARBONS

In petroleum basins, the quantity of organic matter converted to oil in the source bed ranges from a few percent to about 15 percent. The quantity of generated oil that eventually forms commercial accumulations (Hunt 1977a)

Table 7-2

Distribution of C_{15+} Hydrocarbons and Related Organic Matter in Sediments

Sediment	Residual organic matter	Extracted organic matter	
		Asphaltic NSO compounds	C_{15+} hydrocarbons
All sedimentary rocks	94	3	3
Ancient carbonates	74.5	13.8	11.7
Ancient shales	95.7	2.9	1.4
Recent sediments	95.5	4.2	0.3
Cariaco Trench (4 m)	94.6	5.2	0.2
Cariaco Trench (85 m)	98.1	1.7	0.2
Black Sea	93.6	6.1	0.3
Orinoco Delta	94.5	5.0	0.5
Mediterranean Sea	94.8	4.9	0.3
Ancient sediments			
Cherokee, Kansas	96	3	1
Wilcox, La.	92	6.3	1.7
Frontier, Wyo.	92	6	2
Winnepeg, Mont.	95.9	1.4	2.7
Duvernay, Alberta	89	7.3	3.7
Phosphoria, Wyo.	92	3.4	4.6
Woodford, Okla.	89.2	5.8	5
Green River, Utah	72.3	21.4	6.3
Traverse, Mich.	79.4	13.3	7.3
Zechstein, Denmark	73.4	17.4	9.2
Madison, Mont.	68.6	18.9	12.6

Note: Numbers represent weight percent of total organic matter in rock. The extractable fraction is that which is soluble in solvent mixtures such as 70 percent benzene, 15 percent methanol, and 15 percent acetone at their boiling point and is not lost on removal of solvent. Values normalized to 100.

also is in this range. The inefficiency of source beds in generating and releasing oil was recognized by Trask (1936) who estimated that the Santa Fe Springs oil field in California originated from about 4 percent of the organic matter in the sediments from which the oil was derived.

The range of hydrocarbon yield from organic matter is indicated in Table 7-2. In all sedimentary rocks, about 3 percent of the organic matter is converted to C_{15+} hydrocarbons with an equal amount converted to nitrogen, sulfur, and oxygen compounds. The generation of hydrocarbons is more efficient in carbonates, because their organic matter is almost entirely amorphous material derived from plankton. Shales have large amounts of woody and coaly material, which forms fewer hydrocarbons. However, carbonates contain less organic matter than shales. The mean organic carbon content of all shales is about 1 percent, and of all carbonates is about 0.2 percent.

The ratio of hydrocarbons to organic matter in Recent sediments is very low, because only the biogenic hydrocarbons are present. The quantity of both hydrocarbons and asphaltic compounds increases during the catagenetic stage, as shown by the data on ancient sediments. These numbers represent only one sample, and there is no information on the stage of maturation for each rock. Most samples were taken in the vicinity of oil accumulations, however, so in a general way they represent the oil-forming stage of catagenesis. If samples were taken down a continuous section in a petroliferous area, they would show low values until the threshold of intense generation was encountered, after which the C_{15+} hydrocarbon yield could represent 10 percent of the total organic matter.

The color of a rock is a rough, but not always reliable, indicator of its organic content. Figure 7-1 is a plot of the organic carbon vs. extractable hydrocarbon content of rocks from various sedimentary basins. Practically all of the fine-grained rocks fall in the shaded area, which shows an increase in organic carbon and hydrocarbon content with the darkening. Reservoir rocks differ from fine-grained sediments in having much higher hydrocarbon content relative to their organic carbon. Coals differ in having low hydrocarbon content relative to their organic carbon. Also, immature and metamorphosed sediments are low in hydrocarbons.

The organic matter of red beds generally has been destroyed by oxidation. They are plotted in the lower left corner of Figure 7-1. These shales represent environments that accumulated very little organic matter during deposition. For example, some samples of the Ireton and Duvernay shales of Alberta are identical in mineralogical content but differ in organic carbon content. The green Ireton has about 0.1 percent, whereas the black Duvernay has over 5 percent. Most source rocks are in the range of the gray shales, having from 1 to 2 percent organic carbon and several hundred ppm hydrocarbon. Black calcareous shales are very rich in organic matter and hydrocarbons. Some of them, such as the Nordegg Shale of Alberta, are partly oil wet. Limestones and dolomites that are dull brown or dark frequently have the hydrocarbon content of good source rocks. Pure white recrystallized carbonates generally have very few if any indigenous hydrocarbons.

The use of color as a hydrocarbon source indicator should always be backed up with some analytical data. Color is most useful in an area where previous analyses have been made on the formation being studied, so that there is some justification for hydrocarbon estimates. Samples from wildcat areas can be very misleading. Black colors may be caused by a high manganese oxide content. Some light brown carbonates are known to have relatively high indigenous hydrocarbon content.

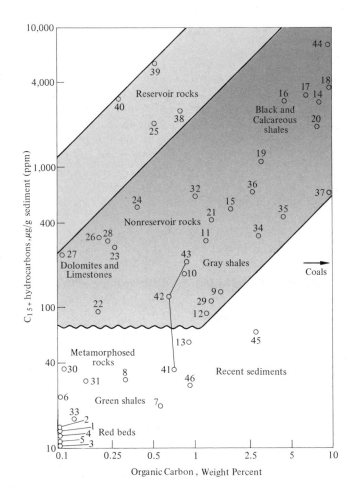

The rocks described in Figure 7-1 will maintain their organic matter content until metamorphosed, because the formation of oil uses up very little material. During metamorphism, the organic matter is converted largely to graphite. The low hydrocarbon content of samples 30 and 31 is typical of samples in the metamorphosed stage. The metamorphosed rocks, green shales, and red beds grouped in the lower left corner of Figure 7-1 have too low a carbon and hydrocarbon content to qualify as source rocks.

Unmetamorphosed Cambrian and Precambrian rocks (samples 29, 32, and 34–37) have carbon and hydrocarbon contents comparable to many known good source rocks. Old rocks with low geothermal gradients are worth investigating, because they could be associated with petroleum accumulations in a suitable depositional environment. Particularly interesting are the

Figure 7-1
Quantity of organic carbon and extractable C_{15+} hydrocarbons in sediments. Oil source rocks usually fall in shaded area above about 70-ppm hydrocarbons. Sediments that are acting more as reservoirs for migrated hydrocarbons than as source rocks usually contain 20–100 percent of the organic carbon in the form of extractable hydrocarbons. At the other extreme, metamorphosed rocks, immature sediments, and coals have very little extractable hydrocarbons relative to organic carbon. *Red beds:* (1) Chugwater, Colo.; (2) Triassic, Sask.; (3) Tertiary, Columbia; (4) Big Snowy, Mont. *Green shales:* (5) Pliocene, Venezuela; (6) Ireton, Alberta; (7) Tertiary, Columbia; (8) Cherokee, Kan., Okla. *Gray shales:* (9) Cherokee, Kan., Okla.; (10) Wilcox, La.; (11) Frontier, Wyo.; (12) Mowry, Wyo.; (13) Pierre, Colo. *Black shales:* (14) Cherokee, Kan., Okla.; (15) Monterey, Calif.; (16) Woodford, Okla. *Calcareous shales:* (17) Duvernay, Alberta; (18) Nordegg, Alberta; (19) Niobrara, Wyo.; (20) Paradox, Utah; (21) Winnepeg, Mont. *Limestones:* (22) Cherokee, Kan., Okla.; (23) Charles, Mont.; (24) Banff, N. Dakota; (25) Madison, Mont. *Dolomites:* (26) Zechstein, Denmark; (27) Madison, Mont.; (28) Phosphoria, Wyo. *Cambrian–Precambrian rocks:* (29) Beetle Creek, Australia; (30) Areyonga Formation, Australia; (31) Animikie Argillite, Minn.; (32) Nonesuch Shale, Mich.; (33) Thomson Slate, Minn.; (34–37) Kuonam Suite, USSR. *Reservoir rocks:* (38) Cherokee Sandstone, Kan., Okla.; (39) Cardium Sandstone, Alberta; (40) Cogollo Limestone, Venezuela. *Neogene formations in Japan:* (41) Nonproductive, Kanto and Miyazaki; (42) Partly productive, Hokkaido; (43) Most productive, Niigata. *Oil shale:* (44) Irati. *Recent sediments:* (45) Lake Maracaibo, Venezuela; (46) Mediterranean Sea. [Data from Baker 1962; Bikkenina and Shapiro 1969; Hunt 1961; Philippi 1957; Powell et al. 1975; Yagishita 1963]

rocks of the Kuonam Suite on the south bank of the Anabar anticline in the USSR (Bikkenina and Shapiro 1969). These samples, numbers 34 through 37, are calcareous and dolomitic clays, marly and silicified in places. It is possible that these samples were comparable to the calcareous shales shown above them in Figure 7-1, but over geologic time they have lost a considerable percentage of their hydrocarbons. They still are hydrocarbon rich compared to most of the samples in Figure 7-1, and they are far richer than the metamorphosed rocks in the lower left corner. Detailed analyses of these samples showed them to contain a complete suite of gasoline-range hydrocarbons in amounts comparable to good source rocks. In the USSR, there appear to be many more petroleum prospects of Cambrian age and older than have been observed in the United States.

Numbers 41, 42, and 43 (connected by a line in Figure 7-1) represent an interesting study by Yagishita (1963) on the organic carbon and hydrocarbon content of several hundred samples of nonreservoir rocks associated with productive and dry reservoirs. Number 41 represents the mean of 126 samples from the nonproductive South Kanto and Miyazaki formations. Number 42 is the mean of 153 samples from the Hokkaido Formation, which has some production and number 43 represents the mean of 415 samples of the most productive Miigata Formation. The organic carbon content of all these formations is roughly about the same but the fine-grained rocks in the most productive formation contain more than five times as much hydrocarbon as the rocks in the nonproductive formations. It is probable that 43 contains the more amorphous type of organic matter that readily gives up hydrocarbons, whereas 41 is either more coaly or more oxidized. The Irati oil shale (44) has dominantly amorphous material.

How much organic carbon is required for a good source rock? Ronov (1958) analyzed several hundred samples of Upper Devonian shale throughout the Russian Platform from Kiev in the west to Ufa in the east. The results are shown in Figure 7-2. Although the Russian Platform has structures and interbedded sands with good porosity and permeability throughout this area, the oil fields are all concentrated in the area near Kuibyshev and Ufa, where the organic carbon content ranges between 0.5 and 5 weight percent. In the Saratov and Kiev areas to the south, where organic carbon contents are around 0.5 percent, some gas is found. There is no Devonian oil or gas in the northern part of the Russian Platform extending from the oil fields in the east to the western boundary of the USSR; organic carbon values are generally less than 0.25 percent. Ronov found that the mean organic carbon content of

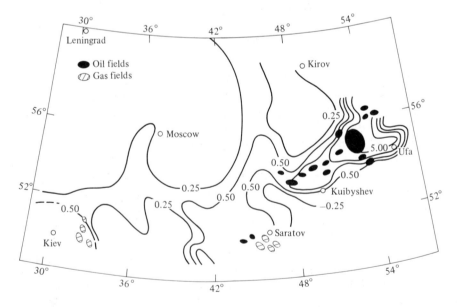

Figure 7-2
Organic carbon content of Upper Devonian sediments from the Russian Platform. The contours represent over 1,000 combined analyses of fine-grained rocks. [Ronov 1958]

fine-grained rocks in petroliferous areas was 1.37 percent for clays and 0.5 percent for carbonates. In nonpetroliferous areas, the means were 0.4 percent for clays and 0.16 percent for carbonates.

Ronov (1958) prepared lithogeochemical maps relating the organic carbon distribution to the depositional environment and its oxidation state. He found that the northwestern and western areas of the Russian Platform represented continental and lagoonal environments where the sediments were more oxidized than in the more marine areas to the east. Ronov determined the ratios of ferric to ferrous iron in the sediments, a measure of the state of oxidation or reduction. The higher this ratio, the more strongly oxidizing is the environment, and the lower the ratio, the more strongly reducing. Ronov found that the Fe_2O_3/FeO ratios in the west were 10 where the organic carbon values were less than 0.25 percent. Going east, there was a gradual decrease in the ratio from 10 to 5 toward the central part of the platform and from 5 to 1 to the east and southeast. In the Volga–Ural petroleum province in the east, the Fe_2O_3/FeO ratio fell below 1, defining this as the most highly reducing environment on the Russian platform. Within the petroliferous provinces in the east, the highest organic carbon contents were in clays deposited in coastal-marine sediment facies. Ronov (1958) concluded, from a statistical evaluation of all of his data, "These figures indicate an existence of a certain minimum of organic substance in the major sedimentary complexes below which the transformations of disseminated organic carbon cannot be conducive to the development of economic accumulations of petroleum. This critical level lies somewhere between the organic carbon averages for the petroliferous and the nonpetroliferous areas, that is, between 1.4 and 0.4 percent, and it is probably closer to the first one of these two figures."

The ratio of ferric to ferrous iron is one of the most sensitive indicators of the oxidizing–reducing conditions that have existed in sediments during deposition and early diagenesis. Iron is so widespread in sediments that its ferric–ferrous state is a dominant contributor to rock color, along with organic matter. MacCarthy (1926) published a table showing the relationship of color to the ferrous/ferric ratio in sedimentary rocks. This ratio (the reciprocal of Ronov's) gives red colors in shales up to about 0.2, yellow and buffs to 0.5, greens and blues to 5, and blacks to 12. Only 2 percent ferric iron will mask all other colors when organic matter is absent. Blue shales usually look gray when dry. Variegated colors occur in sediments because small changes in the ferrous/ferric ratio can produce several colors. Sediment samples from the bottom of the Red Sea show a variety of beautiful red, yellow, orange, and brown streaks over depth intervals of a few centimeters. Variegated shales also occur when red sediments are carried into a reducing environment. Some black or dark brown shales or limestones owe their color more to ferrous sulfides or other heavy metal sulfides or oxides rather than to organic matter.

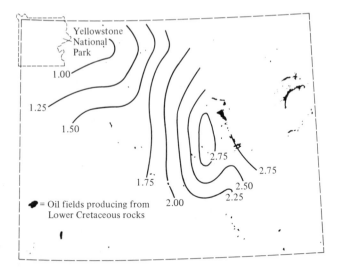

Figure 7-3
Relation of organic carbon content of Mowry Shale to Lower
Cretaceous oil production in Wyoming. [Schrayer and Zarella 1963]

Schrayer and Zarrella (1963) made a detailed study of the organic carbon content of the Mowry Shale in a 35,000-square-mile area (90,650 square kilometers) of Wyoming. The Mowry is considered to be the source of much of the Lower Cretaceous oil in Wyoming. Their data, which are plotted in Figure 7-3, show a general increase in the organic carbon concentration in a southeasterly direction, which is the seaward direction. Organic carbon contents were around 1 percent in the northwest landward direction and reached 3 percent at the southeast limit in the area of major oil production. Later studies (Schrayer and Zarella 1966) showed the total extractable bitumens to also increase from 400 ppm in the northwest to 2,100 ppm in the southeast. Davis (1970) noted that organic carbon is highest in the fine-grained siliceous shale facies of the Mowry.

This and similar studies have convinced many geochemists that 0.4 to 1 percent organic carbon is the minimum for a fine-grained shale to generate sufficient oil for a commercial accumulation. Hunt (1967) has pointed out that fine-grained carbonate rocks generate more hydrocarbon for the same amount of total organic matter, so as little as 0.3 percent organic carbon may be sufficient for some carbonate source rocks.

Some geochemists have contended that the minimum organic carbon content of source rocks should be in the 1–1.5 percent range, because most rocks contain much recycled organic carbon, which is carbonized and cannot produce petroleum. This whole discussion emphasizes that the quantity of organic carbon required depends on its quality. If the organic carbon is largely carbonized material, then it must be subtracted from the organic carbon formed in situ (autochthonous). Shales nearly always have some recycled organic carbon from the continents, along with woody and coaly materials that yield very little petroleum. Carbonates, however, frequently contain only amorphous organic material derived from algae, which has the highest yield of petroleum among the kerogen types. This explains why carbonates often yield more oil than shales for the same organic carbon content. A good example is the Cherokee Formation, studied by Baker (1962). The mean hydrocarbon content for limestones and gray shales in this study was around 100 ppm, even though the limestones contained a mean organic carbon content of 0.19 compared to 1.5 percent for the shales. (See numbers 9 and 22 in Figure 7-1.)

The quantity of nonreservoir oil exceeds the reservoir oil in sedimentary basins of the world by a considerable amount. Table 7-3 shows the estimated volume of oil in reservoir and nonreservoir rocks of the Cretaceous formations of a 15,000-square-mile area (62,550 square kilometers), 1 mile (1.6 km) thick, in the Powder River Basin in Wyoming. The ratio of oil in shales to oil in sands is estimated to be 22 to 1. This ratio varies considerably within formations because of differences in the effectiveness of draining hydrocarbons from the shales. The Niobrara Formation has an average thickness of 450 feet but contains no reservoir sands. The limited amount of production is from fractured shale. The Frontier–Carlisle and older formations contain interbedded sands, resulting in more effective drainage.

Similar calculations show ratios ranging from about 5/1 to 100/1 in sedimentary basins. Coneybeare (1965) calculated a ratio of 20/1 for Jurassic formations in the Surat Basin in Australia.

Table 7-3
Hydrocarbons in Cretaceous Formations, Powder River Basin, Wyoming (15,000 miles^3 or 100,000 km^3)

Formation	In shale (10^6 bbl)	In sand (10^6 bbl)
Steele–Pierre	70,000	500
Niobrara	85,000	3
Frontier–Carlile	25,000	2,500
Mowry–Muddy–Newcastle	6,000	3,000
Thermopolis–Skull Creek–Inyan Kara–Fall River	4,000	2,500
	190,000	∿8,500
	190/8.5 = 22/1	

Source: Hunt 1961.

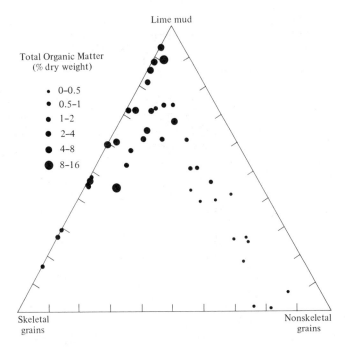

Figure 7-4
Variation in organic matter of different carbonate particle sizes. [Gehman, 1962]

The quantity of organic matter in sediments is closely related to particle size. A sample of Viking Shale from Alberta, Canada, was disaggregated, dispersed in water, separated by centrifuging, and analyzed for organic matter (Hunt 1963b). Organic content of the siltstone size was 1.79 percent; clay 2–4 μ, 2.08 percent; and clay less than 2 μ, 6.50 percent. A similar study by Gehman (1962) on carbonates showed the highest concentration of organic matter in the finest-grained particles (Figure 7-4). The largest circles, representing the highest organic contents, are concentrated in the area of lime mud. The increase in organic content with a decrease in sediment grain size is established at the time of deposition, as discussed in Chapter 4.

The organic content and the soluble (extractable) bitumen of carbonate rocks increases with an increase in the HCl-insoluble mineral residue (Uspenskii and Chernysheva 1951). The bitumen (Table 7-4) includes hydrocarbons plus asphaltic NSO compounds. Since the HCl-insoluble residue of carbonates is frequently clay, this means that the organic matter and bitumen tends to increase with the clay content of carbonates. Hunt (1967) reported that a Florida Bay carbonate mud containing 15 percent clay minerals was found to have 75 percent of its organic matter attached to the clay minerals and only 25 percent to the carbonate minerals.

Table 7-4
Relationship Between Organic Matter and
Insoluble Residue of Carbonate Rocks

Insoluble residue	Organic matter	Bitumen
4.3	0.06	0.015
10.2	0.15	0.021
15.5	0.28	0.034
24.5	0.49	0.034
57.9	0.70	0.046
66.1	0.93	0.051
72.8	2.36	0.052

Note: All values in weight percent.
Source: After Uspenskii and Chernysheva 1951.

TYPE OF ORGANIC MATTER

The term *kerogen* originally referred to the organic matter in oil shales that yielded oil on heating. In recent years, it has been defined as all the disseminated organic matter of sedimentary rocks insoluble in nonoxidizing acids, bases, and organic solvents. Kerogen, which is the precursor of most oil and gas, has three sources: marine, terrestrial, and recycled. The terrestrial kerogen has components similar to coal.

Kerogen and Coal

The classification of the organic matter of sedimentary rocks has been made by (1) palynologists, from the microscopic examination of mineral-free organic residues in transmitted light; (2) coal petrographers, from the microscopic examination of polished rock surfaces in reflected light; and (3) petroleum geochemists, from the changes in the elemental composition of kerogen with maturation. The interrelationships of these three classification systems, which appear in the petroleum literature, are discussed in this section.

Practically all organic matter may be classified into two major types, *sapropelic* and *humic* (Potonie 1908; see Figure 7-5). The term *sapropelic* refers to decomposition and polymerization products of fatty, lipid organic materials such as spores and planktonic algae deposited in subaquatic muds (marine or lacustrine), usually under oxygen-restricted conditions. Sapropelic organic matter such as fats, oils, resins, and waxes have hydrogen/carbon (H/C) ratios in the range from 1.3 to 1.7. Organic-rich sapropelic deposits undergo maturation to form boghead coals and oil shales. A modern

	SAPROPELIC			HUMIC	
KEROGEN (by transmitted light)	Algal	Amorphous	Herbaceous	Woody	Coaly (Inertinite)
COAL MACERALS (by reflected light)		Liptinite (Exinite)		Vitrinite	Inertinite
	Alginite	Amorphous	Sporinite Cutinite Resinite	Telinite Collinite	Fusinite Micrinite Sclerotinite
KEROGEN (by evolutionary pathway)	Types I, II		Type II	Type III	Type III
H/C	1.7–0.3		1.4–0.3	1.0–0.3	0.45–0.3
O/C	0.1–0.02		0.2–0.02	0.4–0.02	0.3–0.02
ORGANIC SOURCE	Marine and lacustrine		Terrestrial	Terrestrial	Terrestrial and recycled
FOSSIL FUELS	Predominately oil Oil shales, boghead and cannel coals		Oil and gas	Predominately gas Humic coals	No oil, trace of gas

Figure 7-5
Classification of organic matter in sedimentary rocks.

freshwater fat-secreting planktonic green alga, *Botryococcus braunii,* is the source of Australia's coorongite, a boghead peat. The Carboniferous equivalent of *Botryococcus b., Pila,* is concentrated in Scottish oil shales (torbanites). The Permian oil shale "tasmanite" in Tasmania is formed from a single-celled green alga. Some boghead coals of the Siberian Jurassic contain only amorphous liptinite (Stach et al. 1975, p. 236).

The term *humic* refers to products of peat formation, mainly land plant material deposited in swamps in the presence of oxygen. Peat has an H/C ratio around 0.9, too low to be an important progenitor of oil. Humic organic matter is derived from plant cell and wall material, which is composed mainly of lignin and cellulose plus the aromatic tannins, which have a high resistance to rotting. The humic category also includes carbonized (fusinitized) organic matter, such as charcoal from fires and other oxidized plant remains. Humification is accelerated by the presence of oxygen and heat (tropical climates).

When the roots, bark, and wood of trees are deposited in forest and reed swamps and swamp lakes, they undergo bacterial and chemical changes that lead to peat formation. As the peat is buried deeper, it changes with time and temperature (coalification) to brown coal, bituminous coal, and finally anthracite. These are the major humic coals. During coalification, the hydrogen/carbon and oxygen/carbon ratios of the humic material changes as

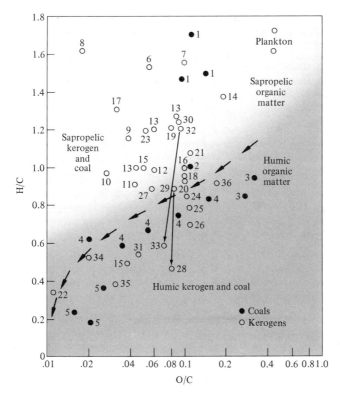

Figure 7-6
Atomic ratios of kerogen and coal. *Coals:* (1) Boghead, (2) Cannel,
(3) Brown, (4) Bituminous, (5) Anthracite. *Kerogen:* (6) Green
River Mahogany Ledge, Utah; (7) Green River Shale, Colo.; (8)
Green River 260 m below Mahogany Ledge, Utah; (9) Duvernay,
Alberta, Canada; (10) La Luna, Colombia; (11); Woodford, Okla;
(12) Madison, Mont.; (13) Toarcian, Paris Basin; (14) Messel,
Germany; (15) Triassic, Spitzberg; (16) Silurian, Sahara; (17) Apon,
Venezuela; (18) Frontier, Wyo.; (19) Monterey, Calif.; (20)
St. Genevieve Limestone, Kentucky; (21) St. Genevieve. Stylolite;
(22) Collier Limestone, Ark.; (23) Cretaceous, Mississippi, offshore;
(24) Cretaceous, Mississippi, bay; (25) Cretaceous, Mississippi, marsh;
(26) Cretaceous, Mississippi, channel fill; (27) Viking, Alberta;
(28) Pierre, Colo., 30 cm from dike; (29) Pierre, Colo., 600 cm from
dike; (30) Bakken, Mont.; (31) Atoka, Okla.; (32) Devonian, Alberta,
1,440 m; (33) Devonian, Alberta, 2,027 m; (34) Precambrian
Dolomite, S. Africa; (35) Phyllite, Texas; (36) Maikop fish bone,
Caucasus. [Data from Combaz 1973; McIver 1967; Forsman and
Hunt 1958; Drozdova and Kochenov 1960]

shown in Figure 7-6. The black circles in the shaded area of this figure
represent the humic coals.

If these same particles of roots, bark, and wood are carried out to sea and
deposited in a sedimentary basin as the 1 or 2 percent organic content of a clay

mud, they contribute to the humic part of the kerogen. Likewise, any spore, pollen, or algal material carried by wind or water into the marine environment contributes to the sapropelic part of kerogen. Organic matter from marine organisms is added to the disseminated land-derived humic and sapropelic material to form the kerogen of marine sediments.

Coal scientists classify coal components based on their appearance under microscopic examination in incident light, using oil immersion objectives with 25 to $50\times$ magnification. Coals are composed of macerals comparable to the mineral components of rocks except that macerals are not crystalline and vary more widely in chemical composition than minerals (for a complete discussion of coal petrology, see Stach et al. 1975). The three major maceral groups in coals are *liptinite* (also called *exinite*), *vitrinite*, and *inertinite* (Figure 7-5). Humic coals usually contain over 60 percent vitrinite, while boghead coals often have over 60 percent liptinite, a high proportion of which is the maceral alginite, from algae. These same coal macerals are recognized in the disseminated form as part of the kerogen of sedimentary rocks, and some petroleum geochemists use these terms to describe kerogen (Kendrick et al. 1978a).

Meanwhile, palynologists in the petroleum industry developed a nomenclature for the kerogen remaining after dissolving most of a rock's mineral matter with hydrochloric and hydrofluoric (HF) acids. They classified the kerogen as *algal, amorphous, herbaceous, woody,* and *coaly* (inertinite). Amorphous material is largely sapropelic organic matter from plankton and other low forms of life. The organic matter of the Green River shales of Colorado and Wyoming and the Nordegg and Bakken shales of the Western Canada Basin contain mostly amorphous kerogen. Amorphous kerogen can be of marine or lacustrine origin, and it is the dominant kerogen in carbonates. Recognizable algal structures of marine or nonmarine origin are classified as algal kerogen. Herbaceous kerogen consist of pollen grains, spores, cuticles, leaf epidermis, cellular structures of plants, and other recognizable discrete cell material. Herbaceous material is mostly of continental origin. Woody kerogen is fibril material with recognizable rectangular, woody structures. Coaly (inertinite) kerogen is recycled material plus plant material that has undergone natural carbonization such as charring, oxidation, moldering, and fungal attack. Charcoal from forest fires is in this category.

Visual kerogen analyses of drill cuttings from wells frequently use the designations Am, H, W, and C to indicate the presence of amorphous, herbaceous, woody and coaly (inertinite) kerogen, respectively (Cernock and Bayliss 1977).

The third kerogen classification system is by Tissot, Durand et al. (1974). It characterizes kerogen as Types I, II, and III, depending on the elemental composition of the kerogen and its evolution path on a Van Krevelen diagram

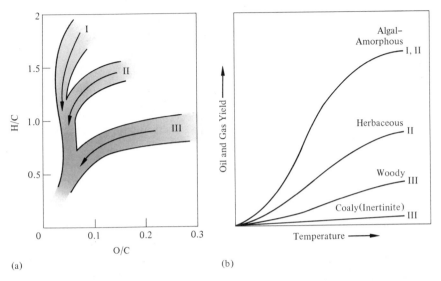

Figure 7-7
Effect of increasing temperature on kerogen type. (a) Change in hydrogen/carbon (H/C) and oxygen/carbon (O/C) ratios. (b) Relative yields of oil and gas.

that plots H/C versus O/C. Figure 7-7a shows the three kerogen types on such a plot, and Figure 7-5 shows the approximate range in the H/C and O/C ratios for these three types. The basic chemical differences are that Type I is mostly normal and branched paraffins, with some naphthenes and aromatics. Type II is predominantly naphthenes and aromatics, whereas Type III contains a high percentage of polycyclic aromatic hydrocarbons and oxygenated functional groups plus some paraffin waxes. The kerogen of oil shales and boghead coals is typical of Type I. The kerogen of the Lower Toarcian shales of the Paris Basin is Type II, and that of the Lower Mannville shales of the Western Canada Basin and Upper Cretaceous shales of the Douala Basin are Type III.

The approximate relationships between these different classification systems are shown in Figure 7-5. The three systems are not completely analogous. Some geochemists use the term *amorphous* for unstructured sapropelic kerogen, while *liptinite* is limited to the structured algal and herbaceous kerogen. The Types I, II, and III are really mixtures, with Type I generally comparable to the algal-amorphous categories, Type II to amorphous-herbaceous-woody, and Type III to woody-inertinite. The systems are similar in showing a division between the humic and sapropelic components of kerogen and coal. This is important, because most oil and gas are generated from sapropelic organic matter, whereas humic materials yield primarily gas. Coal is not normally a source of oil since the vast majority of coal beds in the world are humic. When oil is found coming from humic coal beds, it is usually related to the liptinitic fraction of coal such as waxes and resins.

Sapropelic and humic organic matter in sedimentary rocks may be either marine or terrestrial in origin, but the bulk of the amorphous (unstructured) sapropelic material in carbonates and oil-rich marine shales is believed to be marine or lacustrine. Also, although there is marine humus, it is not differentiated from the vitrinite, inertinite, or woody classifications of Figure 7-5. Most of the humic fraction of kerogen is considered to be terrestrial in origin.

As oil and gas are generated, the hydrogen required to form them comes from the kerogen, so its H/C ratio decreases to a minimum of about 0.45. This corresponds to about nine condensed aromatic rings per plane (see p. 317). Beyond this, no oil and only small amounts of gas are generated as H/C values drop to about 0.3. Algal and amorphous (Type I) kerogen can generate the most oil and gas, because the H/C ratio will drop by 1.25 (from 1.7 to 0.45 in Figure 7-5). Woody (vitrinite) Type III kerogen generates the least oil and gas, because the H/C ratio only drops by 0.55. Kerogen that is oxidized during recycling and inertinite are incapable of forming oil but may yield very small amounts of gas.

A schematic illustration of the relative hydrocarbon yields of the major kerogen types is shown in Figure 7-7b. All of these types yield both oil and gas on heating with a shift from predominately oil in algal-amorphous-liptinite (Type I) to predominately gas in woody-vitrinite-inertinite (Type III). Yields are highest from the algal-amorphous material and lowest from the woody-inertinite material. Unfortunately, the quantity of the latter outweighs the former in sedimentary basins of the world.

Boghead coals and oil shales differ in that the latter usually contains more amorphous liptinite and is also highly diluted with rock minerals. Coal is a readily combustible rock containing more than 50 percent by weight and more than 70 percent by volume of organic matter (Schopf 1956), whereas an oil shale can have much less than this amount. In terms of carbon and hydrogen, the kerogen of oil shales such as the Green River of Utah and Colorado (samples 6, 7, and 8 in Figure 7-6) is similar to the boghead coals (labeled 1 in Figure 7-6). The coals are higher in oxygen.

In most sedimentary rocks, there are few pure sapropelic or humic kerogens; instead, there is a gradation caused by variable contributions of continental- and marine-derived organic matter. The division in Figure 7-6 is not sharp but very gradual. The humic, sapropelic, and several intermediate kerogens were identified in a detailed study of Paleozoic rocks of the middle Volga region by Rodionova and Chetverikova (1962). They found that the humic kerogen is formed in nearshore marine clay–silt–sandstone facies where there is free access to oxygen. The sapropelic type formed in deep water, low-energy clay, and clay–carbonate sediments with little or no oxygen. The sapropelic kerogen was dominant in marls and limestones.

Both sapropelic and humic organic matter is altered by diagenetic and thermal maturation processes comparable to coalification. The H/C–O/C diagram (Figure 7-6) for kerogen shows that both types lose oxygen and hydrogen moving in the direction of the arrows with maturation. Consequently, there are two factors that determine the composition of kerogen: source material (including microbial changes during deposition),* and thermal alteration.

Examples of the source and thermal effects are shown in Figure 7-6. For source effect, compare samples 24, 25, and 26, representing Cretaceous bay, marsh, and channel fill, with sample 23 representing Cretaceous offshore (McIver 1967). The former are humic, the latter sapropelic. The former have a much lower H/C ratio and a somewhat higher O/C ratio than the latter. For thermal effects, compare samples at different depths in the Western Canada Basin. Sample 32 is a Devonian shale at 1,440 m (4,724 ft). Sample 33 is the same formation at 2,027 m (6,650 ft). The latter has lost part of its hydrogen by maturation. Samples 29 and 28 show the drop in H/C ratio caused by the high-temperature baking effect of an igneous dike on the Pierre Shale of Colorado. This shale contains a low hydrogen kerogen largely of continental origin, and the thermal alteration changed it to a fusinitic (charcoal-like) material.

Kerogens capable of generating oil have higher H/C ratios (Duvernay, Sample 9) than kerogens capable of generating only gas (Atoka, Sample 31). Samples 22, 34, and 35, which represent older, more indurated rocks, could have originated either from humic kerogen or from sapropelic kerogen and subsequently lost much of their hydrogen as gas or oil.

Limestone cores frequently show stylolitic partings containing black organic matter. Samples 20 and 21 represent kerogen from the limestone matrix and from the partings. The latter shows a slightly higher H/C ratio but otherwise is similar.

Composition of Kerogen

The kerogens isolated from inorganic minerals by dissolving the latter with HCl and HF acids are fine, brown to jet-black powders resembling coal dust. Degradation studies on these kerogens in recent years have revealed that they contain practically the entire suite of organic structures identified in living organisms. Terpenoid and steroid structures have been found, along with porphyrins, amino acids, sugars, carboxylic acids, ketones, alcohols, olefins,

---

*Lijmbach (1975) has found that the more that plankton are reworked by bacteria during sedimentation the higher is the conversion of the resulting organic matter to oil on heating and the lower the temperature of conversion. Apparently, bacterial reworking of algal and herbaceous substances results in an amorphous biomass high in bacterial bodies that more readily forms oil.

and ether bridges. The most detailed composition studies have been carried out on the kerogen of the Green River oil shale of Colorado (Anders and Robinson 1971, 1973; Gallegos 1973, 1975). This kerogen is predominantly cyclic, with 45–60 percent heterocyclics, 20–25 percent naphthene hydrocarbons, 10–15 percent aromatics, and 5–10 percent normal and isoparaffins. Normal paraffins range in size up to C_{38}. Cycloparaffins contain up to six fused rings with the two- and three-ring structures dominant. Many steranes and terpanes are present. Monoaromatic pentacyclic rings, tetralin, and dihydronaphthalene nuclei were also identified. About half of the monoaromatics are fused to saturated cyclopentyl or cyclohexyl rings.

The different kerogen and coal types in Figure 7-5 have different elemental compositions. Coal chemists refer to liptinite as high in hydrogen, vitrinite as high in oxygen, and inertinite as high in carbon. Likewise, amorphous kerogen is high in hydrogen, and woody is high in oxygen. Composition data for examples of the visual kerogen types of Figure 7-5 are shown in Table 7-5 normalized on a nitrogen and sulfur free basis for comparison. The hydrogen/carbon ratio decreases in the following order: amorphous > herbaceous > woody > inertinite. The potential to form petroleum decreases in the same order.

Table 7-5
Characteristics of Kerogen Types

	Marine or lacustrine origin	Land origin		
	Amorphous	Herbaceous	Woody	Inertinite
Composition Percent				
Carbon	83	82	83	89
Hydrogen	11	8	5	3.5
Oxygen	6	10	12	7.5
H/C ratio	1.6	1.2	0.72	0.47
Biological Markers				
n-Alkanes, odd–even predominance				
$C_{15}-C_{21}$	Odd		None	
$C_{27}-C_{35}$	Even or none		Odd	
Pristane/phytane	Low (<1)		High (>3)	
Pristane/n-C_{17}	Low		High	
Terpenoids	Low		High	
Fatty acids $C_{12}-C_{18}$	High		Low	
$C_{24}-C_{36}$	Low		High	
Hydrocarbon Yields				
Methane	High		Intermediate	
C_2-C_{14}	High		Low	
$C_{15}+$	High		Low	

These kerogen types have subtle chemical differences that also distinguish their origin. The normal alkanes of marine origin show a predominance of odd-carbon chain lengths in the C_{15}–C_{21} range and a slight even-carbon or no predominance in the C_{27}–C_{35} range. In contrast, normal alkanes of continental origin show a strong predominance of odd-carbon chain lengths in the C_{27}–C_{35} range and have essentially no predominance in the C_{15}–C_{21} range. The reason for these differences, as discussed in Chapter 4, are that all plants synthesize predominantly odd-carbon chain lengths, but the land plants synthesize long-chain waxes (C_{27}–C_{35}) to preserve their water content, whereas the marine plants that live in water synthesize short-chain hydrocarbons (C_{15}–C_{21}). This results in the n-alkanes of clays and coals that are mainly land derived having a strong odd predominance in the C_{27}–C_{35} range, whereas the n-alkanes of marine limestones and evaporites show an even or no predominance (Brooks 1970; Powell and McKirdy 1973; Tissot, Pelet, et al. 1975; Dembicki et al. 1975). Since the C_{27}–C_{35} n-alkane odd predominance occurs in higher plants on the continents, it is not present in the lower Paleozoic, as there were no higher plants at that time. Thus, Tissot, Pelet, et al. (1975), in their comparison of the n-alkanes in 842 rock samples, observed little or no odd or even preference in the C_{27}–C_{35} range of Lower Paleozoic samples. The paraffins found were attributed to bacterial lipids. These n-alkanes were frequently associated with iso- and anteisoalkanes. In contrast, a strong odd predominance in C_{27}–C_{31} n-alkanes occurred in the Mesozoic, mostly from Lower Cretaceous samples related to the expansion of angiosperms.

The ratio of pristane to phytane and pristane to n-C_{17} tends to be high in peat swamp environments and low in aquatic environments (Lijmbach 1975). A high pristane/phytane ratio is favored by an oxidizing environment and a low ratio by a reducing environment (Powell and McKirdy 1973).

Certain terpenes, notably the diterpenoidal acids, have long been recognized as major components of tree resins, and they exist in rocks as continental marker compounds. Fatty acids of even predominance in the C_{12}–C_{18} range are predominantly of marine origin, and those in the C_{24}–C_{36} range are mainly of land origin (Simoneit 1975). There are other organic structures that may be more concentrated in one type of kerogen, such as chlorins, steroids, tri- and tetraterpenoids, isoprenoids and polycyclic naphthenes and aromatics. Philippi (1974b) observed that source beds with abundant continental organic matter such as the Eocene Wilcox shales of the Texas Gulf Coast are rich in five-ring naphthenes derived from pentacyclic triterpenes. The Miocene shales of the Los Angeles Basin, which have a kerogen derived from a more marine source, contained mainly four-ring naphthenes derived from steroids. Vandenbroucke et al. (1976) compared the maturation of the land-derived kerogen of the Douala Basin with that of the marine-derived kerogen of the Paris Basin and found the latter to yield more naphthenoaromatics and four-ring naphthenes than the former. The Douala

Basin kerogen yielded mainly paraffin waxes, some di- and tricyclic aromatics, and fewer naphthenes than the Paris Basin kerogen.

In general, paraffin-base waxy crude oils are from a more land-derived organic matter, while asphaltic base crudes are formed from a more marine-type kerogen. A review of the probable environments of origin of some 500 high-wax oils worldwide (Hedberg 1968) shows their occurrence to be genetically related to continental or nearshore depositional facies and a shale–sandstone lithology, with, commonly, an abundance of carbonaceous matter and coal. The high wax content is an inheritance from land-derived organic matter. High-wax oils are not usually associated with far offshore carbonate source sequences, because these contain amorphous kerogen with relatively small amounts of waxes transported from land.

When amorphous kerogen is heated, it yields about twice as much liquid hydrocarbons as the herbaceous kerogen, which in turn yields more liquids than the woody and coaly kerogens (Figure 7-7b). The high oil yield from retorting oil shales is caused by the high percentage of amorphous kerogen most of them have. The hydrocarbons thermally produced from amorphous kerogen are largely naphthenes.

Amorphous kerogen may be marine or lacustrine in origin. An example of the latter is the kerogen of the Green River Formation of the Uinta Basin in Utah. The amorphous kerogen deposited by the Green River and overlying Uinta formations contains primarily naphthenic hydrocarbons plus large amounts of nitrogen and sulfur compounds that have migrated short distances out of the source beds to form asphalt deposits in veins, fissures, and bituminous sands. Some herbaceous and woody kerogen is found in the shore deposits of the Green River Formation but the largest concentration of these clastic types is associated with the basal Green River and Wasatch formations. These had an alluvial shoreline lacustrine facies, compared to the open lacustrine facies of the later Green River Formation. The nearshore deposits of the Wasatch accumulated large quantities of herbaceous and woody organic matter, which resulted in a kerogen that yields primarily waxy hydrocarbons in the C_{15+} range.

This major difference in the type of kerogen and hydrocarbon formed from the Green River and Wasatch formations enabled Hunt et al. (1954) to make one of the earliest chemical identifications of a source rock (see Table 8-7). A liquid gilsonite high in naphthenes and aromatics was clearly associated with the naphthenoaromatic extracts of the Green River, while the waxy Duchesne oil was identified with waxy extracts of Wasatch shales. Although there was a difference in maturation—the Green River kerogen being immature and the Wasatch mature—it was not great enough to mask the large difference in the kerogen types. The relative contributions of amorphous algal-derived kerogen to the Green River and herbaceous plus woody kerogen to the Wasatch was evident in the fine-grained rocks irrespective of the state of maturation.

The different kerogen types thus become important in evaluating both the quantity and the quality of petroleum that will be formed during catagenesis. When the kerogen is primarily the woody–inert type with a low hydrogen content, it forms mainly gas. Dow and Pearson (1975) compared the organic content and type with the depositional environment for over 300 samples from 20 wells and shallow core holes in the Louisiana Gulf Coast. They concluded that the neritic sediments not only were low in kerogen but also contained the woody–inert kerogen more capable of forming gas than oil. The underlying bathyal sediments (Figure 7-8) had adequate amounts of

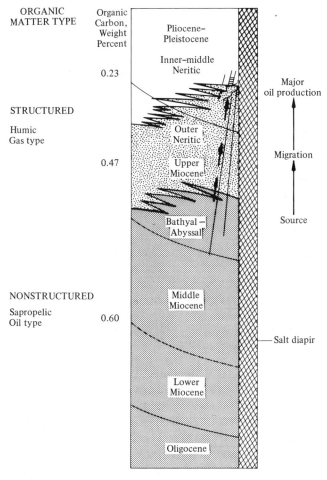

Figure 7-8
Detailed slice of cross section seaward of Southwest Pass, Louisiana.
Diapir-related fault–fracture systems permit vertical migration of
hydrocarbons from bathyal source sediments to neritic reservoirs.
[Dow and Pearson 1975]

organic matter, and it was of the amorphous sapropelic type capable of generating oil. They concluded that the hydrocarbons generated in the bathyal sediments migrated vertically into neritic reservoirs through fault-associated fracture systems, most commonly associated with piercement structures.

MATURATION OF ORGANIC MATTER

The quantity of oil generated by a source rock depends on the ease of degradation of the kerogen and the time–temperature relationship as discussed in Chapter 4. The hydrocarbon yields in Figure 7-7b represent the relative quantities of hydrocarbons produced when each of the four kerogen types are heated in the laboratory. In nature, kerogen is usually a mixture, and the yields depend on the dominant types within the mixture. Microscopic examination of the mineral-free kerogen under transmitted light is useful in determining the importance of that sample in generating oil and gas. When amorphous kerogen dominates, there will be a relatively high yield of naphthenic oil. When woody kerogen dominates, there will be a relatively low yield of waxy oil. Both types will yield increasing amounts of condensate and gas with increasing maturation.

The small amount of soluble bitumen in coal apparently is formed in the high-volatile and middle-volatile bituminous stage of coal maturation (Figure 7-9). The upper curve in this figure represents the ratio of extractable C_{15+} hydrocarbons to organic carbon in several humic coals from the United States. The lower curve represents the weight percent soluble bitumens of coals from the Upper Carboniferous of the Saar district of Germany on an ash-free basis. The Saar coal bitumens are about one-fourth saturated hydrocarbons and three-fourths aromatic and asphaltic compounds. Although oil is generated during the maturation of coal, the amounts are very small compared to the quantity of oil formed during the maturation of amorphous and herbaceous kerogens. Most of the coal bitumens come from the liptinite fraction of coal.

Larskaya and Zhabrev (1964) examined 2,500 rock samples of Mesozoic–Cenozoic age in the western Ciscaspian region for soluble bitumens and compared the yields with formation temperature. Figure 7-10 is a plot of temperature versus bitumen coefficient; that is, the ratio of chloroform-soluble carbon (C_b) to total organic carbon (C_o). The kerogen type also was examined microscopically. Curve A shows the increase in bitumen for the source rocks associated with oil accumulations, presumably those containing amorphous and herbaceous kerogen. Curve B is the increase in bitumen for rocks not containing oil accumulations, but some gas, and Curve C is the increase for rocks with inert organic matter. Microscopic examination of the

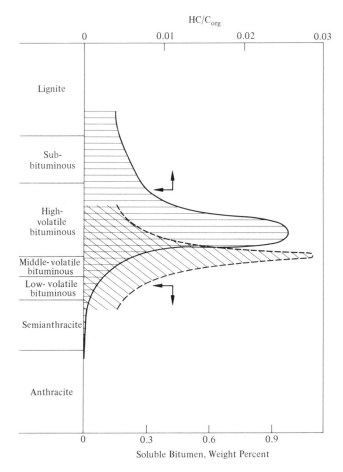

HC/C$_{org}$

Figure 7-9
Curve showing increase in bitumen extracted from high-volatile and
middle-volatile bituminous coals. Upper curve, ratio of hydrocarbon
to carbon in U.S. coals [Hood and Castano 1974]; lower curve,
soluble bitumen in Saar coals. [Leythaeuser and Welte 1969].

dispersed organic matter in samples of Curve C showed they were coaly
particles of coarse pelitic or fine silt size. Samples on Curve A were described
as amorphous brown matter. The authors recognized that wherever the rocks
contained predominantly coaly particles the bitumen coefficient was very
small and changed very little with increasing temperature, whereas the
amorphous matter had a high bitumen coefficient that increased rapidly with
temperature. Their studies clearly showed that bitumen yields of fine-grained
rocks in sedimentary basins are related to kerogen type.

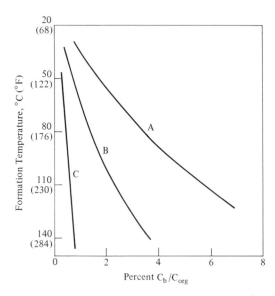

Figure 7-10
Temperature vs. bitumen
coefficient for 2,500 Mesozoic–
Cenozoic fine-grained rocks from
the western Ciscaspian region.
C_b = carbon of soluble bitumen,
C_{org} = organic carbon in rock.
A = mean for oil-bearing rocks,
B = non-oil-bearing, C = coaly
organic matter. [Larskaya and
Zhabrev 1964]

Time–Temperature Relationships

The majority of source rocks are mature; that is, they have entered the
principal phase of oil generation as discussed in Chapter 4. Immature
sediments are not source rocks of commercial petroleum accumulations
except for biogenic gas and possibly a few heavy immature oils. In a depth-
versus-hydrocarbon yield profile of fine-grained rocks, as shown in Figures
4-16, 4-20, and 4-21, the source rocks are located below the threshold of
intense generation within the zone of relatively high hydrocarbon yield.
Because this concept is important in source rock recognition, some additional
examples are shown in Figure 7-11. Figure 7-11a shows the increase in gasoline hydrocarbons with depth of Jurassic–Cretaceous shales from western
Siberia. Figure 7-11b shows the yield of bitumens for Jurassic shales of
eastern Ciscaucasia, and Figure 7-11c shows the increase for shales from the
Carnarvon Basin of Australia. The depth at which these samples would be
considered source rocks starts around 2,500 m (8,202 ft) for (a), about 1,500 m
(4,921 ft) for (b), and 2,000 m (6,562 ft) for (c). These depths correspond to
the temperature range of 50 to 90°C (122 to 194°F), which is the range where
most oil generation becomes significant. If factors such as kerogen type and
quantity are similar at all depths in these examples, then the shallower
samples have the potential to be source rocks, but until they are buried
deeper this potential will not be realized.

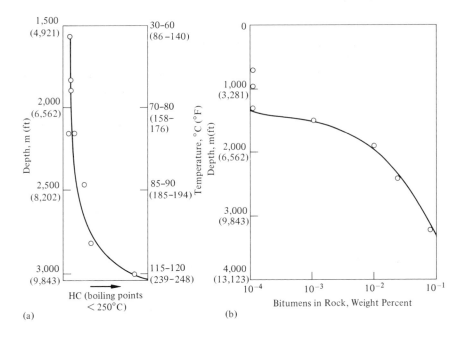

(a)

HC (boiling points
< 250°C)

(b)

Bitumens in Rock, Weight Percent

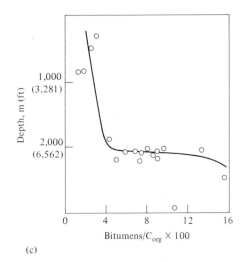

(c)

Bitumens/C_{org} × 100

Figure 7-11
Examples showing the threshold of intense hydrocarbon, or bitumen, generation with depth.
(a) Gasoline-range hydrocarbons in Upper Jurassic–Lower Cretaceous shales of Western
Siberia. [Ivantsova and Shapiro, cited by Vassoevich 1971] (b) Soluble bitumens in Jurassic
shales of E. Ciscaucasia. [Konyukhov and Teodorovich 1969] (c) Bitumens of shales of
Carnarvon Basin. [Shibaoka et al. 1973]

Geothermal History

The Connan time–temperature chart of petroleum genesis (Figure 4-17) was based on continuously subsiding basins. All case histories involving uplift, intense erosion, volcanism, or extensive orogenic activity, or major breaks in sedimentation were not used in constructing the model. Although the model has some shortcomings because of simplification (Connan 1976), it is still a useful way from a practical standpoint to evaluate time–temperature effects on oil and gas genesis. It is not applicable to basins where intermittent subsidence or uplift have delayed exposure of the kerogen to higher temperatures.

When basins are not continuously subsiding, there may be long geological time periods during which organic-rich, fine-grained rocks are unable to generate and transport hydrocarbons to nearby reservoir rocks. The Permian Irati Shale of Brazil, in areas where it is not buried deeply, is an example of delayed generation.

The geothermal history of a basin involves an analysis of the time intervals during which sediments are subjected to various temperatures. It is the best way of evaluating oil and gas generation in a basin provided reasonable paleotemperatures can be reconstructed. Lopatin (1971) was the first to emphasize the importance of geothermal history in his coalification study of the Munsterland I coal hole in the Ruhr district of Germany. This well, which penetrated to 5,956 m (19,541 ft), reached the anthracite stage of maturation at 3,000 m (9,842 ft). Lopatin used the detailed studies of coal petrographers to determine changes in the depth of Munster Coal throughout its geological history. Maturation levels were estimated from vitrinite reflectance values that ranged from 2.5 percent at 3,000 m (9,842 ft) to 5.17 percent at 5,000 m (16,404 ft). The geothermal gradient was 3.4°C/100 m (1.9°F/100 ft) to 3,000 m (9,843 ft) and 3.8°C/100 m (2.1°F/100 ft) in the deeper sections. Uplifting occurred during the Permian, Jurassic, and Cretaceous periods. Lopatin (1971) calculated a time–temperature index of maturation, τ, which was a product of the time of coalification in each 10°C temperature interval and the temperature coefficient of the speed of reactions of maturation in that interval. When this index, τ, was plotted against vitrinite reflectance, it gave a straight line with a correlation coefficient of 0.99.

An example of the application of Lopatin's time–temperature index, τ, to oil and gas generation is shown in Figure 7-12. This is a nomograph developed by Lopatin (1976) to determine $\Sigma\tau$ for various time–temperature intervals. In the upper left corner of the nomograph is a small Lopatin diagram reconstructing the burial history of two sediments A and B over the last 100 million years. Sediment A spent 80×10^6 years in the 50–65°C (122–149°F) temperature range, 5×10^6 years in the 65–100°C (149–212°F) range and 15×10^6 years in the 100–120°C (212–248°F) range. To

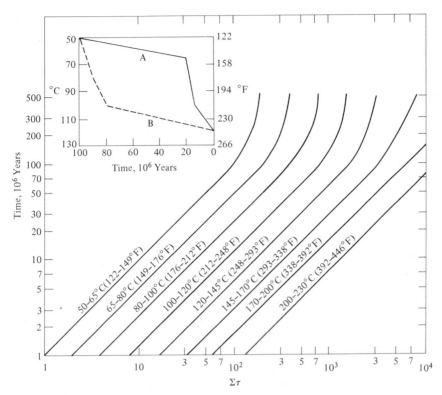

Figure 7-12
Nomograph for calculating the time–temperature index of maturation from the times at which a sediment is at 15–30°C (27–54°F) intervals, from 50 to 230°C (122 to 446°F). [Lopatin 1976]

use the nomograph, read 80 on the ordinate, move horizontally to the 50–65°C line, and move vertically to the abscissa, where $\Sigma\tau = 80$. Likewise, $\Sigma\tau$ for the other two temperature ranges is 15 and 120, giving a total of 215. Sediment B spent 10×10^6 years in the 50–80°C (122–176°F) temperature range, 10×10^6 years in the 80–100°C (176–212°F) range, and 80×10^6 years in the 100–120°C (212–248°F) range. The corresponding $\Sigma\tau$ values from the nomograph are 14, 41, and 630, for a total of 685. $\Sigma\tau$ values for oil and gas generation as determined by Lopatin (1976) are as follows:

70–85	Beginning of principal phase of oil generation
160–190	Maximum phase of oil generation
170–210	Zone of maximum oil migration into reservoir rock
380–400	End of principal oil-generation phase
550–650	Maximum gas generation
1,500–2,000	End of gas generation

This means that Sediment A in Figure 7-12 is well into maximum oil migration but Sediment B is through the maximum gas generation. Both sediments are the same age and are now at the same temperature, but they have had entirely different geothermal histories. Lopatin diagrams of this type were used by Zieglar and Spotts (1978) to evaluate oil and gas generation in the Great Valley of California.

Another approach to the problem of geothermal history was made by Tissot, Deroo, and Espitalié (1975) in their study of the Illizi Basin in Algeria. This basin in the eastern Sahara Desert consists principally of Paleozoic and Mesozoic sediments. The Silurian and Middle Upper Devonian sandstone–shale sequence includes some excellent source rocks. The principal hydrocarbon deposits are in Ordovician, Lower Devonian, Upper Devonian, and Carboniferous sandstones. Tissot et al. used a mathematical model of the genesis of hydrocarbons by kerogen degradation along with laboratory experiments involving both pyrolysis and extraction of the source rocks to calculate the amount of oil and gas generated by increasing temperatures. They correlated the hydrocarbon yield with the geological history as shown in Figure 7-13. Average geothermal gradients were estimated from present-day and paleogradients using vitrinite reflectance.

The top of Figure 7-13 shows the change in depth of burial for the southwest (A), the east (B), and the northeast (C) parts of the basin. In the southwest, the sediments were buried to 3,200 m (10,500 ft) by the end of the Paleozoic. At a geothermal gradient of $40°C/1,000$ m, they would have reached a temperature of about 94°C (201°F) at 1,800 m (5,906 ft), assuming a surface temperature of 22°C (72°F). At 2,700 m (8,858 ft), the temperature would be 130°C (266°F), which is at the end of oil generation and well into gas generation. Note the decrease in oil and increase in gas postulated by Tissot, Deroo, and Espitalié (1975) at the 290×10^6-year point in the second illustration of Figure 7-13. This is at the 2,700 m burial depth (130°C). By the end of the Paleozoic, at a depth of 3,200 m (10,500 ft)—equivalent to 150°C (302°F)—the organic matter in the southwest part of the basin had lost its ability to generate either oil or gas. Much of the gas formed was lost because the Hercynian folds occurred at the end of the Carboniferous. The gas that did accumulate was displaced toward the elevated structures of Djebel Tan Elak and Essaoui Mellene, which were severely eroded during the uplift at the end of the Paleozoic. Both Ordovician and Devonian reservoirs were eroded and invaded by meteoric water during this period. Consequently, the southwest part of this basin is a poor prospect for oil or gas, because both its source capability and the hydrocarbons generated were destroyed by the end of the Paleozoic. Only traces of residual gas have been encountered during exploration in this area. The maturation level of the kerogen has been verified by vitrinite reflectance as being at the end of the gas-forming stage.

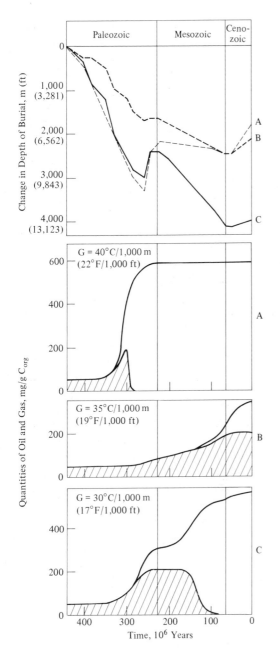

Figure 7-13
History of the depth of burial of the base of the Silurian source rocks and the quantities of hydrocarbon formed during the Phanerozoic in the Illizi Basin, Algeria. The three areas shown are A, southwest; B, east; C, northeast. Shaded area equals oil yield, open area gas yield. G is geothermal gradient. [Tissot, Deroo, and Espitalié, 1975]

In the eastern (B) part of the basin the deposition rate has been slower but continuous through the Mesozoic to a maximum depth of 2,400 m (7,874 ft) at the beginning of the Cenozoic. This is equivalent to a temperature of about 106°C (223°F) for the geothermal gradient of 35°C/1,000 m. Increasing amounts of oil were formed through the Mesozoic, with gas generation being initiated in the Cretaceous. This oil and gas generation occurred after the formation of Hercynian structural traps. Several major oil fields occur in this region, and the gas/oil ratio ranges from 100 to 200 m^3 gas/m^3 oil.

In the northeast area (C), oil and gas generation was initiated in the Late Paleozoic, when the burial depth reached about 3,000 m (9,843 ft), equivalent to a temperature of 112°C (234°F). Erosion was only slight or nonexistent in this area, and as the sediments were buried beyond their original depth during the Mesozoic, the generation of oil and gas continued. During the Mesozoic, the burial reached a maximum of 4,000 m (13,123 ft), equivalent to 142°C (288°F), out of the oil-generating stage and well into gas generation. Vitrinite reflectance values for the kerogen of 1.6 to 1.7 may be compared to the temperature and hydrocarbon stage in Figure 7-49. Meanwhile, the oil that had formed in this area was converted to gas by reservoir maturation during the last 100 million years. Only major gas fields are found in this region.

This example shows how a combination of geological and geochemical data can define the oil, gas, or no production prospects for different regions of a sedimentary basin. The mathematical model of Tissot, Deroo, and Espitalié (1975) utilized the nature of the organic matter, the geothermal gradient, and the reaction time period to determine the principal periods of petroleum genesis and accumulation in the various parts of a basin. The nature of the organic matter was determined by standard laboratory tests on cuttings or cores, as discussed in Chapter 10. The history of burial time was obtained from a synthesis of the geological and geophysical data plus vitrinite reflectance and other maturation techniques to verify the geological conclusions concerning maximum depth of burial. The paleogeothermal gradient was estimated from the present-date gradient and from an evaluation of maturation data such as vitrinite reflectance.

A similar evaluation of the Western Canada Basin showed that the Devonian source rocks were never buried deeper than about 400 m (1,312 ft) for over 200 million years until the Late Mesozoic. During this time, they had the potential to generate petroleum but were not buried deeply enough to enter the catagenetic stage of generation. It is only since the rapid subsidence at the beginning of the Cretaceous that the Devonian source rocks were buried deep enough to yield economic accumulations of oil and gas.

These examples emphasize that petroleum source rocks may lie in a latent state for very long periods of geologic time until buried deeply enough to pass the threshold of intense oil generation. Kerogen is relatively inert in a subsurface reducing environment. It will change very slowly over millions of years at subsurface temperatures in the range of 20–40°C (68–104°F).

This situation means that the exploration geologist must look for structures and traps associated with Paleozoic sediments that have been at low to moderate temperatures during most of their early history, and have been buried deeply only since the late Mesozoic. The concept of late oil generation appears to be much more widespread than was previously believed. New examples of this are sporadically cited in the literature (Webb 1976).

Heat Flow

In Chapter 6, in the discussion on dogleg geothermal gradients, it was pointed out that heat flow equals the product of the geothermal gradient and the thermal conductivity of a sedimentary section. Heat flow is constant for a specific part of a basin, so if the sediments in that area have a low thermal conductivity there will be a high geothermal gradient and vice versa. In comparing different basins with similar types of normally compacted sediments, large variations in heat flow can cause correspondingly large changes in the geothermal gradient. Makarenko and Sergiyenko (1974) compared 403 heat flow measurements in oil, gas, and gas-condensate fields around the world. Heat flow in these productive structures ranged from a minimum of 0.6 to a maximum of 2.8 cal/cm^2 sec. The mean was about 1.4 cal/cm^2 sec. The highest heat flow was in zones of Cenozoic volcanism, while belts of Precambrian folding had low heat flow.

Halbouty et al. (1970), in their study of the factors affecting formation of giant oil and gas fields, recognized that above-normal geothermal gradients accounted for the efficiency of hydrocarbon generation from source rocks in basins such as the Pre-Caucasus, Los Angeles, and Central Sumatra. Klemme (1972, 1975) expanded on this by showing that the yield of hydrocarbons per volume of sediments was higher in basins of high heat flow than those of low heat flow. The effect was greatest for intermediate crustal zone basins that occupied the transition area between continental and oceanic crusts. Typical are the basins associated with the belts along which crustal plates are underthrusting or overriding one another along subduction zones. For example, the West Coast of the United States, the basins behind the Indonesian volcanic arc, the Baku Basin, and southeastern Australia are areas of high heat flow associated with large petroleum accumulations. Klemme's worldwide survey showed considerable variation in geothermal gradients, caused by local hot spots or hot belts through sedimentary basins. The hotter areas contained either shallow oil or deep gas. Klemme recognized that both present and past geothermal gradients are critical in evaluating the potential of a new area to yield oil or gas. Vitrinite reflectance is one of the better maturation techniques for doing this, as discussed in Chapters 7 and 10.

Klemme (1972) also made the interesting observation that in clastic sediments of high geothermal gradients the reservoir accumulations will be larger than those in low-gradient areas because the average porosity is greater.

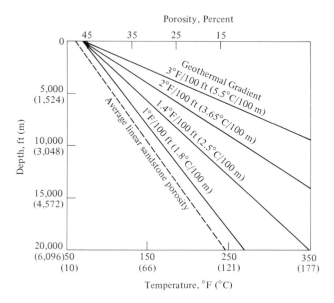

Figure 7-14
Model relating geothermal gradients to porosity in sandstone
reservoirs. [Adapted from Klemme 1972]

This relationship is shown in Figure 7-14, where the average porosity of
sandstones with depth is plotted with different geothermal gradients. At a
gradient of 3°F/100 ft (5.5°C/100 m), a sandstone at 250°F (121°C) will have
a porosity of 38 percent. In an area with a gradient of 1.4°F/100 ft
(2.5°C/100 m), the 250°F (121°C) temperature is reached at a depth of about
13,000 ft (3,962 m), where the average sandstone porosity is only 26 percent.
The shallower field would contain about 50 percent more oil than the deeper
field. Of course, large variations in individual porosities would partially offset
this difference, but the model clearly shows that shallow high-temperature
fields have the potential to contain more hydrocarbons than do the deep fields.

Plate margins have been defined as involving divergent (pull apart),
convergent (colliding) and transform (parallel) movements. Thompson
(1976) emphasized that high earth temperatures are associated with the onset
of divergent movements at a time when shallow water deposition favors
euxinic muds for source beds, carbonate reefs for reservoirs, and evaporites
for seals. He considers the giant El Morgan field to be in a fossil rift, the Gulf
of Suez. Maturation levels of kerogen in sediments associated with this field
indicate that paleotemperatures were higher than present temperatures. The
concept of fossil rifts being prospective for oil and gas accumulation would
apply to rifted margins of extended ocean basins, such as the Sirte Basin of
North Africa and the North Sea Basin.

Convergent oceanic and continental plates result in low heat flow in the trenches but high heat flow along the volcanic ridges. High temperatures behind the ridges are caused by the presence of basalt magmas. The giant oil and gas accumulations previously mentioned behind the Indonesian volcanic arc are in a trend of block fault basins extending up the eastern side of the Bay of Bengal.

Determination of present and past maturation levels is essential in exploration areas of complex tectonics associated with plate margins. The youngest organic matter in a Pleistocene sediment recovered by DSDP drilling on the continental slope adjacent to the Aleutian Trench had a maturation level equivalent to a burial temperature of 68°C (155°F) according to Grayson and La Plante (1973). At a high geothermal gradient of 5.5°C/100 m (3°F/100 ft), this represents a burial depth of about 1,067 m (3,500 ft), but the sample was found at 340 m (1,115 ft). Thompson (1976) postulates that the samples were buried and then uplifted in the complex tectonics of the thrust fault zone in front of the volcanic ridge.

These few examples are cited to emphasize that the exploration geologist must consider the effects of extreme variations in tectonics and temperature when exploring new areas. The thermal alteration of the organic matter is the most sensitive and easily measured indicator of paleotemperatures.

Salt Domes

Salt is the most effective conductor of heat in the sedimentary crust. The thermal conductivity of rock salt has been measured as high as 17 mcal/cm sec°C compared to around 5 mcal/cm sec°C for a typical clayey sandstone. This high conductivity means that salt domes rising from a mother salt bed transfer heat flow rapidly into a basin. The situation is analogous to a copper-bottomed teakettle with copper wires extending into the water being heated. As heat flows to the bottom of the kettle, it is rapidly transferred through the copper wires into the water. Likewise, salt domes act as a mechanism for moving heat from deep in a basin to much shallower levels. Also, when salt forces its way up through a sedimentary column it creates numerous faults through which thermal waters can carry additional heat from great depths. The high conductivity of salt results in its having a low geothermal gradient. The adjacent sediments, like the water in the teakettle, will have a lower thermal conductivity and a correspondingly higher geothermal gradient. This means that high and low geothermal gradients through the same depth interval will exist side by side in the salt dome and adjacent sediments of a basin.

Dzhangir'yantz (1965) recognized this anomalous heat distribution in the vicinity of salt domes of the Emba region, USSR. At the same depth, the temperature above the crests of the salt domes was higher than on the periphery. For example, on the southeast flank of the Kulsary field, the geoisotherms dropped 15–20°C (27–36°F) from the crest toward the area between domes. The geothermal gradient ranged from 1.1–3.5°C/100 m (0.6–1.9°F/100 ft) in the 100–1,000 m (328–3,281 ft) depth range. The magnitude of the gradient decreased with distance from the crest of the dome to its periphery. A similar relationship was noted on other domes in this area northeast of the Caspian Sea.

In the Makat well, the geothermal gradient in 1,200 m (3,937 ft) of mixed clastic and carbonate lithology was 2°C/100 m (1.1°F/100 ft). In 1,000 m (3,281 ft) of salt, the gradient was 0.9°C/100 m (0.5°F/100 ft). This well stood for 8 days before temperature measurements, to ensure equilibrium. Other wells stood for lengths of time extending up to 30 days.

At a depth of 1,000 m, the temperature in Mesozoic sediments within the zone of occurrence of salt structures varied from 31.5 to 52.4°C (88.5 to 126°F). Such a large difference shows that temperature equilibrium is not reached in the vicinity of salt domes even through long geological time periods.

Thus, salt domes not only create favorable structures and stratigraphic traps but also draw heat from the deeper parts of a basin, thereby expediting the conversion of kerogen to petroleum in closely adjacent sediments. This has some importance for exploration, because salt domes are common in many future offshore areas such as the Algero–Provençale Basin in the Mediterranean Sea (Biju-Duval et al. 1974) and the Gulf of Mexico (Worzel et al. 1968). In the latter area, there are over 150 salt domes and knolls covering a band 450 miles (724 km) long and 40 to 125 miles (64 to 201 km) wide, starting 275 miles (443 km) northwest of the Yucatan Peninsula. The term *domes* refers to completely buried structures without surface expression; *knolls* are structures that produce a topographic rise.

In 1968, the Deep Sea Drilling Project drilled Site 2 of Leg 1 on top of one of these knolls in 11,720 ft (3,572 m) of water. An oil-saturated core was encountered at a sediment depth of about 450 ft (137 m). The sediment was recognized as that part of a typical Gulf Coast anhydrite cap rock in which there has been active alteration to gypsum, calcite and sulfur. Carbon and sulfur isotope data indicated that the sulfur came from the oxidation of H_2S and the calcite from microbial oxidation of methane or oil. The core also contained about 400 pollen grains and spores of late Jurassic age, which were probably derived from the underlying diapiric salt (Davis and Bray 1969).

The oil and associated gas were analyzed in considerable detail by several major oil company laboratories, using the most advanced techniques available at the time. The sample was clearly defined as a typical Gulf Coast cap rock crude oil more immature than the average oil.

Figure 7-15 shows the carbon isotopic composition of the hydrocarbon components of the knoll gas compared to the gas from two Gulf Coast

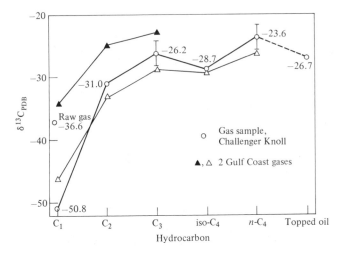

Figure 7-15
Comparison of the carbon isotopic composition of the methane, ethane, propane, isobutane, *n*-butane, total gas, and associated oil in the Challenger Knoll cap rock, with data for two Gulf Coast salt dome gases. [Erdman et al. 1969]

petroleum samples associated with salt domes. Immature gas samples generally show a steep rise in the $\delta^{13}C$ values in going from methane to ethane to propane. The knoll gas clearly is the most immature of the three examples, because it shows the steepest rise. Otherwise, the shape of the curve is normal for gas associated with oil having a marine source. Gases from continental source material generally show a higher concentration of ^{13}C.

The actual depth at which the oil originated is unknown, but it almost certainly came from deeper than the discovery depth. Although the geothermal gradient in the source beds next to the salt dome might be double that in a typical oil-producing basin, the depth of the oil discovery was not great enough to form a mature oil.

The low naphtha content of this oil is characteristic of immaturity. Also, the ratios of individual hydrocarbons indicate immaturity. For example, branched-chain paraffins are abundant in immature oils, whereas in most crudes the normal isomers are present in the highest proportions. Figure 7-16 shows that the branching of the heptane isomers in the Challenger Knoll oil is comparable to Tertiary oils at Spindletop and Stratton Ridge but is unlike the heptane isomer distribution of most Mesozoic and Paleozoic oils.

The Challenger Knoll oil also has a high cyclohexane/cyclopentane ratio, which is believed to be typical of either an immature crude or nonmarine source. Since other data indicate a marine source, this also supports the immature crude concept. Several isoprenoid hydrocarbons including farne-

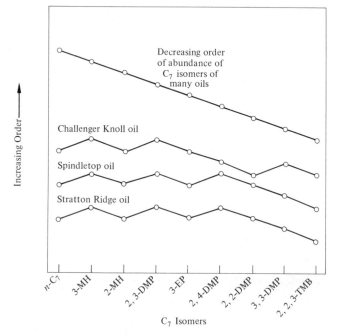

Figure 7-16
Order of abundance of heptane paraffin isomers of Challenger Knoll oil compared to other oils: n-C_7 = n-heptane; 3-MH = 3-methyl-hexane; 2-MH = 2-methylhexane; 2.3-DMP = 2.3 dimethylpentane; 3-EP = 3-ethylpentane; 2,4-DMP = 2,4-dimethylpentane; 2,2-DMP = 2,2-dimethylpentane; 3.3-DMP = 3.3 dimethylpentane; 2,2,3-TMP = 2,2,3-trimethylpentane. [Davis and Bray 1969]

sane, pristane, and phytane were present in the oil. The naphthene content of the gasoline range is unusually high, which would result in a young age by the age-dating method of Young et al. (1977), discussed in Chapter 8.

Three laboratories using four techniques found this oil to be highly aromatic as shown in Figure 7-17. A high naphthenoaromaticity is characteristic of immature oils. For example, a very immature oil, recovered from a sand at 125 ft (38 m) in the delta of the Orinoco, contained about 70 percent aromatic hydrocarbons in the C_{15+} fraction (Kidwell and Hunt 1958). A high-resolution mass spectra analysis of the aromatic fraction obtained by the standard chromatographic separation showed it to contain about 23 percent heterocompounds. Since all of these analyses use similar separation techniques, it is possible that the indicated high aromaticity is caused by a greater concentration of condensed-ring aromatic heterocompounds in immature oils.

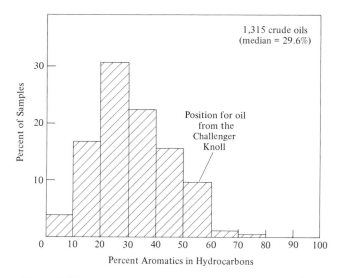

Figure 7-17
Aromatic hydrocarbon content of C_{18+} fraction of Challenger Knoll oil compared to other oils. [Davis and Bray 1969]

The mass spectra analysis of the vanadyl petroporphyrins in the Challenger Knoll oil is typical of an oil of marine origin with a mild thermal history.

The presence of a typical immature oil at shallow depth on a Sigsbee salt knoll has important implications for deep-water drilling beyond the continental shelf. If geothermal gradients are sufficiently high on continental rises and other locations of reasonably thick sediments, vast areas of the oceans are opened up for future exploration.

Gasoline-Range Hydrocarbons

The C_4–C_7 gasoline-range hydrocarbons occur in immature sediments in small amounts, from 10^{-8} to 10^{-6} g/g organic carbon (Hunt 1975b). These hydrocarbons, which constitute about a third of the average crude oil, are formed with the heavier hydrocarbons during catagenesis. Table 7-6 lists the gasoline content of unconsolidated sediments that have not been subjected to temperatures above about 40°C (104°F) and of mature rocks that are well into the catagenetic stage. The gasoline content of the latter is from 100 to several thousand times greater than the former. The mature rocks are well-known source beds associated with commercial reservoirs, except for the last one. The Kuonam outcrop is of interest because it is as rich in gasoline as most established source rocks and it indicates a potential for associated oil

Table 7-6

Quantities of Gasoline Range Alkanes in Immature and Mature Sediments

Sample	Age	Depth, m (ft)	C_4–C_7 Alkanes in 10^{-6} g/g C_{org}
Immature Sediments			
Indian Ocean 33°S, 39°E	Lower Pliocene	414 (1,358)	0.2
Off California 39°N, 127°W	Lower Pliocene	113 (371)	1
Gulf of Mexico 23°N, 92°W	Upper Pleistocene	28 (92)	2
Sea of Japan 39°N, 138°E	Pleistocene Pliocene	200 (656)	2.3
Bengal Basin 9°N, 86°E	Middle Miocene	769 (2,523)	4.4
Off Tasmania 42°S, 143°E	Upper Miocene	28 (92)	15
Mature Rocks			
B-7, Venezuela	Eocene	1,900 (6,234)	1,100
Frontier, Wyoming	Cretaceous	2,100 (6,890)	3,100
Duvernay, W. Canada	Devonian	2,400 (7,874)	6,200
Madison, Montana	Mississippian	1,840 (6,037)	6,800
Wolfcamp, Texas	Permian	3,080 (10,105)	18,700
Kuonam, USSR	Cambrian	Outcrop	1,200 5,500

production. Bikkenina and Shapiro (1969) identified about 50 individual gasoline-range hydrocarbons in these Early to Middle Cambrian rocks from the Anabar anticline. Good source rocks will usually have a gasoline content of at least 10^{-3} g/g organic carbon.

Tissot et al. (1971) noted that the gasoline content of the Toarcian shales increased more than an order of magnitude in the depth interval from 750 to 2,500 m (2,500 to 8,200 ft). Figure 7-18 shows the increase on a semilog plot of the gas through gasoline yields of cores from several wells in the Paris Basin. Gas and gasoline yields increased with depth at a faster rate than the increase in C_{15+} hydrocarbons. The largest increase is in the 1.5–2 km (4,900–6,600 ft) depth interval. A similar large increase in gasoline was previously shown for shale cuttings from a well in Canada (Figure 5-13). In that example, the largest increase occurred in the 1.5–1.7 km (5,100–5,600 ft) depth interval.

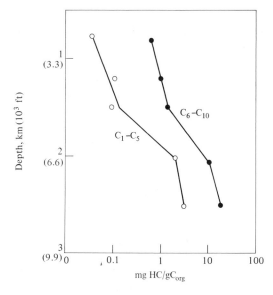

Figure 7-18
Gas and gasoline in cores from Jurassic Toarcian shales of the Paris Basin. Open circles are milligrams of C_1–C_5 hydrocarbons per gram of organic carbon. Closed circles are the same, for C_6–C_{10} hydrocarbons. [Tissot et al. 1971]

In the Los Angeles and Ventura basins, Philippi (1975) observed that the ratio of gasoline-range hydrocarbons to total hydrocarbons in the shales increased from 10 to 26 percent through the temperature interval from about 125 to 150°C (257 to 302°F). Since the geothermal gradient in the Ventura Basin is 2.6°C/100 m (1.5°F/100 ft) compared to 3.9°C/100 m (2.1°F/100 ft) in the Los Angeles Basin, this increase occurred through the 2,700–3,300 m (8,860–10,800 ft) depth range in the latter and 3,800–4,700 m (12,500–15,400 ft) depth range in the former. Clearly, Philippi's data show that the subsurface temperature, rather than the depth, is the controlling factor in the generation of hydrocarbons from source rocks. Philippi's data also show that the generation of gasoline in the 125–150°C range in these basins increases faster than the generation of the total hydrocarbon range.

Maturation Changes in Biological Markers

The thermal breakdown of kerogen to form oil during catagenesis results in significant changes in the biological markers that enable them to be used for source rock evaluation. The predominance of the odd-carbon normal paraffin chains formed biologically is destroyed through the breakdown of larger molecules to form smaller ones and by the dilution of odd-carbon chains with equal amounts of even and odd chains generated thermally. The cholestane structures derived from cholesterol are converted to a variety of naphthene homologs and partially to polycyclic aromatic hydrocarbons. Four- and five-ring condensed naphthene hydrocarbons gradually disappear, while the proportion of one-ring naphthenes increases. The most detailed study of individual hydrocarbon changes in shales with depth was first reported by Philippi (1965) in his comprehensive study of the Los Angeles and Ventura basins of California. This was followed by a similar study of the Paris Basin by Tissot et al. (1971) and of the Douala Basin by Albrecht and Ourisson (1971). These papers laid the foundation for the modern application of biological markers as sensitive time–temperature indicators and as tools for source rock–crude oil correlation.

Odd-Numbered Normal Paraffin Chains

In their study of the hydrocarbons of Recent sediments and ancient source rocks, Bray and Evans (1965) recognized that there was a decrease in the odd/even ratio of n-paraffin chain lengths in going from Recent sediments to ancient sediments to crude oil. They calculated a carbon preference index (CPI), which is a ratio telling how much more odd-carbon than even-carbon n-paraffins there are in a sample (Table 7-7). They found that the n-paraffins in Gulf Coast muds had five times as many odd-carbon chains as even in the C_{24}–C_{33} range. The odd/even ratio in paraffins of ancient shales ranged between 1 and 3, whereas in oils it was 1. They recognized that the transformation of organic matter to petroleum was adding paraffins with an odd/even ratio of 1, thereby reducing the ratio in shales as the shales became more mature. In subsequent studies, they deduced that a good source rock must have generated enough hydrocarbon to reduce the odd/even ratio in its entrained n-paraffins to the crude oil range, which is about 0.9 to 1.3.

Odd/even ratios are currently used for source rock evaluation but there are some problems in interpretation. One problem is variability of the organic source material. In Chapter 4, it was pointed out that marine organisms synthesize odd-carbon chains only in the low-molecular-weight range, not in the C_{24}–C_{33} range. Consequently, their CPI's are very close to 1 (Table 7-7). Sediments composed only of marine source material will have a CPI of 1 at the surface and at all depths. In contrast, the CPI of continental plants ranges up to about 20 and samples with any appreciable contribution from land will have CPI values considerably greater than 1. In actual practice, examples like the Cariaco Trench sample in Table 7-7 are very rare. They require that essentially no herbaceous or woody organic matter be transported to the sediment.

Figure 7-19 depicts schematically how the odd-carbon chains contribute organic matter to the sediments. The n-paraffins in many sediments show an odd-carbon predominance in both the high- and low-molecular-weight ranges. For example, Figure 7-20 shows the quantity of n-alkanes extracted

Table 7-7
Ratio of Odd/Even Carbon Chain Lengths of $C_{24}-C_{33}$ n-paraffins

Continental plants	CPI[a]	Marine organisms	CPI
Barley	7	Sponges	1.2
Maize	5	Coral	1.1
Tree leaves	4	Plankton	1.1
Nearshore sediments		**Deep sea sediments**	
Basins off S. California	2.5–5.1	Cariaco Trench	1.0
Offshore Texas, Louisiana	2.6–5.5		

$$^{a}CPI = \frac{\% \, C_{25}-C_{33} \text{ odd} + \% \, C_{23}-C_{31} \text{ odd}}{2(\% \, C_{24}-C_{32} \text{ even})}$$

Source: Data from Bray and Evans 1961; Koons et al. 1965.

from four samples of the Green River oil shale of Colorado. In the shallowest sample, the C_{17} and C_{31} alkanes are dominant. The C_{17} could be coming from lacustrine plankton or from bacteria. Table 4-6 showed that many bacteria have a high content of the C_{17} alkane. The C_{29} and C_{31} alkanes in this sample are from land plant waxes.

Figure 7-20 also shows that the odd-carbon predominance gradually disappears with increasing depth of burial until it is barely noticeable at 3,000 ft (914 m). This is because of the thermal generation of the complete suite of alkanes from the oil shale kerogen. The depths shown are present day rather than the maximum depths of burial.

The odd–even predominance of the paraffin hydrocarbons in the shales in the Green River and Wasatch formations of the Uinta Basin is sufficiently

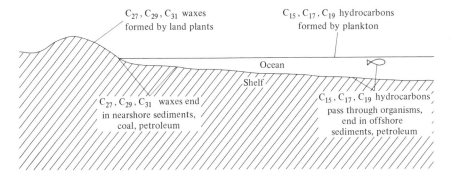

Figure 7-19
The contributions of odd-chain-length n-alkane molecules to sediments by land and marine organisms. [Hunt 1968]

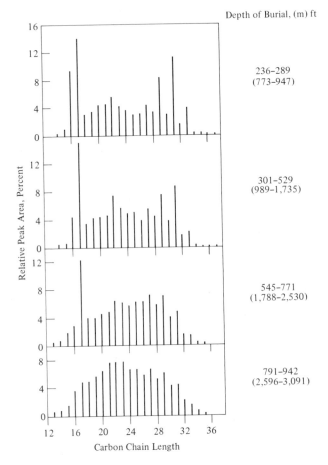

Figure 7-20

Relative concentrations of *n*-alkanes in C_{14}–C_{36} range of paraffins extracted from Green River oil shales of Colorado. [Anders and Robinson 1973]

strong to show up in the waxy oils from that basin (Figure 7-21). The CPI of 1.6 for some Uinta Basin oils is among the highest recorded for oils. Other oils in this basin, such as Red Wash, have CPI values within the normal 0.9–1.3 range. The Kawkawlin oil in Figure 7-21 shows a predominance in the marine odd-carbon range from C_{13} to C_{19}. The John Creek and Pine Unit oils also described by Martin et al. (1963) have an odd predominance in this lower range. When oils show a slight predominance like this instead of the usual 1.0 value, the source bed generally shows a correspondingly higher odd–even predominance than is typical of mature rocks.

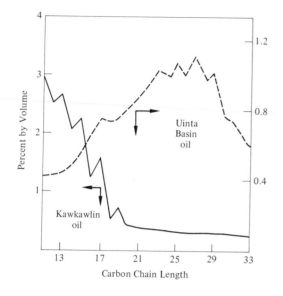

Figure 7-21
Odd-carbon predominance in the low range for the marine Kawkawlin oil of Michigan and in the high range for the nonmarine Uinta Basin oil of Utah. [Martin et al. 1963]

The change in the *n*-paraffin distribution with depth through diagenesis and catagenesis has been followed in several sedimentary basins. Figure 7-22 shows the concentrations of the C_{12}–C_{33} *n*-alkanes in the Douala Basin through the same range as the hydrocarbon yields shown in Figure 4-16. The pattern at 1,500 m (4,912 ft) is typical of the alkanes in sediments near the end of diagenesis before any significant thermal effects. A bimodality exists, because of the respective contributions of marine and continental alkanes. Two changes occur during catagenesis; a gradual disappearance of the odd–even predominance and a shift toward the lower-molecular-weight range because of the synthesis of smaller molecules and the cracking apart of larger molecules. When a source rock has reached sufficient maturity to generate considerable oil, the alkane distribution within the rock will peak between C_{13} and C_{18} and will show a steady decrease in concentration of alkanes with increasing chain lengths. Thus, the profile at 2,375 m (7,792 ft) is mature and corresponds to the peak of oil generation as shown in Figure 4-16. The difference in the pattern at 2,500 m (8,202 ft) is caused by the fact that each of these sediment samples started with different source materials so some will take longer than others to show the maturity profile. Eventually, all profiles will look like those at 2,645 m (8,678 ft) and deeper. The pattern at 3,450 m (11,319 ft) represents a very mature source rock that has passed through the oil generating stage and is well into the gas stage. The change in *n*-alkane distribution shown here is duplicated in the Paris Basin, where the Jouy Shale at 700 m (2,297 ft) shows a pattern similar to 1,500 m (4,921 ft) in Douala, and the Bouchy Shale of Paris at 2,510 m (8,235 ft) shows a pattern

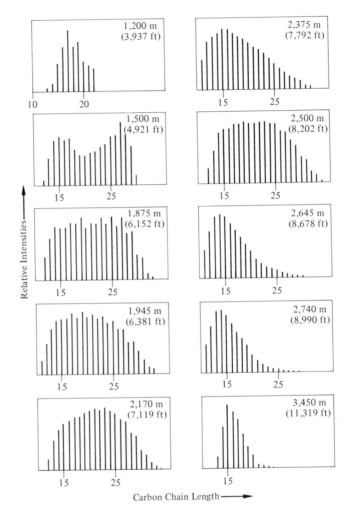

Figure 7-22
Change in *n*-alkane distribution with depth in Upper Cretaceous
sediments of the Douala Basin of Cameroon. [Albrecht 1970]

similar to 2,375 m (7,792 ft) in the Douala Basin. Estimated maximum depths
of burial are quoted for the Paris Basin.

Philippi (1965) noted that the ratio of $2C_{29}$ to $(C_{28} + C_{30})$ in alkanes from
shales of the Los Angeles Basin decreased from 8 to 1 at the threshold of
intense oil generation (Figure 7-23b). As the hydrocarbon yield increased, the
odd–even predominance decreased as shown in Figures 7-23a and 7-23b.
The latter figure represents a different way of calculating odd/even ratios.

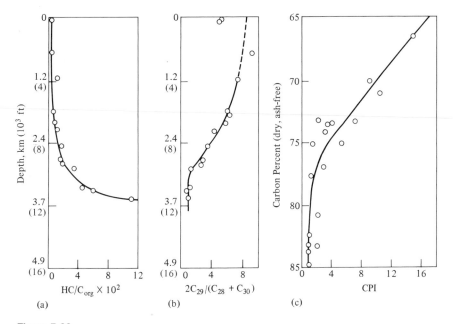

Figure 7-23
Curves showing disappearance of the odd-carbon predominance with increasing temperature (greater depth). (a) Hydrocarbon content of Los Angeles Basin shales. (b) Decrease in ratio of $2C_{29}/(C_{28} + C_{30})$ in these shales. (c) Decrease in CPI of n-alkanes extracted from Australian coals with increase in coal rank. [Philippi 1965; Brooks 1970]

The C_{29} used by Philippi is probably the most common land-derived n-alkane, as discussed in Chapter 4.

Coals contain very high odd/even ratios because their waxes are all land derived. Brooks (1970) noted that the odd/even ratio (CPI) decreased with coal maturation reaching 1 in the high-volatile bituminous coal range at around 80 percent fixed carbon (Figure 7-23c). Waxes that are residing in coals, oils, or sedimentary rocks will show this maturation change with increasing temperature.

The decrease in odd–even predominance can be duplicated in the laboratory and is a technique for evaluating the source potential of a deeply buried formation when only shallow near surface samples are available. Heating the sample to moderate temperatures, 200–300°C (392–572°F) will result in a drop in CPI if the kerogen is able to form oil. Samples of gas-generating kerogen or too little kerogen will show essentially no change in the CPI. Table 7-8 shows the change in odd–even predominance (see glossary for calculating OEP) for various sediments heated in the laboratory by Erdman (personal communication) for 12 days at 250°C (482°F). Heating

Table 7-8
Effect of Accelerated Aging on Typical Organic-Rich Marine to
Semimarine Sedimentary Rocks of Varying Geologic Age and Depth
(Conditions 12 Days at 250°C, or 482°F)

Origin	Geologic age	Depth, m	L_2/L_1 [a]	OEP Before aging	OEP After aging
North Sea	Miocene	1,665	0.80	1.7	1.1
North Sea	Miocene	1,667	0.87	2.1	1.7
North Sea	Miocene	1,680	0.73	12.7	1.1
North Sea	Miocene	1,701	0.30	1.83	1.09
Alaska, N. Slope	Paleocene	3,221	0.44	2.7	1.3
Louisiana OCS[b]	Pleistocene	862	0.27	4.3	1.4
Louisiana OCS[b]	Pleistocene	1,814	0.11	4.0	1.1
Papuan New Guinea	Middle Pliocene	2,203	0.32	3.3	2.2
Papuan New Guinea	Late Pliocene	3,352	0.15	2.6	2.0

[a]L_1 is the amount of lipid in the unheated sample; L_2 is the amount of additional lipid formed by the heat treatment.
[b]Outer continental shelf.
Source: Erdman, personal communication.

was in a sealed pressure vessel in the absence of air. Prior to heating, the sediments were freeze dried, extracted with solvents, dried, and reconstituted with water. After heating, they were freeze dried and reextracted. The *n*-alkane fraction from the extracts was separated by liquid–solid chromatography and urea adduction prior to analysis by gas chromatography.

The North Sea clastics show a significant drop in OEP, indicating them to be potentially good source rocks. The Papuan New Guinea samples formed very little additional lipid and the OEP ratio failed to drop below 2, indicating them to be poor source rocks for oil. The North Sea samples formed much larger quantities of lipids on heating than the New Guinea samples.

Erdman's studies (personal communication) of deep sea drilling samples have suggested that time can be an important factor in lowering OEP values independent of temperature. Figure 7-24 contains Erdman's data for a large number of deep sea samples of varying ages and depths. None of these samples have been subjected to temperatures above about 50°C (122°F). There is considerable scatter in OEP values due to varying mixtures of marine- and land-derived organic matter, but the overall drop in OEP with time is clearly evident. Erdman also grouped the samples in three depth intervals and found that there was no correlation with depth. Erdman concluded from his studies that the OEP in young sediments never exceeds 10 and that it continuously decreases with time irrespective of the temperature.

Nixon (1973) used the odd/even ratio of alkanes extracted from the Cretaceous Mowry Shale of the northwestern United States to evaluate its

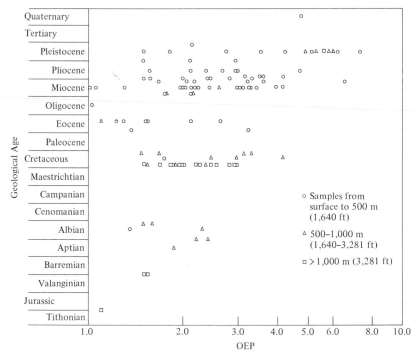

Figure 7-24
Change in odd/even predominance (OEP) with time for sediments obtained in the
Deep Sea Drilling Project. [Erdman, personal communication]

source potential. He found that the total extractable bitumen in the Mowry
Shale was typical of a good source rock in a wide belt extending from eastern
Wyoming north through eastern Montana and the western Dakotas. However, high CPI values indicated inadequate oil generation except where the
Mowry was buried deeper than 7,000 ft (2,134 m). This occurred only in
eastern Wyoming and a small part of the bordering states. A comparison of
the CPI's for the nonsource and oil source parts of the Mowry Shale is shown
in Table 7-9. Since the Muddy–Newcastle crude oils, which are believed to be
sourced in the Mowry, have CPI values in the range from 0.9 to 1.15, Nixon
concluded that the source beds would have to have CPI's around 1.2 or
less. The decrease in CPI through the 1.2 range occurred at about 7,000 feet
(2,134 m).

The Mowry example demonstrates the difference in CPI between immature and mature kerogen, assuming it to be oil generating. If the kerogen
originally is the woody–coaly type it will be capable of generating mainly gas,
so its CPI will not change much with maturation. Thus the Cretaceous Pierre
Shale of Wyoming has alkanes with CPI's around 3, while the alkanes in the

Table 7-9
Mowry Shale, Powder River Basin

Area	Depth, ft (m)	CPI
Nonsource Beds		
Mercer County, N.D.	3,700 (1,128)	2.22
Powder River County, Mont.	5,300 (1,615)	1.99
Fremont County, Wyo.	6,200 (1,890)	1.66
Oil Source Beds		
Niobrara County, Wyo.	7,100 (2,164)	1.14
Natrona County, Wyo.	9,100 (2,774)	1.08
Carbon County, Wyo.	13,500 (4,115)	1.09

Source: Data from Nixon 1973.

immediately underlying Niobrara Shale are 1.3 to 1.5. Both are deep enough to form petroleum but the Pierre will yield mostly gas since its CPI does not decrease with burial as it should with oil generation.

Odd/even alkane ratios in the C_{24}–C_{33} range cannot be used for source rock evaluation in Early Paleozoic sediments because of the absence of land plants. They are of limited value in many carbonates that contain mostly marine kerogen with no odd-carbon predominance or an even predominance in the higher-molecular-weight range.

Changes in Other Hydrocarbon Structures During Catagenesis

The steroids (Figure 4-4) and triterpenoids (hopene in Figure 4-8) contribute four- and five-ring condensed naphthenes to the organic matter of source rocks and eventually to crude oils. These naphthene rings are optically active. At increasing temperatures with deeper burial, the rings are degraded and aromatized to optically inactive structures. This explains why the young Tertiary crude oils of California are more optically active than the old Paleozoic crude oils of Pennsylvania.

Louis (1966) in his early studies of the Toarcian shales of the Paris Basin observed a significant decrease with depth in the optical activity of the bitumens extracted from the shales. The largest decrease occurred at the threshold of intense oil generation. This decrease in shales occurs both from dilution of the naphthenes with newly formed hydrocarbons and their thermal alteration.

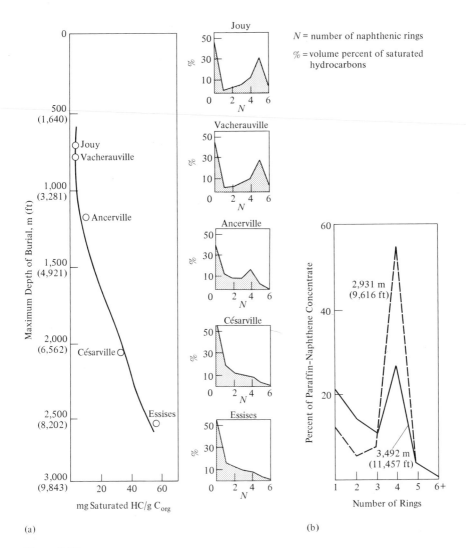

Figure 7-25
Increase in percentage of one-ring and two-ring thermally derived naphthenes and decrease
in biologically derived four- and five-ring condensed naphthenes with depth in saturate hydrocarbons extracted from (a) Paris Basin Toarcian Jurassic shales. [Tissot et al. 1971] (b) Los
Angeles Basin Pliocene shales. [Philippi 1965]

The percentage of naphthene rings in the saturated hydrocarbon fraction
of shale extracts from the Los Angeles and Paris Basins follows this trend with
depth very clearly. Figure 7-25a illustrates the change in the percentage of
naphthene ring structures of different sizes through the maximum depth

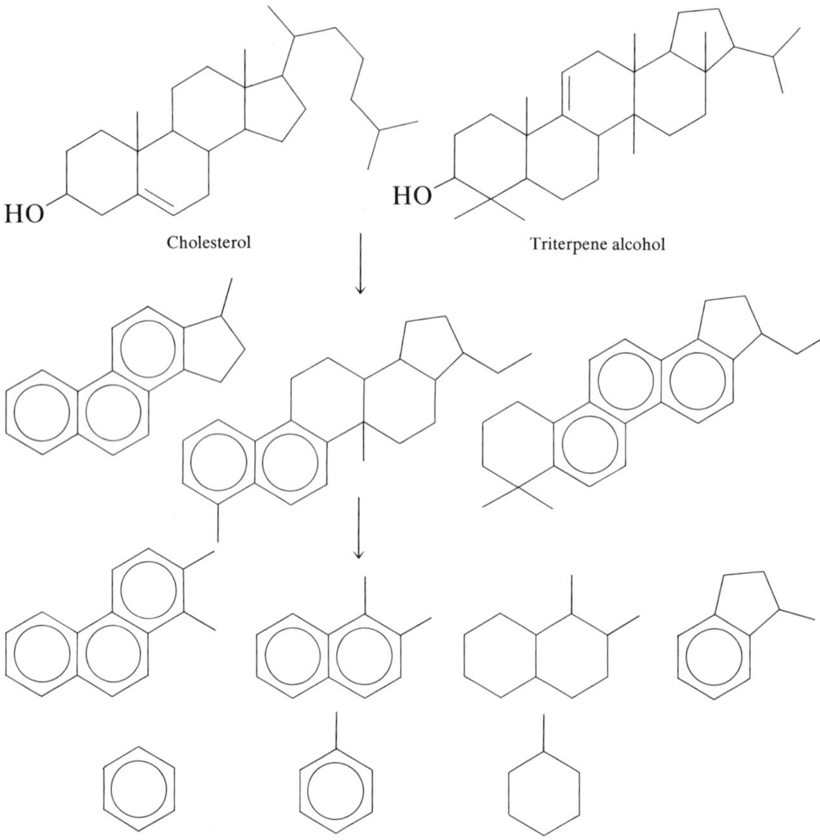

Figure 7-26
Maturation of four- and five-ring condensed naphthene biological markers to one- and two-ring naphthenes and aromatics with increasing temperatures. The condensed napthenes are optically active, but most of their degradation products are inactive. The structures shown, or their homologs, have been identified in shale extracts and crude oils.

intervals from 700 to 2,510 m (2,297 to 8,235 ft) in the Paris Basin. The decrease in four- and five-ring naphthenes and the increase in one- and two-ring naphthenes are evident. Figure 7-25b shows the same change in the Los Angeles Basin. The increase in total saturated hydrocarbons with depth for the Paris Basin suggests this change is largely a dilution effect. However, Tissot et al. (1971) also showed that the ratio of aromatic to naphthene rings in the hydrocarbons extracted from the shales more than doubled in the depth interval shown in Figure 7-25a, indicating some thermal alteration. The kinds of reactions that can occur in the shales are shown in Figure 7-26. Cholesterol and the triterpene alcohol are typical examples of four- and five-ring structures formed by living organisms that may be complexed with the kerogen. These can be aromatized through hydrogen elimination reactions, and some of the side chains can be stripped off to give the types of structures shown in the center of Figure 7-26. Further thermal degradation can result in the small aromatic and naphthenic molecules at the bottom of the figure. The general trend is from four- and five-ring naphthenes to one- and two-ring naphthenes and aromatics.

The molecular structures of steranes, pentacyclic triterpanes, carotanes, and lycopanes (Figures 7-26 and 4-7) are found in many Tertiary shales and oils, but they are generally absent from Paleozoic oils. Many homologs of these structures have been isolated from oil shales since the high hydrocarbon content permits more detailed analysis of individual molecular structures. For example, the Messel oil shale of Germany and the Green River Shale of Colorado, both of Eocene age, are rich in steranes and triterpanes. The older Triassic Serpanio Shale of Switzerland and Pennsylvanian Leo Shale of Wyoming do not contain these structures. The Irati shale of Permian age, which extends for 1,700 km (1,056 miles) in Brazil is old enough to have lost these compounds, but in many areas it was not heated enough to bring about the changes illustrated in Figure 7-26. Consequently, the Irati Shale is rich in terpanes, steranes, carotanes, and isoprenoid branched paraffins as shown in Figure 7-27. The Irati kerogen is the amorphous type derived from algae. Because of shallow burial, it has altered very little in 250 million years. The Irati oil shale is very rich in heavy hydrocarbons (Figure 7-1) even though it is somewhat immature.

Chlorophyll

Chlorophyll, the green coloring pigment of plants, is the most common and widely distributed of the biological markers. The basic tetrapyrrole structure (Figure 4-6) undergoes maturation to form hundreds, possibly thousands, of homologs. The spectrum of chlorins in very young sediments is rather sharp and distinct, but with time and deeper burial a very broad spectrum forms. Baker and Eglinton (1978) have studied the maturation of chlorophyll derivatives using deep sea drilling samples. The major change involves a gradual aromatization of green chlorins to red porphyrins with increasing time and temperature. Free-base porphyrins unattached to metals were found to be a minor constituent of sediments at all depths and ages. As the porphyrins form, they chelate (combine) first with nickel and later at higher temperatures with the vanadyl ($V=O$) group. Most of the porphyrins in Tertiary sediments analyzed by Baker and Eglinton contained only nickel, but in Early Cretaceous samples the ratio of nickel to vanadium was 3. Most of these samples had not been heated above about 50°C (122°F). Vanadyl

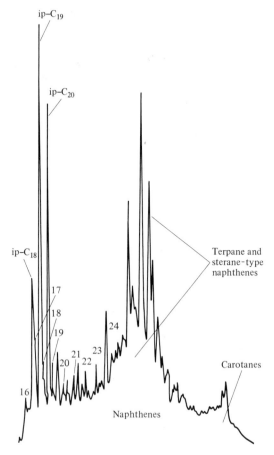

ip–C_{19}

ip–C_{20}

ip–C_{18}

17

18

19

21 23
20 22

24

16

Naphthenes

Terpane and
sterane-type
naphthenes

Carotanes

Figure 7-27
Gas chromatograph of the C_{15+}
paraffin–naphthene hydrocarbon
fraction of the Irati shale.
Isoprenoid (ip-) hydrocarbons C_{18},
C_{19} (pristane), and C_{20} (phytane)
are dominant, plus terpanes,
steranes, and carotanes.

porphyrins are common in Tertiary source rocks and crude oils of petroliferous formations where temperatures are well into the catagenetic stage.

The importance of temperature over time is evident in comparing the older
immature Permian Irati oil shale (Figure 7-27) with the younger, somewhat
more mature, Eocene Green River oil shale. The former contains nickel
porphyrins, the latter vanadyl porphyrins.

Pristane and phytane (Figure 4-6), the isoprenoid hydrocarbons derived
from the phytol chain of chlorophyll, are widely distributed in both
fine-grained sediments and crude oils. The pristane/phytane ratio generally
increases during diagenesis and the early stages of catagenesis. Brooks et al.
(1969) noted that the ratio reached a peak during the maturation of
Australian coals. The peak occurred at a carbon content of about 84 percent
near the end of the high-volatile bituminous coal stage, which is past the peak
of oil generation.

Thermal Alteration Experiments

The analysis of a shale outcrop or shallow core may show it to be rich in sapropelic kerogen but too immature to have generated oil. The question then becomes, "How much oil and gas will this shale generate if buried to higher temperatures?" By heating the sample in the laboratory in an inert atmosphere, it is possible to duplicate nature sufficiently to evaluate the potential of that sample to be a source rock. The process is shown schematically in Figure 7-28 starting with a hypothetical kerogen having a hydrogen/carbon ratio of 1.43. The kerogen can break down to yield small molecules such as toluene, 3-methyloctane, and 3-methylpentane. The kerogen provides hydrogen for these molecules by ring condensation. The process continues with naphthene rings in the kerogen releasing hydrogen as they aromatize. Oil is formed first then condensate, and finally methane. The kerogen continually loses more hydrogen than carbon until it begins to resemble graphite, as shown at the bottom of Figure 7-28.

Gorokhov et al. (1973) demonstrated in the laboratory that marine algae that are the most common organisms of the Precambrian are converted to graphite by heating under increasing temperatures and pressures. Samples were heated under a pressure of 15,000 atmospheres and the first graphite peaks on x-ray diffraction appeared at 875°C (1,607°F). The diffraction lines indicated that in addition to the completely ordered graphite there was some carbon with a disordered structure. Heating at atmospheric pressure did not result in graphitization of the kerogen. Pressures and temperatures higher than normal are required to conduct the experiment in a short time. Under natural conditions, the same results might be obtained at lower temperatures and pressures over millions of years of time.

The hypothetical example of kerogen degradation in Figure 7-28 shows only hydrocarbon structures for simplicity. The typical kerogen contains one oxygen atom for every 10 to 20 carbon atoms and lesser amounts of nitrogen and sulfur. As the kerogen breaks down, it may release nonhydrocarbon lipids (alcohols, ketones, and fatty acids), which can subsequently reduce to hydrocarbons. Also, there may be nonhydrocarbon lipids in the rock that form hydrocarbons independent of the kerogen. In evaluating source rock potential, it is irrelevant whether the hydrocarbons are generated directly from the kerogen or indirectly from the reduction of nonhydrocarbon lipids in the sediment. This is because most studies of bitumen generation such as in Figure 7-11 have shown that the nonhydrocarbons increase exponentially with depth below the threshold of intense generation, as do the hydrocarbons. The quantity of nonhydrocarbon lipids present in a rock during diagenesis is small compared to the quantity formed from kerogen during catagenesis.

Harwood (1977) heated thermally immature mineral-free kerogens at temperatures from 250 to 450°C (482 to 842°F) for several days and mea-

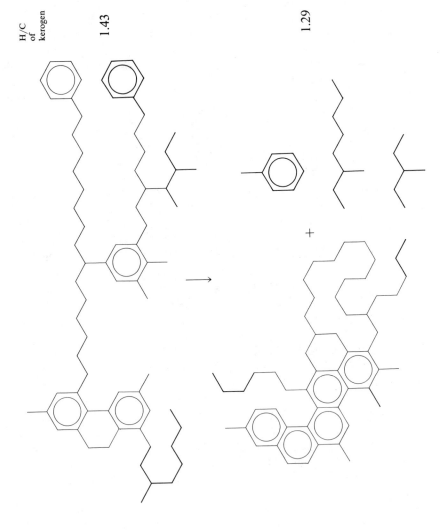

H/C
of
kerogen

1.43

1.29

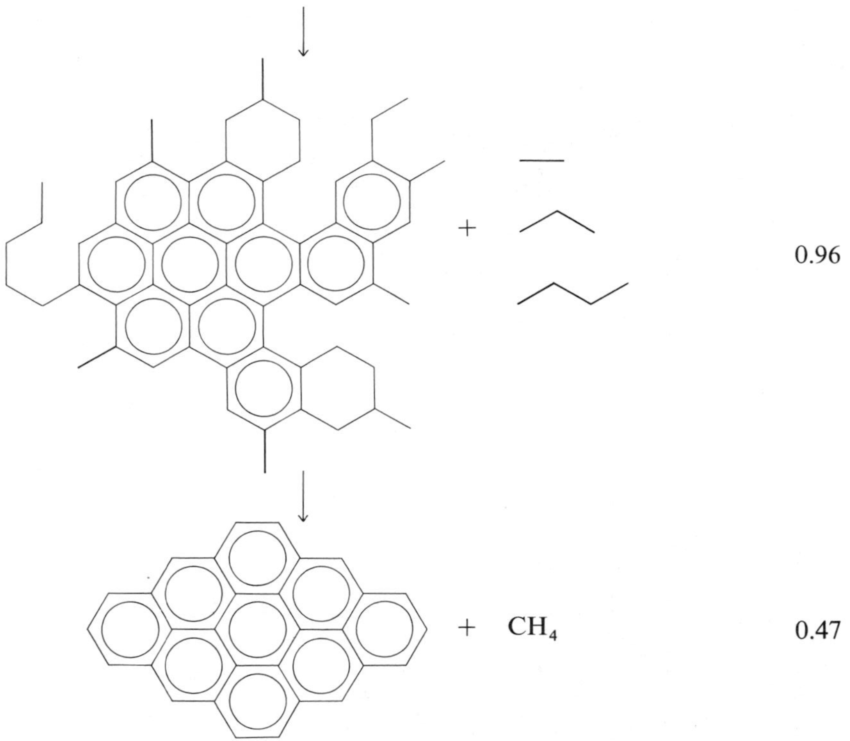

0.96

0.47

Figure 7-28
The thermal alteration of kerogen involves hydrogen disproportionation reactions in which the kerogen loses hydrogen to form gasoline, wet gas, and dry gas in succession. The hydrogen-depleted kerogen condenses and aromatizes to eventually form graphite.

sured the yields of liquids and gases along with changes in kerogen composition. Peak oil generation occurred when the carbon content of the kerogen was between 77 and 85 percent (87 percent for high-hydrogen kerogens). At the end of peak oil generation, the H/C ratio of the kerogen was about 0.8. Vitrinites and liptinites of coals both show their first coalification jump (significant change in chemical and optical properties) at 80 percent carbon content corresponding to the peak oil generation carbon content of kerogen (Stach et al. 1975, p. 198).

Peak gas generation was found by Harwood (1977) to occur between 85 and 89 percent carbon content of the kerogen (up to 92 percent for high-hydrogen kerogens). H/C ratios ranged from 0.8 to 0.4 during peak gas generation. In coals, an 87 percent carbon content corresponds to the second coalification jump, where coals increase their generation of methane.

Figure 7-29 is Harwood's plot of the maximum quantity of bitumen generated (expressed as weight percent of the kerogen) versus the initial H/C ratio for both oil- and gas-forming kerogens. Bitumen in this study is hydrocarbons plus nonhydrocarbons soluble in benzene. From 60 to 70 percent of the bitumen is hydrocarbon, so a pure hydrocarbon curve would be similar with the quantities on the ordinates reduced about a third. The data indicate that only kerogens with H/C ratios above 1 generate appreciable quantities of oil.

Figure 7-30 shows the maximum quantity of hydrocarbon gas generated versus the initial kerogen H/C ratio (Harwood 1977). Most kerogens with an H/C ratio above 1 (the oil-forming kerogens) generate twice as much gas as

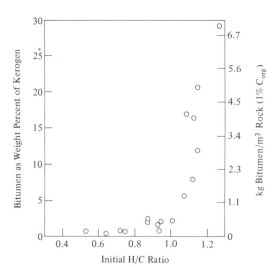

Figure 7-29
Maximum quantity of bitumen (hydrocarbons + NSO compounds) generated versus the initial H/C ratio for oil- and gas-generating kerogens. Kerogens for which the bitumen yields were severely affected by bitumen thermal cracking are not included. [Harwood 1977]

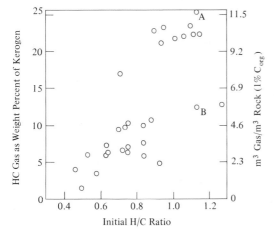

Figure 7-30
Maximum quantity of gas generated versus initial H/C ratio for oil- and gas-generating kerogens. A, Gas formed from kerogen plus gas from cracking of preformed oil of oil-generating kerogen. B, Gas from kerogen only of same sample. [Harwood 1977]

those with ratios below 0.8. This is because oil forms first and then is converted to gas. Point A in Figure 7-30 is the total gas yield from the Permian Cherry Canyon oil source kerogen of West Texas. Point B is the portion of the gas formed directly from the kerogen. The difference between A and B is the gas formed from cracking of pregenerated oil. A comparison of Harwood's data for other oil-forming kerogens shows that most of them generate about the same amount of gas as gas-forming kerogens if gas from the cracking of preformed oil is excluded. In the natural situation, this gas would be included, which means that a large percentage of the gas in sedimentary basins is from the cracking of oil still in the source rocks. It could be half the total gas in basins containing only oil source kerogens and a third in basins with equal amounts of oil and gas source kerogens.

Figure 7-29 also indicates that about 10 percent of the oil source kerogens are converted to liquid hydrocarbons and possibly another 5 percent to nitrogen, sulfur, and oxygen (NSO) compounds. The highest H/C ratio kerogen shows the highest conversion to bitumen, about 28 percent. An average of about 7 percent of gas source kerogens are converted to hydrocarbon gas (Figure 7-30). These data agree with estimates from field studies that about 10 to 15 percent of the kerogen is converted to hydrocarbons during catagenesis (Tissot et al. 1971; LaPlante 1974).

Harwood's (1977) oil source kerogens are in the H/C ratio range typical of liptinitic (amorphous, algal, herbaceous) organic matter while the gas source kerogens have ratios typical of humic, land-derived material. Some humic coals used in his study had hydrocarbon gas yields comparable to the gas source kerogens.

Techniques for Measuring Maturation

The identification of petroleum source rocks requires methods for determining the state of maturation of the kerogen. Has the kerogen passed through the threshold of intense oil generation? Is it in the oil- or gas-generating stage, or has it passed through all of these stages? If a rock has not been heated sufficiently to generate petroleum thermally, it can only contain biogenic methane or a heavy asphaltic biogenic oil, assuming no vertical migration from deeper beds. If a rock has been heated too hot or too long, it can only contain graphite and methane. Geochemists and palynologists have developed various techniques for evaluating kerogen maturation. In the last 20 years, these techniques have reached a level of sophistication that enables them to be used not only for maturation studies but also for defining the provenance of sediments and interpreting the sedimentary and tectonic history of basins.

Carbon Ratio Theory and C_r/C_t

The carbonization or metamorphism of coal is a well-recognized phenomena. Coal ranks based on oxygen content were established early in the nineteenth century (Regnault 1837). Later, Hilt (1873) demonstrated that the non-volatile, or fixed, carbon of coals increased with increasing depth and increasing temperatures. This observation became known as Hilt's law, and it has been confirmed in all coal basins except where local magmatic heating has disrupted the normal geothermal gradient.

The early petroleum geologists noted a relationship between the occurrence of petroleum and the metamorphism of coal as explained by Hilt. This relationship was clearly stated by White (1915) who showed that the oil fields in the eastern United States were in areas where the associated coals had a fixed carbon content of less than 60 percent. Gas accumulations were found in the 60–65 percent range, and there was no oil or gas near fixed carbon values above 70 percent. The analytical technique simply involved weighing a coal sample, heating it in a closed container with a small opening for removal of volatile gases, reweighing, heating with the lid removed to permit oxidation of all carbon, and a final weighing. The difference between the first and last weighings represents the total ash-free organic carbon, because all the organic matter would burn with the lid removed. The difference between the second and last weighings would be the nonvolatile carbon. The fixed carbon, or carbon ratio, was obtained by dividing the nonvolatile carbon by total carbon and multiplying by 100. Isocarb maps were made by joining lines of equal carbon ratios. Thom (1934) published isocarb maps for prospective areas of the United States.

The basic concept of the carbon ratio theory was correct. As coals mature, the stratigraphically equivalent source rocks mature and the threshold of intense oil generation is approximately equivalent to the subbituminous coal rank. Oil generation is completed in source rocks when coals reach the end of the high-volatile bituminous stage, and most methane generation is completed by the end of the low-volatile bituminous stage.

Although the carbon ratio theory was the key to the recognition of source rocks, the technique failed because of insurmountable sampling problems. Surface coal samples could not be used to interpret subsurface sediments, which usually had been exposed to higher temperatures. Suitable coal samples were rarely available from drill cuttings, so there was no way to make an accurate evaluation of carbon ratios down a drill hole.

Gransch and Eisma (1970) solved this problem by determining carbon ratios on the insoluble organic matter of sedimentary rocks. They treated well cuttings samples with organic solvents to remove the soluble bitumen and with hydrochloric acid to remove the carbonate carbon. Aliquot portions of the residue were burned in air to obtain the total organic carbon (C_t) and under nitrogen at a maximum temperature of 900°C (1,652°F) to obtain the volatile organic carbon. The difference between the total organic carbon and the volatile organic carbon equaled the nonvolatile organic carbon or carbon residue (C_r). The ratio of nonvolatile carbon to total carbon (C_r/C_t) is the same as the carbon ratio calculated for coals. This technique enabled carbon ratios to be determined on the kerogen of sediments with depth in any drill hole.

C_r/C_t ratios did not prove as useful as it was hoped because the different types of kerogen (liptinite, vitrinite, and so on) yielded different nonvolatile fractions as they matured. White's carbon ratio technique used coals, which are mostly vitrinite, whereas kerogens are highly variable. C_r/C_t would work best if it were possible to analyze only vitrinite particles or only liptinite particles going down a hole. Other techniques that follow the maturation of a single kerogen type are replacing C_r/C_t.

Exinite Color Changes

Palynologists separate hystrichospherids, spores, and pollen grains from sediments by dissolving the mineral matter with hydrochloric and hydrofluoric acids. These microfossils are used for stratigraphic correlations and age dating. In their studies, palynologists noted that the color of spores and pollen change with depth from light to dark when viewed under a microscope with transmitted light. Eventually they equipped their microscopes to measure light absorption and used this to follow the maturation of spores and pollen. By analyzing the spores and pollen in coals of known rank, they were able to

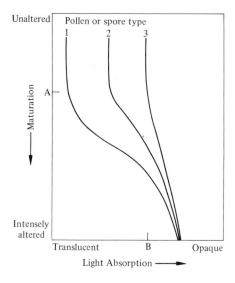

Figure 7-31
Comparison of the range in light absorption of three types of exines. Type I is used for measuring maturation changes in spores and pollen. [Gutjahr 1966]

correlate the light absorption of spores and pollen with fixed carbon (carbon ratios). The technique had all the advantages of the C_r/C_t approach, plus the fact that the kerogen mixture problem could be avoided by measuring light absorption for a single species of pollen grain or spore with depth. Gutjahr (1966), who perfected the technique for subsurface well studies, recognized early that different spores or pollen would give different maturation or rank indications at the same stratigraphic level. Figure 7-31 shows his comparison of the light absorption of three different pollen or spore types. In going from unaltered to intensely altered organic matter, type 1 shows the greatest change, while type 3 shows the least. At maturation level A, the three types have three different levels of light absorption. At light absorption level B, the three types are in three different levels of maturation. Clearly, the technique can only work if the same type or species of pollen or spore is analyzed with depth.

By following only the type 1 microfossils, Gutjahr was able to develop a fairly accurate correlation with carbon ratios. Figure 7-32 is a north–south cross section of Eocene and younger sediments from Houston to the Texas Gulf Coast. The isocarb lines equivalent to carbon ratios of coals from 55 to 90 indicate a high geothermal gradient at the southern end of this section and a lower gradient toward the center. Because the largest commercial oil fields in the world occur between carbon ratios of 50 and 70, the interpretation is that the Lower Miocene and Frio (Oligocene) have the best oil prospects, while the Vicksburg and deeper sections would be limited to gas in decreasing quantities, assuming that there is not extensive vertical migration of oil and gas from source beds.

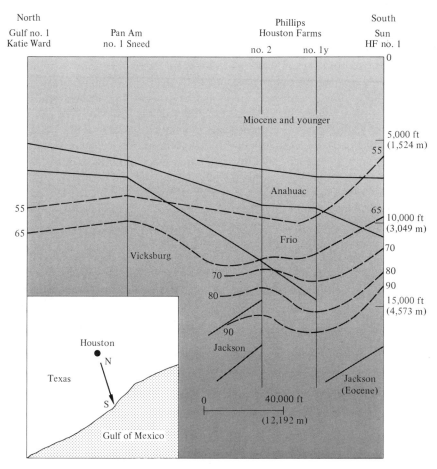

North
Gulf no. 1
Katie Ward

Pan Am
no. 1 Sneed

Phillips
Houston Farms
no. 2   no. 1y

South
Sun
HF no. 1

Figure 7-32
North–south cross section of Texas Gulf Coast, showing increased maturation of kerogen with depth. Solid lines are formation boundaries; dotted lines are isocarb lines of equal fixed carbon, as determined from exine light absorption. Note that isocarb lines cross formation lines. [Gutjahr 1966]

Staplin (1969) developed a relatively simple and rapid technique for evaluating kerogen maturation directly from its change in color. The standard palynological processing technique was modified to eliminate any treatment that might cause partial loss of color. The color designation is restricted to plant particles that are initially yellow, yellow-green, or pale orange. The color designation is made on spores, pollen, plant cuticles, algae, and amorphous organic matter whose colors are in this range. Coaly material is recorded but not used in designating colors. Generally, the lightest-colored material is used for color designations, assuming it is not a contaminant.

The current color scale, which is widely used as an exploration tool along with other maturation indicators, is shown in Table 7-10. The 1–5 indices represent kerogen color changes from light yellow to orange, brown, and black. The yellow is immature, the orange and brown mature, and the black metamorphosed. Only dry biogenic gas and possibly heavy oil will be found in the immature facies, oil and wet gas in the mature facies and only dry gas in the metamorphosed facies. Again, these definitions only refer to source rock characteristics and do not take into account vertical migration of oil or gas. The temperatures shown are only for relative comparisons, because there is a time effect as previously discussed. Some palynologists divide the five indices further into plus and minus values, while others use seven indices.

An application of the kerogen color technique is shown in Figure 7-33. The Middle Devonian carbonates of northeastern British Columbia and the adjacent Northwest Territories yield only dry gas, while shows of oil and wet gas are in the same carbonates to the east in northwestern Alberta. Staplin's (1969) analysis showed that, throughout the dry gas area and westward into the disturbed belt, the spores, cuticles, amorphous debris, and phytoplankton are all dark brown and black in color. To the east in Alberta, they are light yellow. The westward darkening is gradual and is accompanied by a decrease in the quality of preservation. There is also a darkening with depth in all wells except the one furthest east.

The correlation of the maturation index shown in Table 7-10 with the composition of the gas in the fine-grained rock cuttings for a well in the

Table 7-10
Organic Maturation Facies

Approximate temperature, °C (°F)	Maturation index[a]	Kerogen color	Maturity level	Petroleum prospects
30 (86)	1 Unaltered	Light yellow	Immature	Dry gas
50 (122)	1+ Slightly altered	Yellow		Dry gas, heavy oil
100 (212)	2 Moderately altered	Orange	Mature	Oil, wet gas
150 (301)	3 Strongly altered	Brown		Condensate, wet gas
175 (347)	4 Severely altered	Brownish black		Dry gas
>200 (>392)	5 Metamorphosed	Black	Metamorphosed	Dry gas to barren

[a]Also referred to as thermal alteration index (TAI).

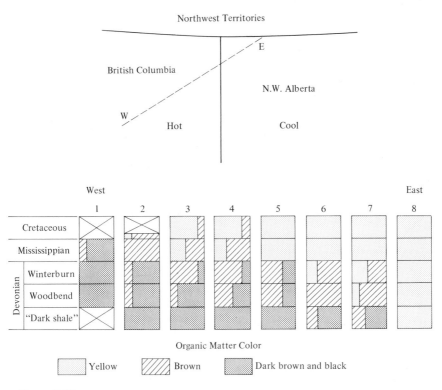

Figure 7-33
Darkening of kerogen color caused by increased maturation with depth and going west in eight wells of the Western Canada Basin. [Staplin 1969]

Northwest Territories of Canada is shown in Figure 7-34. In the Middle Devonian, the kerogen color changes from light brown to dark brown and black (index 3 to 4) at the base of the transition from wet to dry gas in the cuttings. As the kerogen color can be correlated directly with the composition of the cuttings gas, the former can be used on outcrops in rank wildcat areas where canned cutting samples from wells are not yet available. Staplin (1969) used this technique to map the boundary between condensate and dry gas in major sedimentary units from the 60th parallel to the Arctic coast. Unweathered samples must be used.

Oil accumulations in the Jurassic of the Aquitaine Basin are restricted to sediments of orange and light brown kerogen (Figure 7-35, Correia 1969). The dry gas province to the south is associated with source beds containing only dark brown and black kerogen. Color indices are contoured in Figure 7-35.

When spores are subjected to increasing temperatures and pressures in the laboratory, they undergo the color changes observed in nature, along with

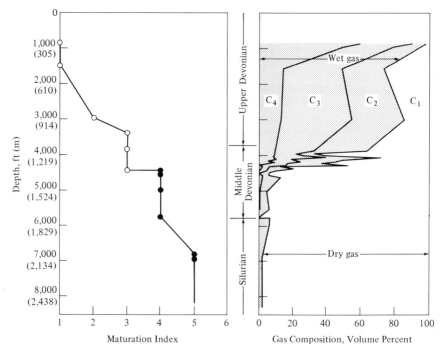

Figure 7-34
Increase in maturation index (kerogen color) with depth and corresponding change in composition of gases from well cuttings of I. O. E. Clare drilled in Northwest Territories, Canada. Index 4 corresponds to end of wet gas. [Evans and Staplin 1971]

size reduction, shape deformation, and ultimate amalgamation and crystallization (Sengupta 1975). The reduction of spore sizes appears to occur about the same time as the blackening of spores and loss of gaseous decomposition products.

Kerogen color is widely used as a rough and inexpensive maturation indicator. Interpretation of the total kerogen color can be difficult in many samples because of the different maturation rates of different species and organisms (Figure 7-31). Color, light transmission, reflectivity, and similar maturation indicators all change at different rates with different kerogen particles. Consequently, color interpretations are more accurate when limited to individual exinite species and when calibrated against cuttings gas analyses or some other maturation indicator. Sophisticated techniques use microscopes with optics capable of measuring the transmission of light through individual spore and pollen grains relative to a blank.

Experience is needed in all of these techniques to understand the problems. For example, recrystallization causes a higher level of maturation in the kero-

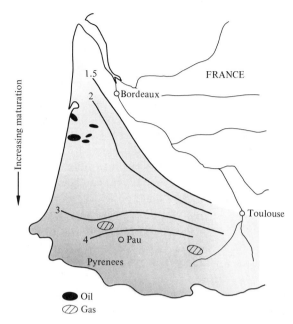

Figure 7-35
Correlation of Jurassic oil and gas provinces in the Aquitaine Basin,
France, with maturation index. Thermal alteration indices (TAI)
1.5–3 are in sediments of the oil province; above 3, in the gas
province. [Correia 1969]

gen of carbonates. Such kerogen is frequently black, while the stratigraphically equivalent shales may contain orange or brown kerogen. Oxidizing
agents and weathering cause substantial changes in translucency. Burgess
(1975) reviewed several methods of determining organic maturation and
compared costs, speed of analysis, quality, applications, and limitations.

Other techniques, such as fluorescence and refractive index analysis of
exinites, are used experimentally as maturation indicators. The wavelength of
the fluorescence maximum in exinite spectra lengthens with maturation.

Maturation techniques involving land-derived plant materials as discussed
above are not applicable to the Pre-Devonian. Conodont coloration has been
correlated with vitrinite reflectance and appears to be useful for both
Pre-Devonian rocks and organic-poor rocks such as red beds (Epstein et al.
1975). Conodonts (fauna made of Ca_3PO_4) change from light brown to black
during maturation. All of these techniques need to be used in combination
with other maturation indicators for proper evaluation of oil and gas
prospects.

Vitrinite Reflectance

The use of vitrinite reflectance as a technique for determining the maturation of argillaceous and calcareous rocks was first described by Marlies Teichmüller in her study of the Wealden Basin (1958). Teichmüller had observed the relationship between coal rank and oil previously discussed by White as the carbon ratio theory. In the Wealden Basin, the oil fields were confined to sediments containing low-volatile to medium-volatile bituminous coal. Teichmüller had been using the reflectance of the vitrinite maceral of coal to measure coal rank. It occurred to her that in sediments where coals were not readily available, as in northwest Germany, it should be possible to measure the reflectance of the small vitrinite inclusions that commonly occur in carbonates and shales. She used this technique to classify successfully petroleum regions that did not possess deposits of coal. Later, in a detailed study (1963) of the coalification process in the Munsterland I bore hole drilled to 5,956 m (19,541 ft), she used the reflectance of vitrinite particles in the shales to extend rank parameters into the deeper horizons where no coal seams occurred.

In 1961, Ammosov and Tan published a detailed study of the relationship of coal rank to vitrinite reflectance and the occurrence of oil and gas. They recognized that coal is rarely encountered in oil- and gas-bearing deposits, so they followed the stages of alteration of sedimentary rocks by the reflection of vitrinite particles in the rocks. Their studies involved the Cis-Caucasus, the Volga–Ural area, the northern Don, the Kuznets, and Irkutsk basins of the USSR. Figure 7-36 contains a semilog plot of vitrinite reflectance with depth for the Don Basin. Several stratigraphic sections were correlated to give a total burial depth of about 11 km (36,089 ft). They concluded that oil accumulations occurred in sediments with vitrinite reflectance values up to about 74 ($R_a + 10$). Condensate occurred to about 90, and above this there was only dry gas or barren formations. The slope of the line in Figure 7-36 is for the lowest geothermal gradient. Here the transition from oil to gas or barren is at a depth of about 7 km (23,000 ft), but it could be as shallow as 2 km (6,562 ft) in a district of high geothermal gradient. The studies of Ammosov and Tan were significant in showing the application of vitrinite reflectance in defining the limits of oil occurrence in several petroliferous basins. Their oil floor in the Don Basin at about 6,401 m (21,000 ft) corresponds to the oil phase-out zone on Landes' chart (1967) for a geothermal gradient of about 2.2°C/100 m (1.2°F/100 ft).

There is more vitrinite in coal than any other maceral. Vitrinite includes telinite, the cell wall material of land plants, and collinite, the substance that fills the cell cavities. The reflectance of these two submacerals is about the same, and they mature at the same rate. The reflectance of light on a polished surface of vitrinite increases with maturation because of a change in

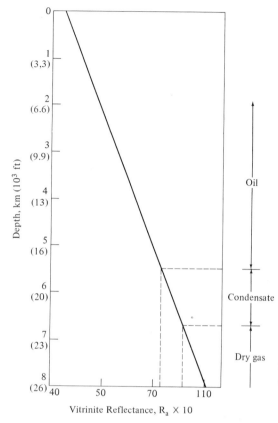

Figure 7-36
Change in vitrinite reflectance measured in air (R_a) with depth, for vitrinite particles from sediments of the Don Basin, USSR. Large oil fields are associated with source beds having reflectance values up to about 74; small oil fields and condensates, to 90; and dry gas or barren, above 90. Oil phase-out is in vitrinite range equivalent to transition from high-volatile to medium-volatile bituminous coals. [Ammosov and Tan 1961]

molecular structure of the maceral. Vitrinite is composed of clusters of condensed aromatic rings linked with chains and stacked on top of each other. With increasing maturity, the clusters fuse into larger, condensed aromatic ring structures, as in Figure 7-28. Eventually they form sheets of condensed rings that assume an orderly structure. The increase in the size of these sheets and their orientation causes increased reflectivity. Condensation of the rings also releases hydrogen, which is utilized to form methane.

Irreversible chemical reactions in which the rate increases exponentially with temperature are responsible for the changes in molecular structure. Consequently, reflectance, which measures these maturation changes, will increase exponentially with a linear increase in temperature.

Ting (1975) demonstrated this relationship in the laboratory when he heated woody lignite particles at a pressure of about 1,000 atmospheres and temperature of 100°C for seven days. Vitrinite reflectance was measured, and the experiment was repeated at 200, 300, and 400°C. The results are shown in

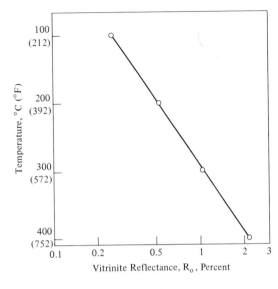

Figure 7-37
Increase in vitrinite reflectance with temperature of woody lignite heated in the laboratory. The linear temperature increase causes an exponential reflectance increase, which plots as a straight line on a semilog graph. [Ting 1975]

Figure 7-37. The exponential increase in reflectance with a linear increase in temperature is shown as a straight line on a semilog plot.

Reflectance measurements must be made only on vitrinite, since the other macerals, like other pollen species, mature at different rates. Figure 7-38 shows the change in reflectance with maturation for the three major coal macerals. At a carbon content of 88 percent, the reflectance varies from about 0.55 percent for liptinite to 1.7 percent for inertinite. There is also a maximum and minimum reflectivity of vitrinite, which differ considerably in the higher maturity ranges, where anisotropy is greatest. Most source rock maturity involves the R_o range from 0.2 to 2 percent. The statistical mean reflectivity of 50 to 100 measurements is used in most computations. This mean reflectivity is indicated by the symbol R_o or $R_{m\ oil}$. If maximum or minimum reflectivities are measured, the symbols R_{max} and R_{min} are used, respectively.

The reflectance technique, which is also discussed in Chapter 10, involves first concentrating the organic matter by removing the mineral matter with hydrochloric and hydrofluoric acids. The organic matter is freeze-dried, mounted in epoxy, and polished; R_o values are determined with a reflecting microscope using oil immersion objectives. From 50 to 100 reflectance measurements are made on a single sample and plotted on histograms as shown in Figure 7-39. This shows a bimodel distribution with R_o peaks equal to 0.8 and 1.6 percent. The mode with the higher reflectance represents recycled vitrinite, while that with the lower reflectance is primary vitrinite.

The correlation of reflectance with other maturation indicators and with oil and gas accumulations has resulted in an empirical definition of the R_o

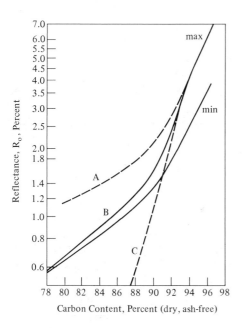

Figure 7-38
Change in reflectance in oil (R_o) for A, inertinite; B, vitrinite; and C, liptinite, with increasing coal rank. [Murchison 1969]

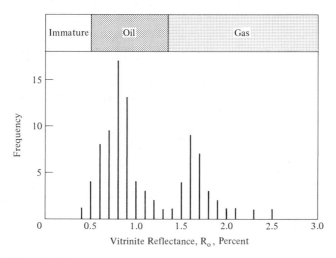

Figure 7-39
Histogram of reflectance of vitrinite particles from a single sediment sample.

numbers representing the limits of oil and gas generation. The lowest value associated with the known generation of oil is about 0.45 percent, and 0.6 percent is generally recognized as the beginning of commercial oil accumulations. Oil generation peaks at a maturation level around 0.8 to 1, and at higher levels the gas/oil ratios increase rapidly with reflectance. The end of oil generation is around 1.3 percent, condensate around 2 percent, and methane around 3.5 percent (Dow 1977a). To date, no major methane accumulations have been reported in which the vitrinite reflectance of associated nonreservoir rocks was higher than 3.5 percent.

Vitrinite reflectance values only indicate the level of maturation of the sample examined. They cannot predict the presence of oil and gas, because these frequently migrate updip along permeable beds or through fracture–fault systems to shallower reservoirs at lower levels of maturation. Consequently, oil is found in northwestern Germany, western Siberia, the Ural–Volga districts, and southern Florida where vitrinite reflectivities are only 0.3 percent R_o, which is believed to be too low for significant generation. The other end of the scale is more clearly defined, because extensive downward migration is less common. No major oil accumulations are associated with reflectivities above about 1.3 percent. The larger gas fields are found in the R_o range from 1.3 to 3 percent.

Vitrinite particles are found in 80 to 90 percent of all well cuttings. Red shales, interbedded red and green shales, and sediments near unconformities that outcrop at the surface contain oxidized vitrinite, which has anomalous reflectance values, frequently in a broad range. Shales and carbonates containing only amorphous kerogen such as the Green River oil shale and the Baaken Shale from Canada have too little vitrinite for accurate evaluation. Pre-Silurian rocks contain no vitrinite, although there is some cell wall material (telinite) in Silurian and Early Devonian rocks that gives a reflectance measurement.

Vitrinite histograms tend to broaden as rank increases. Broadening starts above $R_o = 1$ and becomes very noticeable above 2 and 3. Broadening also occurs in the lower ranks but it is more typical of high rank maturation. Figure 7-40 shows vitrinite reflectance histograms from a well in the Cook Inlet Basin of Alaska. The Eocene histogram gives a mean maximum of 0.64. The Upper Cretaceous sample is bimodel and also has a broad histogram without a clearly defined mean. The latter is a bathyal shale in a turbidite sequence. It has probably been oxidized and reworked.

Vitrinite maturation is not affected by pressure, only by temperature. When the mean R_o values for samples are plotted against depth on semilog paper, they form a straight-line maturation profile, as on the left of Figure 7-41 (Dow 1977a). In a sedimentary cycle of continuous deposition, the line will intersect the surface at about 0.2. Early vitrinite may be pitted, rough, jellied, or colloidal. The lowest measurements obtainable are around 0.18 to

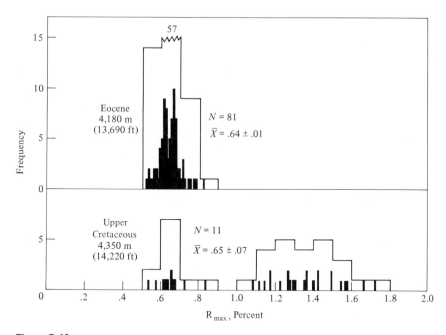

Figure 7-40
Vitrinite reflectance histograms from two samples in a well in the Cook Inlet Basin, Alaska. [Hood and Castano 1974] N = number of readings; $\overline{X}$ = arithmetic mean of maximum reflectance under oil immersion.

0.2. Good histograms are obtained at depths equivalent to 0.3. The slope of the primary vitrinite line will be determined by the geothermal gradient and the rate of deposition. Any change in these will cause a change in the slope. Recycled vitrinite tends to mature more slowly than primary vitrinite so the slope of the line to the right is steeper. Recycled vitrinite is very common, according to Dow (1977a). Extrapolation of the recycled vitrinite line to the surface shows the level of maturity it reached prior to its second deposition. Thus in Figure 7-41 the recycled vitrinite had a maturity of about 0.5 R_0 when deposited. Such information is useful in determining the source area of the sediments with the recycled vitrinite.

On geological cross sections, the oil and gas source rock intervals can be mapped based on reflectance numbers. For example, 0.6 representing the beginning of major oil generation is at about 2,650 m (8,694 ft) on Figure 7-41. The 1.35 point representing the end of oil is at about 4,200 m (13,780 ft). This represents the major oil generation interval in this well. It should be recognized, of course, that vitrinite reflectance is only a maturation indicator. Other geochemical data would be needed to determine if the organic matter in this well is capable of generating oil, gas or nothing. The primary vitrinite

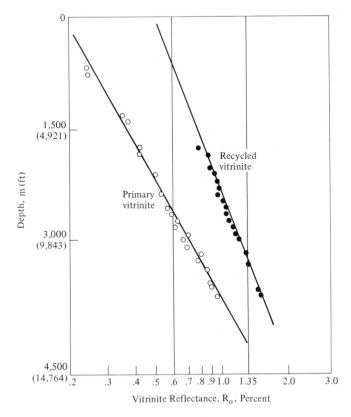

Figure 7-41
Reflectance maturation profile of the kerogen in an offshore Texas
Gulf Coast Miocene well, comparing primary and recycled vitrinite.
[Dow 1977a]

line in Figure 7-41 also shows that sedimentation has been continuous
through this depth interval; otherwise, the line would show a break and
offset.

Dow (1977b) constructed vitrinite reflectance profiles for Cretaceous
through Pliocene and Pleistocene sediments of the Louisiana Gulf Coast
using data from 12 wells, 2 in each producing trend (Figure 7-42). All the
wells had uniform geothermal gradients of about 1.4°F/100 ft (2.5°C/100
m). Initiation of oil generation (0.6 R_o) is around 18,300 ft (5,578 m) or 164°C
(327°F) in the Pliocene, 11,800 ft (3,597 m) or 113°C (235°F) in the
Oligocene, and 8,100 ft (2,469 m) or 84°C (183°F) in the Cretaceous of this
area. Higher temperatures are required in the younger rocks, compared to the
older rocks, to reach equivalent maturities. The variations in slope in Figure
7-42 are caused by differences in burial time. Increasing the geothermal
gradient would decrease the depth of the 0.6 R_o maturation level and vice

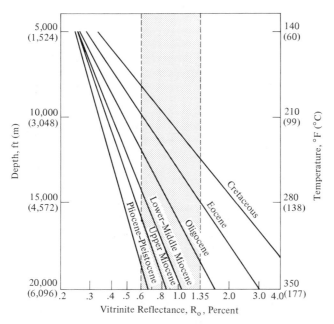

Figure 7-42
Maturation profiles for vitrinite in Cretaceous through Pliocene–
Pleistocene sediments of the Louisiana Gulf Coast. Location of 12
wells sampled is shown in Figure 7-43. Shaded area is oil generation
zone. [Dow 1977b]

versa. Figure 7-43 shows a schematic cross section of the Gulf Coast by Dow
(1977b) with the approximate location of the 0.6 and 1.35 R_o maturity levels,
showing the beginning and end of oil generation, respectively. Gas generation could occur on down to possibly 30,000 ft (9,144 m) in the Lower
Miocene. Figure 7-43 also shows that the oil generation maturity zone from
0.6 to 1.35 R_o becomes thicker and deeper in the younger rocks.

Much of the Louisiana Gulf Coast oil production is from thermally
immature prograding shelf sands overlying the mature slope shale source
rocks. Presumably the oil moved updip along bedding planes to the
reservoirs, although some geologists believe that extensive vertical migration
along deep-seated faults and along fracture systems near piercement domes
also occurred. Some type of vertical migration is the most likely explanation
for any Pleistocene oil accumulations in this area. Pleistocene gas accumulations could be biogenic in origin or could be coming from deep Pliocene–
Miocene source beds. From their calculated ages of hydrocarbons in Gulf
Coast oils, Young et al. (1977) concluded that a general upward migration of
oils occurred in this area.

Vitrinite reflectance defines the zone of catagenesis within which oil and
gas is thermally formed, but it cannot indicate where the oil or gas may be

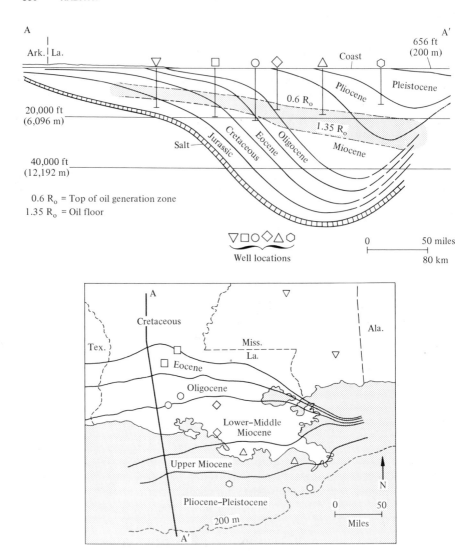

Figure 7-43
North–south cross section through Louisiana Gulf Coast, showing oil generation zone between
vitrinite reflectance levels of 0.6 and 1.35 R_o. [Dow 1977b]

reservoired, and it does not apply to the origin of biogenic hydrocarbon
accumulations, which can occur in shallow immature beds.

Although vitrinite reflectance profiles are used mainly to define zones of oil
and gas generation, they can also be used to interpret some of the sedimentary
and tectonic history of basins. For example, Figure 7-44 (Dow 1977a) shows
the reflectance profile at a major unconformity. Extrapolation of the
reflectance line in the Mesozoic upward until it intersects a line drawn to the

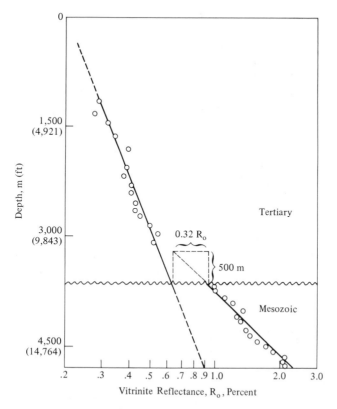

Figure 7-44
Reflectance profile of an Indonesian well showing the effect of a major erosional unconformity on kerogen maturation. The subsidence rate in the Mesozoic is indicated to be slower than in the Tertiary. [Dow 1977a]

unconformity contact of the reflectance line in the Tertiary indicates that about 500 m (1,640 ft) of section has been lost at the Tertiary–Mesozoic boundary. The maturity of vitrinite at the Mesozoic surface at the beginning of Tertiary sedimentation was about 0.32 R_o (0.94 minus 0.62). The different reflectance slopes indicate either different sedimentation rates, or geothermal gradients in the Mesozoic and Tertiary. Dow (1977a) interpreted this example to be caused by a much slower sedimentation rate (longer exposure time) in the Mesozoic.

Intrusives cause a large increase in vitrinite reflectance owing to the high temperatures, as shown in Figure 7-45. Reflectance values above 3 are common in contact metamorphism. Dow (1977a) states that the maturity of the intruded rock is affected to about twice the diameter of the intrusive body, often slightly more above the intrusive than below it. The actual distance depends on the temperature of the intrusive and the thermal conductivity of

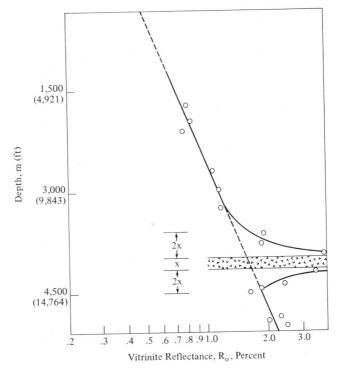

Figure 7-45
Reflectance profile of a Delaware Basin, Texas, well showing effect
of an igneous intrusion on the normal profile. Kerogen in contact with
the intrusive is converted to a graphitic residue. [Dow 1977a]

the intruded rocks. Extrapolation of the reflectance line in Figure 7-45 shows
an R_o value of 0.48 at the surface instead of the normal 0.2, indicating that
possibly 2,000 m (6,562 ft) of strata has been eroded from the surface since
maximum burial took place.

A higher geothermal gradient will cause a faster rate of maturation within a
shorter depth interval. Figure 7-46 shows the change in geothermal gradient
on entering a high-pressure, high-temperature zone. The change in slope of
R_o indicates that oil generation started around 10,000 ft (3,048 m). If the
original gradient of 13°F/100 ft had been maintained in the deeper
high-pressure section, oil generation would not have started until about
14,000 ft (4,267 m). This illustrates the importance of higher geothermal
gradients at depth, causing early generation of oil.

The maturation of vitrinite is an irreversible thermochemical transformation. Consequently, an R_o value of vitrinite cannot decrease. Once the dogleg
pattern in Figure 7-46 is recorded in the sediments, it becomes a permanent

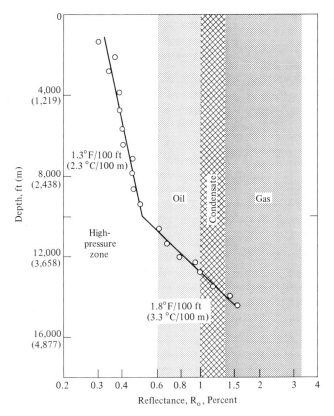

Figure 7-46
Vitrinite reflectance profile showing effect of change in geothermal gradient in high-pressure zone of well in Powder River Basin, Wyoming.

record of the existence of a high-pressure, high-temperature zone. For example, in a shallower part of the Powder River Basin there is a dogleg reflectance curve exactly like this but with the change in slope at about 7,000 ft (2,134 m) and no evidence today of a high-pressure, high-temperature zone. This indicates that a paleo high-pressure zone once existed in this shallower part of the basin.

Elemental Analyses

The maturation of kerogen with increasing temperature and time can be followed from an elemental analysis for the principal elements, carbon, hydrogen, and oxygen. The procedure consists of removing the mineral matter of fine-grained rocks with hydrochloric and hydrofluoric acids and analyzing the concentrated kerogen. In thermal maturation, the oxygen is removed from kerogen mainly as CO_2 and H_2O, the hydrogen as hydrocarbons and H_2O, and the carbon as hydrocarbons and CO_2. Some hydrogen is removed as H_2S, particularly in the final stages of maturation.

By analyzing kerogen with depth and plotting the data on an atomic H/C versus O/C diagram, it is possible to follow the evolution pathway of the kerogen. Figure 7-47 shows the data for five kerogens from oil and gas source rocks plus individual analyses for a group of sapropelic and humic materials. This figure is a modification of those published by Van Krevelen (1961) and Tissot, Durand, et al. (1974). The O/C curve is on a log scale since many kerogens have O/C ratios below 0.1. Liptinite is the oil forming, sapropelic material derived from plant and animal lipids. It includes alginite, sporonite, cutinite, resinite, and amorphous material, as shown in Figure 7-5. These source materials form the oil shales and boghead coals (numbers 1 through 6). The visual kerogen classification for these materials when disseminated in rocks includes amorphous (unstructured) and algal and herbaceous (structured).

Humic materials follow the vitrinite and inertinite evolutionary curves. Marine humus has a somewhat higher H/C ratio than terrestrial humus, but a similar O/C ratio. Marine humus, like its terrestrial counterpart, yields mostly gas during maturation. The visual kerogen types, woody and coaly (inertinite) are in these two humic evolutionary paths.

Data points for the Toarcian shales of the Paris Basin show them to have the highest oil-yielding kerogen (highest liptinite content), but they are not as mature as the other kerogen samples. The vitrinite reflectance (R_o) lines are plotted from Dow (1977a), and they indicate that the deepest Toarcian shales have not yet entered the condensate and gas-forming range. The Mannville and Logbaba shales are very similar in that both contain a high percentage of humic gas-yielding organic matter. The arrows, which show the approximate direction of evolution, indicate that the Mannville may be yielding more CO_2 relative to hydrocarbons during maturation than the Logbaba, because the slope of the Mannville pathway is less steep.

Elemental analyses of the kerogen of cuttings taken in a wildcat well would enable a reasonable estimate to be made of the range of oil and gas source rocks and barren zones by plotting the data on a base chart like Figure 7-47. Such a chart has the advantage of showing both the type of kerogen and its level of maturation. Tissot, Durand, et al. (1974) are using a pyrolysis technique to continue the maturation pathway of a well sample to a graphitic end product. This technique, which is discussed in more detail in Chapter 10, enables a geological formation to be defined as oil source, gas source, or nonsource.

Depth plots of the carbon, hydrogen, and oxygen content of kerogen in shales were utilized by LaPlante (1974) to estimate the depth at which the source beds began to generate hydrocarbons. In a well at South Pecan Lake,

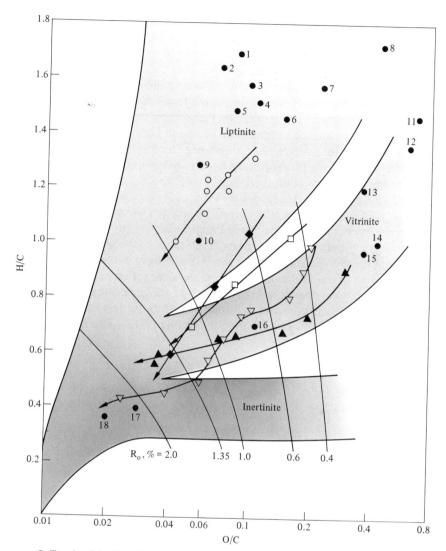

Figure 7-47

Maturation pathways of the dominant kerogen types in terms of their atomic H/C–O/C ratios. Natural organic materials are shown along with maturation changes with increasing depth (temperature) for five oil and gas source rock kerogens. R_o is approximate location of vitrinite reflectance values. (1) Coorongite, Australia; (2) Boghead coal; (3) Green River Shale kerogen, USA; (4) Tasmanite, Australia; (5) Torbanite, Scotland; (6) Kukersite, Estonia; (7) Cutinite; (8) Plankton; (9) *Botryococcus braunii;* (10) Exinite; (11) Wood; (12) Peat; (13) Marine humus; (14) Terrestrial humus; (15) Lignin; (16) Bituminous coal; (17) Anthracite; (18) Fusinite. [Data from Combaz 1975; Durand and Espitalié 1976; McIver 1967; Tissot et al. 1971; Van Krevelen 1961]

Louisiana, he noted that the percent hydrogen in the kerogen increased to a depth of about 10,000 ft (3,048 m) and then decreased. The apparent increase in the hydrogen in the upper section was due to the release of CO_2, and the decrease in the lower section was due to the formation of hydrocarbons. His interpretation is shown in Figure 7-48. Generation of the hydrocarbons was initiated at a subsurface temperature of about 86°C (187°F) in Miocene sediments 23 million years old. About one-third of the hydrogen in the kerogen was used up in generating hydrocarbons in the 10,000–15,000 ft (3,050–4,575 m) interval. At total depth of 15,000 ft (4,575 m) he estimated that 12 percent of the kerogen had been converted to hydrocarbons, 24 percent to CO_2, and 3 percent to water. The kerogen was a low-hydrogen type that generates mainly gas. The South Pecan Lake field produces wet gas that becomes drier at greater depths in the depth range of the generation zone, as shown in Figure 7-48. The composition of the kerogen in these source beds is similar to that of the Mannville and Logbaba shales (Figure 7-47). LaPlante compared this gas-generating kerogen with an oil-generating kerogen from the Permian in Texas, as shown in Table 7-11. Data are also included for oil- and gas-generating kerogens reported by McIver (1967). The key difference

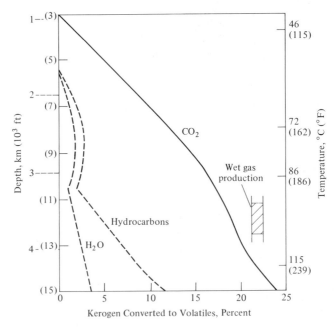

Figure 7-48
Conversion of kerogen to hydrocarbons, CO_2 and H_2O during maturation as estimated from elemental analyses, S. Pecan Lake, Louisiana. [La Plante 1974]

Table 7-11
Composition of Oil- and Gas-Generating Kerogens

	Oil-generating		Gas-generating	
Age	Permian	Cretaceous (Viking)	Miocene	Pennsylvanian (Atoka)
Depth, ft (m)	6,500 (1,981)	5,280 (1,609)	14,500 (4,420)	–
Location	Pecos County, Texas	Alberta, Canada	Cameron Parish, Louisiana	Oklahoma
Wt. % Carbon	80	86	80	89
Hydrogen	8	6.3	5	4.4
Oxygen	10	6.6	13	5.2
Nitrogen	2	1.1	2	1.4
Empirical Molecular Formula	$C_{67}H_{80}O_6N$	$C_{72}H_{63}O_4N$	$C_{67}H_{50}O_8N$	$C_{74}H_{44}O_3N$

Note: A nongenerating kerogen could have 95.5 percent C, 3 percent H, 1.5 percent O, and an empirical formula of $C_{79}H_{30}O$. The H/C ratio would be 0.38 and O/C 0.012. This plots to the left of the fusinite in Figure 7-47.

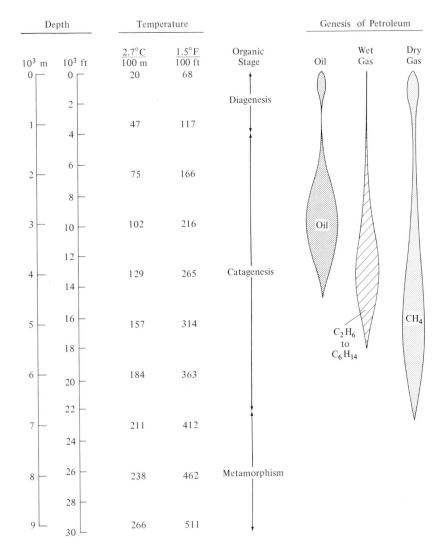

Figure 7-49
Chart of organic maturation R_0 = reflectance with oil immersion objective; CPI = carbon preference index. Maturation data are for an Eocene mixed-kerogen type. [Maturation limits from Dow 1977a; Staplin 1969; Teichmüller 1974]

between oil-, gas- and nongenerating kerogens is in the hydrogen content. Oil-generating kerogens usually have 6 percent or more of hydrogen, gas-generating 3–5 percent, and nongenerating less than 3 percent. The 3 percent hydrogen corresponds to a vitrinite reflectance beyond 3.5, which is approximately the end of methane generation.

	Maturation Indicators				Kerogen		Coal Rank
Vitrinite Ref., R_o, %	Thermal Alteration Index (TAI)	Kerogen Color	CPI C_{24}–C_{32}	Wt. % C	Wt. % H	Atomic H/C	
			5	67	8	1.5	
0.3							Lignite
	1	Yellow		70	8	1.4	
0.4			2				Subbituminous
				75	8	1.3	
0.6			1.5				C (1st Jump)
	2	Orange		80	7	1.1	B High-Volatile
0.8							Bituminous
							A
1.0							
			1	85	6	0.85	(2nd Jump)
1.35	3	Brown					Medium-Volatile Bituminous
				87	5	0.7	
							Low-Volatile Bituminous
2		Dark Brown, Black		90	4	0.5	Semianthracite
	4						
4				94	3	0.38	Anthracite
	5	Black					Meta-anthracite
				96	2	0.25	

The maturation of organic matter in fine-grained sedimentary rocks is summarized in Figure 7-49. The range of oil, wet gas, and dry gas generation relative to temperature and depth are shown for an Eocene mixed-kerogen type. The approximate ranges for maturation indicators such as vitrinite reflectance, the thermal alteration index (TAI), kerogen color, and the carbon preference index (CPI), or odd-even preference of the normal paraffins, are shown from diagenesis through catagenesis and metamorphism. Other data include the carbon and hydrogen composition for a kerogen with a starting atomic H/C ratio of 1.5 and corresponding coal ranks.

A small amount of methane and heavy bitumen is formed biogenically and deposited with fine-grained sediments during the first few hundred meters of burial. Major oil generation begins at an R_o of about 0.6; a TAI of 2; and, for this kerogen, a temperature of about 85°C (185°F), equivalent to a depth of 7,000 ft (2,134 m). Generation of the oil causes a rapid reduction in the H/C ratio shown on the right from about 1.2 to 0.7.

Coincident with the formation of oil from kerogen, Teichmüller (1974) reported a "first coalification jump of liptinites," which consists of several changes, such as the expulsion of liquid oil from fine fissures and holes of liptinite macerals. The release of oil was accompanied by a fluorescence maximum of sporinite and by the formation of granular micrinite, which appears to be a product of disproportionation reactions of certain liptinites. It is interesting that micrinite, which has not been identified in lignite and subbituminous coal, appears to be a secondary maceral formed as a high-carbon residue during oil generation in coals. Teichmüller believes that the hydrogen used to form the liquid oil is released from aromatic structures that then condense to form the high-carbon micrinite, which has an H/C ratio of 0.5 or less. The same thermal degradation reactions are producing oil from both kerogen and coal, but very little is produced from the latter owing to its initial low hydrogen content.

The generation of methane is beginning to reach its peak in the R_o 1.35 percent range (TAI = 3) as oil generation is completed. In comparing with coal, this corresponds to the "second coalification jump," which has been known for some time as the beginning of peak generation of methane from coal. The origin of methane from humic kerogen is essentially identical to its origin from coal. This reaction also causes a marked reduction in the H/C ratio. By the time the kerogen contains only 3 percent hydrogen (R_o = 4 percent, TAI = 4+, color = black) the methane generation has been terminated. Neither meta-anthracite nor a low-hydrogen kerogen is capable of forming CH_4.

It should be emphasized that the depth–temperature intervals for oil and gas genesis shown in Figure 7-49 apply only to a 50 million-year-old mixed-kerogen type buried under the geothermal gradient indicated in the figure. The importance of time is second only to temperature. For example, Figure 7-50 shows the change in vitrinite reflectance with temperature for samples varying in age from 20 to 300 million years, lines A to C respectively. At the same temperature of 100°C (212°F), the Miocene vitrinite has an R_o of 0.68, the Jurassic 0.88, and the Pennsylvanian 4. This means that the Miocene

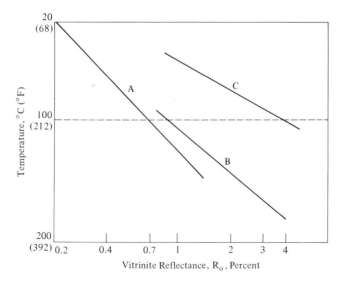

Figure 7-50
Change in vitrinite reflectance, R_o, with subsurface temperature
for A, Miocene, Texas Gulf Coast; B, Jurassic, LaSalle Co., Texas;
C, Pennsylvanian, E. Oklahoma. [Bostick 1973; Dow, 1977a]

kerogen is in the middle of the oil-generating stage, the Jurassic kerogen is into wet gas generation, and the Pennsylvanian kerogen is past the stage of methane generation. The R_o values and the genesis of petroleum are affected by changes in both temperature and time. The Pennsylvanian reservoirs may still contain methane, but it was formed before the kerogen matured to an R_o level of 4.

Time alone is not enough to bring about these changes. Kukersite (see Figure 7-47), the 500 million-year-old organic-rich shale of Estonia, has never entered the oil-generating stage. Heating kukersite in the laboratory yields large amounts of oil that are locked up in the kerogen structure because the rock was never heated in its natural environment.

In Chapter 4, Figure 4-17 was shown as a modified Connan (1974) chart with the times and temperatures of the starting and phasing out of oil and gas generation. Figure 7-51 shows the same chart with vitrinite reflectance lines plotted from 0.5 to 3 percent R_o. These lines enable this chart to be used somewhat more precisely in determining maximum oil, condensate, and gas generation zones, which are around 0.8, 1.2, and 2 percent R_o, respectively. Differences in kerogen composition will cause some variation in these numbers, so interpretations should be made cautiously.

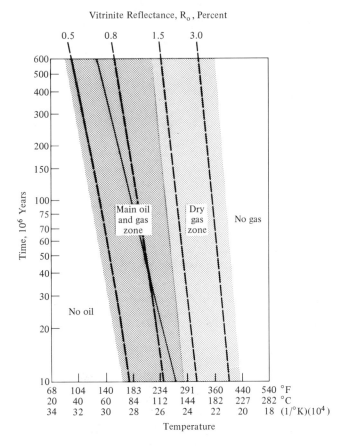

Figure 7-51
Vitrinite reflectance lines plotted on the modified Connan chart of
Figure 4-17.

SUMMARY

1. A petroleum source rock is a fine-grained sediment that in its natural setting has generated and released enough hydrocarbons to form a commercial accumulation of oil or gas.

2. Source rocks are clay or carbonate muds deposited under low-energy, reducing conditions. Shale source rocks typically contain 0.5 to 5 percent organic carbon, while carbonate source rocks sometimes have as little as 0.3 percent.

3. Kerogen, like coal, can be classified as sapropelic or humic. Sapropelic kerogen includes the algal, amorphous, and herbaceous categories. The

corresponding coal macerals in the liptinite group are alginite, resinite, sporonite, and cutinite. Humic kerogen includes the woody, vitrinite, and inertinite categories. Sapropelic kerogens and coals yield much more oil and gas than do humic kerogens and coals.

4. Oil shales, carbonate rocks, and highly calcareous rocks generally contain mostly amorphous and algal kerogen.

5. Paraffin waxes with an odd–even predominance of the n-paraffin hydrocarbons in the C_{27}–C_{35} range are derived from the herbaceous and woody kerogen of continental origin. Liquid paraffins with an odd predominance in the C_{15}–C_{21} range are derived from amorphous and algal marine kerogen.

6. A sedimentary rock must pass through the time–temperature threshold of intense hydrocarbon generation in order to form enough hydrocarbons to yield commercial petroleum accumulations. The rate of hydrocarbon generation from kerogen increases exponentially with a linear increase in temperature. The quantity of oil or gas formed depends primarily on the kerogen type, the temperature, and the time.

7. If the kerogen type is capable of forming oil and gas, the best way of evaluating past generation is by reconstructing the geothermal history of the presumed source sediments. One approach is to determine the time spent by the sediment in 15–30°C (27–54°F) temperature intervals throughout its geological history, and sum the time–temperature indices, τ, by the Lopatin method.

8. Biological markers undergo well-defined maturation changes during catagenesis. The odd predominance of n-alkanes decreases. Large molecules such as four- and five-ring naphthenes predominate in shallower samples, while small one- and two-ring naphthenes increase proportionally in deeper samples. The chlorophyll molecule forms hundreds of porphyrin homologs in which nickel is gradually replaced by the vanadyl group.

9. Laboratory experiments indicate peak oil generation occurs when the carbon content of kerogen is between 77 and 87 percent and peak gas generation when it is between 85 and 92 percent. Oil-generating kerogens have H/C ratios usually above one and a hydrogen content of 6 percent or more. H/C ratios of gas-generating kerogens are usually below 0.8, and hydrogen contents range from 3 to 5 percent.

10. The high earth temperatures associated with the onset of continental rifting and with the volcanic ridges along convergent plates favors petroleum generation. Determination of present and past maturation levels is essential in exploration areas of complex tectonics associated with plate margins.

11. Salt diapirs not only create favorable structures and stratigraphic traps but also act as conduits for heat from the deeper parts of a basin, thereby expediting the conversion of kerogen to petroleum in adjacent sediments.

12. Many oil and gas accumulations in Paleozoic reservoirs are examples of delayed generation. The hydrocarbon potential of these formations remained in a latent state during slow subsidence on broad platforms. The petroleum formed and accumulated only after post-Paleozoic orogenies buried the sediments beneath the threshold of intense oil generation.

13. Kerogen maturation indicators include the color and light transmission of kerogen and exinite particles, the change in gas and gasoline yields with depth from fine-grained rocks, the odd/even ratio of the n-paraffins, vitrinite reflectance, and element analyses.

14. The threshold of intense oil generation in source rocks starts where the wet gas and gasoline yields increase by a factor of more than 10, the kerogen color changes from yellow to orange, the thermal alteration index (TAI) from 1 to 2, the vitrinite reflectance R_0 reaches about 0.6 and there is an appreciable drop in the hydrogen content of the kerogen. The oil phase-out is around an R_0 of 1.35 and a TAI of 3. Generation of wet gas is essentially complete at R_0 of 2 and TAI of $3+$, and of methane at R_0 3.5 and TAI $4+$.

15. Semilog profiles of depth vs. vitrinite reflectance can be used in some instances to interpret sedimentary events such as the quantity of overburden removed, changes in geothermal gradients or depositional rates with time, and paleo high-pressure–high-temperature zones.

SUPPLEMENTARY READING

Alpern, B. (ed.). 1975. *Pétrographie de la matière organique des sédiments, relations avec la paléotemperature et le potential pétrolier.* Proceedings of International Meeting held in Paris, September 15–17, 1973. Paris: Éditions du Centre Nationale de la Recherche Scientifique.

Fischer, A. G., and F. Judson (eds.). 1975. *Petroleum and global tectonics.* Princeton, N.J.: Princeton University Press. 322 p.

Hedberg, H. D. 1964. Geologic aspects of origin of petroleum. *Bull. Amer. Assoc. Petrol. Geol., 48,* 1755–1803.

Momper, J. A. 1975. Time and temperature relations affecting the origin, expulsion and preservation of oil and gas. *Proceedings of the Ninth World Petroleum Congress, Geology.* Vol. 2. London: Applied Science Publishers.

Vassoevich, N. B. 1975. Origin of petroleum. *Geologiya, 30* (5), 3–23.

8

Petroleum in the Reservoir

The petroleum reservoir is the part of the rock that contains the pool of oil or gas. The reservoir rock extends beyond the limits of the pool and is generally considered to be any rock that contains interconnected pores with sufficient permeability to allow oil or gas phase production. Petroleum reservoirs occur in practically every type of rock: sand, siltstones, fractured shales, basement rock, limestone, chalk, and dolomite. The fluids migrating from compacting source beds will channel through practically any coarse-grained rock where permeabilities are the highest. These fluids carry the hydrocarbons, which are then trapped or sieved out at permeability barriers to hydrocarbons. At any one moment during the compaction history of a basin, the fluids moving through coarse-grained reservoir rocks are carrying far more hydrocarbons than are accumulated in the reservoirs. Subsurface waters in petroliferous basins are estimated to contain 100 to 200 times as much dissolved gas as exists in reservoirs. For example, Zor'kin et al. (1974) estimated that the quantity of hydrocarbon gases dissolved in underground waters of the Proterozoic and Paleozoic rocks of the Russian Platform is around 400×10^{12} m^3 (14×10^{15} ft^3) compared to a possible $2-4 \times 10^{12}$ m^3 (70 to 141×10^{12} ft^3) gas in reservoirs.

The quantity of dissolved gas in formation waters increases with depth, particularly in the range where large amounts of gas are being generated by the source beds. Spevak (1972) found 10 to 40 cc hydrocarbon gases per liter (0.06 to 0.22 ft^3/bbl) of formation water in the Jurassic–Lower Cretaceous formations of the East Precaucasus in the subsurface temperature range from 80 to 100°C (176 to 212°F). This concentration more than doubled in the depth range equivalent to a temperature from 100 to 140°C (212 to 284°F).

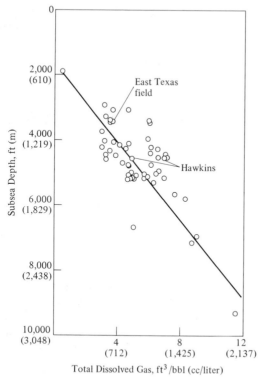

Figure 8-1
Dissolved gas content of Woodbine water, East Texas. [Buckley et
al. 1958]

Buckley et al. (1958) used a pressure core barrel to sample subsurface formation waters at formation pressures. They found dissolved hydrocarbons to be widespread in formation waters throughout the areas sampled, which included 300 wells distributed from New Mexico to Florida although most were in the East Texas, upper Gulf Coast of Texas, and southern Mississippi areas. Formations sampled ranged in age from Jurassic to Upper Miocene. A typical dissolved-gas profile with depth is shown in Figure 8-1. The giant East Texas and Hawkins fields both produce oil from the Woodbine Formation. Woodbine water contains from 3 to 12 ft³ of gas per barrel (500 to 2,100 cc/liter) of water through the subsea depth range from about 3,000 to 9,000 ft (900 to 2,700 m). The Frio sands, which produce oil along much of the upper Texas Gulf Coast, contain about 4 ft³ gas/bbl H_2O at 4,000 ft and 14 ft³/ bbl at 8,000 ft (equivalent to 700 cc/liter at 1,200 m and 2,500 cc/liter at 2,400 m).

At the threshold of intense oil generation, where there is a large increase in the hydrocarbons within the source rock, there is also an increase in the heavier hydrocarbons in formation water. The increase in gasoline-range hydrocarbons of Mesozoic shales of western Siberia was shown in Figure

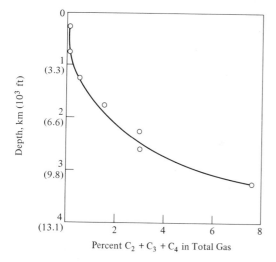

Depth, km (10³ ft)

0
1
(3.3)
2
(6.6)
3
(9.8)
4
(13.1)

2   4   6   8

Percent C₂ + C₃ + C₄ in Total Gas

Figure 8-2
Increase in the percent of ethane-
plus hydrocarbons in the total dissolved gas of reservoir rock waters
associated with Mesozoic source
rocks of western Siberia. [Sokolov
et al. 1972]

7-11a. The corresponding increase in ethane and higher hydrocarbons as a percent of the total dissolved hydrocarbon gas in reservoir rocks associated with these source rocks is shown in Figure 8-2 (Sokolov et al. 1972). Clearly, as soon as wet gas is generated in the source beds, it begins to show up in the associated reservoir waters.

The higher hydrocarbons that form liquid oil also are found at varying concentrations in formation waters. Some of these represent hydrocarbons dissolved from a nearby reservoir accumulation, while others are probably dispersed hydrocarbons that have not yet accumulated in a reservoir. An example of the kinds of analyses obtained on a conventional drill-stem test is shown in Table 8-1 (McAuliffe 1969). The concentration of dissolved gaseous and liquid hydrocarbons in formation waters has been used to some extent to evaluate the petroleum potential of a formation.

How much source rock is required to provide enough petroleum to fill a commercial reservoir accumulation? Smith (1971a) developed a mathematical model in which it was assumed that the contributing source rocks are in contact with the reservoir sands and the quantity of oil transferred into the sands is proportional to the area of contact between the sandstone and source rock. His model indicated that on the average 1.7×10^{-3} volume of oil would be produced by one volume of source rock. This means that a source rock drainage area of 10 km (6.2 mi) by 20 km (12.4 mi) would require only 5 m (16.4 ft) of the source rock to provide 1.7 million m³ (10.7 million bbl) of oil to the reservoir accumulation. Obviously, the larger the drainage area, the smaller the thickness of source rock required for commercial oil.

Table 8-1
Dissolved Hydrocarbon Content of Three Subsurface Brines

Hydrocarbon	g HC/10^9 g H$_2$O in the brines		
	1	2	3
Methane	525	1,825	5.86
Ethane	303	240	0.41
Propane	208	23	0.58
Isobutane	89	1.37	0.31
n-Butane	120	5.10	0.71
Isopentane	82	1.50	0.48
n-Pentane	56	1.40	0.54
Cyclopentane + 2-Methylpentane	34	24.4	0.31
3-Methylpentane	13	0.51	0.08
n-Hexane	17	0.69	0.15
Methylcyclopentane	23	5.19	0.12
Benzene	1,665	10,400	0.86
n-Heptane	–	1.90	0.33
Methylcyclohexane	–	2.76	0.46
Toluene	274	4,680	1.70

Source: McAuliffe 1969.

CHARACTERISTICS OF RESERVOIR OIL

Reservoir oils vary widely in properties from light gasolines to heavy viscous asphalts. Gas associated with the oil may be in a free gas phase above the oil accumulation or entirely dissolved in the oil. In the former case, the pool is called *saturated;* in the latter case, where there is no gas cap, the pool is called *undersaturated.* A condensate pool is a gas-phase accumulation in which the light liquid hydrocarbons are dissolved in the gas. If the pressure of a condensate pool is reduced during production, the liquid hydrocarbons will condense within the reservoir, resulting in a much lower recovery than production in the gaseous phase. Repressuring can return part of the condensate to the gaseous phase.

Statistics on the characteristics of giant fields reveal the habitat of reservoir oil. Table 8-2 compares 100+ million barrel fields in 1956 with 500+ million barrel fields in 1971. In those 15 years, the number of giant fields more than doubled, and they continued to represent about 85 percent of the world's reserves. There was a significant increase in the most productive depth interval from 3,500 to about 7,000 ft (1,067 to about 2,134 m). This is a reflection of deeper wildcat drilling plus deeper production on structures where the shallow fields have been depleted. The range from 6,000 to 8,000 ft (1,829 to 2,438 m) now includes about 40 percent of the total oil in giant

Table 8-2
Characteristics of Oil Fields with More Than 100 × 10⁶ bbl of Oil

	1956	1971
Number of fields	236	546
Percent of world reserves	83	85
Richest depth interval	∿3,500 ft (1,067 m)	∿7,000 ft (2,130 m)
Current basin position	Percent of Total Fields	
Shelf	68	70
Basin deep	11	10
Mobile belt	21	20
Lithology	Percent of Total Fields	
Sand	66	60
Carbonate	34	40

Source: Data from Knebel and Rodriquez-Eraso 1956; Gardner 1971; and Moody 1975. Fields of USSR and socialist countries not included. 1971 figures are for >500 million bbl.

fields. During the coming decades as drilling goes deeper, particularly outside the United States, the peak may shift to 8,000 or 10,000 ft (2,438 or 3,048 m) but probably not much deeper. The 12,000–15,000 ft (3,658–4,572 m) depth represents the oil generation phase-out in most sedimentary basins so fields at deeper horizons would probably produce only gas.

The present average depth and deepest pay for the 198 fields with over 500 million bbl of recoverable oil are shown in Figure 8-3 (Moody 1975). The average reservoir depth is around 7,000 ft (2,130 m).

In classic geosynclinal basins, about two-thirds of the number of giant oil fields are on the shelf side, where there is a favorable conjunction of reservoir beds and source facies. There are more oil accumulations in sands than in carbonates but the total quantity of oil in the two lithologies is about the same (Moody 1975). Carbonate reservoirs tend to be larger.

The most productive era is the Mesozoic. However, if the Middle East is excluded, the Cenozoic has the most production. The pattern of ultimate recovery versus time is shown in Figure 8-4. There is a sharp increase at the end of the Triassic, which some have attributed to the splitting apart of continents and changes in the earth's heat flow patterns at that time. The breakup of continents would increase the number of continental margins, create narrow seaways, and restricted basins with widespread deposition of evaporites. There is also the possibility that the Caledonian and Hercynian orogenies destroyed most of the oil fields of the Paleozoic era. The growth of the biomass over geological time could cause some of the difference in Figure 8-4, but some biologists maintain the biomass in the Palezoic era may not have been much different from now.

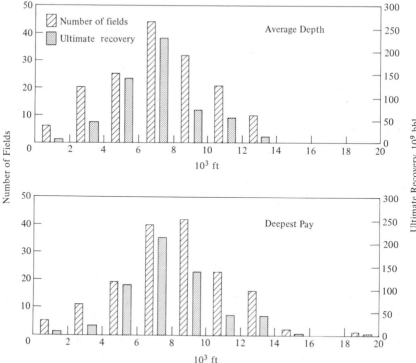

Figure 8-3
Average depth and deepest pay of giant oil fields (over 500 × 10⁶ bbl) in the world. [From
J. D. Moody, "Distribution and Geological Characteristics of Giant Oil Fields," in *Petroleum
and Global Tectonics,* eds. Alfred G. Fischer and Sheldon Judson. Copyright © 1975 by
Princeton University Press. Reprinted by permission of Princeton University Press.]

The world recoveries for the Paleozoic also may appear low because of a
lack of Paleozoic testing in much of the world. Figure 8-5 shows the oil
recovery-age pattern for giant fields in the United States, where drilling has
been intense. Here the Paleozoic era contains 38 percent of the recoverable
oil, the Cenozoic 35 percent, and the Mesozoic 27 percent. The Paleozoic
pattern in the rest of the world may look more like this in the next 30 years.

One characteristic of reservoir oil that is observed worldwide, is the
decrease in specific gravity (increase in °API) with depth. The maturation of
crude oils, which is discussed in the next section, occurs in both source and
reservoir rocks. The combination of lighter oils being generated from kerogen
with depth and of reservoir oils maturing to lighter oils with depth causes an
inevitable progression toward higher °API gravity oils. Figure 8-6 shows the
change in the percentage of oil fields with depth and the gravities of the oil.
These curves, which were calculated from data in the 1975 International

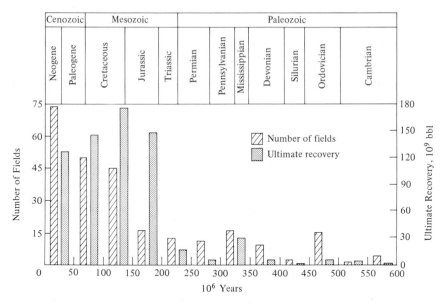

Figure 8-4
Age of giant oil fields in the world over 500 × 10⁶ bbl. [From J. D. Moody, "Distribution
and Geological Characteristics of Giant Oil Fields," in *Petroleum and Global Tectonics*, eds.
Alfred G. Fischer and Sheldon Judson. Copyright © 1975 by Princeton University Press.
Reprinted by permission of Princeton University Press.]

Petroleum Encyclopedia should be looked on as showing the general trend
worldwide. There is considerable variation in any one producing area, so this
cannot be used for predicting gravity changes with depth except in a very
rough manner. For example, in the 2,000–3,000 ft (610–914 m) depth range
worldwide, the °API gravities vary between 12 and 43. In the 11,000–12,000
ft (3,353–3,658 m) depth range, they vary from 20 to 53. Some extreme
anomalies include the Eocene Barracouta field of Australia with a 63°API
gravity oil at 4,700 ft (1,433 m), the Boscan field with 10° API gravity crude at
7,500 ft (2,286 m) in Venezuela, and the Ragusa field of Italy with 20° API
gravity crude at 12,460 ft (3,798 m).

There are many explanations for such anomalies. For example, biodegradation and washing by meteoric waters can make an oil heavy at depth.
Eroded reservoirs will contain a heavy asphalt residue from weathering. They
can be reburied and the asphalt will start maturing toward a crude oil. Oils
may be lightened by gas deasphalting, and condensates may escape through
vertical permeability to shallow reservoirs, causing wide variations in
depth–gravity patterns. Despite these anomalies, all crude oils are relent-
lessly moving toward condensate and eventually gas as they are exposed to
higher temperatures at depth.

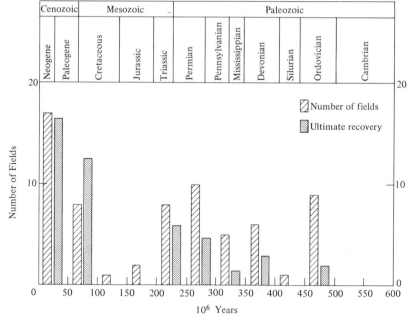

Figure 8-5
Age of giant oil fields over 500 × 10⁶ bbl in the United States. [Moody et al. 1970]

   The combination of source and reservoir maturation results in the number of oil and oil-plus-gas fields gradually diminishing with depth. In the 6,000–10,000 ft (1,829–3,048 m) depth range of most basins, there is a shift from an oil to a gas majority in reservoirs. Beyond the 12,000–14,000 ft (3,658–4,267 m) range, less than a fourth of the reservoirs contain oil, and below 20,000 ft (6,096 m) it is only a few percent. Deep drilling will not change this picture very much, because the basic cause is the effect of increased temperature rather than a lack of sufficient deep tests for oil.

THERMAL MATURATION

In petroleum reservoirs, the hydrocarbons in the crude oil and gas are being heated by earth temperatures that cause them to undergo slow but continuous changes in molecular structures toward a state of equilibrium. The equilibrium mixture that finally forms depends on the free energy of the system at constant temperature and pressure. Reactions usually proceed in the direction of a lower free energy. The most stable isomers of a hydrocarbon molecule are those with the lowest free energy.

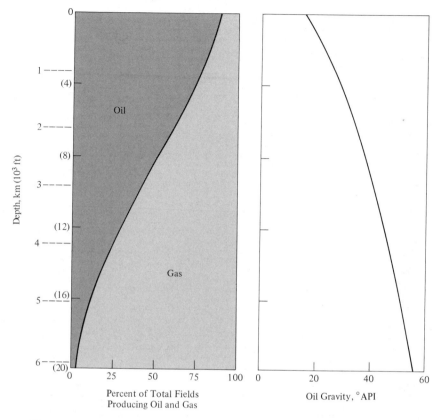

Figure 8-6
Decrease in the percentage of commercial oil fields with depth and the corresponding increase in °API gravity.

Five of the isomers of heptane are shown in Table 8-3 with data on their free energy and concentration in crude oils. The isomers are shown in order of increasing free energy; that is, decreasing stability. The 2,3-dimethylpentane, the most stable of the heptane isomers, has a free energy of formation of 2.4 kcal per carbon atom, and at 127°C (261°F), it would represent 29 percent of an equilibrium concentration of the heptane isomers. In contrast, 3-ethylpentane, with the highest free energy, would represent only 1.5 percent of the total isomers at equilibrium. The total in the last three columns does not add up to 100 because only five of several heptanes in crude oil are being compared.

The last two columns in Table 8-3 show the actual concentrations of these isomers in typical crude oils. Natural mixtures are nowhere near equilibrium. The n-heptane dominates and 2,2-dimethylpentane, which is second highest

Table 8-3
Comparison of Equilibrium and Actual Heptane Isomer Distributions in Crude Oil

Hydrocarbon	Free energy of formation in kcal/C atom	Equilibrium concentration at 127°C (257°F)	Actual concentration in 18 crude oils	Actual concentration in N. Sahara oils
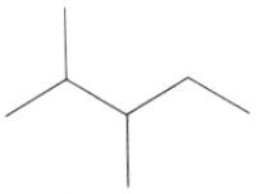	2.40	29	6	5
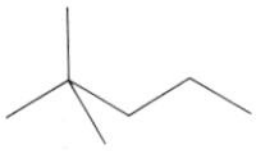	2.47	13	0.5	0.5
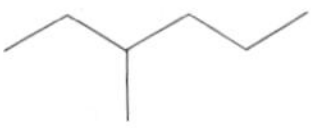	2.51	12	19	22
	2.62	8.5	56	52
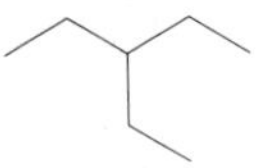	2.75	1.5	2.5	2.5

Note: Concentrations as percent of total heptane isomers in liquid phase (Tissot 1966).

in equilibrium concentration, is the lowest in actual crude oil concentration. The reason for this disparity is that *n*-heptane is more common in the source bed than the other isomers. More *n*-paraffins than 2,3-dimethylparaffins are being cracked from kerogen during catagenesis. Also, this hydrocarbon mixture does not represent a closed system since its origin. Instead, it has been involved in transport and accumulation mechanisms plus various losses after accumulation, all of which may alter the mixture.

Different types of hydrocarbons have much larger differences in free energy than the isomers of one type. Figure 8-7 shows the thermal stability of aromatic, naphthene, and paraffin hydrocarbons in terms of free energy of formation in kilocalories per carbon atom. The zero line corresponds to the free energy of the elements carbon and hydrogen. This figure is only valid at unit activity of the compounds and at a pressure of 1 atmosphere. At higher pressures, the relative differences are essentially the same, but absolute values change somewhat. The figure shows that at the low temperatures of shallow drilling, the aromatic hydrocarbons are the least stable and the low-molecular-weight paraffin hydrocarbons are the most stable. At the high temperatures encountered in deep drilling, only the hydrocarbon gases are stable, and at very high sediment temperatures only methane is stable. This is usually the only hydrocarbon found in high-temperature reservoirs. At temperatures

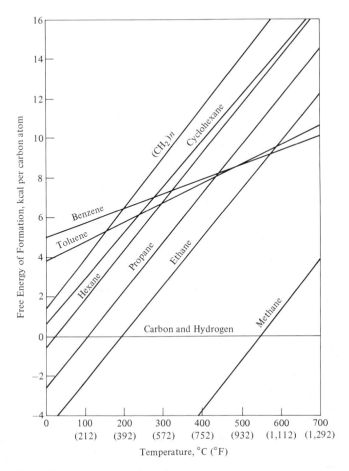

Figure 8-7
Thermal stability of hydrocarbons with increasing temperature. [Hunt 1975a; data from National Bureau of Standards, C-461, 1947]

well above those encountered in sediments, benzene and toluene become more stable than methane (note difference in the slope of the lines in Figure 8-7).

The following conclusions result from Figure 8-7: First, paraffins are the most stable hydrocarbons at the low temperatures comparable to sedimentary rocks, whereas aromatics are the most stable at very high temperatures. Second, the stability of paraffins increases (free energy decreases) as the number of carbon atoms in the molecule decreases. Methane is stable at temperatures up to 550°C (1,022°F), considerably higher than any sedimentary rocks. Third, the naphthenes are intermediate in stability between the aromatics and paraffins.

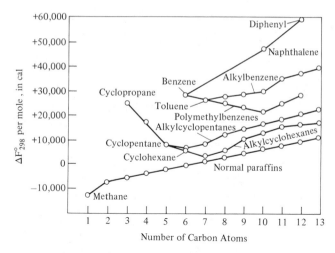

Figure 8-8
Thermal stability of hydrocarbons at 25°C. [Reprinted with permission from Andreev et al., *Transformation of Petroleum in Nature*, International Series of Monographs in Earth Sciences, v. 29. Copyright © 1968, Pergamon Press, Ltd.]

The last statement is more evident from Figure 8-8 where the free energy in calories per mole at 25°C (77°F) for individual hydrocarbons in each group are shown. It follows that all hydrocarbons should tend to move toward the paraffins and eventually the smallest paraffin, methane. It has the lowest free energy of all the hydrocarbons shown in Figure 8-8.

The free-energy data indicate that a ratio of naphthenes to paraffins (n/p) in petroleum will decrease with increasing age or depth of burial (temperature) of the reservoir. Kartsev (1964) calculated naphthene/paraffin ratios for the gasoline through lubricating oil fractions of 67 crude oils from the USSR. The average ratios are plotted in Figure 8-9. The influence of temperature (depth) on this ratio is evident from the slope of the line. The deepest samples within each geological era have the highest paraffin content (lowest n/p ratio). The influence of time is evident from the displacement of the lines to a more paraffinic ratio with increasing age. The change in the n/p ratio is logarithmic with a linear increase in depth (temperature), at least through the Mesozoic. The Paleozoic represents very long-term low-temperature reactions that do not appear to show an exponential change in n/p ratio with time.

Kartsev also noted that 65 oils from various parts of the world had n/p ratios averaging 2 for the Cenozoic, 1.25 for the Mesozoic, and 0.6 for the Paleozoic. Computing the ratio for the gasoline range only of 16 crude oils, Kartsev obtained 1.35 for the Cenozoic, 0.9 for the Mesozoic, and 0.4 for the Paleozoic. The gasoline fraction is less influenced by biogenic input, so it is a

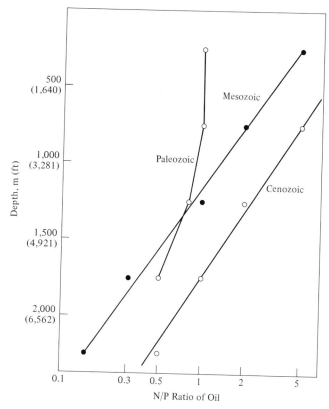

Figure 8-9
Change in the naphthene/paraffin ratio of the up to 550°C (1,022°F) boiling range (lubricating oil) of 67 crude oils of the USSR. Data averaged for 500 m depth intervals for each era. [Kartsev 1964]

truer measure of the n/p ratio resulting from catagenesis. Kartsev stated that his results agreed completely with what is predicted by thermodynamic considerations and by laboratory experiments.

The thermal alteration of reservoired oil to more and smaller paraffin molecules requires a source of hydrogen. There is very little free hydrogen in most reservoirs, but fortunately there are thermodynamically favored reactions that can make hydrogen available, namely the condensation of aromatics. The free energy of formation of benzene at 25°C (77°F) is 4,930 cal per carbon atom. In going from benzene with one ring to naphthalene with two, phenanthrene with three, and pyrene with four, there is a steady decrease in free energy to 4,015 cal. This decrease continues with further ring condensation all the way to graphite. Consequently, the formation of

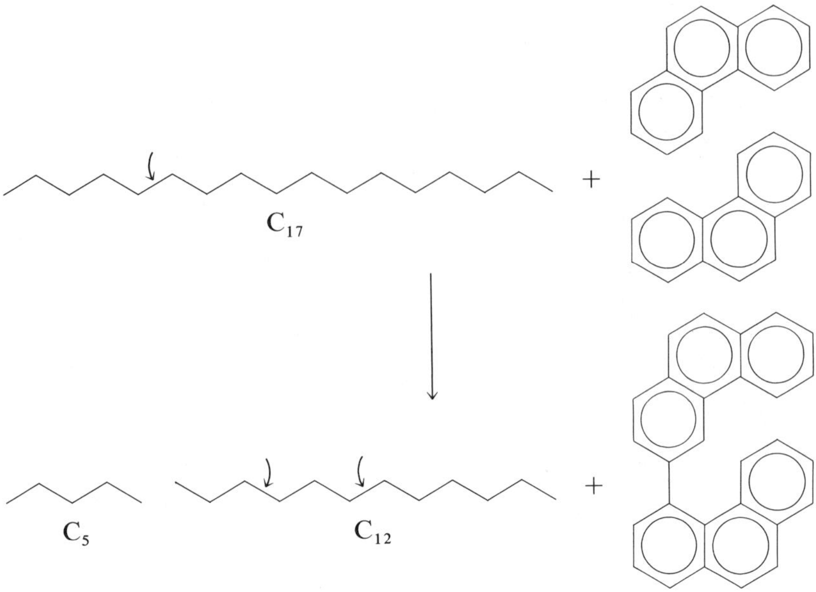

Figure 8-10
Formation of hydrocarbon gases and a pyrobitumen residue from disproportionation
reactions of oil molecules during the thermal alteration of petroleum in a reservoir.
[Connan et al. 1975]

polycondensed aromatic hydrocarbons to form an asphaltite or pyrobitumen
in the reservoir is favored.

This disproportionation scheme is schematically illustrated in Figure 8-10.
Phenanthrene shown at the top of this figure is a katacondensed polycyclic
aromatic hydrocarbon. Its rings are fused in a linear series. Katacondensation
releases four hydrogen atoms with each new added ring. Simple linking of
rings only releases two hydrogen atoms, as in the example shown of two
phenanthrene molecules being linked instead of condensed. The hydrogen
can be used to form the C_5 and C_{12} n-paraffins. Further rupture of the long
hydrocarbon chains to small C_3, C_5, and C_6 molecules occurs utilizing
hydrogen from pericondensation of the aromatics. Pericondensed polycyclic
aromatics are formed by fusing of rings on all sides and this can release more
than four hydrogen atoms. Eventually the end products in Figure 8-10 are
methane, ethane, and propane, plus a large pericondensed, polycyclic

aromatic molecule that is beginning to approach the structure of a pyrobi-tumen. The combination of kata- and pericondensation plus aromatization of naphthenes and linking of rings provides enough hydrogen to form the hydrocarbon gases.

Some process like the one just described has been occurring in reservoirs throughout geological time since the first oil began accumulating, as evidenced by the precipitation of black asphaltites and pyrobitumens in reservoirs along with gases. These bitumens have been identified in many reservoir cores. They are not always present, because the gases may migrate from their area of origin in the reservoir. Also, many gases form directly from kerogen in source rocks. There is no way, except possibly by a detailed geological–geochemical (isotope) study, that a hydrocarbon gas mixture could be identified as a thermal alteration product from oil within a reservoir or as a direct generation product from the kerogen of a nearby source rock.

Connan et al. (1975) observed that the saturated hydrocarbons (paraffins plus naphthenes) in pooled oils of the Upper Jurassic–Lower Cretaceous fields in the Aquitaine Basin of France increased with depth, being about 8 percent at 1,400 m (4,593 ft) and 40 percent at 3,400 m (11,155 ft). There was also a shift from large molecules to small molecules, with the n-alkanes peaking at C_{19} around 2,500 m (8,202 ft) and at C_{17} around 3,400 m (11,155 ft). The hydrogen for this increasing saturation probably came from the asphaltene fraction of the oil. Connan simulated this change in the laboratory by heating a shallow, immature crude oil from the Aquitaine Basin at 300°C (572°F) for several months. The aromatic/saturate hydrocarbon ratio in the crude oil changed from about 3.6 to 1 over a 12-month period and the n-alkane peaking decreased from C_{18} to C_{16}. The solid bitumen content of the crude oil increased from near 0 to about 40 percent.

Reservoir Bitumens

The wet gas and condensate formed during maturation of reservoir oil causes a natural deasphalting of the hydrogen-poor asphaltic fraction. Propane deasphalting, which is a widely used refinery process for removing asphaltenes from residuum, was discussed in Chapter 3. Its counterpart in nature is reservoir deasphalting, which probably occurs slowly over a long period of time. The process would result in precipitation of solid bitumen as it is formed. As the large aromatic molecules in oil condense to form asphaltenes and these condense further to pyrobitumens, there is a gradual increase in the solid phase until nothing is left in the reservoir but condensates, gases, and solid bitumens. Rogers et al. (1974) have analyzed and described the bitumens from Western Canada Basin oil reservoirs in detail. They considered reservoir bitumens to have two sources: (1) thermal alteration of oils to dry gas and solid bitumen and (2) deasphalting of heavy oils after solution of large amounts of wet gas. They distinguished these two processes from the change in $\delta^{13}C$ in the bitumens. Figure 8-11 shows these two processes for reservoir bitumens in the Western Canada Basin. The Triassic oil, which has a much higher H/C ratio than its corresponding bitumen, has about the same $\delta^{13}C$ value. Thermal cracking to form methane gives off carbon with a more negative $\delta^{13}C$ value than the starting product. Consequently, as methane is formed from the bitumen, the latter becomes richer in ^{13}C, changing in $\delta^{13}C$ from -30.4 to $-28.4 \%_{00}$. Triassic reservoirs in the west, which are at a higher level of maturation than those in the east, contain dry gas and a bitumen rich in ^{13}C.

The Devonian Leduc oil is similar in $\delta^{13}C$ to the bitumens in three Leduc reservoirs. These bitumens range in H/C ratio from 0.8 to 0.48, but their ^{13}C content is about the same. These different bitumens were originally dissolved

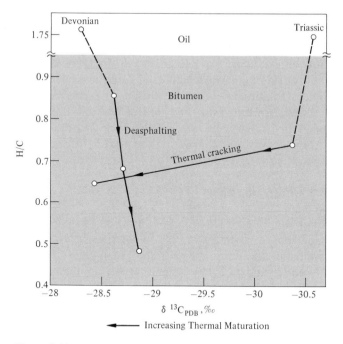

Figure 8-11
Change in properties of reservoir bitumens with deasphalting and
thermal cracking. Cracking forms CH_4 with a low ^{13}C content leaving
a reservoir bitumen with a higher ^{13}C content. Deasphalting does not
significantly alter the ^{13}C in bitumens. [Data from Rogers et al. 1974]

in the reservoir oils but deasphalting resulted from gas injection. The
precipitated bitumens showed very little change in $\delta^{13}C$ from the deasphalting process.

Rogers et al. (1974) also were able to interpret the thermal history of a
reservoir from the properties of the bitumen, such as the H/C atomic ratio,
the solubility in CS_2, and a discriminant function score. The latter was based
on several properties of the bitumen, including elemental analyses. They
found that mature oils had reservoir bitumens with an H/C ratio greater than
0.58 and a solubility in CS_2 greater than 2 percent. The bitumen H/C ratio in
wet gas or condensate reservoirs was 0.53 to 0.58 and below 0.53 was limited
to bitumens of metamorphosed dry gas areas. The H/C ratios of the reservoir
bitumens generally decreased westward into the deeper, hotter foothills belt
of the Western Canada Basin. In the USSR, the Shebela gas condensate
reservoir contains a pyrobitumen with an H/C ratio of 0.54, correctly fitting
the Rogers et al. classification (Zaritskaya and Zaritskiy 1962).

Reservoir bitumens precipitating from crude oil can be distinguished from
coal particles (Figure 8-29) on the basis of their atomic H/C ratios and their

atomic (N + S)/O ratios. Bitumens have been reported in the dry gas reservoirs of the Ellenburger Formation of the Brown–Bassett field in the Delaware–Val Verde Basin (Holmquest 1965). They also have been described in the Jurassic Smackover reservoirs of the Mississippi interior salt basin as an infusible asphalt with a high fixed-carbon ratio (Parker 1973). The more infusible pyrobitumens associated with dry gas are formed by thermal maturation, whereas many fusible asphalts result from oil and condensate deasphalting. Discussions with field geologists indicate that reservoir bitumens are more common than has been reported in the literature.

Change in Oil Gravity with Depth

The maturation of oil in a reservoir shows up as an increase in °API gravity (decrease in specific gravity) with depth. A typical example is shown in Figure 8-12, where the data for four oil fields in the Volgograd region of the USSR are plotted. Gabrielyan (1962) also noted that the paraffin content increased with depth and the sulfur content was highest in the most shallow samples.

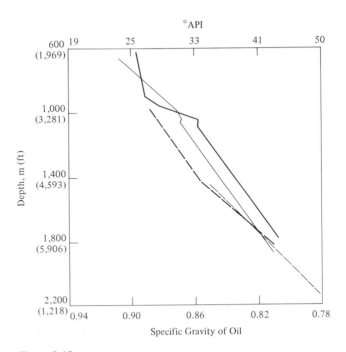

Figure 8-12
Decrease in specific gravity of oil with depth from pools of the Bakhmet'ev, Zhirnov, Archedin, and Klenov fields of the Volgograd region, USSR. [Gabrielyan 1962]

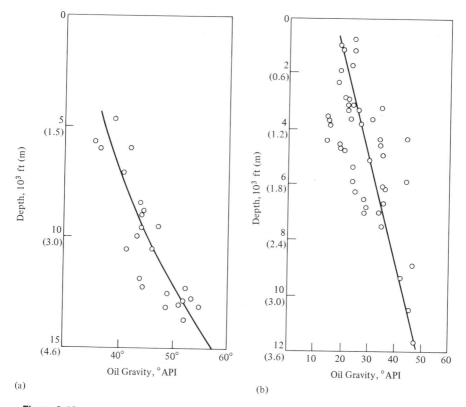

Figure 8-13
Increase in °API gravity with depth. (a) Ordovician Ellenburger reservoirs in Delaware–Val Verde Basin. [Holmquest 1965] (b) Pennsylvanian Tensleep reservoirs in Wyoming. [Hunt 1953]

Makarenko and Sergiyenko (1970) observed the same change in the eastern Ciscaucasia, where the specific gravity decreased from 0.89 to 0.81 (equivalent to the change from 26 to 42°API) in the subsurface temperature range from about 30 to 130°C (86 to 266°F). In this same interval, the paraffin hydrocarbons in the oil increased from 35 to 56 percent, naphthenes decreased from 47 to 31 percent, and the aromatics decreased from 18 to 13 percent. This caused the naphthene/paraffin ratio to change from 1.34 to 0.55.

The change in °API gravity with depth for oil accumulations in the Ordovician Ellenburger and Pennsylvanian Tensleep reservoirs are shown in Figure 8-13. The °API gravity of the Ellenburger oils increases about one degree for each 486 ft (148 m) of burial. Increases for the Tensleep and the previously shown Volgograd oils are one degree API for 333 ft (102 m) and

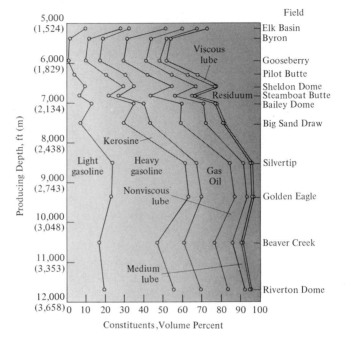

Figure 8-14
Composition of crude oils in the Tensleep Formation of Wyoming.
Some shallow oils have been degraded, while deep oils have matured
to light crudes. [Hunt 1953]

Table 8-4
Compositional Differences Between Deep and Shallow Tensleep,
Wyoming, Oils

Constituent	Volume % difference between Riverton Dome (11,700 ft) and Pilot Butte (6,300 ft)
Light gasoline C_4–C_7	+ 16.3
Heavy gasoline C_8–C_{10}	+ 20.0
Kerosine C_{11}–C_{12}	+ 3.5
Gas oil C_{13}–C_{20}	+ 2.6
Nonviscous lube oil C_{21}–C_{30}	− 3.1
Medium lube oil C_{31}–C_{35}	− 5.0
Viscous lube oil C_{36}–C_{40}	− 5.3
Residuum >C_{40}	− 29.0

Source: Data from Hunt 1953.

for 215 ft (66 m) respectively. The Western Canada Basin oils increase one degree API for each 270 to about 500 ft (82 to about 152 m) of burial. Most reservoir oils increase one degree API (decrease 0.005 in specific gravity) for every 200 to 400 ft (61 to 122 m) of depth.

The Tensleep oils of Wyoming are of particular interest because the geological evidence indicates that for approximately 100 million years between the Permian and Cretaceous these oils were believed to be within 1,000 ft (305 m) of the same depth of burial. They were probably identical in °API gravity and composition at the beginning of the Cretaceous. The Laramide orogeny caused intense deformation in the Wyoming area, resulting in some Tensleep reservoir sands being buried to 11,000 ft (3,353 m) while others were uplifted to within 1,000 ft (305 m) of the surface. The major differences in the composition of Tensleep oils today as shown in Figure 8-14 probably occurred since the Laramide orogeny. Reservoir maturation caused the deepest oils to become the lightest. They have over 50 percent gasoline. The largest difference in gasoline content (10 to 60 percent) is in the 6,000–9,000 ft (1,829–2,743 m) producing depth. It is interesting to note that there is very little difference in the kerosine, gas oil, and lubricating oil percentages with depth. The largest difference is in the decrease in residuum and the increase in gasoline.

Two of the Tensleep oil reservoirs, Riverton Dome and Pilot Butte, are only 32 miles (52 km) apart at a depth difference of 5,367 ft (1,636 m) in the Wind River Basin of Wyoming. Table 8-4 shows the compositional differences that exist today between these two nearby oil reservoirs. All fractions lighter than gas oil with less than 20 carbon atoms per molecule have gained in volume, while the heavier oil fractions have lost volume. Reservoir maturation of the Tensleep oils appears to have cracked the hydrocarbons containing more than 20 carbon atoms into smaller molecules, which become part of the gasoline, kerosine, and light gas-oil range. The largest contribution of cracked products comes from the residuum.

Tensleep oils at depths of less than 3,000 ft (914 m) have gravities below 25°API. Many of these are in contact with meteoric waters, which cause degradation through oxidation, microbiological decomposition, and solution of the lighter components. This has not affected the deeper Tensleep oils.

The importance of reservoir maturation to exploration is apparent when it is realized that many deep horizons were not tested in the past because they were believed to contain uneconomic heavy oil comparable to that in shallow reservoirs. The Tensleep is an example of a formation where deep tests were not planned because the Tensleep oil was generally known to be heavy when the first pools were discovered. Today there are many Tensleep and Phosphoria oils being produced with gravities higher than 40° API from fields deeper than 10,000 ft (3,048 m).

Oil and Gas Phase-Out Depths

In 1967, Landes published a paper on the eometamorphism (early metamorphism) of petroleum. At that time, the conversion with depth of oil to condensate and eventually to methane was a recognized phenomenon, and various maturation techniques generally related to coal metamorphism were used to define a petroleum floor. Landes developed a chart, based on well data, that showed the depth of the oil phase-out zone for geothermal gradients varying between 1 and 2°F/100 ft. Landes concluded that the commercial oil floor ranged from about 14,000 ft (4,267 m) at a geothermal gradient of 2°F/100 ft (3.65°C/100 m) to 27,500 ft (8,382 m) where the gradient is 1°F/100 ft (1.8°C/100 m).

Landes recognized that time as well as temperature is involved in reservoir maturation, so to keep the chart simple he placed the oil phase-out zone at the maximum depth where oil could occur in the youngest sediments. Figure 8-15 is a modified form of Landes' original chart, in which the approximate oil phase-out zone is shown for the Cenozoic, Mesozoic, and Paleozoic eras. The oil phase-out temperatures were estimated from well data, the time–temperature chart (Figure 4-17), and the maturation data presented in Chapter 7. The curves in this figure may be used in a very general way to indicate the transition horizons where the gas/oil ratios increase sharply. These curves add to Landes' original chart a time dimension that is admittedly rough but does recognize the shallower oil phase-out floor in Paleozoic compared to Cenozoic sediments. This figure also reemphasizes the concept of exploring in young, hot or old, cold basins. A well drilled to 8,000 ft (2,438 m) where the geothermal gradient is 1°F/100 ft (1.8°C/100 m) would find oil in the Paleozoic, but at 2.75°F/100 ft (5°C/100 m) oil would occur mainly in the Cenozoic at that depth. There would be gas in Paleozoic reservoirs.

Charts like Figure 8-15 are only useful for gross evaluations because they do not include the geothermal history of the hydrocarbons, as do the Lopatin and Tissot methods discussed in Chapter 7. For example, Tissot, Deroo, and Espitalié (1975) presented evidence for the late generation of Silurian oil at Hassi Messaoud in Algeria. Geochemical studies confirmed that the oil originated from lower Silurian source rocks and migrated into Cambro-Ordovician reservoir beds. The oil has a gravity of 49°API and produces from a depth of 3,400 m (11,000 ft). The geothermal gradient is 2.35°C/100 m (1.3°F/100 ft), which places the reservoir near the Paleozoic oil phase-out in Figure 8-15. If this oil was formed in the Mesozoic, however, it is well within the oil range.

Tissot's evidence is shown in Figure 8-16, where the change in reservoir depth and the quantity of oil and gas generated are both plotted against time. From 400 to 300 million years ago, the Hassi Messaoud reservoirs were never buried deeper than about 1,100 m (3,609 ft), which was too shallow to form

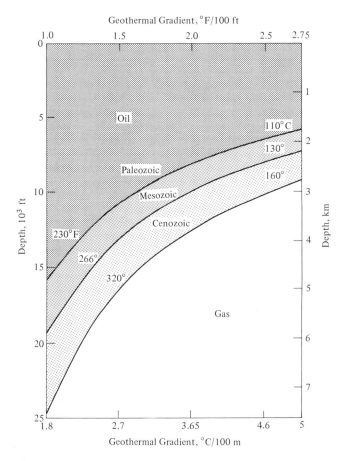

Figure 8-15
Estimated depth of oil phase-out zones for different geothermal
gradients and time periods.

anything but a small amount of early biological asphaltic oil. From the end of
the Carboniferous to the Permian, the reservoirs were uplifted and exposed,
destroying any early biogenic hydrocarbons. In the subsequent period of
deposition, an evaporite bed was deposited in the area, which effectively
sealed in the immature kerogen still present in the Silurian source beds.
Burial of these beds to depths greater than about 2,000 m (6,562 ft) initiated
the generation of large amounts of petroleum and gas that accumulated in the
Hassi Messaoud reservoirs. This generation occurred during the Cretaceous
or later. Consequently, although the oil and gas were generated in Silurian
source beds and are reservoired in Cambro-Ordovician rocks, they were
formed during the Cretaceous or later periods. The Mesozoic evaporites

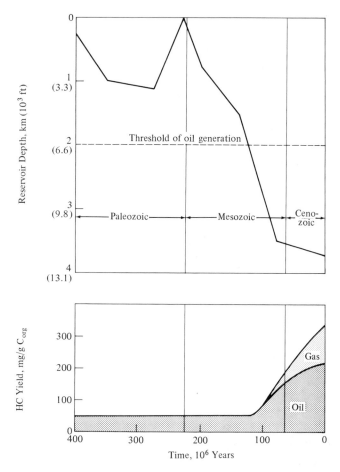

Figure 8-16
History of the depth of Cambro-Ordovician reservoirs and the
quantity of hydrocarbons formed in Silurian source rocks during
geologic time in the Hassi Messaoud, Algeria, area. [Tissot, Deroo,
and Espitalié 1975]

provided a perfect seal for the freshly generated hydrocarbons. The Silurian
source rocks were protected from erosion in the low parts of the basin and in
synclinal structures in the Late Paleozoic.

The phase-out of gas in source rocks can be recognized from its disappearance in analyses of cuttings. It is more difficult to predict a phase-out depth
for reservoir gas. Because methane is thermally indestructible at sedimentary
rock temperatures, some geologists maintain that the only limiting control on
deep methane production is reservoir porosity. If this is true, then deep gas
production would be limited to fracture porosity and to carbonate rocks,
which have enough internal strength to maintain porosity to very great

depths. In contrast, sands lose most of their porosity in deep horizons. Rittenhouse (1973) noted that the average porosity of Miocene sands in the U.S. Gulf Coast decreased from about 32 percent at 6,000 ft (1,829 m) to 18 percent at 16,000 ft (4,877 m). This is 1.4 percent for each 1,000 ft (305 m) of additional burial. Sand grain rearrangement, modification, and solution, plus the filling of pores with all types of crystalline materials, caused the decrease in porosity.

Chemical reactions may create a methane floor in some areas. Orr (1974) has pointed out that any sulfate, such as gypsum or anhydrite, is capable of reacting with hydrogen sulfide to form elemental sulfur. Sulfur and H_2S can then react to form polysulfides, which are powerful oxidizing agents. If a methane reservoir acquires sulfur and polysulfides, the methane can be destroyed, yielding H_2S and CO_2. This reaction is more apt to occur in carbonate reservoirs, which frequently contain H_2S in deep basinal facies because of the absence of iron (see discussion in Chapter 5). However, any reservoir in contact with sulfate brines or $CaSO_4$ has an inexhaustible supply of sulfur for oxidation of hydrocarbons.

Methane and the heavier gaseous hydrocarbons all produce H_2S when treated with sulfur at elevated temperatures and pressures in the laboratory. The presence of increasing amounts of H_2S and sulfur during deep drilling would not be encouraging for finding methane (Hunt 1975a).

Methane also could be lost through diffusion or migration along deep-seated fractures or faults. Considering the factors that could destroy or remove methane, and the decrease in porosity, it is unlikely that very many profitable methane accumulations will be found at depths much beyond 30,000 ft (9,144 m). Those that are profitable will probably be in areas with geothermal gradients less than 1.8°C/100 m (1°F/100 ft). Lower temperatures would preserve hydrocarbons longer, postpone the generation of H_2S from nearby source beds, and slow up some of the processes that reduce reservoir porosity.

Carbon Isotope Changes During Maturation

In a gas–oil mixture, the gas-phase hydrocarbons generally contain from 11 to 22 ‰ less ^{13}C than the oil (Silverman 1964), because when methyl, ethyl, or propyl groups split off from larger molecules they tend to leave behind the heavier carbon isotope, ^{13}C. Silverman determined the $\delta^{13}C$ value for several fractions of a crude oil and found that the gas and gasoline fractions boiling up to about 93°C or 200°F (through C_6) showed a deficiency in ^{13}C, compared to the higher boiling fractions, which were very close to the whole crude. The heavy lube oil and residuum fractions were slightly enriched in ^{13}C, indicating that they were cracking off the light hydrocarbons forming the gas and gasoline.

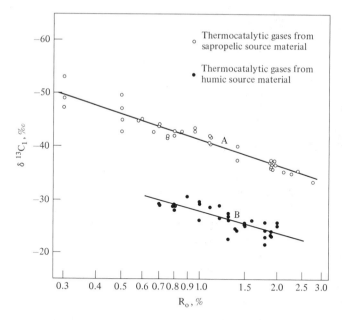

Figure 8-17
Variation in $\delta^{13}C$ of methanes from natural gas deposits plotted against the rank of maturity (vitrinite reflectance) of their source rocks. Methanes from sapropelic source rocks are along line A, while those from humic source rocks are along line B. [Stahl 1977]

In the discussion of Figure 5-11, it was pointed out that cracking at low temperatures yields a methane more deficient in ^{13}C than does cracking at high temperatures. Also, coals and related continental deposits yield methane with a higher ^{13}C value than marine deposits. The relationship between source material, maturation, and the ^{13}C content of methane is shown in Figure 8-17. In this figure, the vitrinite reflectance R_o of sediments is plotted against the $\delta^{13}C$ value of methane originating from those sediments. Thus, the humic deposits (line B) form methane with a $\delta^{13}C$ around $-30\,\%_0$ from organic matter at a maturation level of R_o of 0.8 percent. In more deeply buried sediments at an R_0 of 2 percent, the methane formed has a $\delta^{13}C$ of $-25\,\%_0$. Methane from sapropelic sediments (A) follows a similar change with maturation, but the methane is more deficient in ^{13}C by about $14\,\%_0$, compared to the methane from humic organic matter (Stahl 1977).

Laboratory experiments generally duplicate the maturation changes observed in nature. For example, heating the gasoline fraction of a crude oil at 165°C (329°F) for about three weeks produces methane, by cracking, that is $10–11\,\%_0$ lower in ^{13}C than the gasoline. The gasoline goes up about $2\,\%_0$ in ^{13}C.

Changes in Hydrocarbon Ratios with Depth

The decrease in the naphthene/paraffin (n/p) ratio of crude oils with time and increasing temperature (depth in Figure 8-9) is accompanied by changes in other hydrocarbon ratios. Young, immature oils contain more isoalkanes, whereas old, mature oils contain more of the n-alkane homologs. Consequently, ratios of i-C_4/n-C_4 and i-C_5/n-C_5 decrease with depth. This is related to the increased formation of n-alkanes in deeper horizons.

Among the naphthenes, the ratios of methylcyclopentane to cyclohexane and of methylcyclopentane to methylcyclohexane both decrease with depth. In fact, the broader ratio of total cyclopentanes to cyclohexanes generally decreases with depth. The four-, five-, and six-ring naphthenes derived from biological carbon structures decrease considerably with respect to the one-ring naphthenes. Total naphthenes decrease with respect to paraffins plus aromatics. Naphthenes are believed to play a dual role in maturation, some molecules splitting open to form paraffins, while others yield hydrogen to the paraffins by forming aromatics. More mature oils generally have more aromatics in the heavier fraction, most of which were derived from aromatization of the naphthenes.

Heterocompounds—that is, those containing nitrogen, sulfur, and oxygen in their molecular structure—generally decrease with depth. Heavy immature oils contain large quantities of these compounds compared to the light mature oils. More details on the nature of these reactions, along with their verification by laboratory experiments, are described by Andreev et al. (1968).

Optical Activity of Petroleum

The ability of crude oil fractions to rotate a beam of plane-polarized light was discussed in Chapter 4, because optical activity is one of the many proofs of the biological precursors to petroleum molecules. The optical activity of oil has been known for a long time but it was not until the 1930s that a very practical use was found for this property. The automobile industry in the United States was expanding rapidly at that time, and the high-performance engines required good lubricating oils. The best lube oils in the country were refined from Pennsylvania-grade crude oil because of its high paraffinicity. Going west, the crude oils became increasingly naphthenic and aromatic, with a corresponding reduction in lubricating oil properties. Today, additives are used to offset this difference, but in those early years Pennsylvania crude sold for a considerable premium over West Coast crudes. Consequently, it was not unusual for an enterprising promoter to sell cans of West Coast lubricating oils with Pennsylvania-grade labels on them.

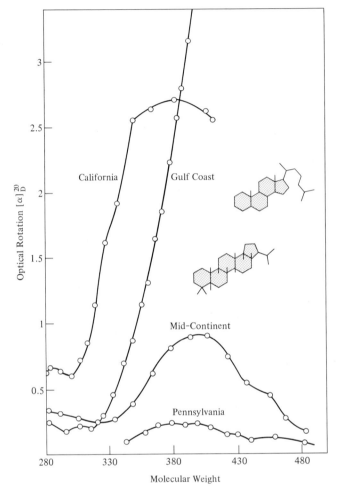

Figure 8-18
Optical rotation vs. molecular weight for the neutral lubricating oil
fraction of various U.S. oil refinery blends. $[\alpha]_D^{20}$ means rotation at
20°C, sodium D light. [Reprinted with permission from Fenske et al.,
Optical Rotation of Petroleum Fractions, Industrial and Engineering
Chemistry. Copyright by the American Chemical Society, 1942]
Cholestanes (top structure) and the hopanes (lower structure) are
among the hydrocarbon structures causing optical activity. These
are steranes and triterpanes, respectively.

The Pennsylvania Grade Crude Oil Association asked Professor M. R.
Fenske (1942) of Pennsylvania State University to develop a method to easily
distinguish Pennsylvania lubricating oils from those of other producing areas.
Because the optical activity of crude oil is concentrated in the lube oil fraction,
this was the obvious property to use to distinguish the oils.

Figure 8-18 shows the change in specific optical rotation with the molecular weight of neutral lubricating oil fractions from refineries in various parts of the United States. The Pennsylvania oil clearly has the lowest optical rotation, the Gulf Coast and California oils having the highest. The Pennsylvania Grade Crude Oil Association was able to use data of this type to settle court cases involving the labeling of cans as Pennsylvania-grade crude oil.

Oakwood et al. (1952) concentrated the optical activity of petroleum and was able to obtain specific rotations as high as +36.8. He determined that the rotation was caused by four- or five-ring naphthene hydrocarbons. Subsequently, Whitehead (1971) and his colleagues established, through mass spectrometry and nuclear magnetic resonance spectrometry, that the optical activity of petroleum in this high-molecular-weight range was caused by a mixture of steranes and triterpanes. Cholestane, whose structure is the upper one shown in Figure 8-18, is one of a group of steranes derived from the sterols, which are widespread in all living organisms. The lower structure is a typical pentacyclic triterpane derived from natural triterpenoids. Many triterpenoid homologs have been identified in crude oils and petroleum source rocks. The triterpenoids are widely distributed in plants.

Amosov (1951) measured the optical rotation of the lube oil fraction of several USSR crudes. He tentatively concluded that this fraction of all Tertiary oils had optical rotations above 1°, whereas the Mesozoic and Paleozoic oils were below 1°. In a later study (1955), Amosov compared the optical rotation of the total oil without resins, asphaltenes, and paraffins. These values would be lower than a narrow lube oil range, but they would show the same difference. Table 8-5 contains Amosov's results, and it is clear that there is a decrease in optical rotation with increasing age. Optical rotation is a useful tool in crude oil correlation, because it is a relatively simple measurement and it varies widely, depending mainly on the sterane–pentacyclic triterpane content of the oil.

Table 8-5
Change in Optical Rotation of Crude Oils with Age

Age of oils	Number of oils	Optical rotation + [α]D[a]
Tertiary	86	0.63
Cretaceous	18	0.28
Jurassic	20	0.20
Carboniferous	23	0.24
Devonian	21	0.18
Silurian	14	0.12

[a]The + [α]D means that the monochromatic light of the D line (589.3 mm) of sodium is rotated clockwise (+). [α] is the specific rotation.
Source: Data from Amosov 1955.

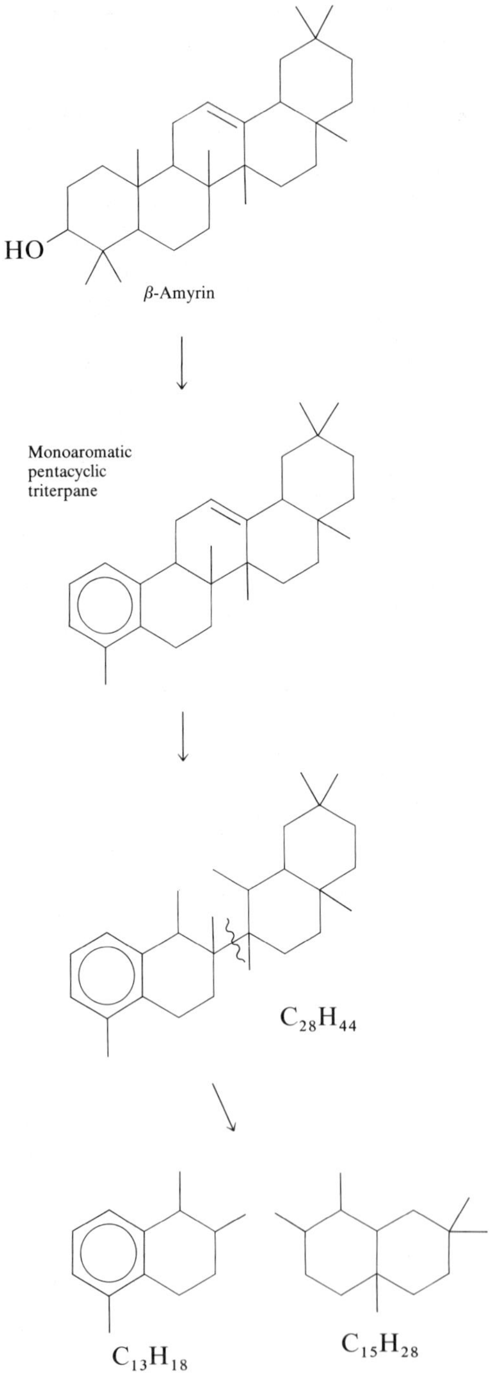

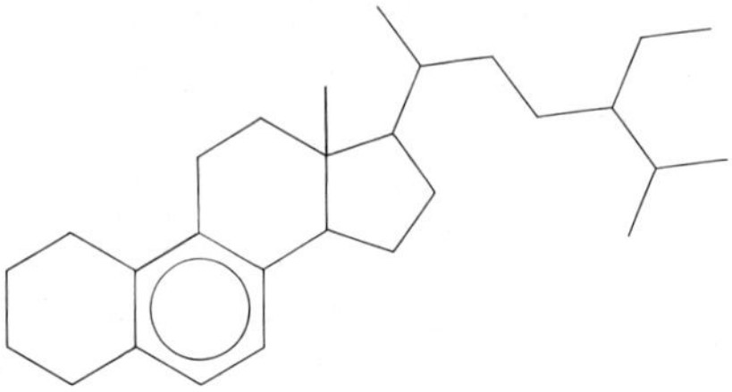

β-Amyrin

Monoaromatic sterane

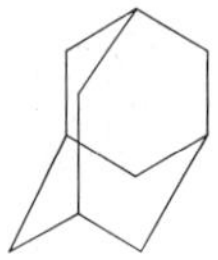

Monoaromatic
pentacyclic
triterpane

Adamantane

$C_{28}H_{44}$

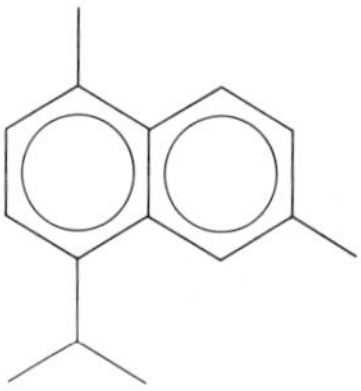

Cadalene

$C_{13}H_{18}$ $C_{15}H_{28}$

Changes in Biological Markers of Crude Oils

Heavy immature crude oils contain proportionately larger quantities of isoprenoids, terpenoids, and other biological generation products than do light, more mature crude oils. Pristane and phytane are found in higher concentrations in immature oil. Naphthenic crude oils of Rumania with specific gravities between 0.95 and 0.91 (18 to 23°API) contain from 209 to 325 ppm of adamantane (Landa and Hala 1958). More deeply buried oils with gravities of 0.85 to 0.79 (35 to 47°API) contain no adamantane. Although the source of adamantane (Figure 8-19) is not well defined, it is similar in structure to the terpenoids.

Bendoraitis (1974) analyzed several crude oils from the Eocene of southern Texas and found they contained a range of bi-, tetra-, and pentacyclic hydrocarbon molecules. Mass spectrometry identified several pentamethyl-decahydronaphthalenes, isoprenoid alkanes, and aromatic tetra- and penta-cyclic hydrocarbons presumably derived from the triterpenes. The sesquiter-pene, cadalene, also was identified as a prominent constituent. Bendoraitis proposed a maturation scheme for the formation of the various hydrocarbons he identified, as shown in Figure 8-19. The β-amyrin would lose the OH and CH_3 groups and hydrogen to form a monoaromatic pentacyclic derivative. This would undergo ring opening and subsequent cleavage to yield the C_{13} and C_{15} ring compounds shown at the bottom. More mature crude oils do have higher concentrations of the C_{13} and C_{15} cyclic hydrocarbons, while less mature oils have more of the pentacyclic triterpanes. Structures such as those in Figure 8-19 are useful in making precise correlations of different crude oils and their source rocks.

DEGRADATION PROCESSES

Thermal maturation improves the commercial value of an oil accumulation by increasing the quantity of light hydrocarbons and the paraffinicity, while reducing the percentage of asphaltic compounds high in nitrogen, sulfur, and oxygen. Degradation processes have the opposite effect. They reduce the economic value of an oil by destroying the paraffins, removing the light ends, and oxidizing the remaining fractions of the oil. In extreme cases, degradation processes can completely destroy the oil accumulation. Whereas increased thermal maturation increases the °API gravity of an oil, an increase in degradation lowers the gravity.

Figure 8-19
Biological marker hydrocarbons in petroleum and a postulated source of the C_{13}, C_{14}, and C_{15} cyclic compounds. [Bendoraitis 1974]

There are two ways in which degradation processes are actually beneficial. First, if uplifting and erosion exposes a large oil accumulation, degradation processes can create an asphalt seal that will effectively trap a large percentage of the oil before it is destroyed. Asphalt seals are not as effective as most types of cap rock. Consequently, there is a complete range of examples in nature from effective asphalt seals over oil reservoirs to examples where asphalts or asphaltites are the only remaining evidence of a former oil accumulation.

The second beneficial effect can occur when a crude oil that is paraffinic in the gasoline range undergoes mild bacterial degradation resulting in removal of the n-paraffins. Paraffins in the gasoline range, as discussed in Chapter 3, lower the octane rating for spark ignition engines. If bacteria destroy these n-paraffins, the octane rating is improved and the gasoline fraction is more valuable.

Biodegradation

Sediments, soils, and waters contain a wide variety of microorganisms that can utilize hydrocarbons as a sole source of energy in their metabolism. Paraffins, naphthenes, and aromatics, including gases, liquids, and solids, are all susceptible to microbial decomposition. There are more than 30 genera and 100 species of various bacteria, fungi, and yeast that attack one or more kinds of hydrocarbons. Also, microbial populations are highly adaptable and can alter their metabolic processes depending on the hydrocarbons available. Microbes are widely distributed in nature. They are beneficial in causing the relatively rapid disappearance of petroleum from natural seeps and spills. They are destructive in causing the deterioration of asphalt-base highways, asphalt-coated pipelines, and the contamination of stored gasoline and jet aircraft fuels. Military jet planes have been known to crash owing to the clogging of fuel systems using microbially altered fuels.

Several reviews have been written on the biological oxidation of hydrocarbons (ZoBell 1946a; McKenna and Kallio 1965). The general mechanism of attack is shown in Figure 8-20. Hydrocarbons are oxidized to alcohols, ketones, and acids. For example, n-butane is oxidized to methylethylketone and butyric acid. The organism *pseudomonas methanica*, which grows at the expense of methane, also will oxidize ethane, propane, and butane in a mixture yielding the corresponding acids. Long-chain paraffins are oxidized at the terminal of carbon atoms to yield di-acids. Naphthene and aromatic rings are oxidized to di-alcohols on adjacent carbon atoms, as shown in Figure 8-20. The order in which the hydrocarbons are oxidized depends on a variety of factors, but in general small molecules up to C_{20} are consumed before large ones, and within the same molecular weight range the order is usually n-paraffins first, followed by isoparaffins, naphthenes, aromatics, and polycyclic aromatics. Single-ring naphthenes and aromatics are attacked before

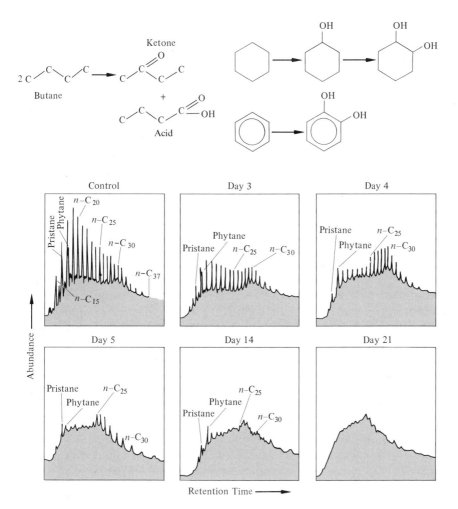

Figure 8-20
Microbiological oxidation of hydrocarbons showing conversion of n-butane to ketone and acid and conversion of cyclohexane and benzene to alcohols. Gas chromatograms show disappearance of n-paraffin peaks first in the C_{16}–C_{25} range and later in entire range during incubation of Saskatchewan crude oil with a mixed microbe population at 30°C. Original oil is shown along with changes over a 21-day period. [Jobson et al. 1972]

isoprenoids, steranes, and triterpanes. Seifert and Moldowan (1979) found diasteranes and tricyclic terpanes to survive heavy biodegradation, so they can be used as source fingerprints in biodegraded oils.

Jobson et al. (1972) treated crude oil samples with pure and mixed bacterial cultures and observed considerable degradation in 21 days of incubation. A North Cantal (Saskatchewan) oil changed in specific gravity from 0.827 to

1.046 (from $40°API$ to $5°API$). Thirty percent of the paraffin–naphthene fraction was destroyed, along with a few percent of the aromatic hydrocarbons. Figure 8-20 shows the change in the gas chromatograph analyses of the North Cantal Saskatchewan oil over the 21-day period. Note that at 4 days it has lost a substantial percentage of the n-paraffins with chain lengths shorter than C_{25}. At 5 days, even these longer chains are being attacked, and at 14 days most of the n-paraffins have disappeared.

Similar experiments and field observations indicate that after oxidation of n-alkanes the single-branched alkanes are attacked, followed by the double-branched hydrocarbons. Condensed-ring naphthenes, especially with six rings, are not readily attacked. Thiophenes and other sulfur compounds are concentrated in the oil because the microbes do not alter them.

The bacterial alteration of oil in its reservoir can occur if reservoir temperatures do not exceed about $82°C$ ($180°F$) and if oxygen, inorganic trace nutrients, and water are present. It was once thought that oxygen is not present in the subsurface, but modern analytical techniques indicate concentrations of 1 to several ppm dissolved oxygen are not unusual. The requirement is that the subsurface waters be in hydrodynamic connection with the surface as shown in Figure 8-21. In this hypothetical example, the oil seep on the far left and the oil accumulation in the deep aquifer with moving water could both be subject to bacterial degradation. The oil trapped in the zone of stagnant water could not be degraded, assuming that it has not had a connection to the surface since accumulation.

An example of an oil field that has been partially degraded by meteoric water is Montana's Bell Creek field (Winters and Williams 1969). The field produces oil ranging in gravity from less than 32 to over $42°API$. The heaviest crude as shown in Figure 8-21 is in the central area, where water resistivities are 2.2 ohms. This water contains about 8 ppm oxygen, and the invasion of microbes has resulted in a complete elimination of n-paraffins out beyond C_{25}. The $35°API$-gravity oil at the northern end of the field has the n-paraffins missing up to about C_{17}, while the $40+°API$-gravity oil at the southern end of the field has not been altered. Reservoir temperatures are about $35-41°C$ ($95-105°F$). Microorganisms thrive within this range. The reservoir, which is located near the edge of the basin, has an average depth of about 4,500 ft (1,372 m). The altered oils from the middle and updip parts of the field contain three times as much nitrogen as the unaltered oil to the south. This is more than would be expected from the increase in nitrogen compounds because of removal of paraffins. Microbial activity appears to be introducing nitrogen into the crude oil. Specific optical rotation of the aliphatic fraction of the altered oils is about twice that of unaltered oils. This is more than the paraffin loss would lead us to expect and indicates optically active compounds are being added to the oil by the microorganisms.

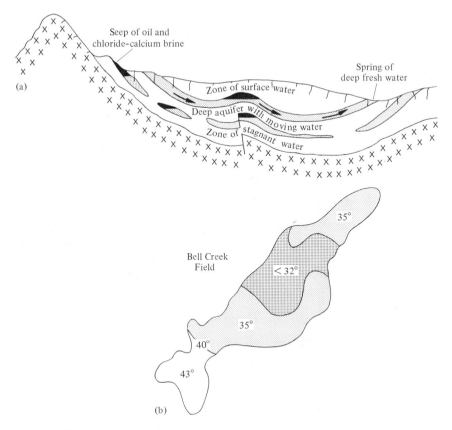

Figure 8-21
(a) Microbial alteration of hydrocarbons will occur at seep outlets and in deep, meteoric
water aquifers where microbes and dissolved oxygen can penetrate. Stagnant connate water
zones are preserved. [Courtesy of P. A. Dickey] (b) °API gravities of Bell Creek, Montana,
crude oils. Meteoric (2.2 ohm) water in center of field has degraded the oil, while connate
(0.25 ohm) water at southern end has unaltered oil. [Winters and Williams 1969]

Harwood (1973) believes that subsurface waters must contain at least 8
mg/liter of dissolved oxygen in order for aerobes to oxidize a major part of an
oil accumulation.

Philippi (1977) plotted the ratio of normal heptane to total heptane
paraffins against reservoir temperature for several oil fields where there was
evidence of bacterial degradation. Such a ratio would be low in bacterially
degraded oils and relatively high in unaltered oils because of preferential
consumption of normal alkanes by bacteria. In the Los Angeles Basin, San
Joaquin Valley, and Louisiana Gulf Coast, Philippi (1977) found a sharp drop

in this ratio from about 0.5 to 0.05 at reservoir temperatures below about 77°C (171°F). In southern Sumatra and southeastern Borneo, the drop occurred at about 54°C (130°F). Philippi (1977) concluded from all his studies that the cutoff temperature for microbial degradation averaged 66°C (150°F) with the maximum around 82°C (180°F). He also observed an increase in optical activity with biodegradation.

Oxidation of some hydrocarbons by anaerobes in the absence of dissolved oxygen may occur if there is sulfate ion in the formation waters. Davis (1967) made a detailed study of the biodegradation of oil in reservoirs of the Eocene Wilcox Carrizo sands of south Texas. Meteoric waters move through these sands at a rate of 15 to 30 m/year (50 to 100 ft/year). An oil accumulation trapped in a small faulted structure of the Carrizo contains only 3.5 percent paraffins in the saturates, most of the oil being naphthene and aromatic hydrocarbons. The situation is somewhat analogous to the upper part of Figure 8-21. Sulfate ion is found throughout the aquifer but methane, ethane, and hydrogen sulfide are found only downstream from the oil accumulation. Davis found both facultative bacteria and sulfate reducers in the aquifer. The hydrocarbon oxidizing ability of the latter were tested in the laboratory with radioactive methane and ethane. After about a month of culture, both the carbon dioxide formed by bacterial metabolism and the bacteria bodies themselves were radioactive indicating that methane and/or ethane were utilized by the sulfate-reducing bacteria. Other studies have indicated that anaerobic sulfate-reducing bacteria can degrade oils in a manner similar to aerobic strains but require more time (Dostalek et al. 1957; Bailey et al. 1973).

The northeast edge of the Williston Basin in Saskatchewan contains a number of Mississippian Mission Canyon reservoir oils that have been altered by bacterial degradation. In the area of saline brines to the southeast in the Williston Basin, the oils are about 36°API gravity and 1 percent sulfur, whereas in the fresher water area to the northwest the oils drop to 15°API gravity and increase to 3 percent sulfur. A comparison of the gas chromatograms of the saturated hydrocarbons of Stoughton (southeast) and High Prairie (northwest) oils are shown in Figure 8-22. The n-paraffins in the High Prairie oils have been biodegraded, leaving only the isoprenoids, pristane and phytane. The ratio of pristane to n-C_{17} and phytane to n-C_{18} increased steadily from the unaltered crudes in the southeast to the altered crudes in the northwest. The proportion of saturate hydrocarbons decreased from 47 to 19 percent, while the nitrogen, sulfur, and oxygen compounds in the crude increased 37 percent and the asphaltenes increased 100 percent. One- and two-ring hydrocarbons were attacked more readily than the larger polycyclic hydrocarbons (Bailey et al. 1973).

Data on microbial alteration have been published for oils from the MacKenzie Delta, the Hackberry fields of Louisiana, and the Moonie field of Australia. This is only an indication of the thousands of oil accumulations that

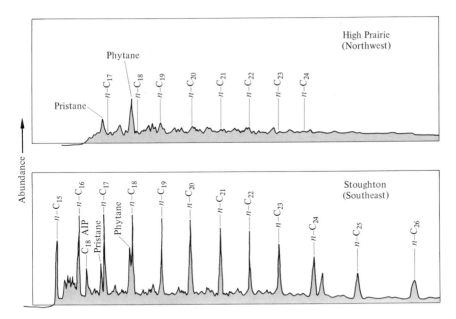

Figure 8-22
Gas chromatograms of saturated fraction of Mission Canyon oils of Williston Basin, comparing biodegraded High Prairie oil with unaltered Stoughton oil. C_{18} AIP is an 18-carbon anteisoprenoid alkane. [Bailey et al. 1973]

have been altered by bacteria. It has been estimated that as much as 10 percent of the world's petroleum reserves has been destroyed by bacteria and an additional 10 percent altered. When published analytical data shows paraffinic oils of a formation to change to naphthenic in contact with meteoric waters, bacterial degradation may be suspected. For example, the Devonian oils of northwestern Pennsylvania are paraffinic throughout, with low correlation index (see Chapter 11) values for oils of the third Franklin sands (Figure 8-23). Two oils from the Franklin first and second sands are highly naphthenic in the lower boiling cuts, as shown by the correlation indices in the 35–50 range. These oils are associated with relatively fresh water entering the outcrop of their producing sand. In this case, microbial alteration caused an increase in the antiknock value of the gasoline range by selectively removing the low-molecular-weight n-alkane hydrocarbons. In terms of economic value, these first and second sand crudes are ideal in having a high-octane (low normal paraffin) gasoline and a high-viscosity-index (high n-paraffin) lubricating oil.

It should be emphasized that oils in shallow, low-temperature reservoirs are not always microbially altered. In Kiowa County, Oklahoma, an unaltered oil is produced at 300 ft (91 m) in granite wash by a fault zone. The oil

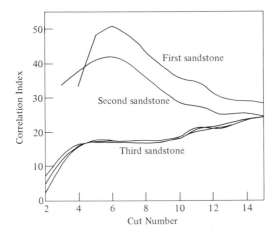

Figure 8-23
Correlation index curves for
U.S. Bureau of Mines Hempel distillation of Devonian oils, Venango
District, Pennsylvania. [Dickey
et al. 1943] Low-molecular-weight
paraffins have been microbially removed from crudes of the first
and second sands, leaving a crude
with naphthenic high-octane gasoline (cuts 2 to 6) and a paraffinic
lube oil (cuts over 13). Oil in
the third sand is paraffinic in the
entire range (low CI).

presumably migrated up the fault from its source in the Viola Limestone and
has not been altered in the process. Philippi (1977) found many oils were not
biodegraded at reservoir temperatures lower than 80°C (176°F) in the basins
he studied.

Chemical and Physical Degradation

Water washing, inspissation, and oxidation are other mechanisms by which
crude oil can change within the reservoir. Asphalt seals at the outcrop of seeps
and asphalt mats at the oil–water interface of pools in contact with meteoric
water (Figure 8-21) have formed by a combination of these processes and
microbial alteration.

Asphalt seals are much more important than most geologists realize. In the
San Joaquin Basin of California, the giant Coalinga field, with over 600
million barrels of recoverable oil, and the Kern River field, with more than
700 million barrels, are both sealed in by thick asphalt covers. The giant
Lagunillas field of Venezuela with reserves measured in the billions of barrels
is sealed by an asphalt outcrop. The combination of microbial and other
degradation processes has caused the change in gravity as shown in Figure
8-24. Fresh water covers a belt 5 to 10 km (3 to 6 miles) wide from the outcrop
to the shallowest oil with 12°API gravity. The oil becomes lighter downdip,
being about 20°API 1,500 m (4,921 ft) from the outcrop. In the deepest
sections, gravities rise to 36°API. Within both Eocene and Oligo-Miocene
formations in the Bolivar coastal fields, heavy- and medium-gravity oils
alternate in successive sandstones. Heavy oils are found near bottom or edge
waters and near overlying waters whose freshness indicates meteoric origin.

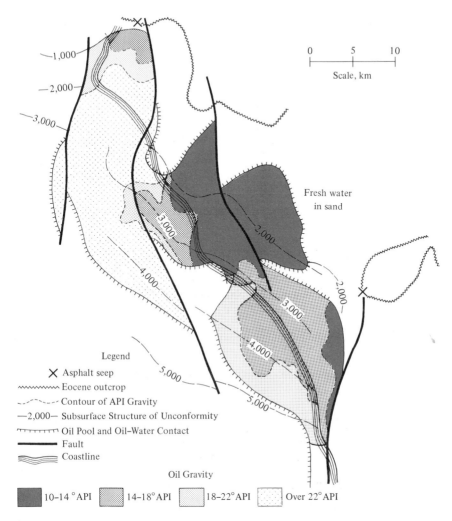

Figure 8-24
Asphalt seal at Lagunillas field, Venezuela. Fresh water overlies heavy oil, which increases in °API gravity downdip. Oil–water contact on southwest edge drops more than 300 m in 20 km because of post-Miocene tilting. [Dickey and Hunt 1972]

Formation waters in parts of the Bolivar coastal fields are so fresh that those from depleted oil reservoirs are used for drinking.

In the northern part of the South Sumatra Basin, the Palembang Formation contains an asphalt seal at the outcrop. Downdip at a subsurface depth of 100 m (328 ft) is a heavy asphaltic oil with no gasoline and 70 percent residuum. At 250 m (820 ft) depth, the oils contain 17 percent gasoline and 49 percent residuum, and at 500 m (1,640 ft) depth 63 percent gasoline and 14 percent residuum. The deepest oils run from 40° to 50°API gravity, compared to 8 to 10°API near the surface. Clearly, this is a case of water washing, biodegradation, and loss of gas and the light ends near the outcrop (Dufour 1957).

The Seria field of Borneo, with more than a billion barrels of recoverable oil, contains an asphaltic, biodegraded, nonwaxy crude at about 300 m (984 ft) an intermediate, nonwaxy crude at about 600 m (1,969 ft), and a light, waxy, unaltered oil from 2,000 to 3,000 m (6,562 to 9,843 ft). Oil gravities are 19, 26, and 37°API respectively. In the Balikpapan trend of the East Borneo oil fields, heavy oil without gasoline is found at shallow depths with a lighter paraffinic oil rich in gasoline in the deeper horizons. In the La Brea-Pariñas field of Peru, a well may penetrate a low-pour-point, nonwaxy crude in one sand followed by a high-pour-point, waxy crude in another; followed by another nonwaxy crude. In such cases, where each sand body acts as an individual reservoir, biodegradation can influence one oil in which there has been meteoric water invasion and leave other oils untouched. Whenever high-pour-point and low-pour-point crudes are in juxtaposition, biodegradation is the probable explanation.

Formation waters flowing by an oil reservoir will remove hydrocarbons by solution up through C_{15} and probably higher, considering the geologic time periods involved. When oxidation by a sulfate ion or dissolved oxygen is added to this, an asphalt mat is created at the oil–water interface. Amosov and Kozina (1966) observed that the oils in the Sakhalin reservoirs of the USSR became heavier approaching the oil–water contact. They found a direct correlation between the specific gravity of the oil and the bicarbonate content of the formation water. The oils became lighter in going from the bottom to the crest of the pool, but only about one-fourth of this distribution was attributed to gravity segregation. Most of the difference was caused by the degrading effect of the meteoric water at the oil–water contact. In the Permian oil fields of the Urals, Yarullin (1961) noted asphalt mats of 30 to 80 m (98 to 262 ft) in thickness at the oil–water contact of the pools. They attributed these to oxidation of the oils by the formation waters, which were high in sulfate ion.

The Burgan field in Kuwait and the Hawkins field in East Texas also contain asphalt mats at the oil–water interface. The °API gravity of an oil accumulation will decrease updip toward its crest or downdip toward its oil–water contact, depending on which end is being degraded by contact with meteoric water.

The Athabasca heavy oil deposits of eastern Alberta, Canada, probably represent the largest accumulation of biodegraded oil in the world. Current evidence indicates that the oil was formed in Middle or Late Cretaceous time and migrated updip to form a wide area of bituminous sands at the outcrop.

Detailed gas chromatographic–mass spectrometric analyses of Lower Cretaceous oils of Alberta and the Athabasca heavy oil have fairly well established
that the latter was a conventional oil that has undergone extensive biodegradation and water washing (Deroo et al. 1974; Hunt 1976; Rubinstein et al.
1977).

If an oil is physically, chemically, and biologically degraded near the
surface, can it be reburied to great depths and undergo thermal maturation to
form a light crude? Connan et al. (1975) carried out thermal maturation
experiments in the laboratory by heating both biodegraded and unaltered
immature oil samples in an inert atmosphere at 300°C (572°F) for extended
periods of time. The immature, unaltered oil underwent normal maturation
resulting in an increase in saturated hydrocarbons from 10 to 30 percent. The
hydrogen was obtained from disproportionation reactions, which caused a
gradual increase in the insoluble pyrobitumens from a trace to about 30
percent of the oil over a 12-month period. In contrast, the biodegraded oil
showed no significant change over a four-week period, after which about half
the oil was converted to an insoluble pyrobitumen, the remaining oil showing
a decrease in saturated hydrocarbons and a corresponding increase in
aromatics. The reason for this difference is that the paraffins shown on the left
side of Figure 8-10 are not present in the biodegraded oil. The polycyclic
naphthenes and aromatic rings in the biodegraded oil would tend to be
converted directly to pyrobitumens, plus some mono- and bicyclic aromatic
rings. The interpretation is that once an oil is intensely biodegraded it cannot
be converted to a normal oil by thermal alteration without some additional
paraffinic input from adjacent source beds.

Sulfurization and Desulfurization of Oil and Gas

The threshold of intense hydrogen sulfide (H_2S) generation from the
decomposition of organic sulfur compounds in source beds is in the range of
130 to 150°C (266 to 302°F). Small amounts of H_2S are formed earlier at
lower temperatures from relatively unstable compounds. This H_2S is highly
soluble in both water and oil. Goncharov et al. (1973) state that under conditions corresponding to depths of 2,000 to 3,000 m (6,562 to 9,843 ft) the
solubility of H_2S in fresh water at a partial pressure of 50 kg/cm^2 (711 psi)
approaches 50 m^{3}gas/m^3 of water (281 ft^3/bbl).

Hydrogen sulfide is a catalyst for the reduction of sulfate by organic
compounds such as hydrocarbons. Toland (1960) demonstrated that sulfate
ion will rapidly oxidize hydrocarbons in the presence of H_2S. For example,
reaction 1 in Figure 8-25 converts m-xylene to isophthalic acid in 89–100
percent yields at 325°C (617°F) in one hour. Reaction 2 converts 90 percent
of the methane to CO_2 in 70 min at 325°C (617°F). Although these reactions

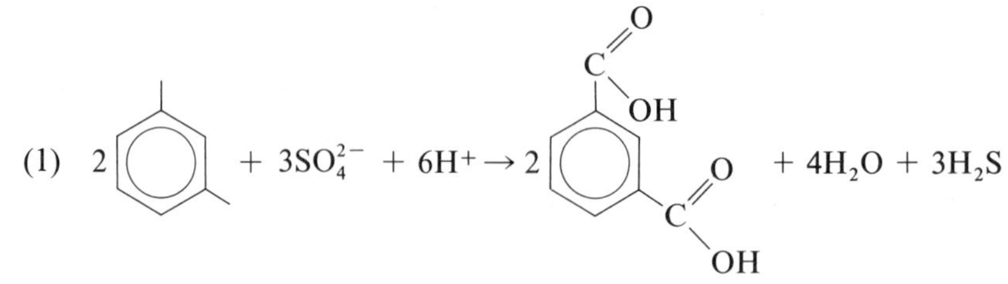

(1)

(2) $CH_4 + SO_4^{2-} + 2H^+ \rightarrow CO_2 + 2H_2O + H_2S$

(3) $SO_4^{2-} + 3H_2S + 2H^+ \rightarrow 4S^0 + 4H_2O$

(4) $H_2S + S^0 \rightleftharpoons H_2S_x \rightleftharpoons 2H^+ + S_x^{2-}$

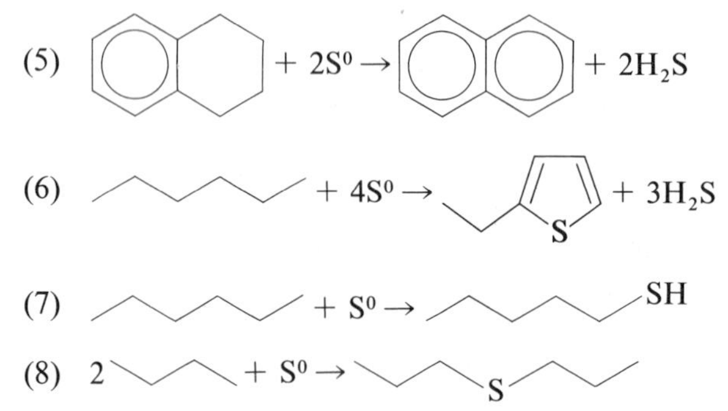

(5)

(6)

(7)

(8)

(9) $CH_4 + 4S^0 + 2H_2O \rightarrow 4H_2S + CO_2$

Figure 8-25
Chemical degradation reactions involving the oxidation, sulfurization, or dehydrogenation of hydrocarbons including methane from the reactions of sulfate ion, free sulfur,
polysulfides, and hydrogen sulfide in the reservoir.

are at higher temperatures in the laboratory than in the field, the reaction
rates would probably be high enough, at the reservoir temperatures of 130 to
150°C (266 to 302°F) to result in considerable oxidation of the hydrocarbons
over relatively short geologic time periods.

Elemental sulfur has been identified in many crude oils and has been
encountered in reservoirs at depths to 30,000 ft (9,144 m). When elemental
sulfur reacts with H_2S, as in equation 4 (Figure 8-25), it forms polysulfides,
H_2S_x. Polysulfides are powerful oxidizing agents that can convert saturated
hydrocarbons completely to carbon dioxide at elevated temperatures. Under
the mild conditions typical of petroleum reservoirs, the sulfurization of
hydrocarbons is favored. Reactions 5–9 in Figure 8-25 show the kinds of
conversions that will occur in the presence of H_2S, sulfur, and polysulfides.
Reaction 5 involves the conversion of tetralin to naphthalene. In reaction 6, a

thiophene is formed, in 7 a thiol, and in 8 a sulfide. Reaction 9 involves the conversion of methane to CO_2. These are all chemical degradation reactions that reduce the hydrocarbon content of the oil accumulation and increase its sulfur compounds, many of which are water soluble. An exception is reaction 5, which simply represents aromatization of a naphthene ring.

All of these reactions have been carried out in the laboratory and similar types of reaction products have been identified in crude oil. The proof that such reactions occur in the reservoir rests on sulfur isotope measurements (Orr 1974). The sulfur of the dissolved sulfate in reservoirs is about 15 ‰ heavier isotopically than is the sulfur in immature crude oils. As these reactions occur, the sulfate sulfur becomes part of the sulfur in crude oil and the associated H_2S. Consequently, the sulfur in oil and H_2S becomes isotopically heavier in the more mature oils subjected to higher temperatures, eventually approaching the isotopic value of the dissolved sulfate as a limit. The isotopically heavy sulfur is preferentially introduced into lower boiling compounds, such as the thiophene, thiols, and sulfides of reactions 6 through 8 in Figure 8-25. With increasing maturity, the ^{34}S, which is the heavy sulfur isotope, is introduced into the entire crude but most rapidly in the low boiling fractions. Orr calculated that 64 percent of the organic sulfur in the Manderson oil of the Big Horn Basin of Wyoming was derived from the sulfate sulfur of the formation waters. A major conclusion of Orr's study was that a large percentage of the organic sulfur compounds in mature crude oils of Paleozoic age in Wyoming formed after the oil was accumulated.

There is some evidence that carbonates catalyze these reactions based on the experiments of Valitov (1974). He found that when gasoline and sulfur are heated at 175°C (347°F) it takes 2 hours and 40 minutes for hydrogen sulfide to appear. In the presence of limestone, however, hydrogen sulfide evolved within 35 minutes.

The chemical reactions that may create a methane floor are shown in equations 2 and 9 in Figure 8-25. Undoubtedly some methane has been oxidized in reservoirs by sulfur, but the importance of these reactions in creating a methane floor cannot be evaluated until more information is available on reaction rates and catalytic effects over geologic time periods.

AGE DATING OF RESERVOIR OIL

Geologists and geochemists have always been intrigued with the idea of dating oil. This would simplify recognition of the probable source rocks, particularly for oils that have migrated long distances. Unfortunately, oils do not contain the atomic clocks that have been used successfully in dating rocks. Spores and pollen have periodically been suggested for age dating but their diameters, which are mostly over 15 microns, prevent their migration with liquid or gaseous hydrocarbons. Particles of that size are unable to migrate through shales or siltstones although they may be released by the recrystallization of carbonates or the solution of evaporites.

The possibility of dating oils based on maturation changes in specific hydrocarbon structures became a reality with the development of precise analytical techniques such as gas chromatography–mass spectrometry, combined with computer readout systems. In 1967, Reznikov reported a method for estimating the ages of oil accumulations based on the changes in the percentage of naphthenes, paraffins, and aromatics in the gasoline-range hydrocarbons. It was previously mentioned that naphthenes probably undergo disproportionation to yield paraffin and aromatic hydrocarbons. Reznikov (1967) developed a linear relationship between the hydrocarbon composition of crude oils and the product of the oil's reservoir temperatures and ages. He calibrated his method with a suite of oils believed to be indigenous to reservoirs in which they were found. The age of an unknown oil was estimated by measuring the oil's hydrocarbon composition and its reservoir temperature and solving the equation for time.

Young et al. (1977) refined Reznikov's method, using a more sophisticated approach for estimating the thermal histories of the beds. They also extended the method to the C_{15+} hydrocarbons in order to have confirmation of the age with two different hydrocarbon ranges. In the light gasoline range (C_5–C_7), they measured the concentrations of 10 naphthenes, 17 paraffins, and 2 aromatics. The basic disproportionation reaction by Reznikov is 4 naphthene = 3 paraffin + 1 aromatic. The age of the oil is determined by the extent to which paraffins and aromatics have been formed from the naphthenes. Only naphthenes are analyzed for the C_{15+} age dating. The basic equation is four-ring + three-ring naphthenes = one-ring naphthenes. The extent to which the latter increases over the former is a measure of the age.

Hydrocarbon ages calculated by the Young et al. method are generally in good agreement with geologically interpreted ages of the oils. The agreement is better for oils from clastic reservoirs than for oils from carbonate reservoirs. The problem with carbonates may be that the calibrations used in setting up the method do not have enough indigenous oils from Tertiary carbonate reservoirs. Other problems include difficulties in estimating paleo geothermal gradients, in calculating the amounts of missing sections at unconformities, and in trying to evaluate breaks in depositional rates. Thus a reservoir buried at a shallow depth during most of its history will not give the same age as one buried steadily deeper during its history. Also, condensates need a separate calibration, because there may be phase separation of some of the hydrocarbons during different stages of burial.

Despite these problems, the method shows considerable promise and has already been useful in solving some of the age-old questions regarding the origin of various reservoired oils. For example, some geologists have felt that

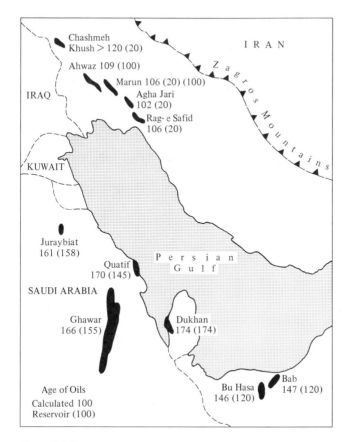

Figure 8-26
Calculated ages, in 10^6 years, of gasoline range (C_5–C_7) hydrocarbons
in oil reservoirs of the Persian Gulf. [Young et al. 1977]

the oils found in Miocene reservoirs of the Bolivar coastal fields of the
Maracaibo Basin of Venezuela are not indigenous but have migrated across
an unconformity from Eocene source beds. The calculated ages of most of
these oils are Eocene, which supports this concept. There are a few oils
however, with ages indicating them to be indigenous to the Miocene. The
dating method also indicates that Cretaceous reservoired oils in the Maracaibo Basin are indigenous.

Dunnington (1967a) has long maintained that some of the Miocene
reservoired oils in the Middle East originated in Cretaceous or Jurassic source
beds. The calculated ages support this hypothesis. Figure 8-26 shows the
calculated ages for the C_5–C_7 hydrocarbons in several Persian Gulf reservoirs. The detailed data for these and additional reservoirs are listed in Table

Table 8-6

Calculated Ages of Hydrocarbons in Middle East and Offshore
Louisiana Oil Fields

Field	Depth		Estimated temperature		Assigned reservoir age, 10^6 years	Calculated age, 10^6 years	
	ft	m	°F	°C		C_5–C_7	C_{15+}
Middle East							
Chashmeh Khush	11,300	3,444	265	129	20	>120	
Marun	10,700	3,261	265	129	20	106	
Agha Jari	8,700	2,652	230	110	20	102	
Rag-e Safid	8,350	2,545	235	113	20	106	
Ahwaz	11,100	3,383	260	127	100	109	
Marun	11,600	3,536	275	135	100	107	
Bu Hasa	7,800	2,377	230	110	120	146	
Bab	8,400	2,560	252	122	120	147	
Ghawar-D	7,400	2,256	215	102	155	166	
Dukhan	7,200	2,195	219	104	174	174	
Quatif D	7,375	2,248	226	108	145	170	
Juraybiat C	8,550	2,606	250	121	158	161	
Louisiana							
Main Pass Block 290	6,670	2,033	151	66	6	9.1	10.7
South Pass Block 62	7,905	2,409	167	75	3.6	12.3	11.6
West Delta Block 30	3,095	943	100	38	2.0	5.2	2.2
West Delta Block 30	12,346	3,763	243	117	9	12.7	14.2
Ship Shoal Block 72	9,060	2,762	218	103	4.3	17.4	18
Eugene Island Block 208	5,045	1,538	123	51	1.3	15.5	10.7
Eugene Island Block 18	9,350	2,850	204	96	4.3	19.8	18.7
S. Marsh Island Block 6	9,110	2,777	202	94	3.4	11.2	8.3
Vermillion Block 164	6,765	2,062	157	69	2.7	14.1	20.0

8-6. Chashmeh Khush, Marun, Agha Jari, and Rag-e Safid are examples of
Miocene reservoired oils that originated in the Cretaceous. The Marun field
contains a Miocene reservoir at 10,700 ft (3,261 m) and a Cretaceous reservoir
at 11,600 ft (3,536 m). Both oils have calculated ages of about 106 million yrs.
The oils of Abu Dhabi, in the lower right corner of Figure 8-26, have Late
Jurassic calculated ages. The Saudi Arabian oils range from Early to Middle
Jurassic in age.

The Late Tertiary of the offshore Louisiana Gulf Coast is a section in which
there has been considerable controversy regarding the origin of the Plio-
Pleistocene oils. The chemically calculated ages clearly indicate that these oils
originated in older source beds. For example, the Pleistocene reservoired oil

at Eugene Island, Block 208 (Table 8-6) has a calculated age of Middle Miocene. The Pliocene reservoired oil at Eugene Island has a calculated age of Early Miocene. The other Pliocene reservoired oils in Table 8-6 also have Miocene calculated ages. Young et al. (1977) found that the average difference between the assigned reservoir age and the calculated C_5–C_7 age was 8.7 million years, while it was 5.3 million years for the calculated C_{15+} age of 70 offshore Louisiana oils. They also estimated that these oils originated at depths from 6,000 to 11,000 ft (1,829 to 3,353 m) deeper than their present reservoirs. It is interesting that these chemically calculated ages confirm the vitrinite reflectance interpretations shown in Figure 7-43. Young et al. are in general agreement that there has been considerable upward migration of hydrocarbons in the Gulf Coast and the oils in the sediments younger than Miocene are rarely indigenous.

Hydrocarbon ages calculated from the C_{15+} fraction are sometimes low, because of biodegradation. An example of this is the West Delta Block 30 oil at 3,095 ft, or 943 m (Table 8-6). The calculated age of 2.2 million years is low, because the one-ring naphthenes, which should increase with time, have been depleted along with paraffins by microbial oxidation.

Figure 8-27 shows the distribution of paraffins and naphthenes for West Delta Block 30 oils at 3,095 ft (943 m) and at 12,346 ft (3,763 m). The latter oil, which is too deep to be altered, shows 40 percent paraffins, while the shallow oil has no paraffins. The one-ring naphthenes also have been depleted considerably in the shallow oil. When nonthermal processes such as this alter the oil, the dating technique fails. This emphasizes that the application of such a method should be supplemented with other geological and geochemical techniques for evaluating causes of compositional changes.

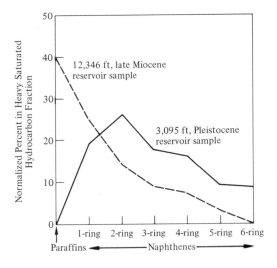

Figure 8-27
Relative amounts of paraffins and one- to six-ring naphthenes in two West Delta Block 30, offshore Louisiana oils. [Young et al. 1977]

NATURAL ASPHALTS AND RELATED SUBSTANCES

Reservoir rocks, both at depth and exposed at the surface, often contain intervals of solid bitumens as thick as many oil reservoirs. Some of these result from the physical, chemical, and biological degradation of normal crudes, but others are immature heavy asphalts that probably never existed as light oils. Some immature asphalts also form vein fillings in fractured rocks exposed near the surface. An example of this is the gilsonite of the upper Green River Formation in the Uinta Basin of Utah (Hunt et al. 1954). This gilsonite probably entered the vein openings as a heavy immature liquid and then solidified and hardened over geologic time as the formation was uplifted and exposed. Its maximum depth of burial probably did not exceed 1,500 m (about 4,900 ft) which was not deep enough for generation of a light crude oil. Both liquid gilsonite and liquid wurtzilite (liverite) were found in the Uinta Basin during the early years of exploration in that area. These asphalts are found in veins that terminate directly in the source beds. Source and reservoir are in juxtaposition with some bitumen in the fracture porosity of the source bed. A mass spectrometer analysis of liquid gilsonite is shown in Table 8-7. About 70 percent of it consists of condensed polynuclear aromatic and heterocyclic ring compounds. This high aromaticity is typical of immature oils, but some of this hydrocarbon-type distribution can also be attributed to the lacustrine algae source of the Green River kerogen. It contrasts markedly with the ozocerite wax derived from the kerogen of the basal Green River–Upper Wasatch formations. The latter are alluvial shoreline lacustrine sediments with herbaceous and woody organic matter high in paraffin wax. These sediments reached depths of 3,000 m ($\sim$10,000 ft) and more, which was sufficient to generate a highly waxy crude oil subsequently produced in the Duchesne field. During drilling of the Duchesne field, an oil show was encountered in the upper Green River Formation that was generally similar to liquid gilsonite and distinctly different from ozocerite (Table 8-7). The ozocerite, which is found in exposed sections of the Wasatch Formation, is very similar to the residue of the waxy oil produced at Duchesne.

Solid, relatively pure bitumens found in drill holes or in outcrops have been characterized and classified on the basis of their physical and chemical properties (Abraham 1960; Hunt 1963a; King et al. 1963). Figure 8-28 is a classification modified from these authors. The autochthonous coals, formed in place, are differentiated from the allochthonous bitumens and pyrobitumens that have migrated from their source, as petroleum has. Further classification of the bitumens is based on their solubility in organic solvents such as CS_2, the pyrobitumens being relatively insoluble. Bitumens are further differentiated by their fusibility, the asphaltites being relatively infusible. The pyrobitumens include unmetamorphosed, highly polymerized compounds that have relatively high hydrogen/carbon ratios, and metamor-

Table 8-7
Composition of Uinta Basin Liquid Gilsonite

	Wt. %
Paraffin hydrocarbons	3
Naphthene hydrocarbons	
Noncondensed rings	6.4
Condensed, 2-ring	4.6
3-ring	2.7
4-ring	1.1
5-ring	0.2
Aromatic hydrocarbons	
Alkylbenzenes	11.1
Indanes	8.1
Indenes	5.0
Naphthalenes	8.5
Acenaphthenes	8.0
Acenaphthylenes	4.1
Tricyclic aromatics	3.2
Nitrogen, sulfur, oxygen compounds	34
	100

Comparison of Produced Oil Residues with Liquid Gilsonite
and Ozocerite

	Duchesne field oil show at 3,000 ft (914 m)	Liquid gilsonite	Duchesne field oil at 7,596 ft (2,315 m)	Ozocerite
Paraffin and naphthene hydrocarbons	30	18	82	81
Aromatic hydrocarbons	36	48	11	10
Nitrogen, sulfur, oxygen compounds	34	34	7	9

phosed residues that have very low hydrogen/carbon ratios. Coals are divided between the sapropelic deposits, largely derived from plankton, and the humic deposits, largely derived from higher terrestrial plants.

From left to right across the chart there is a decrease in solubility, volatility, and hydrogen content and an increase in the fusing point, refractive index, and molecular weight of the substance. Asphalts are most apt to be associated with active seeps. Asphaltites and pyrobitumens are the carbonized residues left from seeps that have long since dried up. Anthraxolite is the last stage of metamorphism for bitumens comparable to anthracite for coals.

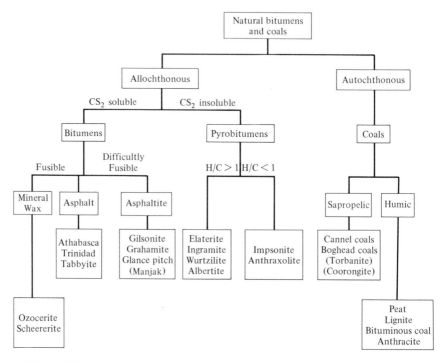

Figure 8-28
Classification of natural bitumens and coals. The bitumens generally have more nitrogen plus sulfur than coals, and less oxygen.

The principal difference in the chemistry of bitumens and coals is in their nitrogen, sulfur, and oxygen content. Coals derived from humic and sapropelic materials are generally higher in oxygen than in nitrogen plus sulfur (Hunt 1978). Bitumens derived from the organic matter of fine-grained rocks usually have more nitrogen and sulfur than oxygen. This is apparent in Figure 8-29 where the H/C ratio is plotted against the (N + S)/O ratio for coals, bitumens, pyrobitumens, petroleum asphaltenes, and reservoir bitumens. Although the coals and bitumens overlap in their H/C ratios, they are separated in the (N + S)/O ratios. The sapropelic coals are comparable to the asphalts, asphaltites, and petroleum asphaltenes in their H/C ratios, but are much lower in the (N + S)/O ratios. The humic coals are comparable in H/C ratio to reservoir bitumens of the Western Canada Basin reported by Rogers et al. (1974), but they are also lower in the (N + S)/O ratio. Only the impsonite and anthraxolite are comparable to the humic coals in both ratios. They can be differentiated by other properties. Both impsonite and anthraxolite devolatilize at lower temperatures than coals of corresponding rank. Impsonite, like most bitumens, contains more vanadium and nickel than the coals.

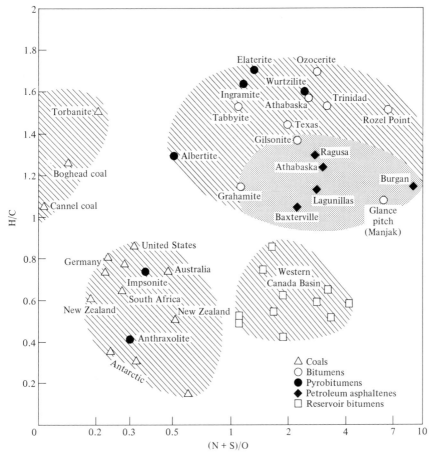

Figure 8-29
Atomic ratios of bitumens and coals. [Hunt 1978]

Figure 8-29 also shows that the native asphaltites, such as gilsonite, grahamite, and glance pitch, are similar in their H/C and (N + S)/O ratios to the asphaltenes of petroleum. This suggests that the conversion of asphalt to asphaltites that occurs in nature is similar to the process of asphaltene coagulation as the oily medium of an oil is removed.

Mineral Waxes

Ozocerite is the most important native wax. It is found in Galicia, the Uinta Basin of Utah, Rumania, several areas in Russia, Jordan, Wales (hatchettite), and Switzerland (scheererite). The H/C ratio of ozocerites varies between about 1.7 and 1.99, indicating them to be almost a pure mixture of paraffins and naphthenes.

Ozocerites in general represent the natural residuum of high-wax oils exposed to the surface. Although ozocerites are most frequently found in bituminous nonmarine shales or in beds associated with cannel coals or lignites, it is possible for such waxes to be derived from the inspissation of almost any high-wax petroleum.

Asphalts and Asphaltites

Asphalt deposits occur worldwide, most of them around structural uplifts where oil-producing sandstones crop out or fractured formations reach to the surface. Breached and fractured reservoirs are a common source of asphalt deposits. Total asphalt in the world in shallow sands and in the pure state is unknown, but is probably on the order of 3×10^{11} MT (metric tons). The bulk of this is in the asphalt sands of Alberta (8×10^{10} MT), and eastern Venezuela (9.3×10^{10} MT). Large bitumen deposits in the Permian sediments of the Tataria area of Russia have been described by Akishev et al. (1974).

It is not clear how much of the 3×10^{11} MT of asphalt originated from heavy or light oil, but it is interesting to speculate that if all of it had come from an average 35°API gravity crude oil, it would represent 15×10^{11} MT of petroleum. This is 1.5 times the estimated 10×10^{11} MT of oil in reservoir rocks of the earth's crust. Obviously, enormous amounts of oil have been lost because of tectonic activity and lack of adequate seals over geologic time. Oklahoma is one of the major oil-producing states in the United States, but the reservoir sands of Oklahoma leak all along the Arbuckle and Wichita uplifts to the south. The whole southern area of Oklahoma has asphalts in both liquid and solid forms occurring as springs, seepages, and rock impregnations all associated with formations that are oil-producing downdip away from the uplifts.

Asphaltites are native bitumens with fusing points above 110°C (230°F). On an H/C versus (N + S)/O diagram (Figure 8-29), the asphaltites are grouped with the petroleum asphaltenes. Some asphaltites are indurated bitumens formed by the polymerization and gelation of asphalt components, while others are thermal alteration products formed by increasing temperatures. In most areas where asphaltites have been found, there are also some asphalt deposits. Examples are the liquid gilsonite and gilsonite of the Uinta Basin and the grahamite and asphalts of southern Oklahoma, Peru, Argentina, Cuba, Trinidad, and Mexico. In Vera Cruz, Mexico, deposits of pure asphalt change over to glance pitch, and glance pitch veins outcrop as grahamite. Asphalt, glance pitch, and grahamite also occur in Cuba. Asphalt and glance pitch occur in the Dead Sea area. Asphalt–asphaltite combinations are common in many areas of the USSR, such as in the southern Urals and the Dzhungar depression.

Pyrobitumens

The pyrobitumens are infusible and relatively insoluble compared to the asphalts and alphaltites. There are three groups: the bitumen polymers, elaterite and wurtzilite; the metamorphosed bitumens, impsonite and anthraxolite; and a group that appears to be more indurated forms of asphalts and asphaltites, ingramite and albertite.

Elaterite and wurtzilite are products of a highly unsaturated, unstable organic matter. Olefinic structures are common in organisms (Figure 3-5). Normally, when these materials are buried, they undergo diagenetic reactions that hydrogenate the olefins to saturated paraffin hydrocarbons. In some instances, this process is delayed, and the unstable olefins polymerize to eliminate their double bonds. An example is the formation of rubber from isoprene. Many tree resins are olefinic and become hardened polymers on exposure to the atmosphere. Studies on the origin of various elaterites (Uspenskii and Gorskaya 1951) indicate that it originates as a viscous, unsaturated bitumen that quickly polymerizes to the rubbery elaterite on exposure to meteoric waters or surface weathering. Uspenskii mentions the Mirzaanite elaterite from Georgia, which is accompanied by a viscous bitumen that can be stretched into filaments when rubbed between the fingers. The same property has been observed in the elaterite of Derbyshire, England. Uspenskii states that all elaterites found to date originated under the direct influence of surface agents, either in strata immediately underlaying the surface or in cavities of mineral veins affected by surface waters. Wurtzilite appears to be a more indurated polymer with an origin similar to elaterite. The structures of these compounds might be interpreted from liverite, a liquid form of wurtzilite found in the Uinta Basin. The saturate fraction of liverite contains only 1 percent paraffin, 35 percent noncondensed naphthene rings, and the rest condensed napththenes containing two- to six-ring structures. The Uinta Basin wurtzilites and several of the elaterites are distinctly different from other bitumens and pyrobitumens in having a high sulfur and nitrogen content (Figure 8-30). Sulfur analyses are made after removal of pyrite and elemental sulfur, which is associated with some samples. Uspenskii and Gorskaya (1951) report an elaterite found in the western Ukraine capable of being vulcanized by treatment with elementary sulfur, similar to natural rubber. They concluded from this that the difference in solubility and fusibility of various elaterites depended on their increasing degree of vulcanization with sulfur as follows: elaterite → thioelaterite →

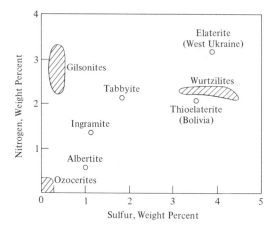

Figure 8-30
Nitrogen and sulfur content of bitumens and pyrobitumens. All samples except the elaterites are from the Uinta Basin, Utah. Elaterites and wurtzilites have the characteristics of vulcanized rubber.

wurtzilite. This is analogous to the series: raw rubber → vulcanized rubber → ebonite. The same conclusion was reached by Hunt (1963a) in suggesting that wurtzilite consists of multiple units of naphthene rings crosslinked with sulfur bridges.

Imponite and anthraxolite represent the final stage in the induration of bitumens. The imponites and anthraxolites from the metamorphosed Ordovician rocks of Quebec appear to have reached their stage by the effects of temperature and pressure. King (1963) found that about half the samples had a crystalline structure, which is not the case with imponites found in less metamorphosed rocks, such as exist in southern Oklahoma.

SUMMARY

1. In 1971, the average reservoir depth of known giant fields with more than 500 million bbl of recoverable oil was around 7,000 ft (2,130 m). In classic geosynclinal basins, about two-thirds of the giant fields are in reservoir rocks located on the shelf side. Giant fields contain about 85 percent of the world's reserves.

2. The combination of source and reservoir crude oil maturation results in the number of oil fields gradually decreasing and gas fields increasing at depth. Below 12,000 to 14,000 ft (3,658 to 4,267 m) less than one-fourth of the reservoirs contain oil, and below 20,000 ft (6,096 m) all but a few percent contain gas.

3. Thermodynamic calculations predict that n-paraffins are the most stable type of hydrocarbons in the gas phase at sedimentary basin temperatures. Their stability increases as the number of carbon atoms in the molecule decreases, with methane being the most stable.

4. The maturation of crude oil in the reservoir is believed to involve several hydrogen disproportionation and cracking reactions with large molecules giving up hydrogen to permit increased formation of low-molecular-weight paraffins. The large molecules condense to polycyclic aromatic hydrocarbons, which eventually form asphaltites or pyrobitumens in the reservoir.

5. Thermal maturation in the reservoir causes the gravity of most crude oils to increase one degree API (decrease 0.005 in specific gravity) for every 200 to 400 ft (61 to 122 m) of depth. Reservoir maturation appears to crack the hydrocarbons containing more than 20 carbon atoms into smaller molecules that become part of the gasoline range.

6. The oil floor in a sedimentary basin can vary between about 5,700 ft (1,738 m) and 25,000 ft (7,620 m) depending on the geothermal gradient, the depositional rate, and the time of oil formation. Most basins have a gas floor caused either by decreased reservoir porosity or by chemical destruction of the hydrocarbons or their migration to shallower horizons.

7. During crude oil maturation, the percent of naphthenes decreases with respect to paraffins plus aromatics, and the percent of four-, five-, and six-ring naphthenes decrease, while one-ring naphthenes increases. The decrease in percent of steranes and triterpanes with maturation causes a marked decrease in the optical activity of crude oil.

8. Bacteria degrade crude oil by oxidizing the hydrocarbons. Bacterial degradation of oil can occur if reservoir temperatures do not exceed about 82°C (180°F) and if oxygen, inorganic trace nutrients, and water are present. Anaerobic sulfate-reducing bacteria also can degrade oils in a manner similar to aerobic strains but require more time. It has been estimated that 10 percent of the world's crude oil reserves has been altered by microbes.

9. Asphalt seals at seep outcrops and asphalt mats at the oil–water interface of pools in contact with meteoric water are formed by a combination of water washing, inspissation, and chemical and microbial oxidation.

10. Laboratory experiments indicate that a crude oil that has been physically, chemically, and biologically degraded cannot be converted to a normal crude oil by reservoir thermal maturation without some additional paraffinic input from source beds.

11. Laboratory experiments indicate that both sulfate and elemental sulfur in the presence of H_2S are capable of oxidizing hydrocarbons, including methane.

12. Sulfur isotope studies have shown that a large percentage of the organic sulfur compounds in mature Paleozoic crude oils of Wyoming were formed from the sulfur in dissolved sulfate of the formation water after the oil was accumulated.

13. A method for the age dating of reservoir oil has been developed based on the ratio of naphthenes to paraffins plus aromatics in the gasoline fraction and the ratio of three- and four-ring naphthenes to one-ring naphthenes in the C_{15+} reaction.

14. Pyrobitumens and asphaltites can be distinguished from coals in having a higher atomic ratio of $(N + S)/O$ and usually a higher vanadium plus nickel content.

SUPPLEMENTARY READING

Andreev, P. F., A. I. Bogomolov, A. F. Dobryanskii, and A. A. Kartsev. 1968. *Transformation of petroleum in nature.* London: Pergamon Press, 466 p.

Levorsen, A. I. 1967. *Geology of petroleum: The reservoir and reservoir dynamics.* Pp. 47–494. San Francisco: W. H. Freeman, 724 p.

Van Nes, K., and H. A. Van Westen. 1951. *Aspects of the constitution of mineral oils: Genesis of mineral oils.* Pp. 14–66. New York: Elsevier, 484 p.

IV

APPLICATIONS

9

Seeps and Surface Prospecting

Four principal requirements for petroleum accumulations are (1) petroleum source beds, (2) adequate reservoir rock porosity and permeability, (3) structures or traps, and (4) an impermeable cover during and since accumulation. Visible oil and gas seeps are important in an exploration program, because their presence implies that the first requirement has been satisfied and possibly others, if it is a major seep. Many large seeps represent tertiary migration; that is, migration from an accumulation that has been disturbed by tilting of strata, changes in depth of burial, or development of new avenues of escape to the surface, such as a fracture–fault system.

A seep can be defined as visible evidence at the earth's surface of the present or past leakage of oil, gas, or bitumens from the subsurface. This definition does not include microseeps or invisible seeps, which are generally thought of as moving by some mechanism of primary migration such as diffusion or solution in subsurface fluids. These are treated separately, in the discussion of geochemical prospecting.

SEEPS

The importance of seeps tends to be minimized in this era of increased use of highly sophisticated instrumentation and decreased use of ground surveys. Nevertheless, many, if not most, of the important oil-producing regions of the world were detected or discovered through surface oil and gas seeps.

Oil seeps were reported in the earliest recorded history. The use of asphalt as a building material from seepages in the Middle East dates back to 3,000 BC. In 1875 BC, there was an oil trade and complaints even then about the shortage of supply (Owen 1975). Burning gas wells have existed in the Baku area since several centuries before Christ.

The numerous explorers that crossed the United States in the early nineteenth century reported many seeps. For example, W. P. Blake (1855, p. 433) of the U.S. Geological Survey gave the following account of his journey through California in 1853: "It is an interesting fact which I believe is not generally known that there are numerous places in the coast mountains south of San Francisco where bitumen exudes from the ground and spreads in great quantity over the surface. These places are known as *tar springs* and are most numerous in the vicinity of Los Angeles." He goes on to say, "I am informed by Lt. Trowbridge of the U.S. Engineer Corps that the channel between Santa Barbara and the islands is sometimes covered with a film of mineral oil, giving to the surface the beautiful prismatic hues that are produced when oil is poured on water." Now, over a hundred years later, the reaction of the public to oil on water is quite different from that of these early explorers.

The Drake well, completed in 1859 in Pennsylvania, and a well completed the same year in Ontario, Canada, were both drilled on or near oil seeps. The first successful well in Texas in 1865 was drilled near an oil seep. As late as 1949, every oil field in Iran was associated with surface oil or gas seeps (Link 1952).

Weathering of Seeps

When petroleum leaks to the surface, it undergoes a series of chemical and physical changes that considerably alter its appearance and composition. Gas or condensate seeps are difficult to recognize unless they occur underwater, where they appear as bubbles. Most light hydrocarbons evaporate into the atmosphere and are quickly dissipated by air circulation. There is an indirect indicator of gas seepage known as "paraffin dirt," which is found associated with gas seeps in Louisiana and Texas along the Gulf of Mexico. Paraffin dirt is a yellow, elastic material resembling art gum in color and physical appearance. The name is a misnomer, because it contains almost no hydrocarbons. It is composed almost entirely of microbial cells, including large amounts of polysaccharides. Large numbers of methane-, ethane-, propane-, and butane-oxidizing bacteria are present in paraffin dirt samples. This material is believed to be a metabolic by-product of bacterial degradation of the gases issuing from the seep. Carbon-14 analyses show that about 85 percent of paraffin dirt organic matter is composed of ancient carbon coming from the subsurface (Davis 1967, p. 102).

Crude oil leaking to the surface undergoes the following simultaneous changes:

1. *Evaporation of the more volatile hydrocarbons.* In the first two weeks after an oil comes to the surface, it loses the hydrocarbons up through about C_{15} equivalent to a boiling point to 250°C (482°F). In subsequent months, additional hydrocarbons are lost, up through about C_{24}.

2. *Leaching of water solubles.* The most soluble nitrogen, sulfur, and oxygen compounds, along with some lighter aromatics, may be leached out by groundwaters.

3. *Microbial degradation.* As discussed in Chapter 8, hydrocarbons leaking to the surface are subject to microbial attack. The *n*-paraffins, some isoparaffins, and naphthenes are oxidized in varying amounts, depending on the suitability of the environment for microbial activity.

4. *Polymerization.* This change involves the combining of some of the intermediate to larger molecules to form very large complex structures after the elimination of water, carbon dioxide, or hydrogen.

5. *Auto-oxidation.* Many constituents of petroleum absorb sunlight and oxygen, resulting in formation of oxidized polymers. Seeps can take up as much as 6 percent oxygen on long exposure to air and sunlight.

6. *Gelation.* The formation of a rigid gel structure may develop over time with some types of seeps.

All of these reactions lead to thickening or solidification of the original oil. The crude is gradually converted from a liquid oil to an asphalt, an asphaltite, and eventually a pyrobitumen. Consequently, unless a seep is supplied by a continuous flow of fresh oil, it will ultimately harden to a black bitumen deposit. Table 9-1 lists seep samples collected near Lake Maracaibo. Sample 1 is at the site of an active, flowing oil seep. Samples 2 and 3 represent more weathered bituminous materials successively farther from the active flow site, and sample 4 is the farthest away (Dickey and Hunt 1972). The decrease in solubility of these samples in organic solvents coincident with an increase in fixed carbon and a decrease in hydrogen content is evident.

The increase in oxygen content for several samples of gilsonite from the Uinta Basin is shown in Figure 9-1. This comparison includes samples of

Table 9-1
Weathering of Seep at Mene Grande, Venezuela

Sample numbers[a]	1	2	3	4
Solubility in *n*-heptane (wt. %)	78	43	5	0
Solubility in CS_2 (wt. %)	99	78	52	0
Carbon ratio[b]	6	10	28	82
H/C atomic ratio	1.63	1.52	1.36	0.6

[a]Higher numbers are more weathered samples farther from the seep source.
[b]Ratio of nonvolatile organic carbon to total organic carbon.

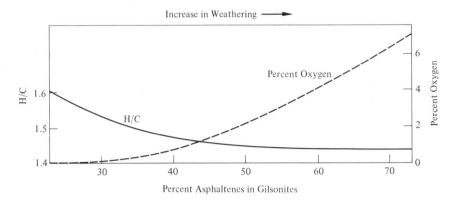

Figure 9-1
Weathering of gilsonite samples from Uinta Basin, Utah. Ten samples, going from liquid gilsonite on left to highly weathered gilsonite on right, show increase in oxygen content of gilsonites with weathering. [Hunt 1963]

liquid gilsonite from deep in a mine plus both fresh unweathered and highly weathered solid forms of gilsonite. The more weathered samples have less hydrogen, more oxygen, and a higher asphaltene content. The nitrogen and sulfur content of these samples showed no change over this entire range (Hunt 1963a). These general observations on seep weathering are important in order to evaluate properly the relationship of a surface seep to possible subsurface petroleum accumulations.

Classification of Seeps

Seeps may be classified for field operations into three groups as follows: (1) active or live seeps, which may be gas, light oil, heavy oil, or mounds of sticky black asphalt; (2) inactive or dead seeps, which are generally asphaltites or pyrobitumens not connected to any liquid material; (3) false seeps, which are materials that may have the appearance of active or inactive seeps but actually are in no way related to a subsurface hydrocarbon accumulation.

Active seeps are important in rank wildcat areas, and detailed surface and subsurface studies should be undertaken in the area of such seeps to determine their relationship to the lithology, stratigraphy and structure. Active seeps should be analyzed for their hydrocarbon range, n-paraffin content, asphaltenes, H/C ratio, and oxygen content. Such data will indicate the degree of weathering and the probability of the seep coming from an actual underground reservoir. Detailed gas chromatography–mass spectrometry analyses will assist in correlating the seep with known crude oils or source rock extracts in the area.

Inactive seeps generally leave a residue of asphalt that matures with time to an asphaltite such as grahamite or to a pyrobitumen such as impsonite and anthraxolite.

The largest known grahamite vein in the world occurs near Tushkahoma, Oklahoma, in a fault in shaley sandstone. The vein is a mile (1.6 km) long with a thickness up to 25 ft (7.6 m). The asphaltite contains pieces of wall rock that fell into it before it solidified. This and other Oklahoma grahamite deposits represent ancient oil seeps that have long since become inactive. They are all located along the structural uplifts in southern Oklahoma, where oil-producing sands outcrop and active seeps are common. Impsonite is also found in Oklahoma, filling fissures caused by faulting around structural uplifts.

Evaluation of Seeps

It is best to collect unweathered samples of seeps and send them to an analytical laboratory for proper identification and interpretation. A few observations can be made in the field and should be included with any samples sent in for analysis. For example, is there any evidence of liquid material associated with solid bitumens? Does the seep melt or become tacky in the hot sun? Will enough of the bitumen dissolve in lighter fluid to cause a coloration? If not, it is probably a pyrobitumen or coal.

Seep samples for laboratory analysis should be collected in glass or plastic bottles rather than in metal containers, because the latter may introduce trace elements that are useful in determining the validity of a seep. Also, brines collected from active seeps frequently cause corrosion and leakage of metal cans. Gas seeps bubbling through water may be collected by inverting a water-filled bottle over the seep, allowing the gas to displace the water.

Laboratory tests can properly classify a field sample into one of the categories shown in Figure 8-28. Asphalts are more soluble and have a lower melting point than do asphaltites. Pyrobitumens are infusible and practically insoluble in all organic solvents. Pyrobitumens can be distinguished from coals by their lower oxygen content, their higher nitrogen plus sulfur content, and higher vanadium plus nickel content.

Rocks colored black with manganese dioxide or metallic sulfides may be mistaken as seeps. Arnold (1959) mentions a manganese-stained conglomerate in Kern County, California, that has been misidentified as an oil sand. He cites other types of false seeps, including a variety of deceptive colors and stains of inorganic and organic origin.

Buried and abandoned containers of refinery products may be mistaken for seeps when the containers rust through and leak. A 23°API gravity oil was submitted for analysis from a presumed seep area in Missouri. The oil contained no hydrocarbons up to the lubricating oil range and no residuum. Although the front end might be lost by evaporation, it was unusual to find no residuum. A trace metal analysis showed the oil to contain 400 ppm iron and 40 ppm zinc, whereas the concentration of these elements in a typical midcontinent oil is 4 ppm and 0 ppm respectively. This analysis identified it as a lubricating oil that had passed through a refinery operation.

A particularly unusual sample was an oil seep with the viscosity of kerosine but with a specific gravity of 0.985 (about 12°API). It was identified as picric acid, an explosive, dissolved in kerosine. Samples of kerosine containing DDT also have been identified as false seeps. As kerosine is used worldwide as a solvent for insecticides and all types of organic compounds, it is not unusual to find it seeping from abandoned storage containers.

There are natural light oil seeps that do not contain a residuum. Many years ago, natural seeps were found in Cuba that included only the gasoline—kerosine range. The seeps could be used to run automobiles without any processing.

Geology of Seeps

Seepages are most numerous on the margins of basins and in sediments that have been folded, faulted, and eroded. Link (1952) categorized seeps into five types depending on their origin as follows:

1. Seeps emerging from homoclinal beds, the ends of which are exposed where these beds reach the surface.

2. Seeps associated with beds and formations in which the oil was formed.

3. Seeps from large petroleum accumulations that have been bared by erosion or the reservoirs ruptured by faulting and folding .

4. Seeps at the outcrops of unconformities.

5. Seeps associated with intrusions such as mud volcanoes, igneous intrusions, and piercement salt domes.

The first type of seep, shown in Figure 9-2, is usually small in volume but persistent in activity. The Ordovician Trenton Limestone produces oil downdip in deeper parts of the Michigan Basin.

Examples of the second type are the bituminous sandstones and vein bitumens of the Green River Formation of the Uinta Basin, Utah, shown schematically in Figure 9-3. Heavy asphaltic oil generated in the Green River shales feeds into fissures on minor structures and into sands interbedded with the shales. Link also cites a Type 2 seep in Costa Rica where a rich shale is crushed and shattered by folding and faulting, releasing indigenous high-grade free oil that accumulates in the fractures at the outcrop.

$\triangle$ = Oil seep

Manitoulin Island, Ontario

Figure 9-2
Type 1 seepage where the oil-bearing Trenton Limestone is exposed on the northern edge
of the Michigan Basin. [Link 1952]

Enormous quantities of oil have undoubtedly been lost over geologic time
from the erosion of reservoirs on anticlines as shown in Figure 9-4 (Type 3).
Such seepages indicate good prospects for finding oil in nearby anticlines
where the same formations have not been eroded. Seepages along faults were
discussed in Chapter 6 (Figure 6-28). The correlation of subsurface faults and
gas seeps has been reported in many areas.

The Masjid-i-Sulaiman field of Iran (Figure 9-5) is on an anticlinal fold
where there is no evidence of major faulting. Seepages occur just at the crustal
turnover, where minor fracturing in the cap rock and overlying beds have
allowed passage of the oil upward from the Asmari Limestone reservoir.

Link (1952) cites some of the Uinta Basin, Utah, seeps as being associated
with unconformities (Type 4). The Athabasca oil sands probably represent
the world's largest seep at the outcrop of an unconformity.

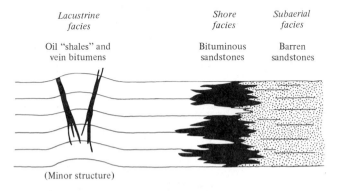

*Lacustrine
facies*

Oil "shales" and
vein bitumens

*Shore
facies*

Bituminous
sandstones

*Subaerial
facies*

Barren
sandstones

(Minor structure)

Figure 9-3
Type 2 seepage where the bitumen is in contact with the source beds.
This is typical of the asphaltic sands and vein bitumens of the
Green River Formation, Uinta Basin, Utah. [Hunt 1963]

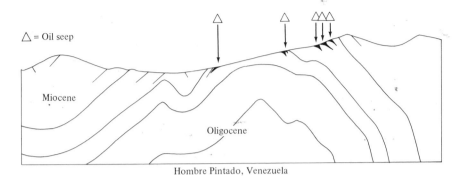

Figure 9-4

Type 3 seepage of an eroded anticline reservoir at Hombre Pintado, Venezuela. [Link 1952]

The numerous oil and gas seepages associated with piercement salt domes in the Gulf Coast are in the Type 5 category. These seepages were known to the early Indians and the first white settlers. In 1901, the prolific Spindletop field was discovered from drilling at a salt dome oil and gas seep. The reported abundance of seepages in the Bay and Bayou areas of Louisiana is partly caused by the ease of recognition of gas seepages in water-covered areas. Many of these salt dome seeps include H_2S gas. Deposits of commercial sulfur from H_2S oxidation occur in the cap rock of several domes, including Spindletop. The intrusion of salt domes is accompanied by deformation that results in fracturing and faulting of both the intruded and overlying sediments. This results in appreciable localized seepage of oil and gas, particularly around the shallowest domes.

Oil and gas seeps associated with mud volcanoes as discussed in Chapter 6, also are in the Type 5 category. Mud volcanoes are generally in geologically

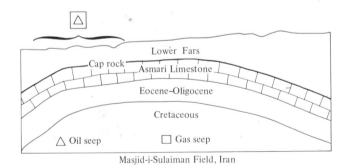

Figure 9-5

Type 3 seepage of oil from the Asmari Limestone at Masjid-i-Sulaiman field in Iran. [Link 1952]

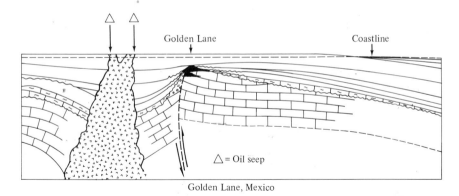

Golden Lane, Mexico

Figure 9-6
Type 5 seepage of oil possibly formed by heating of source rocks penetrated by igneous dike near Golden Lane production of Mexico. [Link 1952]

complicated areas, where the expelled sediments are frequently much older and more deeply buried than the surrounding sedimentary rocks.

Figure 9-6 shows a Type 5 seepage caused by an igneous intrusion penetrating an organic-rich sedimentary section. Such seepages may not be directly associated with oil accumulations, because the seepage could be oil distilled out of source beds by the high temperature of the intrusion. The seeps would indicate the presence of good source rocks, and in this example the prolific Golden Lane production of Mexico has originated in the same source beds penetrated by the igneous dike.

Young sediments in tectonically active areas produce the most seeps. The small Tertiary basins of California have hundreds of oil seeps. Seeps are also numerous in small intermontane basins such as the Magdalena Valley of Columbia, the Maracaibo Basin, and Indonesia. Seeps are numerous on the mobile sides of basins such as in the Mesopotamian geosyncline, the Monagas Basin of eastern Venezuela, and the eastern foothills of the Andean mountain chain from Columbia to Cape Horn. Seeps are particularly numerous along the margins of basins, where unconformities and oil-producing formations come to the surface. A typical example is Oklahoma, as shown in Figure 9-7. Few seeps occur above productive units buried under large thicknesses of nonproductive strata in the central and western part of the state. Seeps are common in northeastern Oklahoma, where the oil-bearing rocks come to the surface on the flanks of the Ozark uplift, and in southern Oklahoma, where the producing sands are sharply upturned by the Arbuckle and Wichita uplifts. Because these are Paleozoic formations, there has been sufficient time and temperature for part of the asphalt to be converted to the asphaltite grahamite and the pyrobitumen imponite.

Another example of seeps occurring along the edges of a basin or where structural uplifts have exposed the oil-bearing stratigraphic sequences is in

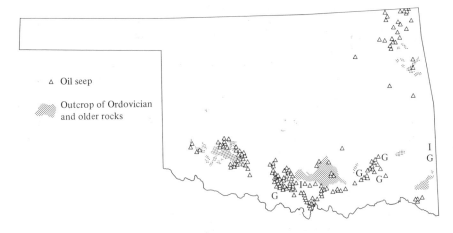

Figure 9-7
Oil seeps and asphalt-saturated sands of Oklahoma. G indicates the asphaltite grahamite; I indicates the pyrobitumen imponite, both of which represent advance states of asphalt maturation. [Dickey and Hunt 1972]

the Lake Maracaibo area, as shown in Figure 9-8. There are over 200 oil and gas seepages in western Venezuela, many of which occur along the flanks of the Venezuelan Andes and the Perija Mountains. The seepages range from a few barrels of oil to asphalt lakes that cover several square kilometers. Along the northwest edge of the Venezuelan Andes, small seeps issue from fractured igneous rocks that have been thrust basinward over Tertiary or Cretaceous sediments. Seeps are associated with Cretaceous, Miocene, and Eocene sediments. The Mene Grande field, El Mene, La Paz, and the Bolivar coastal fields were all drilled because of seeps. Note that there are no seeps in the central, thicker part of the basin southwest of the Bolivar coastal fields, even though prolific oil deposits have been discovered in the central part of the lake.

The rupturing of cap rocks and seepage of oil through small fractures and faults is common in earthquake areas along the edges of crustal plates where continents are in collision. For example, in the Middle East (Figure 9-9) the collision between the Arabian plate and the Eurasian plate regularly jolts Iran and Iraq, resulting in several seepages near the plate boundary. Such boundaries are areas of earthquake activity, which causes the fracturing and faulting that has permitted the upward escape of oil and gas from Cretaceous reservoirs. Probably most of this oil was safely accumulated in reservoirs under an impermeable cap rock 65 million years ago prior to the collision of the African–Arabian plates with Eurasia. In the southern and western parts of Saudi Arabia, where there is practically no earthquake activity, there are very few seeps, even though this area contains some of the largest oil fields of the Middle East, such as Abqaiq-Ghawar. Worldwide, there is a correlation

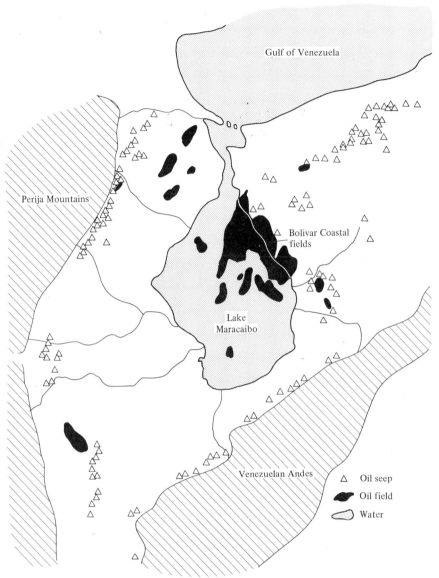

Figure 9-8
Oil seeps of the Maracaibo Basin, Venezuela. [Link 1952]

between seeps and earthquake activity, with the majority of the seeps being close to plate boundaries where such activity is the highest. For example, the western coast of South America has seeps in Ecuador, Peru, and Chile that follow the earthquake belt resulting from the subduction of the Nazca plate beneath South America. Trinidad, southern California, southern Alaska, the

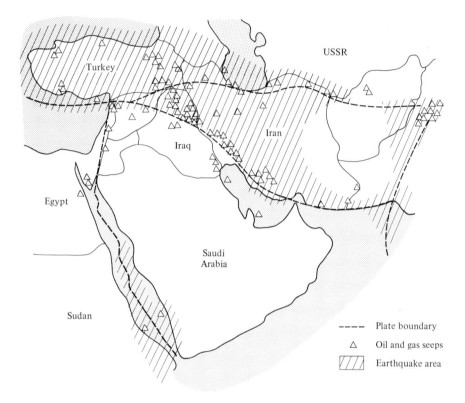

Figure 9-9
Relationship of plate boundaries and earthquake activity to seepage areas of the Middle East. Type 3 seeps involving vertical migration through ruptured cap rocks are common in petroliferous areas near plate boundaries.

Philippines, Indonesia, and Burma are other areas where numerous seeps are related to the earthquake activity of plate boundaries. The hand-dug wells of the Burmese oil fields, which were yielding oil nearly a century before the Drake well, are located on seeps near the Indo-Australian–Eurasian plate boundary. Exploration for oil and gas in wildcat areas near plate boundaries should emphasize the detection and analysis of seeps.

Oil seeps have periodically been reported in coal mines of England, Australia, and other areas. Hydrocarbons in the oil range are generated in small amounts during the maturation of coal, so some seepages would be expected to be associated with coal measures. It is unlikely, however, that a commercial accumulation of oil would form unless there were organic-rich, fine-grained source rocks associated with the coals. Coals can generate commercial accumulations of methane but only trivial amounts of the higher hydrocarbons.

Underwater Seeps

Active seeps are often found offshore on the continental shelf from formations that produce oil on the nearby land area. The seaward extension of the oil-rich Ventura Basin in southern California is the source of the prolific Coal Oil Point seep area, which is reputed to release 50 to 70 bbl of oil a day from several seepage vents. These seeps were observed by the early Spanish explorers and have been a chronic source of beach contamination in the southern California area. Vernon and Slater (1963) clearly identified the source of the seeps in the Santa Barbara area by mapping the asphalt mounds on the sea floor, which were particularly abundant near Point Conception. The asphalt mounds range up to 100 ft (31 m) in diameter and 8 ft (2.4 m) in height. They are irregularly distributed along an east–west trend of faulted anticlines. The asphalt becomes denser than seawater through the loss of gas, light hydrocarbons, and the accumulation of sediment material. Some of the vented asphalt escapes to the surface, but much of it contributes to the growth of the mounds, which are encrusted by marine organisms.

Landes (1973) has summarized the reports of offshore seeps including areas such as the Gaspé Peninsula of Quebec, the U.S. Gulf Coast, the Gulf of Paria near Trinidad, the Gulf of Suez, the Red Sea, the Arctic Coast of Alaska and Canada, and the South China Sea.

Underwater oil and asphalt seeps can be positively identified by divers or remote control underwater cameras. Gas seeps are more difficult to identify, and there is always some question as to whether the gas is biogenic in origin. In the late 1950s, the Atlantic Refining Company developed a gas sniffer for detecting gas seeps in water-covered areas. Basically, the system is similar to mudlogging, except that seawater instead of mud is analyzed for hydrocarbons at a higher level of sensitivity than in the conventional mudlogger. As an exploration vessel moves along the water surface, it continually pumps in seawater from an intake line that is weighted to move at varying distances beneath the vessel. The seawater is stripped of its hydrocarbons and discharged. The hydrocarbon gases are conducted by a helium stream into a gas chromatograph. The sensitivity of the system will vary with the level of instrument sophistication, but most systems can recognize 0.5 ppb of an individual hydrocarbon in the seawater.

Dunlap and Hutchinson (1961) made a statistical comparison of known seeps and known petroleum accumulations in south Louisiana in order to emphasize the importance of marine seep detection. Their data indicated that an exploratory well located near a seep in south Louisiana will have a better chance of finding production than one located only by geological and geophysical data. Also, they concluded that fields found within four miles of seeps will be a third larger than fields further away. They assumed that systems of faults and fractures provided leakage paths for the oil and gas from major zones of accumulation to the surface. They claimed that the Gulf Coast has many faults with dips of 60° or greater, which would result in a horizontal displacement of only about a mile (1.6 km) at a depth of 10,000 ft (3,048 m). Because the average field diameter in south Louisiana is about a mile and a half, it would be possible to encounter production by drilling directly on seeps in this area.

Although considerable publicity attended the introduction of the gas sniffer technique, there are a lot of problems in clearly pinpointing an ocean bottom seep from the analysis of hydrocarbons in the overlying seawater. For example, the hydrocarbon concentration in coastal waters is not constant. In some areas, the water is supersaturated with methane, so the gas is lost to the atmosphere, whereas in other areas the water is undersaturated and acts as a sink for atmospheric methane. Vertical hydrocarbon profiles in the ocean generally show a high in the vicinity of the thermocline. Methane concentrations seem to vary with seawater density, salinity, oxygen content, and biological activity. Seasonal changes in the thermocline cause changes in the depths of the methane maximum.

Higher hydrocarbons, such as butane, have been reported at concentrations of a thousand times open-ocean values in shipping lanes (Brooks and Sackett 1972). Perturbations in higher hydrocarbon concentrations also occur where rivers discharge into the ocean. Seawater samples taken 5 km from the west coast of Africa in the vicinity of the Congo River outflow show an increase in propane from 0.01 to 0.3 ppm and in methane from less than 1 to 60 ppm.

Despite these problems, some companies have installed underwater seep detector systems in their geophysical survey ships. In 1970, Jeffrey and Zarrella reported on the use of a seep detector on the *Gulfrex*. They stated that the system could recognize true petroleum and natural gas seepages even in areas where extraneous sources of high hydrocarbon backgrounds existed. Generally, they did not consider a seep valid unless it could be verified by a second visit to the area. In 1974, the Gulf Oil Corporation launched the *Hollis Hedberg*, a sophisticated research vessel for geophysical and geochemical surveys. The *Hedberg* has three gas sniffers, which were initially towed at the surface, at 200 ft (61 m), and at 600 ft (183 m). Methane through the butanes are measured simultaneously in three gas chromatographs. From the concentration measurements, an aerial plot of the hydrocarbon plume can be constructed on a computer, permitting accurate location of the seep on the sea floor.

All oil companies and service laboratories that have used gas sniffers report a definite correlation between subsurface faults and surface seeps. Off the coast of West Africa, the concentrations of methane and the higher hydrocarbons rise to several hundred ppm within 20 to 30 m (66 to 98 ft) of a fault zone, and then fall to background about 100 m (328 ft) away. Since most surveys include both geophysical and geochemical equipment, it is a simple matter to correlate the hydrocarbon plumes from seeps with seismic anomalies.

Clusters of gas bubbles venting from the sea floor can be detected by the use of 3.5 and 12 kHz acoustical reflection techniques. Bubbles reflect the pulse and show up as a dark vertical line against the clear background of sea-water. Sieck (1973) has correlated these bubble clusters with gas-charged sediment cones below the sea floor. These gas-charged cones, which extend to sediment depths of several hundred feet, create volcano-shaped mud lumps on the sea floor, where they vent into the sea water. High-resolution data can be obtained by combining several seismic systems, such as side-scan sonar, fathometer, tuned transducer, and a sparker or air gun system.

A critical question that is not answered by any of these underwater techniques is whether the gas has a biogenic or thermal origin. Biogenic methane is of little value from an exploration standpoint unless there is reason to believe that commercial accumulations of biogenic gas exist in the area. The presence of high concentrations of propane and butane are generally taken to indicate that the gas is petrogenic in origin. Experience is required to determine what concentration is high, because some propane and butane are formed in Recent sediments. A more accurate way is to use composition data plus carbon-13 analyses, providing enough gas can be collected for such studies. Brooks et al. (1974) collected two gas seeps at sea floor depths of 40 to 50 m (131 to 164 ft) in the Gulf Coast and found them to have $\delta^{13}C$ values of about $-60\ \%_{00}$. This is in the range considered to be of biogenic origin. Later (1975), Sackett reported that 16 seep gases obtained on topographic highs in the 50–100 fathom line off the Gulf Coast had $\delta^{13}C$ values from -40 to $-65\ \%_{00}$. Gas from four producing wells in the Gulf Coast had $\delta^{13}C$ values of -42 to $-50\ \%_{00}$. Sackett concluded that some of the seeps were from biologic sources and others from petroleum reservoirs. Later Bernard, Brooks, and Sackett (1976) reported that only one of these seep gases was unequivocally of thermal origin. This suggests that sniffers probably see a lot of biogenic gas in the Gulf Coast.

Another direct method of determining the probable source of an underwater seep is to analyze sediment samples on the sea floor for contained gases. Carlisle et al. (1975) analyzed sea floor sediment samples in an area where the geophysical data showed the presence of gas bubbles in the water and a subsurface fault. The results are shown in Figure 9-10. Over one of the gas seeps in the vicinity of the fault, the C_2–C_4 hydrocarbon concentrations exceeded that of methane and the C_5–C_7 hydrocarbons increased ten times over background. The large increase in these higher hydrocarbons indicates a petrogenic source for this gas seep. The high concentration of gases in the surface sediments is caused by movement up the fault and a spreading out of the gas into the surface sediments.

Bernard et al. (1977) developed a geochemical model for distinguishing biogenic from petroleum-related gases in marine sediments. Their model

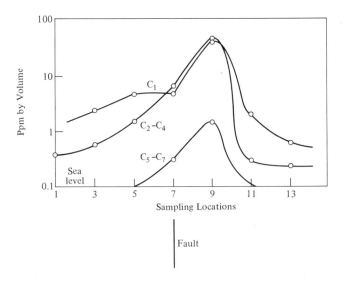

Figure 9-10
Hydrocarbon gases in underwater sediment samples from offshore
Texas–Louisiana in proximity to known gas vents and a fault.
Concentrations shown in volumes of hydrocarbon per million
volumes of sediment. [Carlisle et al. 1975]

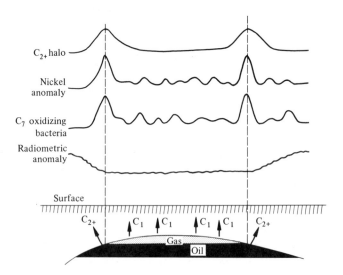

Figure 9-11
Geochemical anomalies presumed to exist over an oil field because
of the vertical migration of hydrocarbons.

included the observations that the composition of thermal gas can be altered during migration through sediments and by mixing with biogenic gas. They considered thermal gas to have a $C_1/(C_2 + C_3)$ ratio of 10 and a $\delta^{13}C$ value in the -35 to -50 ‰ range. Migration could change the $C_1/(C_2 + C_3)$ ratio to 10^3. Biogenic gases have $\delta^{13}C$ values in the -55 to -85 ‰ range and $C_1/(C_2 + C_3)$ ratios from 10^3 to 10^5. Gases with ratios below 10^3 but with $\delta^{13}C$ values more negative than -50 ‰ were considered to be mixtures of biogenic and thermal gases.

Gas sniffing in the water column followed by sediment analyses where sniffing anomalies appear can be effective in mapping underwater seeps. Such studies must be integrated with the geological and geophysical data (subsurface structures, faults, and so on) of the area to have value in prospecting. Where sufficient gas is available, a carbon isotope and composition analysis is useful in defining its source.

All of the techniques discussed above are aimed at locating and identifying the source of the visible seeps. The so-called invisible or microseeps are much more difficult to interpret and utilize as a prospecting technique.

GEOCHEMICAL PROSPECTING

The term *geochemical prospecting* refers to a direct method for finding petroleum that is based on the supposition that some hydrocarbons in an oil or gas accumulation migrate vertically to the surface directly over the reservoir. The technique consists of taking samples of soil at a predetermined depth on a grid pattern and analyzing for hydrocarbons or other materials that are affected by the presence of hydrocarbons. The method is designed to identify invisible seeps, such as those seeps resulting from hydrocarbon diffusion to the surface. Surface prospecting has involved several types of analysis, such as (1) the free hydrocarbon gases in the pore spaces of the soil; (2) hydrocarbon gases adsorbed on soil particles; (3) the fluorescence of soil samples, caused by the presence of high-molecular-weight hydrocarbons; (4) the analysis of soil bacteria that thrive on certain kinds of hydrocarbons; and (5) trace metals and radioactive elements in the soil whose adsorption is affected by the presence of hydrocarbons.

If the mechanism of upward migration is diffusion, the hydrocarbons should show a maxima over the field and a broadening of the anomaly for deeper fields, as with gravity anomalies. Some anomalies do show a maxima, but many show a halo effect, like a doughnut ring, with the field in the center. Various hypotheses have been proposed for these halos. Figure 9-11 shows one. If gases migrate vertically upward, the ethane-plus fraction would originate mainly from the oil, and the methane from the gas cap. This would cause an ethane-plus halo having the configuration of the oil accumulation.

Bacteria that would oxidize only heptane would also show this halo effect. Likewise, nickel and radioactive elements would show a halo. Another idea is that bacteria would consume the hydrocarbons in the more concentrated center section, with less bacterial activity around the edges. Neither of these explanations is very plausible. A more likely explanation is that the halos are reflections of topography and surface geology.*

In the late 1930s, several companies were organized to carry out geochemical prospecting as a direct oil finder. Soil surveys showed many anomalies, and enough of these were drilled to enable proponents of the technique to point to specific discoveries resulting from geochemical prospecting. The method was developed on a purely empirical basis, and failures were explained as caused by improper techniques or inadequate evaluation of the geological factors. Eventually, this led to a thorough investigation of the technique by several major oil companies. Also, a theoretical analysis of the principles involved was carried out by several geochemists.

In 1958, Buckley et al. reported the analysis of gases in formation waters from 300 wells along the U.S. Gulf Coast. The gases were taken with a pressure core barrel. An objective of this study was to determine if the upward diffusion of hydrocarbons caused a hydrocarbon enrichment in sands overlying petroleum accumulations. They found that formations were enriched in hydrocarbons in close proximity to an oil or gas accumulation but overlying formations were sometimes undersaturated with gas and sometimes saturated. For example, in the series of water sands over the Hawkins dome in East Texas, an upper Nacatosh water sand contained more hydrocarbon in solution than a lower sand directly over the productive area. This means that the oil accumulation at the Hawkins dome could not have hydrocarbons migrating upward by diffusion, because the lower Nacatosh sand would have to have a higher concentration than the upper one.

Juranek (1958b) pointed out that the time to reach steady-state diffusion varies with the square of the depth, so that hydrocarbon gases in near surface sediments are more important in causing anomalies than those at depth even when the former are in much lower concentrations. He concluded that the geochemical method is a structure prospecting rather than a direct oil-prospecting technique. Kawai (1963) calculated the diffusion of methane from gas reservoirs in the South Kanto region of Japan. He concluded that migration by diffusion is so slight that the Japanese fields lost very little gas in

---

*Hedberg (personal communication) believes that the halo effect may often be caused by a reduction of permeability by cementation immediately above a structurally high petroleum accumulation. Cementation occurs because the upward movement of waters in this area would not only help to bring petroleum into the trap but would also tend to cause a mineralization of the overlying strata through which water (not oil) would have been escaping during oil accumulation, as in the case of "cap rock."

the last several million years. Migration caused by ground water flow or by buoyancy of gas bubbles is much larger than that by diffusion. Saraf (1970) calculated that it will take 285 million years for 1 percent of a methane accumulation at a depth of 3,000 m (9,843 ft) to migrate upward by diffusion to a depth of 150 m (492 ft) below the surface. It will take 11 million years for 1 percent of an accumulation at 300 m (984 ft) to migrate within 15 m (50 ft) of the surface. These calculations assume that some adsorption of the methane occurs during migration. Because the surface prospecting method is based on some surface adsorption of hydrocarbons, subsurface adsorption would also be expected to occur.

A calculation that is more relevant to gas production in the United States was made by Smith et al. (1971). They calculated the hydrocarbon losses by steady state diffusion from a 2.8×10^8 m³ (99×10^8 ft³) gas accumulation at a depth of 1,740 m (5,709 ft). The other factors in the calculation were consistent with known reservoirs in the United States. Their data, shown in Table 9-2, indicate that it will take 140 million years for methane and 170 for ethane to reach the surface. The rate of loss of methane and higher hydrocarbons at the surface would be less than the quantities of hydrocarbons formed during diagenesis in near surface sediments of most reducing areas (see Chapter 5). This calculation assumes that diffusional losses are being replaced by source rocks in order to maintain the accumulation at a constant size and composition. Since this may not be occurring, it means that the quantities of gases migrating to the surface would be even less. Smith et al. also calculated that a 3 m (10 ft) thick aquifer at a depth of 870 m, or 2,854 ft (half-way between surface and accumulation) and with a lateral water velocity of 1 m (3.3 ft) per year would move the vertically diffusing gases laterally 190 km (118 miles). If it is further assumed that 150 m (492 ft) of the reservoir cap rock has a porosity of only 4.5 percent, the rest of the section being 15.7 percent, then the rates of loss of gas to the surface are decreased by 50 percent.

The Smith et al. model did show that in vertical, water-filled joints and faults the diffusion coefficients of the various gases would be increased by a

Table 9-2
Losses of Reservoir Gas by Vertical Diffusion

	Relative molar composition	Time to reach surface, 10^6 years	Rate of loss to surface, cm³/m²/year
Methane	100	140	2
Ethane	7.5	170	0.16
Propane	4.3	230	0.06
Butanes	2.1	270	0.02

Note: Calculated by Smith et al. (1971) for steady-state diffusion from a 2.8×10^8 m³ accumulation at a depth of 1,740 m, temperature of 70°C and pressure of 172 atm. Overburden porosity assumed to be 15.7 percent.

factor of about 39. This would explain observed hydrocarbon anomalies over faults, although it is still possible the observed hydrocarbons are entering the faults from near surface source beds.

Clearly, vertical diffusion from petroleum accumulations buried at depths such as 1,700 m (5,600 ft) is not the mechanism causing surface geochemical anomalies. More likely these anomalies are caused by the diffusion of gases from decaying organic matter in the first few hundred feet of burial plus some hydrocarbons migrating by other mechanisms. The late V. A. Sokolov, who is generally recognized as the father of geochemical prospecting, always felt that the most intensive migration of gases resulted from filtration, buoyancy, and other mechanisms through faults, fractures, fissures, and other permeability channels. Diffusion supplemented these processes. However, these other mechanisms would show an erratic surface distribution rather than a pool outline. This means that surface prospecting might be useful as a regional prospecting tool where vertical migration occurs, as in offshore areas of the U.S. Gulf Coast where the Pliocene and Pleistocene contain hydrocarbons from deeper formations. The technique would not be able to outline an accumulation. Considerable experience would be needed even for regional prospecting to sort out the strong background variability in near surface hydrocarbons. Hitchon (1974a) has emphasized that an understanding of the composition and flow of near surface fluids is an important factor that is rarely studied in surface prospecting.

Periodically, papers are published stating that geochemical prospecting has not had a fair chance. It has been condemned because of inadequate good research, failure to combine with geological and geophysical studies, and failure to use modern, highly sophisticated analytical equipment. The truth is that enormous sums of money have been spent by most of the major oil companies using the best scientific talent available in order to perfect geochemical prospecting, without success. Many government geological surveys and oil companies throughout the world have tried with discouraging results to recognize oil fields at the surface. Since there is no scientific basis for the concept of direct vertical migration, it is not surprising that surface prospecting cannot outline accumulations.

A few of the many investigations will be cited here. Humble Oil intensively investigated the Blau method, named after their chief of geophysical research. It was found that chemical reagents such as sodium peroxide added to soil samples gave color reactions that presumably outlined subsurface oil accumulations. The color reactions were later found to be caused by differences in humic and fulvic acids originating in surface samples. Humble Oil's most significant contribution was development of the pressure core barrel, which proved that many formation waters overlying oil and gas accumulations were considerably undersaturated in gaseous hydrocarbons (Buckley et al. 1958). The Humble and Carter research laboratories investigated surface geochemical prospecting for about ten years and concluded that there was no correlation between hydrocarbon anomalies at the surface and the presence or absence of subsurface oil and gas accumulations.

Imperial Oil collected and analyzed over 9,000 soil samples and concluded that surface hydrocarbons do not occur as a halo around oil fields. In fact, there appeared to be no general diffusion of hydrocarbons from reservoirs to the surface. They did find that fluorescent highs are closely associated with photolinears. International Petroleum examined over a thousand soil samples from the Middle Magdalena Valley of Colombia. A definite relationship was found between soil fluorescence and faulting in rocks beneath the Upper Cretaceous unconformity. Surface samples within 300 m (984 ft) horizontally of the traces of seismic faults on the unconformity showed significantly higher fluorescence than those from the rest of the area. Creole Petroleum Corporation conducted a large-scale research project to determine if tonal anomalies as seen on air photos may be indirectly related to subsurface hydrocarbons. The conclusion was that the anomalies are related to textural or compositional differences in the soil or vegetation cover as contrasted with the surrounding area.

Pan American (now Amoco) drilled a thousand 9-ft holes in the Powder River Basin of Wyoming for geochemical surface studies. Preliminary experiments indicated that they could map some of the existing producing fields. However, nothing ever came of the program, and there was no significant increase in Pan American's wildcat success ratio compared to other companies.

D. B. Sikka (1959) made a radiometric survey over the Redwater oil field of Alberta and other areas in Canada. He concluded that radiometry is useful in delineating fault and fracture systems. For example, radiometric highs across the St. Lawrence River near Assumption, Quebec, were associated with faults underneath the river. Later, Imperial Oil conducted radiometric surveys over 4,000 square miles (10,360 square kilometers) covering 28 producing fields in various stages of development in western Canada. They found a beautiful correlation with soil types. Sandy soils and bogs are radiometric lows, while loams tend to be relatively higher. The radiometric patterns were not constant with time, and it was inferred that they were dependent on moisture conditions in the soil. The contoured radiometric survey maps showed no relationship to the oil fields. Areas of low and high radioactivity had about the same number of producers. Various statistical tests applied to the data indicated that the radiometric anomalies were about what would be expected from contouring random numbers.

The Mobil Oil Company investigated geochemical prospecting for several years combining hydrocarbon studies in soil samples with microbiological analyses. They concluded that if microseeps exist they are of relatively small aerial extent and low population density, which makes it impossible to outline oil accumulations (Stevens 1962). Mobil conducted carefully controlled tests over known production areas of several basins and concluded that only a small percentage of the soil gas samples contained ethane and higher hydrocarbons exceeding background values. In the Denver Basin, Mobil analyzed soil samples in a grid surrounding the discovery well in each of six prospects before any other wells were completed. No relationship was found between ethane concentrations and the area of subsequent production.

At the Third International Geochemical Conference in Budapest in 1962, the Polish, Czechoslovakian, and Hungarian petroleum geochemists all reported that they could not find oil with surface prospecting methods. In the Precarpathian region of Poland, a correlation was found between hydrocarbon anomalies and the presence of fault zones. The conference concluded that areas with complicated tectonics involving faulting and thrusting give unfavorable results in surface prospecting because the fault zones are preferential migration paths. Another complication, pointed out by Mogilevskii (1963) is that surface geochemical anomalies frequently correlate with biological activity, because microorganisms consume hydrocarbons in some areas but not in others. Biological activity varies with soil nutrients, introducing another complicating factor.

The few areas where surface hydrocarbons seem to show some relationship to subsurface accumulations involve a suitable combination of an arid climate and highly permeable beds above the accumulation. An example is the Fort Polignac Basin in the Sahara Desert where complete absence of vegetation, a low water table, and permeable sediments enabled some hydrocarbons to migrate to a surface structure, where subsequent drilling established a gas reservoir at 2,250 m or 7,382 ft (Pomeyrol et al. 1961). Even in this instance, the data did not outline the size of the accumulation.

The USSR conducted more research on geochemical prospecting, under the able leadership of the late V. A. Sokolov, than any other country. In 1962, about 10 percent of the geochemical field parties were engaged in surface geochemical surveys, mostly on a research basis. Most successful correlations with subsurface accumulations occurred in arid areas such as around Baku and Turkmenia. In addition to soils, water from springs and shallow wells were analyzed for traces of hydrocarbons. Despite many years of research, there is still considerable disagreement in the USSR concerning the value of soil prospecting. Sokolov himself recognized that more hydrocarbons were observed at the surface than could be explained by simple diffusion. Sokolov felt that these excess hydrocarbons were not coming from shallow sediments but rather were coming from several other mechanisms of migration, such as (1) the filtration of gas and oil caused by pressure differences within the interconnected pores and fractures in the rocks, (2) the buoyancy of gas and oil in the water contained in porous and fractured rocks, (3) the transfer of free and dissolved gas and oil by underground waters, and (4) the squeezing of gas and oil along with the water from connecting sediments (Sokolov et al. 1963). The problem is that the intensity of vertical migration by any of these mechanisms does not lend itself to a simple calculation, so there is no way of estimating how much gas moves vertically by these processes. All of these processes would follow erratic migration paths to the surface and would not be related to the configuration of the oil or gas accumulation. In one study made by the Russians, the gas content and microorganisms in soils and water above an artificial gas reservoir were analyzed before and after filling the reservoir. Over a period of a few months, a water sand located 300 m (984 ft) above the reservoir reached a gas content about ten times its initial value, and the concentration of hydrocarbon oxidizing bacteria increased significantly. This is much too rapid to be attributed to diffusion. The cause was believed to be a fault or fracture zone over the reservoir.

There seems to be no question that diffusion and other migration processes increase hydrocarbon concentrations in rocks in close proximity to oil or gas accumulations. Studies of hydrocarbon concentrations and various individual hydrocarbon ratios in the first few hundred feet of sediments above accumulations have frequently shown abnormal values that appear to be caused by migration processes. In many instances, however, this migration does not reach the surface in concentrations high enough to offset the interfering effect of hydrocarbons generated from organic matter in shallow near surface sediments. Also, the fastest mechanisms of migration follow erratic pathways to the surface that do not outline subsurface accumulations. In other words, surface geochemical prospecting can be useful as a regional tool to show the general area of oil or gas accumulations, providing information is available on the geology, the effect of near surface diagenetic hydrocarbons, and the local and regional fluid flow systems in the sediments. Surface prospecting cannot, however, show the outline of a subsurface pool in a way that will enable a drilling location to be made, except in very rare cases.

Probably the most valuable use of surface prospecting is in identifying low levels of seepage associated with faults, fractures, unconformities, and intrusions such as piercement salt domes, mud diapirs, and igneous intrusions. These were discussed in the previous section on visible seeps, but they also may occur as invisible low-level seeps. Lowland areas, which can act as major discharge regions for subsurface fluids, may carry traces of hydrocarbons from subsurface accumulations. Chalk beds and continental deposits such as red beds sometimes have sufficient permeability for vertical movement of pore fluids carrying hydrocarbons. Some geologists believe that microfractures providing vertical permeability occur in many relatively impermeable rocks. Although there is disagreement over the extent and importance of microfractures, it does appear that hydrocarbons do migrate upward with subsurface pore fluids in a wide range of concentrations from the visible to the invisible in many different geological situations.

If surface prospecting is not looked on as a direct oil finder that can outline oil and gas fields but instead is considered an auxiliary technique to conventional geological and geophysical exploration methods, then it does have merit in many areas. Such an application, however, would require a thorough understanding of the active groundwater flow regime and the known links between seeps and hydrodynamic systems, as discussed by Hitchon (1974a). Used in this way, it can assist in differentiating structures and areas that are alive with hydrocarbons from those that are essentially dead. It can be a useful auxiliary tool, but not a panacea, for finding oil and gas.

SUMMARY

1. Visible oil and gas seeps are important in exploration, because they indicate the presence of source rocks or reservoired oil or gas. The most important oil-producing regions of the world were detected through visible oil and gas seeps.

2. When crude oil leaks to the surface, it undergoes evaporation, leaching of water solubles, microbial degradation, polymerization, auto-oxidation, and gelation. These reactions lead to thickening or solidification of the oil. A liquid oil is gradually converted to an asphalt, asphaltite, and eventually a pyrobitumen.

3. Seeps in the field may be classified as (1) active or live, (2) inactive or dead, and (3) false. Seeps in rank wildcat areas should be analyzed in detail and their relationship determined to the lithology, stratigraphy, and structure.

4. Seeps are most numerous on the margins of basins and in sediments that have been folded, faulted, and eroded. Seeps originate in five types of geological situations: (1) emerging from homoclinal beds, (2) direct leakage from a source bed, (3) rupture of a reservoir by erosion, faulting, or folding, (4) migration along unconformities, and (5) migration with intrusions such as mud volcanoes, igneous intrusions, and piercement salt domes.

5. Worldwide, there is a correlation between seeps and earthquake activity, with many seeps being in areas of plate boundaries, where such activity is the highest.

6. Underwater seeps may be monitored by divers, underwater cameras, or gas sniffers. Sniffers involve pumping water from near the sea floor and analyzing it for hydrocarbons. The technique requires considerable experience, because there are many extraneous sources of hydrocarbons in seawater that are not related to subsurface accumulations. Many underwater seeps are biogenic in origin.

7. Vertical diffusion of hydrocarbons from subsurface petroleum accumulations is not the mechanism that causes surface hydrocarbon anomalies. These anomalies are probably caused by the diffusion of gases from decaying organic matter in the first few hundred feet of burial and by hydrocarbons migrating to the surface by mechanisms other than diffusion. Buoyancy is the most likely driving mechanism.

8. Surface geochemical prospecting cannot outline oil or gas accumulations at depth except in rare cases. It can be useful as an auxiliary prospecting tool providing that data on subsurface geology, near surface hydrocarbons, and fluid flow systems in the sediments is available. It is most useful where intrusions, fault or fracture systems, or permeable beds are providing vertical pathways of migration by processes other than diffusion. In such regions, it can assist in differentiating structures and areas that contain hydrocarbons from those that are barren.

SUPPLEMENTARY READING

Hitchon, B. 1974. Application of geochemistry to the search for crude oil and natural gas. In A. A. Levinson (ed.), *Introduction to exploration geochemistry.* Calgary, Canada, pp. 509–545.

Hobson, G. D., and E. D. Tiratsoo. 1975. Surface exploration for petroleum. In *Introduction to petroleum geology.* Beaconsfield, England: Scientific Press, pp. 123–149.

Kartsev, A. A., V. A. Tabasaranskii, M. I. Subbota, and G. A. Mogilevskii. 1959. *Geochemical methods of prospecting and exploration for petroleum and natural gas.* (Chevron Research Corporation, trans. P. A. Witherspoon and W. D. Romey, eds.). Los Angeles: University of California Press. (Original Russian edition published 1954.)

10

Subsurface Prospecting

Too often a dry hole is plugged, abandoned, and forgotten, with little effort being made to determine why the hole was dry. Considering that rank wildcats now cost in the millions of dollars, it is imperative that some funds be allocated prior to drilling for a careful assessment of the geological and geochemical parameters of the sedimentary section drilled. It has been the practice of major companies for several decades to drill stratigraphic tests in new areas primarily to obtain subsurface information. The COST (Coastal Offshore Stratigraphic Test) wells are recent examples of industry-supported deep stratigraphic tests involving detailed geological and geochemical studies of the well samples. Similar test wells are drilled in the USSR.

The kinds of geochemical analyses that can be run on subsurface samples and their value in exploration are discussed in this chapter. These techniques use different approaches to answer the same question, namely, has oil or gas been generated in sufficient quantities to form economic accumulations? Some of these techniques are so similar that it is not necessary to use all of them on a wildcat well. At the same time, it is equally dangerous to rely on only one methodology, because any geochemical approach is subject to anomalous situations, as are many geological methods. Also, subsurface geochemical data should not be interpreted in the absence of geological information. Geochemistry is not a black box, and geologists and geochemists need to work together in the interpretation of the data, if it is to have any value in exploration. Subsurface techniques have been most valuable where the geologists discuss the problem and the questions to be answered with the geochemists, who suggest the best analyses for that type of problem.

GAS AND GASOLINE

In some areas, the subsurface sediments seem to be alive with hydrocarbons. Oil or gas shows are encountered in several permeable beds. Background gas builds up in the circulating mud. In other areas, the sediments are dead. The mud logger shows nothing, and there is not the slightest evidence of hydrocarbons from drill-stem tests. This concept of a well being alive or dead with respect to hydrocarbons can be followed up by mapping hydrocarbon live areas from data such as mud logs.

Gas Logging

Most hydrocarbon gas in the subsurface exists either in the free form, dissolved in formation water or adsorbed on sediment particles. Hydrocarbon mud logging was first used commercially by the oil industry in 1939 to detect pay horizons with rotary drilling. As the drill bit breaks the rock into small fragments or cuttings, the gas present is released into the mudstream. As the mud comes out of the hole a portion is run into a mud–gas separator. This may be simply a cylinder containing baffles to spread out the mud or a stirring unit to beat up the mud and release the entrained gas. The gas–air mixture is then fed into a gas chromatograph, where the individual hydrocarbons, methane through pentanes, are analyzed. Some systems using a simple hot wire detector make only two measurements, the methane and the ethane-plus fraction, which includes all higher gaseous hydrocarbons.

The technique was designed specifically to test reservoir rocks for hydrocarbons during drilling. In fact, it is still the practice in some areas to use the logger only when penetrating prospective pay horizons, but most companies log continuously.

In the USSR, continuous mud logs and intermittent core samples are taken of both reservoir and nonreservoir rock. The cores are hermetically sealed on recovery and the gas is desorbed in the laboratory by heating to 70°C (158°F). The gas logs they obtain represent a combination of data from both mud and core analyses. This enables Soviet geochemists to look not only at the reservoir rocks but also at the fine-grained source rocks throughout the sedimentary section. In the late 1950s, V. A. Sokolov and B. P. Yasenev noted that in sedimentary rocks immediately above productive oil or gas fields there was frequently more gas than in rocks from dry hole areas. For example, in Figure 10-1 the gas logs are shown for two wells drilled in the Kum–Dag region of the USSR. Well 41 encountered a commercial petroleum accumulation at a depth of about 780 m (2,556 ft). Well 42, which was located around 1,200 m (3,937 ft) from 41, was a dry hole. A higher concentration of gases is

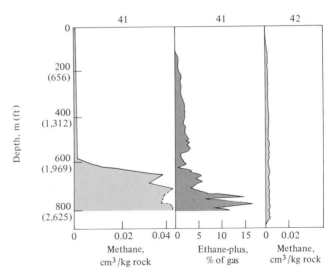

Figure 10-1
Gas logs (mud logs) of two wells in the Kum–Dag region, USSR.
Both methane and ethane-plus increase above the pay horizon
at 780 m in Well 41. Only traces of methane were found in Well 42,
which was a dry hole. Note that the increase in Well 41 does not
extend to the surface. [Yasenev 1962]

noted in Well 41 throughout the section, but the largest increase occurs at about a depth of 580 m (1,900 ft). Here the methane and ethane-plus increases substantially, even though it is 200 m (656 ft) above the oil accumulation (Yasenev 1962).

Another example is shown in Figure 10-2 of three dry holes (1, 2, 5) and two wells (3 and 4), which encountered oil production (Yasenev 1959). Wells 1 and 2 drilled on the Ivanov structure penetrated Cretaceous and Jurassic sediments and bottomed in the Upper Paleozoic. There was almost no gas in this entire section and no production was encountered. Well 5 was drilled in the Peschano–Umet area beyond any oil production. It contained a little more gas than the Ivanov wells but considerably less than Wells 3 and 4, which encountered an oil accumulation in the Jurassic. Other examples from Yasenev's paper are summarized in Table 10-1. The data are shown as the average methane or ethane plus content in the entire rock section drilled. It is clear that in these examples the gas content of subsurface sediments above oil or gas fields is five to ten times greater than those from nonproductive areas.

There are two possible sources for the high gas yields shown in Wells 41, 3, and 4 of Figures 10-1 and 10-2. They could be gases migrating upward from the petroleum accumulations, or they could be syngenetic gases formed in

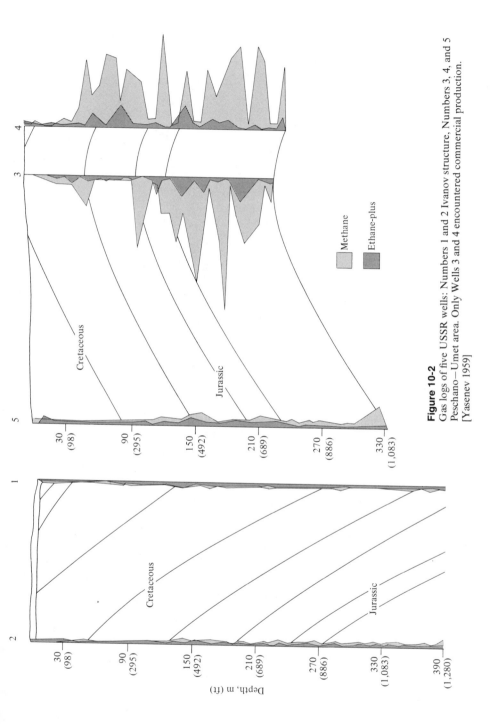

Figure 10-2
Gas logs of five USSR wells: Numbers 1 and 2 Ivanov structure, Numbers 3, 4, and 5 Peschano—Umet area. Only Wells 3 and 4 encountered commercial production. [Yasenev 1959]

Methane
Ethane-plus

Depth, m (ft)

Cretaceous
Jurassic

Table 10-1

Gas in Subsurface Sediments Overlying Petroleum Accumulations

Structure	Well number	Above oil or gas field		Nonproductive area	
		C_1	C_{2+}	C_1	C_{2+}
Ivanov	1	—	—	0.09	0.04
	2	—	—	0.8	0.03
Peschano–Umet	3	1.2	0.26	—	—
	4	1.3	0.22	—	—
	5	—	—	0.22	0.10
North Stavropol	4	1.6	0.40	—	—
	5	—	—	0.40	0.03
Staro–Il'sk	2	0.57	0.91	—	—
Mukhanov–Kuybyshev	67	1.63		—	
	68	1.47		—	
	189	—		0.28	
Kungur–E. Chernon	1	—	—	0.40	0.28
	2	4.6	3.5	—	—
	3	2.6	2.4	—	—

Average gas content of sediments in cm³/kg rock[a]

[a]Averages calculated for methane (C_1) and ethane through pentanes (C_{2+}) for entire section drilled. Only total gas was available for wells 67, 68, and 189 (Yasenev 1959).

place from the organic matter disseminated in the sediments. Yasenev believed that most of the gas resulted from vertical migration, because the gas yields seemed to die out both vertically and horizontally away from petroleum accumulations. Also, he found no direct relationship between gas yields and the organic matter of a rock. There is evidence from other examples around the world that the vertical migration of methane by diffusion and other processes does show up for distances of a few hundred meters above some accumulations. However, over distances of thousands of meters the shale gas that shows up on mud logs is mostly syngenetic. The gas analysis of cuttings, which is more precise than mud logging, does show a linear relationship between methane yield and organic carbon content.

From an exploration standpoint, the source of the gas is less important than the fact that it is there in large quantities. Wells 41, 3, and 4 are alive with hydrocarbons, whereas Wells 42, 2, 1, and 5 are essentially dead. Mud logging has the advantage over surface prospecting of being able to high-grade different stratigraphic units in terms of their hydrocarbon potential. As a general rule, hydrocarbon accumulations are found in the reservoir rocks of stratigraphic sections where the nonreservoir rocks contain relatively large amounts of gas. In gas-prone areas, both the coarse- and fine-grained rock will contain primarily methane. In oil-prone areas, the ethane-plus fraction comprises a substantial percentage of the gas throughout the section. These concepts are discussed not only in Yasenev's papers but also in a book by Sokolov and Yurovskii (1961) that covers all the problems associated with the application of hydrocarbon mud logging in petroleum exploration.

More recently the geochemists of the Soviet Union have been correlating data from mud log and core analyses of shallow (200 to 800 m; 656 to 2,625 ft) test wells drilled in prospective areas to identify anomalous increases in thermal gas in the subsurface (Geodekian and Stroganov 1973). Such studies are made prior to deep exploration drilling in order to quantify zones of high gas concentration. Core samples are taken at 5m (16 ft) intervals and analyzed for adsorbed hydrocarbons. Areal maps are drawn for specific subsurface geological reference horizons on which methane and ethane-plus hydrocarbon concentrations from the core and mud log analyses are plotted. Wells are drilled over regional fault zones (flexure–rupture zones), across wedge-outs of productive formations, across the strike of monoclines and in other structural areas where gas migration may occur.

In Chapter 9, it was pointed out that seepage is common along homoclinal beds, unconformities, fracture–fault systems and around intrusions such as salt domes or igneous rock. In effect, the Soviet technique involves identifying such seepage in the subsurface. For example, in the South Tiub–Karagan area near a discontinuous fault, which appeared to be an exit channel for gases of deep origin, the average methane concentration in the 400–800 m (1,313–2,625 ft) depth range was 11.4 cm^3/kg of rock (26,200 ppm by volume) with a maximum of 39 cm^3/kg (90,000 ppm). This compares with an average background methane concentration of 0.11 cm^3/kg (253 ppm by volume). In the South Mangyshlak area, the methane concentration in the drilling mud at a 600 m (1,969 ft) reference horizon was 1.16 cm^3/liter (1,160 ppm), compared to a background level of 0.16 cm^3/liter (160 ppm). A gas accumulation occurs here at 1,900 m (6,234 ft). Cores of reference horizons over productive areas of South Mangyshlak had 250 times as much methane as those over nonproductive areas. Geodekian and Stroganov (1973) found that core data gave the best results in some areas and drilling mud in others. They also found that variations in the gas composition patterns of the reference horizons often indicated the direction of gas migration.

In the United States, mud logs are still used primarily to evaluate reservoir production and only incidentally to indicate the presence of source rocks. One problem is that on most loggers the mud trap recovers only about 15 percent of the gas in freshwater mud and less than 5 percent in oil emulsion mud. Higher efficiencies can be obtained with loggers using steam or CO_2 to drive the gases out of the mud. Steam will recover about 80 percent of the butanes in a 3 percent diesel oil emulsion mud whereas conventional traps recover only a few percent.

Despite these problems, mud log maps have been compiled to show broad regional trends and stratigraphic differences in the quantity of gas in the subsurface. Since thousands of logs are available from commercial logging companies, this is a source of information that can be obtained at relatively low cost. The procedure consists of reducing logs to a common gas scale and plotting average gas readings for each stratigraphic section from each well. Gas shows also are noted.

When such a study is made for the Western Canada Basin, it clearly shows a relation between the quantity and composition of gas in nonreservoir rock and the nearby presence of oil or gas accumulations. The shallow, immature, Lower Cretaceous rocks of northeastern Alberta show dry gas throughout the mud logs, and the associated reservoirs contain only dry gas or heavy oil. In the deeper, more mature Mississippian system, there is a band of wet gas extending from the Peace River area south to the international boundary of western Alberta. All the major oil production is in this wet gas band. West of this band, the sediments become deeper and hotter with wet gas giving way to dry gas. Mud logs of the deeper Devonian section in central and northern Alberta show high wet gas yields through almost the entire interval. The wet gas of the nonreservoir rocks surrounds the oil-producing reefal limestones that extend from the Peace River area to Calgary. Further northwest, in British Columbia, the mud logs show only dry gas associated with the prolific gas accumulations of that area. From the Devonian to the Cretaceous throughout British Columbia and Alberta, the shale gas is wet in oil-producing areas and dry in gas-producing areas.

In contrast, mud logs from a large area immediately west and north of Regina, Saskatchewan, show no gas through the entire section from the Cambrian to the Cretaceous Mannville. Clearly, this is a dead area with no potential for commercial hydrocarbon accumulations regardless of other geological factors. If no oil or gas is being generated or is migrating anywhere in the section, there can be none in the reservoirs. South and east of Regina, the mudlogs show a sharp increase in wet gas yields, with particularly high yields in the area of prolific Madison oil production. Even here, there is little or no shale gas in the deeper, lower Devonian, Ordovician, and Silurian sections, where there is also no production.

Mud logs become most useful for wells where samples are not available. For new well locations, it is better to can samples of cuttings taken at intervals down the hole and have these analyzed for gas content by a service laboratory. Laboratory procedures extract all of the gas in the cuttings and give a more precise reading of changes with depth than mud logs. Figure 10-3 shows the changes in composition of the gaseous hydrocarbons analyzed in canned cuttings from four wells along an east–west cross section through the Western Canada Basin. The changes are comparable to the results on the mud logs just discussed. Upper and Middle Devonian strata are shown in this section, with the datum being the top of the Woodbend Group. From Well 1 to Well 4, a

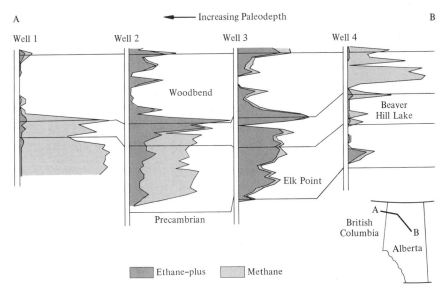

Figure 10-3
Cuttings (C_1–C_4) gas log cross section for Upper and Middle Devonian strata of Western Canada Basin. [Bailey et al. 1974]

lateral decrease in subsurface paleotemperatures results from decreasing paleodepths of burial. Well 1 was subjected to the highest temperatures, estimated to exceed 125°C (257°F). No gas is shown in the Woodbend of Well 1, and no significant production is associated with this formation in British Columbia. The lower Beaver Hill Lake and Elk Point formations have a high yield of dry gas, and dry gas production is significant in associated Beaver Hill Lake reservoirs. In Well 2, the ethane-plus fraction increases and reaches a maximum in Well 3. This well represents the mature part of the basin in central Alberta, which contains prolific Devonian reservoirs. Well 4 further east is in the immature dry gas stage associated only with heavy oils.

Similar correlations show up in other areas. Mud logs from the Paradox Basin show higher shale gas yields near production than in the surrounding barren areas. In the Gulf Coast, relatively high shale gas readings are obtained in the Tuscaloosa Formation of Louisiana and Mississippi. Going east, the gas yields of the Tuscaloosa decrease to almost zero in Georgia and northern Florida. Drill stem tests of Tuscaloosa sands from Mississippi to Georgia show the same decrease in the dissolved gas of formation waters. Whereas the Mississippi sediments are alive with hydrocarbons, the Georgia sediments are essentially dead.

The ratios of hydrocarbons in mud log gas or cuttings gas may be used to discriminate between potential oil and gas production providing the hydrocarbons have not fractionated excessively in the recovery or sampling process.

Pixler (1969) used hydrocarbon ratios in mud logging to determine whether a reservoir will produce oil, gas, or water. He used a steam still for quantitative recovery of gases from drilling mud. The basic concept is that wet gas containing ethane through pentanes is indicative of oil production, while dry gas (methane) indicates only gas production. Concentrations of individual gases, or ratios of gases, are plotted against each other to graphically predict the type of production. Some typical plots used are C_1 versus C_3 or C_2/C_1 versus C_3/C_1. Potential oil source rocks are grouped on the graph where the ratios or concentrations of higher hydrocarbons to methane are relatively high.

Core and Cuttings Analysis

Side wall cores are normally considered less contaminated than cuttings, because they represent a single horizon rather than a mixture of cuttings plus cavings from up the hole. Since side wall cores are generally taken after a well is drilled to total depth, however, the formations are subjected to the high pressure of drilling mud. If the mud has diesel oil or crude in it, the oil can contaminate the side walls. This is particularly true of friable Tertiary sediments. Consequently, cuttings are generally preferable to side wall cores for hydrocarbon surveys. The usual technique is to collect 250 to 500 cc ($\frac{1}{2}$ to 1 pint) of cuttings from the shale shaker. If the shaker is not providing enough cuttings, a hose can be attached to the bottom of the mud flow line before it reaches the shaker and fed into a small box placed over the mud pits. The cuttings should not be allowed to dry out. If oil is used in the drilling mud or if the cuttings are heavily coated with mud, they should be washed off. Normally, however, it is preferable not to wash the samples. The cuttings are placed in a No. 2 or quart-size tin can and filled to within about $\frac{1}{2}$-inch of the top with water. A bactericide such as zephrin chloride is added to the can to stop the bacterial formation of methane and the can is sealed. Depending on the type of lid used, the cans may be sealed with a crimping device or hammered on with a rubber mallet and secured with clips. After labeling, it is a good idea to invert the can with the factory seal on top to prevent any loss of gas. The samples are then shipped to a company laboratory or a service lab for detailed analyses of the hydrocarbons.

Samples are usually taken at 50–100-ft (15–30-m) intervals, but the number actually analyzed will depend on the information desired. It is best for the geologist to pick a limited number of samples down the hole for analysis and then fill in where more information is needed. All samples picked for analysis should be examined microscopically to eliminate those containing high percentages of cavings. Both the sampling and analysis can be done at the well site if the data are urgently needed. Some companies have designed portable analytical laboratories for complete well site analyses.

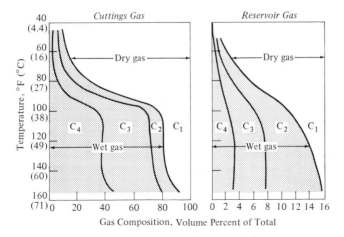

Figure 10-4
Composition of Cretaceous gas in source and reservoir rocks vs.
subsurface temperatures, Western Canada Basin. [Evans and
Staplin 1971]

Examples of the use of cuttings gas analysis in exploration are shown in
Figures 10-4 through 10-7. On the left side of Figure 10-4 is shown the
composition of Cretaceous shale gas on the eastern or updip side of the
Western Canada Basin. Dry gas occurs to a depth range equivalent to
present-day temperatures up to about 90°F (32°C), after which there is a
rapid change to wet gas (Evans and Staplin 1971). The gas in the associated
Lower Cretaceous Mannville reservoirs also changes from dry to wet at the
same temperature interval. This change is accompanied by an increase in
°API gravity and the producibility of Lower Cretaceous Mannville oils.
Consequently, the depth at which this change occurs is used to predict the
depth at which a Mannville oil will be producible. Since these sediments are
believed to have been buried deeper in the past, the paleotemperatures were
probably about 70 to 100°F (21 to 38°C) higher than they are today.

A similar change from a dry gas to wet gas facies in a shale was previously
shown in a well from the Beaufort Basin (Figure 5-13). In that example, the
dry-to-wet conversion occurs at about 5,000 ft (1,524 m), equivalent to a
present-day temperature around 100°F (38°C). Gasoline range hydrocarbons in the shale increased a hundredfold with depth, the largest jump
occurring at a depth equivalent to the wet gas increase.

Figure 10-5 shows on the left the change in shale gas composition from wet
to dry at a present-day subsurface temperature of about 200°F (93°C) in the
Devonian of northern Alberta and northeastern British Columbia. A similar
change is observed in the reservoirs. Primarily dry gas is found in the
deeper, hotter area of northeastern British Columbia and wet gas in the

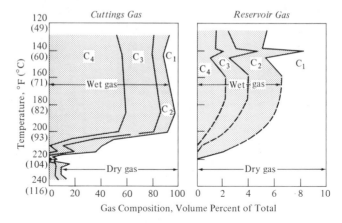

Figure 10-5
Composition of Devonian gas in source and reservoir rocks vs.
subsurface temperatures, Western Canada Basin. [Evans and
Staplin 1971]

shallower reservoirs of Alberta. Paleotemperatures are estimated to be up to
150°F (66°C) higher (Evans and Staplin 1971).

Changes in shale gas composition will be observed laterally within the same
formation if there are significant changes in temperature. As shown in Figure
10-6, the Devonian carbonates and evaporites and the Beaver Hill Lake shales
are all rich in wet gas. The overlying "shale unit" is barren. Going westward,
the gas in the fine-grained rock cuttings changes from wet to dry. This is due

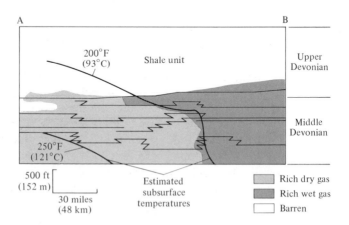

Figure 10-6
Gas facies in fine-grained Devonian shales and carbonates of
northeastern British Columbia and northwestern Alberta, Western
Canada Basin. [Evans and Staplin 1971]

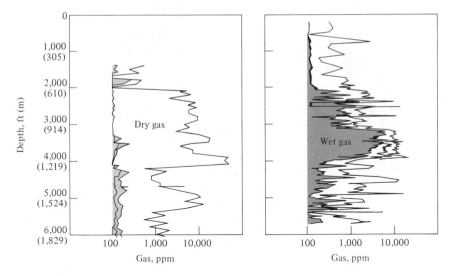

Figure 10-7
Cuttings gas logs from two wildcat wells, Western Canada Basin: 100,000 ppm equals 0.3 ft³ gas/ft³ cuttings. [Evans and Staplin 1971]

to the fact that the western section is at a higher temperature than the east. This is an interesting example because the Imperial Oil geochemists found that the wet to dry conversion occurred just west of the town of Rainbow in northern Alberta. Their data indicated that the Rainbow area to the east would be oil prone, whereas that to the west in British Columbia would be gas prone. Shortly after this, a major oil field was discovered at Rainbow. The transition from wet gas to dry gas in the Middle Devonian of the Western Canada Basin is shown as line D in Figure 5-15. The cuttings gas data show this change with depth in a well drilled in the northwest territories west of line D (Figure 7-34).

An example of cuttings gas logs from two wildcat areas are shown in Figure 10-7. Cuttings from the well on the left recorded only dry gas through the entire section, which defines this as a gas-prone area. The log on the right shows a high yield of wet gas particularly through the 3,000–4,000 ft (914–1,219 m) interval. Suitable reservoirs or traps associated with these intervals could contain oil.

Formation Waters

The distribution of hydrocarbon gases in formation waters was discussed in the first part of Chapter 8. Since there is 100 to 200 times as much dissolved and disseminated gas in subsurface waters as there is accumulated in

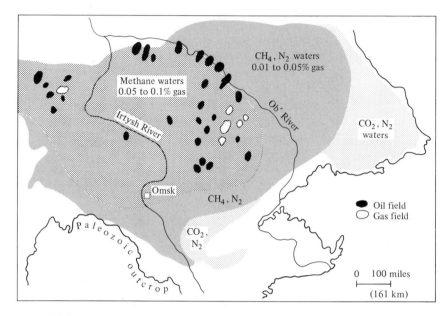

Figure 10-8
Volume percent of gas in the Lower Cretaceous formation waters in the West Siberian lowland. [Kartsev 1963]

reservoirs, it would be expected that regional enrichments would occur in oil- and gas-productive areas. Such enrichments do occur in a general way. The main producing sands of the U.S. Gulf Coast contain from 1 to 14 ft^3 gas/bbl water (178 to 2,492 cc gas/liter) in the oil- and gas-producing regions (Buckley et al. 1958). In nonproducing areas, such as southern Georgia, there is no gas in the waters.

An interesting example of the proximity of gas-rich waters and petroleum production is shown in Figure 10-8. Starting in the mid-1950s, Russian geochemists published analyses of formation waters collected during early wildcat drilling of the West Siberian lowland. Studies by Gurevich (1954) and Bars and Nosova (1962) are cited by Kartsev (1963). In a more recent article, Zimin et al. (1973) cite the work of Gurevich (1956), Torgovanov (1960), Rozin (1966), and Kruglikov (1964). The waters were analyzed for organic carbon, dissolved gases, and several organic and inorganic compounds. The highest methane content of the formation waters was in a 500 square kilometer (192 square mile) area west of the Ob River. To the east and south, the methane content decreased and there was an increase in CO_2 and nitrogen in the dissolved gases. Other geochemical data such as the presence of organic carbon also indicated that this was a favorable area for exploration. About ten years after these data were published, several giant oil and gas fields were found in the favorable area.

In the Lacq–Meillon Basin of Southern France, Coustau (1977) was able to define the most attractive zones for prospecting from the anomalous concentrations of dissolved gases in subsurface waters. Background concentrations were less than 0.2 (volume gas/volume water), but increased to about 1.0 near the Antin–Maubourguet Ridge, where an oil reservoir was found.

This regional enrichment pattern raises the question as to whether hydrocarbon concentrations in waters can be used to determine the proximity of petroleum accumulations. Gas concentration in Tuscaloosa waters within a few feet of the Hub field of Mississippi is about 10 ft^3/bbl H_2O (1,780 cc/liter). In the same sand, about 1 mile (1.6 kilometer) from the Hub field, the gas concentration drops to background levels of about 3 ft^3/bbl, or 534 cc/liter (Buckley et al. 1958).

In a study of the Dnieper–Donets Basin, London (1964) noted that the area of high hydrocarbon concentration in the water related to petroleum accumulations rarely extended more than 1 to 2 km (0.62 to 1.2 miles) beyond the field limits. These examples show that dissolved gas concentrations have limited use for pinpointing accumulations, although they are still useful on a regional basis. Background concentrations of dissolved gases in petroleum-producing zones are higher than in nonproducing zones.

Depth and age have an effect on background concentrations of gases. It was pointed out in Chapter 5 that biological methane forms early and thermally generated methane forms late. Stroganov and Shubbota (1973) noted that in the Central Asian Platform gas saturations are high in Eocene formation waters, low in Cretaceous, and high in Jurassic. The average data are shown in Table 10-2. The high values in the Eocene are probably caused by biological methane since the depths are too shallow for thermal generation. Most of the

Table 10-2
Gas Concentrations in Formation Waters

Source of samples	Age in 10^6 years	Depth range, m	Average liter gas/liter H_2O	Average saturation coefficient[a]
Central Asian Platform				
Eocene formations	40–60	400–500	0.65	0.8
Cretaceous formations	70–135	300–1,800	0.2	0.15
Jurassic formations	135–180	1,300–4,000	1.0	0.45
Alpine basins	Present to 180	—	4.0	0.7
Hercynian basins	230–400	—	0.3	0.3
Caledonian basins	400–500	—	0.08	0.1

[a]The saturation coefficient is the ratio of the actual quantity of gas present to the maximum amount that will dissolve at the conditions of temperature, pressure and water mineralization. Average concentrations are in liters of gas per liter of water.
Source: Data from Stroganov and Shubbota 1973; Mukhin 1960.

Cretaceous samples are also too shallow, but the Jurassic is well within the depth range of gas generation, assuming a normal geothermal gradient. No temperature data were given, but the Jurassic sediments may be in a higher temperature regime than the Cretaceous, because their data indicate that at the same depth intervals the Jurassic contains more gas than the Cretaceous. Some Jurassic waters are completely saturated with hydrocarbons. The authors believe that the Jurassic sediments are the main gas source rocks. Gas accumulations of the Central Asian platform are mainly in the Jurassic and Cretaceous, rarely in the Cenozoic. Data in Table 10-2 were obtained from stratigraphic tests drilled well beyond known pool limits, so there should be no effect of local diffusion of hydrocarbons. The Cretaceous sediments frequently have a high nitrogen content, which is characteristic of shallow, stable platform gas, as discussed in Chapter 5.

Mukhin's (1966) comparison of gas saturations in various basins indicates that there may be a steady loss of gas with time. His data on gas in the middle Caspian Basin, as representing the Alpine system, show much higher gas concentrations than in Hercynian basins, which, in turn, are higher than in Caledonian basins. As hydrocarbon generation ceases in a basin, there are gradual and continual gas losses by diffusion and other mechanisms of migration. This suggests that some old sediments may be very poor prospects for gas exploration. However, there are many old sediments that were not buried deep enough to generate gas during most of their history. If deep burial has occurred only since the Mesozoic, there would be good prospects. An example of this is Hassi Messaoud in Algeria which was previously discussed.

Dissolved Liquid Hydrocarbons

Low-molecular-weight liquid hydrocarbons partition between oil and water phases, depending on their solubility. As oil accumulates, part of the more soluble hydrocarbons diffuse into the aquifer away from the accumulation. This causes a regular decrease in hydrocarbon concentration with distance from the oil reservoir. Zarrella et al. (1967) proposed using benzene, a highly soluble hydrocarbon, as an indicator of proximity to oil accumulations. They found that formation waters associated with production had generally higher benzene contents than those in barren areas. Benzene concentrations varied considerably owing to differences in the composition of associated crude oils, the salt content of the brines, and the temperature–pressure conditions. Analyses of aquifers separated from oil horizons by shales showed only traces of benzene, indicating that there was no vertical migration of this hydrocarbon. In a horizontal direction, benzene concentrations reached background levels generally within a few miles of an oil accumulation. This is comparable to the decrease in methane going away from accumulations, and it means that benzene would have application as an indicator of oil fields mainly when in close proximity to an accumulation.

Benzene may have some value as a regional indicator, as shown in Table 10-3. The highest benzene values are in waters coproduced with condensate. Waters associated with dry gas have no benzene. Background levels in the vicinity of most oil accumulations range from about 0.5 to 5 mg/liter. These numbers are significantly higher than the zero values from nonproductive areas, because concentrations as low as 0.001 mg/liter can be determined. However, Kortsenshteyn (1968) showed that there are productive areas with very low benzene values in the formation waters, such as South Mangyshlak, east of the Caspian Sea. Jurassic waters from oil- and gas-producing horizons in this area contain 0.01 to 1 mg/liter of benzene, with the mean around 0.15. Since benzene concentration varies widely in productive horizons, some experience would be needed in a specific area in order to use it effectively as a regional prospecting tool.

Benzene, the xylenes and dissolved hydrocarbon gases can be useful in delineating step-out wells after a discovery has been made. Production wells that are drilled while developing a field should analyze formation waters and cores from the water leg for dissolved and adsorbed hydrocarbons. Computer

Table 10-3
Benzene Content of Formation Waters

Water source	Formation	Area	Benzene content in mg/liter H$_2$O
Coproduced with oil	Pennsylvanian	New Mexico	10.7
oil	Leduc	Alberta	6.0–4.8
gas	Viking	Alberta	0
condensate	Eagle Ford	Mississippi	18.6
From productive horizons in vicinity of accumulations but not coproduced			
oil	Leduc	Alberta	1.7
oil	Wolfcamp	W. Texas	1.0
oil	Tertiary	Louisiana	0.8
oil	Jurassic	S. Mangyshlak, USSR	0.001–1.0
oil	Jurassic–Cretaceous	Ciscaucasia, USSR	1–20
Nonproductive areas			
	Viking	Saskatchewan	0
	Cretaceous	Georgia	0

Source: Data from Zarella et al. 1967; Kortsenshteyn 1968.

mapping of the compositional trends can assist in determining where to drill successive step-out wells. Such comparisons are made on a relative basis since absolute concentrations of these hydrocarbons will vary widely as mentioned above.

C_{15+} HYDROCARBONS

The range of hydrocarbons in fine-grained sediments that can be used to characterize source potential extend from C_1 to about C_{40}. The value of the gas and light gasoline range (C_1–C_7) in characterizing gas-prone, oil-prone, and nonproductive areas has been discussed. The C_{15+} range analysis was the first technique used extensively for source rock characterization, and it is still useful for evaluation of source quality, source type, and degree of maturation. This fraction contains nearly all of the biogenic hydrocarbons, which are particularly sensitive to maturation changes with increasing temperature. Some research has been done on extracting the C_8–C_{14} hydrocarbons particularly for studying the origin of the substituted aromatics, but it has not been used extensively as a source indicator.

The C_{15+} hydrocarbon analysis involves grinding the fine-grained rock to a specified particle size and extracting with solvents. Trask (1942) used carbon tetrachloride (CCl_4) as a solvent for his source rock studies in the 1930s (see Chapter 7). Chlorinated hydrocarbons such as chloroform ($CHCl_3$) and trichloroethane (CH_3CCl_3) have continued to be widely used in Europe for extracting bitumens from rocks and coals. In the United States, a popular solvent has been the benzene–methanol–acetone triple mixture introduced in the early 1950s by Hunt and Meinert (1954, see Chapter 7). This solvent mixture was designed to remove the hydrocarbons in the presence of much larger amounts of water at a temperature below 100°C (212°F). In the last 25 years, various modifications have been introduced such that each laboratory uses a slightly different technique for the extraction. Acetone was eliminated some years ago, because it tended to add polymers and other impurities to the bitumen extract. Grinding is done carefully to avoid heating, and the procedure may be carried out under a nitrogen atmosphere to minimize oxidation. Some laboratories prefer a ball mill extraction to the soxhlet extraction, because channeling sometimes occurs with the latter. Extraction with an ultrasonic probe was introduced to reduce the extraction time from several hours to a few minutes. Evidence developed, however, that some probes cracked the organic matter to yield smaller molecules during the extraction.

After extraction, the solvents are removed by evaporation, usually at 40°C (85°F) or less. The residue remaining is referred to as *bitumen*. It contains the hydrocarbons plus asphaltic nitrogen, sulfur, and oxygen compounds. The hydrocarbon range may start as low as C_{11} or C_{12}, depending on the method of solvent removal. Most data reported in the literature indicate that more complete recoveries start around C_{15}. The bitumen is further separated by liquid chromatography into the saturates (paraffins plus naphthenes), aromatics, NSO compounds, and asphaltenes. Changes in hydrocarbon recoveries with depth in a well may be plotted as bitumen, hydrocarbons, or saturates. Examples of bitumen and C_{15+} yields with depth were shown in Figures 4-16, 4-18, 4-20, 4-21, 4-22, and 4-24 and Figures 7-9, 7-10, and 7-11.

In using the C_{15+} analysis as shown in these figures to determine the depth of maximum hydrocarbon production, it is best to have all samples analyzed by exactly the same procedure. None of these extractions are quantitative. Consequently, comparisons on a relative basis are meaningful, providing the extraction technique always gives proportionally the same recoveries. Also, the liquid chromatography separations of saturates from aromatics and aromatics from nonhydrocarbons are not quantitative. Aromatic fractions often contain small amounts of acids, esters, and other nonhydrocarbons. Periodic monitoring of extracts by mass spectrometry is advisable, to ensure that separations are adequate for the problem being studied.

A typical data sheet for the C_{15+} analysis is shown in Table 10-4. Three samples are listed in order of maturity. DSDP Sample 280A is from the deep ocean south of Tasmania in Eocene sediments that have never been buried

Table 10-4
Composition of Bitumen Extracted from Sediments

Samples	DSDP 280A-17-6	Irati Shale	Douala Basin[a] Shale (103-29)
Organic carbon, % C_{org}	1.8	9.3	1.32
Total bitumens, mg/g C_{org}	53	174	100
C_{15+} Hydrocarbons, mg/g C_{org}	8.6	83	71
Saturates	3.5	35	46
Aromatics	5.1	48	25
Nonhydrocarbons, mg/g C_{org}	44.4	91	29
Saturated hydrocarbons			
n-Paraffins, %	7	—	23
Isoprenoid, %	2.4	—	4
Naphthene + other branched, %	90.6	—	73
Ratios			
CPI	1.0	—	1.1
Saturate/aromatic	0.7	0.6	1.8
Hydrocarbon/nonhydrocarbon	0.2	0.9	2.5
Pristane/phytane	2.4	1.3	2

[a]From Albrecht et al. 1976.

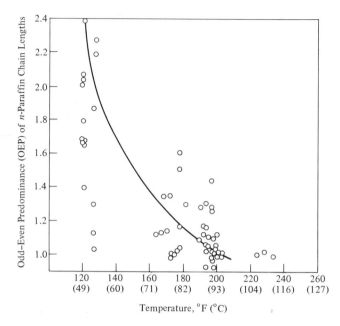

Figure 10-9

Change with temperature in odd/even predominance of *n*-paraffins in the C_{15+} fraction of hydrocarbons extracted from North Sea Basin Tertiary clastics. [From J. G. Erdman, "Geochemical Formation of Oil," in *Petroleum and Global Tectonics*, eds. Alfred G. Fischer and Sheldon Judson. Copyright © 1975 by Princeton University Press. Reprinted by permission of Princeton University Press.]

deeper than about 400 m (1,312 ft). The Irati Shale from Brazil is Permian but also has not been buried deeply enough for maximum generation of hydrocarbons. The Cretaceous Douala Shale from Cameroon is from 2,041 m (6,696 ft) equivalent to peak hydrocarbon generation in this basin.

The total extract (bitumen) recovered relative to organic carbon is two to three times higher in the shales than in the DSDP unconsolidated mud. Both shales have generated some hydrocarbons, but the Irati is still in the stage of a heavy asphaltic oil. This is indicated by the hydrocarbon/nonhydrocarbon ratio, which is low in the Irati and higher in the Douala Shale extract. Also, the saturate/aromatic ratio is higher in the more mature Douala.

The ratio of odd to even *n*-paraffin chains discussed in Chapter 7 is computed from the gas chromatograms of the saturate fraction of the C_{15+} extract. The carbon preference index (CPI) and odd–even predominance (OEP) are methods for calculating these *n*-paraffin ratios. These ratios can be used in oil prospecting, providing there has been some contribution of land-derived waxes in the $C_{27}-C_{33}$ range to the sediment being analyzed.

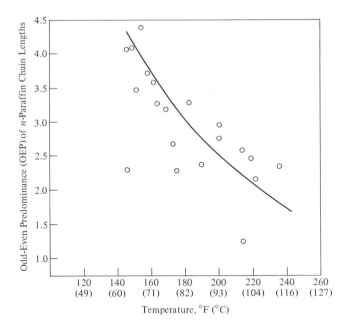

Figure 10-10
Change with temperature in OEP for *n*-paraffins extracted from
Papuan Basin Tertiary clastics. [From J. G. Erdman, "Geochemical
Formation of Oil," in *Petroleum and Global Tectonics,* eds. Alfred
G. Fischer and Sheldon Judson. Copyright © 1975 by Princeton
University Press. Reprinted by permission of Princeton University
Press.]

Figures 10-9 and 10-10 are examples of the application of the odd/even
ratio in exploration (Erdman 1975). These examples represent sediments in
the mature stage, but only one contains organic matter capable of generating
oil. Figure 10-9 shows the OEP plotted against subsurface temperatures in
North Sea Tertiary clastics. There is a dramatic drop in OEP from about 2.3 to
1.0 in the temperature range from 120 to 200°F (49 to 93°C). The cluster of
samples in the 190 through 200°F (88 through 93°C) range are from source
beds associated with the prolific North Sea oil production. The scatter of low
OEP values around 120°F (49°C) represent samples with mostly marine
organic matter. Figure 10-10, which represents similar data from clastics in
the Gulf of Papua shows a decrease in OEP from about 4 to 2.4 at
temperatures around 230°F (110°C). Because this ratio is far above 1 even at
high temperatures, it indicates the organic matter in the Papuan shales is
incapable of generating oil although it still could be a source for gas.
    It is possible to further check the field data by heating cuttings or core
samples in the laboratory and measuring the drop in OEP values owing to

induced maturation. This procedure was discussed in Chapter 7, where Table 7-8 shows the change in OEP values before and after heating for some of the immature North Sea and Papuan Basin clastics (Erdman, personal communication). Three of the North Sea samples decrease in OEP to 1.1, which is comparable to the produced oil. Neither of the Papuan samples decrease in OEP to the oil range (1 to 1.3, depending on the oil). The Alaska sample also shows a drop in OEP indicative of oil formation.

Maturation indicators such as vitrinite reflectance cannot recognize these differences in oil-generating capability, because they respond mainly to temperature differences. This emphasizes that subsurface indicators such as OEP are needed to define the nature of the organic matter as well as the state of maturation.

The C_{15+} fraction also contains individual hydrocarbon molecules that can be analyzed by combined gas chromatography-mass spectrometry techniques for more sophisticated correlation and maturation studies. These are not routine prospecting tools but may be used for special problems. For example, the kerogen of immature rocks contains cyclic diterpanes, which change to aromatics during the oil generation stage. Most diterpanes are gone at a vitrinite reflectance level (R_o) of 1.0 percent. Likewise, the pentacyclic triterpenes of immature rocks change to terpanes before an R_o of 1.0 and then aromatize at R_o levels prior to 1.6 percent.

The C_{15+} analysis can be run on cores, cuttings, or unweathered outcrops of fine-grained rocks. Much of the data in Figure 4-24 was obtained from unweathered outcrops. The effects of weathering on organic matter have been investigated by Leythaeuser (1973) and by Clayton and Swetland (1978). Both studies showed that weathering effects are highly variable, but a 50 percent loss of C_{15+} hydrocarbons and total bitumen can occur to depths up to 3 m (10 ft).

Cuttings are easily contaminated with pipe dope and oil used in the mud during drilling. The mud should be thoroughly washed off of cuttings for C_{15+} analyses, although not for gas analyses, as previously discussed. Periodic analysis of oils and greases used on the drilling rigs is recommended for recognition of contaminants in the C_{15+} range.

KEROGEN

Prospecting techniques involving kerogen are directed toward answering the questions, (1) Is there enough of the right kind of organic matter to form oil or gas in commercial quantities and (2) What is the stage of maturation of the organic matter? Is the kerogen in the pre-, peak, or post-hydrocarbon-generating stage? The techniques may involve analysis of the whole rock or separation of the kerogen from the mineral matrix prior to analysis.

Elemental Analyses

Changes in the distribution of the elements carbon, hydrogen, and oxygen in kerogen with depth can be used to interpret the oil- or gas-generating potential of the source rocks, as discussed in Chapter 7. Procedures for removal of the kerogen from the rock mineral matter have been described by McIver (1967) and LaPlante (1974). The key to the generation of oil or gas is the hydrogen content. Harwood (1977) observed from a large number of source rock analyses that the kerogen atomic H/C ratio was below 0.8 for dry gas generation and above 1 for oil generation.

The separation and analysis of kerogen is important for a research project, but it is not a good technique for the rapid evaluation of well cuttings. The process is expensive and time consuming. Kerogen separations are tricky, because the amorphous organic matter of kerogen is easily oxidized. The kerogen must be separated and stored in a nitrogen atmosphere, particularly if the rock is extracted with organic solvents prior to kerogen isolation. Some kerogen analyses reported in the literature have artificially high oxygen contents, because oxidation occurred during separation or storage.

In recent years, other, more rapid techniques, such as pyrolysis of the whole rock, have been correlated with the elemental analysis of kerogen, so that elemental analysis is largely used as a monitoring procedure in subsurface prospecting.

Pyrolysis

Heating kerogen or a whole rock in the absence of oxygen to yield hydrocarbons is termed *pyrolysis*. The high-temperature pyrolysis of oil shales to yield shale oil was developed in the eighteenth century and is still used for commercial oil production in a few countries. High-temperature pyrolysis, however, cracks not only the kerogen to yield hydrocarbons but also the hydrocarbon products to yield gases and low-molecular-weight aromatic hydrocarbons. In 1962, Hunt (1963b) reported the low-temperature pyrolysis of kerogen as a technique for distinguishing gas- and oil-yielding kerogens. The technique involved heating 300 mg kerogen samples in a helium atmosphere for ten minutes at temperatures ranging up to 400°C (752°F). The ratio of methane to higher hydrocarbons from the kerogen of the Precambrian Gunflint Chert was 15, while for the Monterey Shale of California it was 0.1. The latter was a sapropelic kerogen, the former humic.

In the last 15 years, pyrolysis–FID (flame ionization detector) has become a technique for both characterization of the type of organic matter and the level of maturation. In some laboratories, it has replaced all extraction techniques, because it is faster, involves less manpower, and uses very small samples. Basically, the technique consists of heating a 1–300 mg sample of kerogen or whole rock in a stream of helium at a rate of 5–40°C/min, up to a temperature around 600–800°C (1,112–1,472°F). The hydrocarbon products are monitored with a hydrogen flame ionization detector. Free hydrocarbons in the sample are released in the temperature range up to about 200–300°C (392–572°F). The exact temperature varies with the temperature programming rate. Hydrocarbons from cracking of the kerogen are obtained at higher temperatures. In some geochemical literature, the release of the free hydrocarbons by low-temperature heating is called *thermal distillation*. The term *pyrolysis* is reserved for the higher-temperature cracking process.

The French were the first to publish details on the use of pyrolysis as an exploration technique (Bordenave et al. 1970; Giraud 1970). Since then, there have been several articles describing improvements and modifications to obtain different types of data (Barker 1974a; Tissot, Durand, et al. 1974; Claypool and Reed 1976; Leventhal 1976; Harwood 1977). Some techniques include thermogravimetric analysis; that is, the measurement of weight loss during heating and determination of the CO_2 evolved.

An example of the type of curves obtained is shown in Figure 10-11 from Barker's (1974a) paper. The first peak represents the free hydrocarbons that are thermally distilled during heating up to 250°C (482°F), while the second peak represents the hydrocarbons obtained from the pyrolytic breakdown of the kerogen in the rock. The small first peak for the shallow sample indicates it has not yet entered the oil-generating stage, while the large second peak indicates it is potentially capable of producing oil. If the second peak were predominantly methane, it would indicate a gas source potential, whereas higher hydrocarbons would indicate an oil source potential. The strong first peak in the deeper sample shows that this sample has entered the oil-generating stage, and reservoirs in this zone would be prospective. The first peak for samples that have not reached the petroleum-generating stage is usually small, as shown in the shallow sample. Recent sediments, for example, show a very small first peak. If cuttings samples are run at intervals down a well, the first peak will show a large increase, going through the hydrocarbon-generating range and then a decrease as the generating capability of the kerogen becomes exhausted. The peaks will also show a change from heavy hydrocarbons to condensate to dry gas with depth as hydrogen is used up in the kerogen. Another indication of maturation is the shift in the maximum of the second peak toward higher temperatures. In the discussion on activation energies in Chapter 4, it was pointed out that the different types of kerogen require different activation energies to yield hydrocarbons. As the kerogen in a rock matures, higher temperatures are required to form hydrocarbons. Going down a drill hole, as shown in Figure 10-11, the second peak will continually shift toward higher temperatures. This comparison can only be made on a relative basis, using the same procedure. Changes in the heating

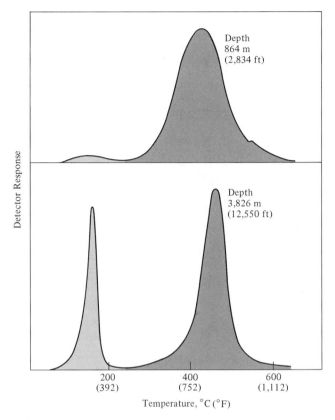

Figure 10-11
Yield of hydrocarbons as indicated by detector response with
increased heating of Miocene shale samples from a Gulf Coast well.
[Barker 1974a]

rate will cause a shift in the absolute temperature maximum, but the relative
changes will remain the same.

Barker kept his cuttings samples under water from the moment they were
brought to the surface in order to avoid loss of free light hydrocarbons.
Claypool and Reed (1976) oven-dry their samples for two hours at 105°C
(221°F) in order to remove light hydrocarbons up to about the C_{15+} range.
Using a slightly different technique, they have found that the first peak obtained up to 400°C (752°F) is directly proportional to the concentration of
extractable C_{15+} hydrocarbons in the rock. The second peak in the 400–
800°C (752–1,472°F) range is approximately proportional to the organic
carbon content of the rock. Claypool and Baysinger (1978) use the temperature maximum of the second peak to indicate maturity. Immature samples

usually peak below 470°C (878°F), but this will vary depending on the kerogen type, the lithology, and the design of the pyrolysis system. Each investigator must calibrate his or her own equipment for determining maturation peaks. Tissot, Durand, et al. (1974) have noted that the CO_2 evolved during pyrolysis is a measure of the humic (gas source) nature of the kerogen versus the sapropelic (oil source) nature.

The hydrocarbons obtained in peaks 1 and 2 of Figure 10-11 may be trapped and analyzed in detail on a gas chromatograph (GC) to determine the individual hydrocarbons present. This is useful when more information is needed on the molecular weight range distribution of the hydrocarbons for correlation purposes. Mass spectrometry can be used for more positive identification of GC peaks. Although pyrolysis techniques are very useful for well surveys, they are less desirable than extraction techniques for monitoring biogeochemical marker compounds or for age dating, because they do not always form the same products that are formed by natural diagenesis. It is possible, however, to use pyrolysis in correlation studies (see Chapter 11).

Analysis of Cuttings and Cores at the Well Site

Most geochemical techniques can be developed for well site operation. The C_{15+} hydrocarbon extraction, when run on samples of 1 to 5 g, can be set up in a well site laboratory. This can include separation of the extracts into saturates, aromatics, and NSO compounds plus gas chromatography of the saturates including odd/even n-paraffin ratios.

Espitalié et al. (1977) adapted the French pyrolysis technique for field operation. The H/C–O/C diagram of Tissot, Durand, et al. (1974) was developed from the analysis of kerogen samples for hydrogen, oxygen, and carbon (Figure 10-12). In this diagram, three types of kerogen are shown. The liptinitic type typical of oil shales and boghead coals are Type I in Figure 10-12 (see also Figures 7-5 and 7-47). These Type I kerogens have a very high H/C ratio and low O/C ratio. Carbonate kerogens, such as those in the Cretaceous limestone of the Persian Gulf, are Type I. At the other extreme are the Type III kerogens, with a low H/C ratio and high O/C ratio. These are the same as the woody and coaly kerogens of Figure 7-5 and the vitrinite and inertinite curves of Figure 7-47. They are derived from humic materials. In intermediate position are the Type II kerogens, which contain liptinite with a lower H/C ratio than does Type I. From an exploration standpoint, Types I and II give the highest yields of oil and gas, while Type III yields mostly gas. The state of maturation is indicated by how far along the curve the sample is located, the most mature being in the lower left corner (see also Figure 7-47).

Espitalié et al. (1977) wanted to obtain the same information directly from the whole rock sample without going through the tedious and time-consum-

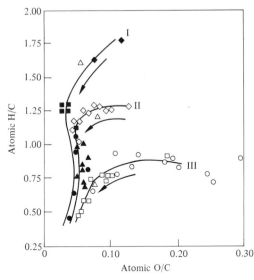

Atomic H/C (y-axis): 2.00, 1.75, 1.50, 1.25, 1.00, 0.75, 0.50, 0.25

Atomic O/C (x-axis): 0, 0.10, 0.20, 0.30

I, II, III

♦ Algal kerogens (*Botryococcus*, and so on)
◇ Lower Toarcian, Paris Basin
▲ Silurian, Sahara–Libya
● Upper Paleozoic–Triassic, Spitzbergen
□ Upper Cretaceous, Douala Basin
■ Cretaceous, Persian Gulf (Oligostegines limestone)
○ Lower Mannville shales, Western Canada Basin
△ Others

Figure 10-12
Classification of kerogens into high-H/C liptinite (I), low-H/C liptinite (II), and vitrinite, inertinite (III) on an atomic H/C–O/C diagram. [Espitalié et al. 1977]

ing kerogen isolation and elemental analysis. They assumed that the oxygen content of kerogen would be proportional to the CO_2 liberated during pyrolysis and that the hydrogen content of kerogen would be proportional to the hydrocarbons liberated. They pyrolyzed ground samples (50 to 100 mg) of rocks and pure minerals with programmed temperature ($20°C/min$) up to $600°C$ ($1,112°F$). The CO_2 from the kerogen in rocks came off mostly in the range up to $400°C$ ($752°F$), while the CO_2 from pure carbonates came off mostly at higher temperatures: calcite above $600°C$ ($1,112°F$), dolomite above $500°C$ ($932°F$), and siderite above $400°C$ ($752°F$). This indicated that carbonates would not interfere with organic CO_2 evolution. Espitalié et al. also found that the yield of organic CO_2 was proportional to the O/C ratio of the kerogen in the rock, as shown in Figure 10-13. This figure is a plot of the atomic O/C ratio in isolated kerogens versus the milligrams of CO_2 per gram of organic carbon coming from the whole rock up to temperatures lower than carbonate decomposition by their procedure. The mg CO_2/g C_o is called the *oxygen index*, and it can be plotted in place of O/C on an $H/C–O/C$ diagram.

Rock samples that are not extracted prior to analysis give two hydrocarbon peaks, as shown in Figure 10-11. The quantity of free hydrocarbons in the

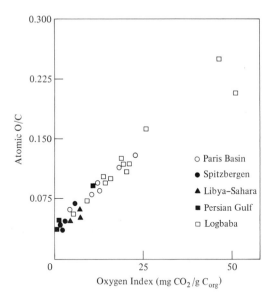

Figure 10-13
Correlation of O/C atomic ratio of kerogen with organic CO_2 evolved from pyrolysis of the corresponding rock. [Espitalié et al. 1977]

first peak is divided by the quantity of total hydrocarbons in both peaks (free and cracked hydrocarbons) to give the *production index*. This is a measure of the hydrocarbons available for accumulation. Espitalié et al. (1977) also found that the yield of cracked hydrocarbons (second hydrocarbon peak) correlated with the H/C ratio of the kerogen. The mg HC/g C_o was called the *hydrogen index,* and it was plotted in place of H/C on an H/C–O/C diagram.

The level of organic maturation was estimated from the hydrogen–oxygen index diagram, the production index, and the temperature at the top of the second hydrocarbon peak. Families of samples showing a change in maturity would show a drop in the oxygen index during early catagenesis, followed by a drop in the hydrogen index in the principal zone of oil and gas formation. A natural variability in the hydrogen content of kerogen, though, can obscure the intepretation of the hydrogen index if the other two maturation indicators are not present to support it. The production index—ratio of $P_1/(P_1 + P_2)$—always increases in the mature zone if kerogen is capable of generating oil or gas. The temperature maximum for the cracked hydrocarbons in the second peak may be used to interpret the kerogen as immature, mature, or metamorphosed, the temperature range depending on several factors, as previously stated.

It should be emphasized that pyrolysis results, like extraction results, are interpreted on a relative basis, because procedural differences will not give the same numbers on the same samples. The interpretations should be the same, however, assuming that each technique is evaluated with known standard samples.

For many years, the value of oil shales has been determined by a modified Fischer assay, which involves heating 100 g of crushed shale at 500°C (932°F) for one hour. Yields are quoted in barrels of oil per ton of rock or kilograms of oil per MT (metric ton) of rock. Espitalié et al. (1977) noted that the area of the second hydrocarbon pyrolysis peak, obtained in pyrolyzing an oil shale, correlated very well with oil yield as measured by the Fischer assay (Figure 10-14). The pyrolysis can be carried out in 20 minutes with a 50–100 mg sample of oil shale, so it has a distinct advantage over the Fischer assay for rapid surveys.

An example of the maturation of kerogen with depth as monitored by pyrolysis is shown in Figure 10-15 (Espitalié et al. 1977). The first and second hydrocarbon peaks, corresponding to the two peaks in Figure 10-11, are the free (P_1) and cracked (P_2) hydrocarbons. These are Tertiary core samples from West Africa analyzed through the 800–2,770 m (2,625–9,088 ft) depth. The free hydrocarbons (P_1) are seen to increase steadily with depth, while the kerogen-complexed hydrocarbons, available by cracking (P_2), are steadily decreasing. The production index $P_1/(P_1 + P_2)$ increases accordingly. It should be noted that the peak of P_2 in Figure 10-15 is steadily shifting to higher temperatures (compare with Figure 10-11).

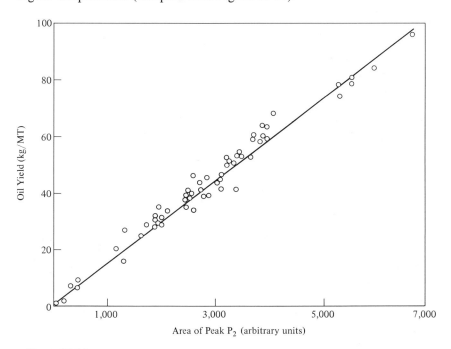

Figure 10-14
Correlation of Fischer assay oil yield from oil shales with quantity of hydrocarbons cracked by pyrolysis. [Espitalié et al. 1977]

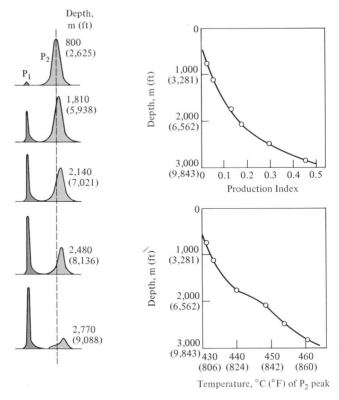

Figure 10-15

Relative hydrocarbon yields from pyrolysis of cores at different levels of maturation. Free hydrocarbons (P_1) increase with depth, while hydrocarbons available from kerogen (P_2) decrease with depth. The production index $P_1/(P_1 + P_2)$ and the P_2 peak temperature increase with increasing maturity. [Espitalié et al. 1977]

The versatility of pyrolysis-FID is such that it is replacing the C_{15+} extraction and elemental analysis of kerogen for well surveys in many laboratories. Pyrolysis can be readily adapted to field or ship operations. A unit using the Espitalié et al. (1977) technique has monitored down-hole cores in several wells drilled by the *Glomar Challenger* for the Deep Sea Drilling Project of the National Science Foundation.

Heacock and Hood (1970) developed a crude but simple technique known as *pyrolysis–fluorescence* (PF) for rapid evaluation of organic richness in well cuttings. Their technique involved heating a 0.1 g sample of cuttings in a test tube until sample and tube became red-hot. The tube, which is open at the unheated end, is held horizontal in order to condense any vapors given off on the colder end of the tube. After becoming red-hot, the sample is cooled, and

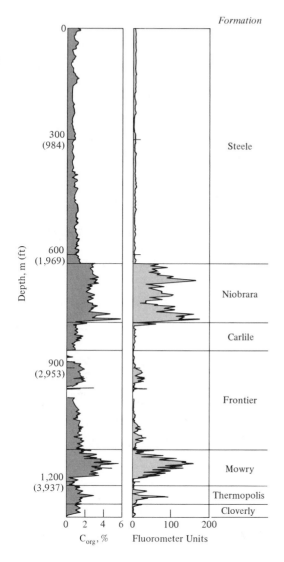

Figure 10-16
Comparison of organic carbon and pyrolysis–fluorescence for Cretaceous sediments, Powder River Basin, Wyoming. [Heacock and Hood 1970]

3 ml of 1,1,1-trichloroethane is added. The solution is transferred to a clean test tube, and the amount of soluble fluorescing material is measured with a fluorometer, modified by addition of a 99 percent-opaque neutral-density filter. Samples with PF readings greater than 20 are diluted with trichloroethane to avoid self-absorption effects. PF values are calculated as scale readings per 0.1 g of sample in 3 ml of solvent using the appropriate dilution factor. Figure 10-16 shows the PF analyses of cuttings from a well in the Powder River Basin of Wyoming along with organic carbon analyses

(Heacock and Hood 1970). The most thermally reactive kerogen is in the Niobrara, Mowry, and Thermopolis formations. These would be the best oil source rocks by the PF interpretation. Parts of the Frontier Formation also contain reactive kerogen. The entire Steele section is unreactive, even though organic carbon contents exceed 1 percent. Pyrolysis-fluorescence is the easiest and fastest technique available for low-cost well site evaluation of cuttings as to whether they are alive or dead in their oil-generating capability.

Another measure of organic richness is effective carbon (C_{eff}). A cuttings sample is heated in a flowing stream of nitrogen from room temperature to 300°C (572°F) at 25°C/min. This removes the volatile free hydrocarbons. Heating is continued to 650°C (1,202°F) to crack available hydrocarbons from the rock kerogen. C_{eff} is calculated as 85 percent of the hydrocarbons generated in the 300–650°C (572–1,202°F) range (Kendrick et al. 1978b). The ratio of C_{eff}/C_o indicates the fraction of the kerogen that is thermally convertible to hydrocarbons. C_{eff} and the hydrogen index of Espitalié et al. (1977) are simply different ways of expressing the cracked hydrocarbons of Barker's (1974a) second peak (Figure 10-11).

Pyrolysis sounds simple, but it is sometimes difficult to obtain consistent high-quality results. Geologists should work with a group that has considerable experience with well samples that have been analyzed by other techniques as well as pyrolysis. Slight modifications in the technique will result in different yields. The inorganic matrix also affects the yields to some extent. Another problem is that the samples are so small that they may not be representative of the section analyzed unless they are an aliquot of a large homogeneous sample. Preventing contamination and loss of free hydrocarbons is also more difficult with small samples. Nevertheless, the advantages of pyrolysis so far outweigh the disadvantages that it is probable this will be the dominant source rock technique in the years to come.

Electron Paramagnetic Spin Resonance (EPR or ESR)

In the discussion of asphaltenes in Chapter 3, it was mentioned that the condensed aromatic structures of these compounds contain free-radical sites with unpaired electrons. Asphaltenes and kerogen both contain these free-radical sites. The kerogen structure is analogous, in that as small groups of carbon atoms and hydrogen atoms crack from the kerogen to form petroleum they leave behind some free radical sites because of the unpaired electrons. The complexity of the kerogen structure shields and stabilizes these sites through geologic time. With increasing time and temperature, the number of free radicals increases in the kerogen structure. Thus, on the left side of Figure 7-28 the increasing ring condensation and aromatization can cause an increase in free-radical sites. Eventually, high temperatures destroy the free-radical sites. Electron paramagnetic spin resonance (EPR or ESR) is a technique for measuring the number of free radicals. The kerogen is placed in a magnetic field of known strength, and microwaves are passed through at a critical fixed frequency. As the free electrons in the kerogen resonate, they alternately absorb and emit microwave energy. The magnetic field is varied in a known manner, and the amount of microwave energy absorbed can be directly related to the number of free electrons.

Pusey (1973) proposed EPR (or ESR) as a technique for estimating maximum paleotemperatures in the subsurface. He measured the EPR of kerogen isolated from well samples in continuously subsiding Tertiary basins. The EPR values were calibrated against well temperatures obtained from long-term shut-in tests. The calibration curves were then used to determine paleotemperatures of unknown samples from their EPR measurement. Later, Pusey extended his calibrations to ten basins of Tertiary to Mesozoic age and four of Paleozoic age.

The EPR technique has not been widely adopted, because there is a lot of scatter in the data, and many geochemists consider it a bulk analysis that is only indirectly related to temperature. Further investigations have shown that other factors, such as variations in the type of organic matter, environment (oxidizing or reducing) at time of burial, and time, also affect the spectroscopic parameters. Thus direct estimates of temperature can be obtained only in special cases. EPR measurements on pyrolyzed kerogen in some cases show no change until a threshold temperature of about 100°C (212°F) is reached. The EPR signal then suddenly widens and collapses over about a 20°C range. The change may be due to rapid graphitization of the kerogen with elimination of free-radical structures. This appears to occur at the high-temperature end of the oil generation scale around a vitrinite reflectance, R_o of 1.2 percent. EPR is really a kerogen structure indicator rather than a temperature indicator.

Kerogen Color

The kerogen color technique developed by Staplin (1969) and his colleagues was described briefly in Chapter 7 as a maturation technique. The procedure involves treating the cuttings samples first with hydrochloric acid after which all calcium ion is removed to avoid formation of calcium fluorides. The sediment remaining is digested in cold hydrofluoric acid. After centrifuging and washing, a zinc bromide flotation acid wash is used to float the kerogen. Boiling should be avoided in all steps, and no oxidizing agents or bases such as KOH (potassium hydroxide) should be used, because they destroy part of the organic matter and cause a partial loss of color.

A microscope slide is prepared from the organic residue and is scanned several times for such things as membrane cuticles of uniform thickness. Reference sets of color standards are normally used, and care must be taken to

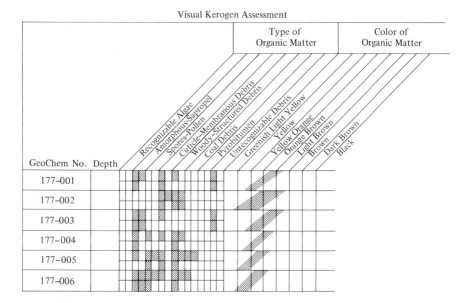

Figure 10-17
Data sheet for recording kerogen color and other characteristics from visual examination of microscope slides. [GeoChem Laboratories]

note natural pigments, thickness of material and reworked organic matter. The least colored material is the maturation indicator, because it is indigenous. Recycled material is darker. Figure 10-17 shows a typical data sheet for recording the visual information. Samples 1, 3, and 6 in this table are dominantly amorphous, while 2 and 4 are woody, and 5 is herbaceous. The latter contains large amounts of spore material. The kerogen color is in the yellow–orange range, indicating only slight maturation with a rating of 1 + and a maximum temperature around 50°C (122°F). These are all deep sea sediment samples in Figure 10-17, and the maturation level corresponds to the estimated bottom-hole temperatures.

Some microscope systems are set up with ultraviolet excitation to measure the fluorescence of the kerogen, which also changes with thermal alteration. The use of kerogen color by Staplin (1969) in mapping oil and gas regions in the Western Canada Basin was discussed in Chapter 7. Burgess (1974) has reported the examination of 4,600 well samples ranging in age from Ordovician to the Holocene from basins all over the world. Sediments with kerogen color of orange to brown (indexes 2 and 3) were generally associated with reservoirs containing oil or wet gas. The dark brown to black kerogens (4 and 5) were associated with either dry gas or barren reservoirs.

Gray and Boucot (1975) discuss some of the problems in spore and pollen color interpretation. Although pressure is not an important factor in the

Visual Kerogen Assessment

State of Organic Matter								Maturation Index													Depositional Environment							
Particle Size				Preservation				Unaltered	Slightly Altered	Moderately Altered	Strongly Altered	Severely Altered	Metamorphosed								Offshore Marine	Nearshore Marine	Restricted Nearshore–Marsh	Lacustrine	Continental	Photograph Coordinates		
Finely Disseminated	Fine	Medium	Coarse	Excellent	Good	Fair	Poor	1	1+	2–	2	2+	3–	3	3+	4–	4	5										Remarks
																												10 × 10
																												X
																												X
																												X
																												X

generation of hydrocarbons from kerogen, there does seem to be some change in palynomorph color because of shear pressure in tightly folded formations. Spore maturation in coals seems to be faster than in associated shales. Radioactive material in sediments also affects spore and pollen color. These and other factors indicate that some experience is needed for the proper interpretation of maturation from kerogen color.

Vitrinite Reflectance

The kerogen concentrate obtained from the acid treatment of the rock discussed above can be used for vitrinite reflectance measurements as well as for visual kerogen studies.

Dow's technique (1977a) involves taking a portion of the washed kerogen and freeze-drying it. In the extraction and drying technique, the sample must not be heated or exposed too long in air, as this will change the reflectance values. The black freeze-dried powder is mixed with a slow-curing epoxy resin and is placed in a press to make small pellets. Three of these pellets are polished with aluminum oxide on a polishing wheel and immediately sprayed with a thin film of plastic to prevent oxidation of the exposed vitrinite particles. The three polished pellets are placed under a microscope and scanned at a magnification of 625–650x in oil immersion. Some training is required to recognize vitrinite particles and distinguish them from other coal macerals. Reflectance measurements are made on between 50 and 100 vitrinite particles from one sample. In sophisticated systems, a computer printout plots the data as a histogram. The population is recorded on the ordinate, and the reflectance value on the abscissa. At the conclusion of the analysis, the computer calculates the mean, median, mode, and standard deviation. The statistical mean reflectivity is reported because rotation of the microscope stage is not required. In coal petrography, maximum reflectivity rather than the mean is generally reported, and there is some difference in the higher maturity ranges where anisotropy is largest. The term R_o refers to reflectivity measurements obtained with oil immersion objectives, whereas R_a refers to reflectivity measurements in air. The latter are reported in much of the Russian literature.

The statistical mean reflectivity of a sample is plotted on a semilog depth–R_o plot as shown in Figure 7-41. Dow (1977a) emphasizes that R_o values are most useful if the data are taken objectively and analyzed subjectively. All data are recorded, but the operator is trained to record for each sample items that can affect the R_o values. For example, R_o values are generally lower if the sample contains a large amount of cavings, rough vitrinite, or mud contamination. Oxidized vitrinite and recycled material will show higher R_o values. The best R_o data are used to draw a least-squares, best-fit line, the slope of which depends on the geothermal gradient and the rate of deposition.

Vitrinite, which has been oxidized in the subsurface, frequently is observed along major unconformities, in formations in contact with meteoric waters, and in red shales and interbedded red and green shales. Oxidized vitrinite is an indication that the sediments have either been eroded or exposed to meteoric waters at some time in their geologic past.

In analyzing cuttings down a hole, there may be gaps in the appearance of vitrinite because organic-rich shales with mostly amorphous kerogen have little primary vitrinite. Also, carbonates that contain mostly marine amorphous kerogen have no vitrinite unless there are some clastics in the carbonate. Anhydrites and other evaporites rarely have vitrinite. However, if enough data can be obtained on good vitrinite particles at intervals down the hole, the straight line can be extrapolated through the barren or unusable sections. The technique is not applicable to most Pre-Devonian rocks because of the absence of land plants.

Histograms of mean reflectivity tend to broaden going up in rank, so the errors are greater (Figure 10-18). Fortunately, the standard deviation is reasonable in the lower rank ranges where oil and gas generation occur. There is less standard deviation using maximum reflectance values at high rank, but this requires rotation of the stage. Reflectance values as high as R_o equals 8 percent have been observed in deep Texas wells, but there is no known production associated with R_o values higher than 3.5 percent. Examples of

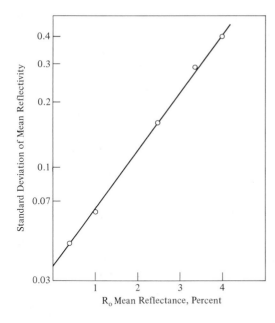

Figure 10-18
Broadening of vitrinite histograms, as indicated by the logarithmic increase in standard deviation of the mean reflectivity with a linear increase in R_o values.

vitrinite reflectance data for several wells were shown in Chapter 7, and Figure 7-49 summarizes the R_o values for oil, condensate, and dry gas production.

Vitrinite reflectance has the advantage over other maturation techniques of covering the entire temperature range from early diagenesis through catagenesis into metamorphism. It is the most useful subsurface prospecting tool for determining the present and past stages of maturation. It does not, however, define the oil- or gas-generating capability of a sediment. A technique such as pyrolysis, cuttings gas analysis, or C_{15+} extraction must be used with vitrinite reflectance to define the most prospective horizons.

PETROLEUM PROXIMITY INDICATORS

Soviet geochemists methodically analyze water samples from drill-stem tests or those associated with oil fields, along with surface water samples of lakes, springs, rivers, and creeks, for various organic and inorganic constituents such as phenol, naphthenic acids, alcohols, organic sulfur, ammonia, nitrogen, boron, iodine, and bromine as well as the conventional ions, chloride, bicarbonate, sulfate, sodium, potassium, calcium, and magnesium. Kartsev (1963), in his book on the hydrology of oil and gas fields, discusses the use of several of these analyses in predicting favorable horizons for oil and gas accumulations. The content of phenols in the waters of oil fields of the

Northwest Caucasus ranges from 1 mg/liter (or ppm) to more than 10 mg/liter, whereas in nonproducing areas it is generally less than 1 mg/liter. During early drilling in the West Siberia area, a series of dry holes showed a strong increase toward the north in phenol and other organic concentrations. The geochemists predicted oil north of the dry holes, and subsequent drilling proved them to be correct. In the Devonian brines of the Volga–Ural petroleum province, the ammonia content is never less than 100 mg/liter, whereas in the barren province near Moscow the Devonian brines contain less than 60 mg/liter. Kartsev recognized that some geochemical constituents are formed by the action of oil and gas on subsurface water, whereas others form from sediments under conditions favorable for the generation and accumulation of petroleum. The first would be direct proximity indicators, comparable to the hydrocarbon gases diffusing from an accumulation, while the second would be indirect indicators for regional prospecting.

The problem with many of these indicators is that they can originate either from the sediments or a petroleum accumulation and the chemistry of the aquifers can considerably alter their concentrations. Aliev et al. (1966) concluded that high concentrations of iodine, bromine, ammonium, and boron are not caused by petroleum accumulations but by geological conditions that are favorable both for the formation of oil and the accumulation of these chemicals. Bogomolov et al. (1970) concluded that the highest ammonium concentrations are in connate waters of low pH. Also, ammonium tends to be higher in carbonates and evaporites than in clay–sand sequences, possibly because of the fixation of ammonia by the clay. Shtogrin (1974) concluded that high ammonia concentrations in formation waters can result from the destruction of an oil accumulation.

Zinger and Kravchik (1973) found that the concentration of C_3H_7OH and higher alcohols ranged from 1 to 8 mg/liter in the vicinity of oil and gas accumulations and from 0 to 1 in nonproductive background areas. They also investigated the organic sulfur compounds of subsurface waters and concluded that these were generally higher in the vicinity of petroleum accumulations. Hitchon and Horn (1974) carried out a statistical study using discriminant analysis on a suite of 438 formation waters from Alberta, Canada, which had been analyzed for Cl^-, Br^-, I^-, HCO_3^-, SO_4^{2-}, Ca^{2+}, Mg^{2+}, and Na^+. They found that producing and nonproducing areas could be distinguished by I and Mg in the Paleozoic and by Cl^- and Na^+ in the Mesozoic.

The indicators discussed in this section have not been used in the United States as subsurface prospecting tools because (1) drill-stem tests are rarely taken on wildcat wells, (2) when water samples are taken, they are usually too infrequent with depth and too few aerially compared to other samples, and (3) there is still controversy over the value of these indicators in subsurface prospecting compared to other well-established methods. Their continued use in the USSR is probably due to the fact that hydrogeochemical indicators have been most valuable in prospecting for ores and minerals other than petroleum.

Probably the most useful proximity indicators in the United States are benzene and other soluble hydrocarbons, as previously discussed. They can be helpful in outlining a pool and indicating whether a dry hole may be close to an accumulation.

SUMMARY

1. Oil and gas accumulations occur in sedimentary sequences that are alive with hydrocarbons. Prospecting should be directed toward those areas and formations where the shale gas readings on the mudlog are high and where core and cuttings analyses show that the fine-grained rocks have a relatively high hydrocarbon content.

2. Mapping gas yields from mud logs can show prospective areas for oil and gas. In the Western Canada Basin, a band of wet shale gas extends from the Peace River area south to the border of Alberta and the United States. All major oil production lies in this wet gas band. West and northwest of this band, the mud logs show dry gas and associated reservoirs contain only gas. Near Regina, Saskatchewan, mud logs show no gas in the entire section from the Cambrian to the Cretaceous. No commercial production is in this area.

3. Subsurface formations can be defined as having an oil, gas or no potential from analyses of the gas–gasoline in wet cuttings. At the depth equivalent to the threshold of intense oil generation, there is a large increase in the wet gas and gasoline content of the cuttings. At greater depths, the change from oil to gas production in reservoirs corresponds to the change from wet gas to dry gas facies in the cuttings. In the Western Canada Basin, the hot lines are defined as the change from wet to dry gas facies in cuttings going west into the foothills of the Rocky Mountains.

4. The concentration of hydrocarbon gases in subsurface formation waters may be used as a regional indicator of producing versus nonproducing areas, but it is less effective for pinpointing accumulations.

5. The C_{15+} hydrocarbon extraction is used extensively for evaluating source rock quality, source type, and degree of maturation. It provides the most detailed information on hydrocarbon type distribution and changes in biogenic hydrocarbons such as steranes, pentacyclic triterpanes, and the odd/even n-paraffin ratio. It can be run on cores, cuttings, or unweathered outcrops of fine-grained rocks.

6. Pyrolysis–gas chromatography of a whole rock or kerogen concentrate differentiates the free hydrocarbons from those formed by the thermal

degradation of the kerogen. It can provide a material balance of the petroleum-generating process through quantitative measurement of the weight loss and all gases evolved. It can give much of the same information available in the gas and gasoline cuttings analysis and the C_{15+} extraction. It has the advantage of being faster, of being less expensive, and of using very small samples. Pyrolysis–fluorescence is a simple, rapid technique for routine surveys of large numbers of samples at the well site.

7. The elemental analysis of kerogen is a useful technique for spot-checking data from other procedures. As a technique for continuous logging of cuttings with depth, it is expensive and time consuming. Kerogen concentrates from solvent-extracted rocks are very susceptible to oxidation. This has caused artificially high oxygen contents to be reported.

8. Kerogen color is a useful maturation technique applicable to both unweathered outcrops and well samples. It has been used for mapping favorable oil and gas regions and formations. It is most effective when incorporated as part of a normal palynological operation, because experience is needed for valid intepretation of the color.

9. Vitrinite reflectance is the most useful technique for determining the present and past stages of maturation of sedimentary organic matter. It can be used on unweathered outcrops and library samples of old cuttings in addition to fresh well samples. It covers a greater temperature range than the color techniques, and it provides geological information such as the quantity of overburden removed and changes in geothermal gradients with time. It does not define the oil- or gas-generating capability of a sediment, so it must be used with extraction or pyrolysis techniques.

10. Chemical components of formation waters such as phenol, alcohols, ammonia, organic sulfur compounds, iodine, and bromine have been used with some success as regional indicators of petroleum in the USSR. They are not used in the United States because of the scarcity of formation water samples and preference for other less controversial prospecting techniques. Some soluble hydrocarbons have been used in the United States as petroleum proximity indicators.

SUPPLEMENTARY READING

Hitchon, B. (ed.). 1977. Application of geochemistry to the search for crude oil and natural gas. *Journal of Geochemical Exploration, 7* (2), Special Issue, 293 p.

Sokolov, V. A., and G. G. Grigor'ev. 1962. *Metodika i rezul'taty gazovykh geokimicheskikh neftegazopoiskovykh rabot* [Methods and results of gas geochemistry in the search for oil and gas]. Gosudarstvennoe Nauchno-tekhnicheskoe Izdatel'stvo Neftyanoi i Gorno-Toplivnoi Literatury Moscow, 403 p. (In Russian.)

11

Crude Oil Correlation

The correlation of crude oils with each other and with the oils from their presumed source rocks can be a valuable tool for assisting the exploration geologist in locating new production and extending existing trends. When two or more pools are discovered in a relatively unexplored area, the question arises as to both the lateral and vertical extent of the production trend. Are there one or several types of oils present in a particular rock series? Are any of these oils related to shallower or deeper accumulations? Can the known seeps in the basin be related to subsurface pools? Can an unconformity oil be related to the oils or source rocks of underlying beds? Can the pathways of secondary migration (permeable rock to reservoir accumulation) or tertiary migration (movement of a pool) be followed by changes in crude oil composition? If crude oil–source correlations could define the source of each oil in a basin, drilling could focus on prospects related to the defined sources.

Modern analytical techniques are so definitive that it is a simple matter to establish the similarity or dissimilarity of two oil samples, whether from reservoirs, source rocks or seeps. Much more difficult is the problem of interpreting how an oil may change in moving from source to reservoir or within the reservoir rock or how two crude oils of the same origin may undergo different physical and chemical changes after accumulation. A major objective of correlation procedures is not simply to compare oils but to compare molecules within oils that may be least or most affected by environmental factors such as water washing, biodegradation, or thermal alteration.

Crude oil samples should be collected in glass bottles or vials with aluminum-foil-lined caps. Hydrocarbons can diffuse out of some plastic containers, and metal containers can introduce trace elements.

ANALYTICAL METHODS

Refinery laboratories estimate the market value of a newly discovered crude oil by determining the °API gravity, viscosity, pour point, and sulfur content. The crude also is distilled to determine the volume of gasoline, kerosine, gas oil, and residuum. More detailed analyses, such as for paraffin, naphthene, aromatic hydrocarbons, and nitrogen, vanadium, and nickel content, are made to evaluate performance in various distillation and catalytic cracking units. Although many analytical procedures for petroleum were developed in refinery laboratories, they needed considerable modification to apply them to geological problems. Thus, the refinery chemist works with several liters of oil from the pipeline, whereas the exploration geochemist may have a fraction of a milliliter extracted from a source rock.

Distillation

Historically, crude oils have been classified into two main groups, paraffin base and asphalt base, depending on the content of paraffin wax or asphalt in the petroleum residuum. Within each group, correlations were made on the basis of the distillation curve. A problem with this early classification was that a high precentage of all crudes were mixed base; that is, they did not belong distinctly to either the paraffin or asphalt base type. Also, simple distillation did not define the chemical composition of the oil fractions, which is essential for valid classification.

In the late 1920s, the U.S. Bureau of Mines adopted the Hempel method of distillation for crude oil correlation. This involved taking ten fractions at atmospheric pressure in 25°C (45°F) intervals from 50 to 275°C (122° to 527°F) and five fractions at 40 mm pressure from 200 to 300°C (392 to 572°F). Later, Harold M. Smith (1940) of the U.S. Bureau of Mines developed the correlation index (CI) as a means of defining the chemical character of the distillation fractions. Smith correlated the specific gravity, G, at 60°F (16°C) with the average boiling point of the fraction in degrees Kelvin (K) in the following empirical formula:

$$CI = \frac{48,640}{K} + 473.7 \, G - 456.8$$

The values of the CI are 0 for the normal paraffins, 51.4 for cyclohexane, 100 for benzene, and above 100 for polycyclic aromatics. The value for methylnaphthalene is 125.

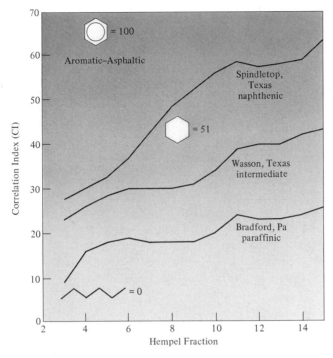

Figure 11-1
Base of a crude oil as defined by the correlation index (CI), which
is plotted against the Hempel distillation cut number. The oils shown
are typical naphthenic, intermediate, and paraffinic crudes.
[McKinney et al. 1966]

Figure 11-1 shows how the composition of a crude is revealed by the
correlation index. Values below 25 indicate a paraffinic crude, and successively higher values indicate intermediate, naphthenic, and aromatic-
asphaltic crudes. The higher boiling fractions have higher CI's because there
is an increase in ring structures with molecular weight, as discussed in
Chapter 3. Usually, CI curves follow the general trends shown in Figure 11-1,
but a distinct advantage of this technique is that it recognizes the hydrocarbon
type within each crude fraction. Thus, in Figure 8-23 the Devonian oil of the
third Venango sand was paraffinic throughout, whereas the oil in the first and
second sands was naphthenic-aromatic at the front end, because of the
microbial destruction of paraffins.

Literally thousands of correlation index curves have been determined by
the U.S. Bureau of Mines for crude oils from all over the world. Geologists
and petroleum engineers have used these data extensively in their operations.
The correlation of Utah and Colorado crudes and of Pennsylvanian crudes by

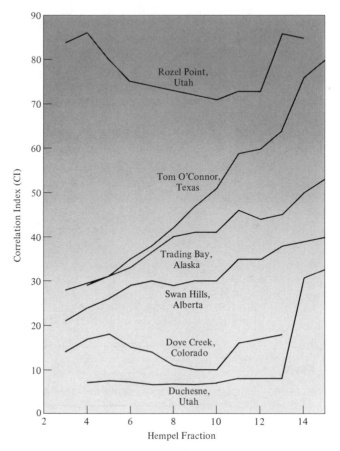

Figure 11-2
Correlation index curves for a variety of crude oils. [Wenger et al. 1957; McKinney et al. 1966; Wenger and Morris 1971; Blasko et al. 1972; Dooley et al. 1973]

CI was previously shown in Figures 6-27 and 8-23, respectively. Figure 11-2 shows examples of CI curves of crudes from the United States and Canada. The Dove Creek and Duchesne crudes are paraffinic, although the latter contains a high percentage of naphthenes in the higher cuts. The Swan Hills oil is intermediate, while the Tom O'Connor crude is highly naphthenic, with aromatic hydrocarbons concentrated in the higher cuts. High CI's can also result from the sulfur and nitrogen compounds in the heavy fractions but this oil only has 0.16 percent sulfur, indicating that the high CI's are primarily caused by aromatics. The Rozel Point crude is an aromatic asphaltic oil with 12 percent sulfur. The high CI values through its entire boiling range are partly caused by sulfur compounds. All of the oils shown in Figure 11-2 are readily distinguished by the correlation index curve. Correlation indices cannot be used in crude oil–source rock correlations, because too large a sample is required. Also, distillation techniques are being replaced by more rapid, definitive analyses such as gas chromatography. Nevertheless, the huge library of correlation index values determined by the U.S. Bureau of Mines over the last 35 years represent a valuable source of information for both petroleum research and operations.

One shortcoming of the CI method, or of any method based on the physical properties of petroleum, is that it does not define the proportions of the different hydrocarbon ring and chain structures present. CI values of about 50 may be caused by the presence of mixtures of naphthenes, aromatics, and neutral resins with paraffinic side chains of variable lengths. The Waterman ring analysis was developed for a more definitive chemical description of distillation fractions. It utilized the refractive index, density, molecular weight, and elemental analysis to define the percentage of hydrocarbon structural groups in the boiling range generally above gasoline (Van Nes and Van Westen 1951). The advent of mass spectrometry, which measured ring structures directly, resulted in abandonment of this technique.

Chromatography

The separation of components of a mixture by differences in the way they become distributed (partitioned) between a mobile and stationary phase is called *chromatography*. The components must differ in their attractions for the two phases in order to become separated.

The chromatographic technique was discovered by the Russian botanist, Tswett, who used a two-phase liquid–solid system to separate the pigments of green leaves. He passed a petroleum ether extract of the leaves through a vertical column of dry, powdered sugar and observed a separation into two green pigments (chlorophylls A and B) and several yellow pigments (carotenoids). The procedure was called *chromatography* because of the separation of these colored substances, but it works equally well with colorless material.

Gas Chromatography (GC)

In gas chromatography, hydrocarbon molecules are partitioned between a moving gas phase and a high-molecular-weight stationary liquid phase inside a column. The procedure is as follows: The hydrocarbon mixture to be analyzed is introduced into a moving stream of inert carrier gas, such as helium or nitrogen, which flows through a long (1–30 ft, or 0.3–9 m) column. This may be a capillary column coated with a thin film of the liquid phase or may be a $\frac{1}{8}$-inch diameter or larger column filled with finely divided powder, which is coated with the liquid phase, or substrate. Each hydrocarbon molecule in the unknown mixture differs in the length of time during which it stays dissolved in the liquid phase. Generally, small paraffin molecules are retained the shortest time and large molecules the longest. For example, passing CH_4, C_4H_{10}, and C_8H_{18} with helium through a column coated with nonpolar silicone oil would cause the methane to be retained for the shortest length of time and the octane the longest. The time from injection of a particular hydrocarbon to the time when it reaches the detector at the end of the column is the retention time. With a particular substrate liquid, it is possible for two hydrocarbons to have exactly the same retention time. However, they can usually be separated by using columns of different substrates. The detector at the end of the column determines the quantity of each hydrocarbon, which is then plotted by a recorder or computed by a digitizer that records the quantity on tape. GC units are used routinely with thermal conductivity detectors on many well logging units to monitor the methane and ethane-plus down the hole.

The gas chromatograph is superior to a distillation column in separating close boiling hydrocarbons. The efficiency of distillation columns such as shown in Figure 3-6 is based on the number of plates, each plate representing one distillation of a mixture from the liquid to vapor state, as in Silliman's original flask. The column shown in Figure 3-6 has 16 plates. When distillation columns are packed with helices instead of plates to increase the vapor–liquid contact, the efficiencies are calculated as theoretical plates.

Early distillation columns, such as those used by API Project 6, were high (15 ft, or 4.6 m) glass columns packed with metal helices with efficiencies equivalent to 200 theoretical plates. Modern spinning band columns have efficiencies of about 1,000 plates, but a high-resolution gas chromatograph will have efficiencies of several thousand plates. This difference is better understood when it is realized that a modern GC unit can determine the quantities of over 100 hydrocarbons in the C_1–C_{12} range of a crude oil in about two hours. It would take hundreds of man-hours to do this by distillation.

GC uses very small samples. It cannot process enormous volumes of material as is required in refinery operations where distillation is still the principal method for separating petroleum components. GC can be modified to separate small amounts of pure hydrocarbons by using a *preparative* gas chromatograph. However, the volume is still much less than with a good spinning band distillation column.

There are so many GC column designs and substrates that it is best to consult a specialist in order to solve a particular analytical problem or when setting up an analytical facility. Modern gas chromatographs are designed with interchangeable columns and detectors to handle almost any problem. For example, Martin and Winters (1963) determined 78 hydrocarbons in the

C_3–C_{10} range in 11 crude oils using a coiled, 500 ft (152 m) capillary column coated with 1-octadecane to separate saturates and a coiled, 800 ft (244 m) capillary column coated with polyethylene glycol to separate aromatics. The columns were placed in parallel with a stream splitter to introduce the sample. An 8 ft prefractionator column coated with silicone rubber (G.E. SE-30) on Chromosorb W was used to prevent hydrocarbons above C_{10} from entering the columns. Later, Merchant (1968) analyzed both saturates and aromatics in the C_5–C_{12} range on one column. He introduced about 0.5 μl of sample into a 200 ft stainless steel capillary column coated with a mixture of squalane and Kel-F40. He identified 90 components in a synthetic blend of C_5–C_{12} hydrocarbons in 80 minutes. A more detailed discussion of the applications of gas–liquid chromatography to geochemistry is given by Douglas (1969). See also McNair and Bonelli (1969).

The composition of crude oil above C_{10} is so complex that the typical GC column does not separate all the hydrocarbons. Oil samples in these high-molecular-weight ranges can be given preliminary separation by adsorption chromatography. Also, narrow fractions from the GC column can be analyzed by mass spectrometry. The most promising development in GC is the improved high-resolution glass capillary column, which is capable of direct separation of highly complex mixtures in the C_{10+} range. Grob (1975) and Grob et al. (1975) analyzed diesel oil in tap water and the major components of cigarette smoke by this technique.

Liquid—Solid (Column) Chromatography

Liquid–solid (column) chromatography is comparable to the original Tswett technique in that the liquid phase is a solvent and the stationary phase is usually a column of silica gel or alumina. The procedure involves placing a concentrated hexane solution of the oil sample on the top of the column and then passing hexane through to elute the paraffin–naphthene fraction and to leave the aromatics and heterocyclic compounds behind on the column. The latter are sequentially eluted by successively passing through cyclohexane, chloroform, and methanol. Further separation of the paraffin–naphthene fraction is by urea and thiourea adduction. This involves forming crystalline complexes (adducts) with some of the compounds, such as n-paraffins, leaving others in solution. Schiessler and Flitter (1952) have listed the various compounds that adduct with urea and thiourea at 25°C (77°F). The adductions are more quantitative at low temperatures such as at -25°C or -13°F (Speers and Whitehead 1969). Molecular sieves also are used for the separation of n-paraffins from branched and cyclic paraffins (O'Connor et al. 1962; Blytas and Peterson 1967). Separation by chromatography, urea adduction, and molecular sieves must be monitored by other techniques such as mass spectrometry, because the separations are not quantitative. Bes-

Table 11-1
Mobile and Stationary Phases in Column Chromatography

Relative eluting power of mobile solvents			Relative activity of stationary adsorbents	
Increasing Eluting Power or Increasing Polarity	Hexane			Cellulose
	Cyclohexane			Sugar
	Carbon tetrachloride			Calcium sulfate
	Benzene		Increasing Activity	Silica gel
	Chloroform			Magnesium oxide
	Diethyl ether			Alumina
	Acetone			↓ Activated charcoal
↓ Methanol				

tougeff (1967) found that the use of the molecular sieve separation of *n*-paraffins in the Hassi Messaoud and Quatar crude oils yielded a product containing only 26 and 16 percent of normal paraffins, respectively, apparently because of the inclusion of naphthene rings with long hydrocarbon side chains.

Table 11-1 compares the eluting power, or ability of a given liquid to move a polar compound down a column, of various common solvents used in separating crude oil. Crude oil contains both nonpolar paraffin hydrocarbons and polar nitrogen, sulfur, and oxygen compounds. Consequently, the elution is started with the most nonpolar solvents and gradually changed to the more polar. Solids used for the stationary phase also are listed in Table 11-1, in order of increasing activity. High-activity solids are used to separate the less polar compounds.

Petroleum fractions boiling above 400°C have large numbers of nitrogen, sulfur, and oxygen compounds that can be partially separated by using special adsorbants, such as (1) anion and cation exchange resins to remove acids and bases, and (2) heavy metals for complexing other compounds. Jewell et al. (1972) were able to characterize several heavy ends of crude oils by combinations of these adsorption chromatographic techniques.

Gel permeation chromatography (GPC) is useful for separating molecules in a specific high molecular weight range from the complex heavier fractions of petroleum. Cross-linked polymer gels contain openings in which molecules of a specific size range are trapped while larger molecules pass on through. By selecting different solvents and gels, it is possible to control the size of molecules to be retained. Baker and Smith (1973) separated chlorophyll derivatives in the 500–1,500 molecular weight range from sediment extracts by passing a tetrahydrofuran solution of the extract through Sephadex LH-20. McKay and Latham (1973) used GPC and fluorescence spectrometry to identify polyaromatic ring systems containing five to eight rings.

In thin-layer chromatography (TLC), the stationary phase is a thin coating of an adsorbant on a support such as a glass slide. The mixture to be separated is applied as a spot to the bottom of the slide, and the eluting solvent is allowed to move up the plate by capillary action. Separation occurs on the plate. TLC has the advantage of allowing rapid analysis of very small samples.

Mass Spectrometry (MS)

A mass spectrometer identifies compounds by their molecular weight or the mass of ionized molecular fragments. When hydrocarbon molecules are introduced into a mass spectrometer, they are vaporized and bombarded with electrons, which ionize some of the molecules. Depending on the voltage used, the parent molecule is ionized. Some ionized molecules break into daughter ions, which form a characteristic pattern for that molecule. In effect, this involves the cracking of a large molecule into several small ones, all of which are ionized. The ions are directed at high velocity past a magnet, which spreads them out depending on their mass. A detector scans and records the quantities of all masses present. Samples smaller than a nanogram (10^{-9} grams) can be analyzed for masses up to 2,000 in seconds. Since the resulting spectrum may contain peaks of several hundred ions in different quantities, the task of interpretation is difficult. Consequently, the MS data are best interpreted when analyzing a single peak of a pure compound from a gas chromatograph. Identification of unknowns is provided through a comparison of the mass spectra of the sample with known spectra. The most advanced spectrometers have their output fed into a computer that contains memory banks of all known compounds. The computer then prints out the most probable identifications of the various compounds.

In gas chromatography–mass spectrometry (GCMS), the gas chromatograph is used for separation of hydrocarbon mixtures. The GC peaks are fed directly into the mass spectrometer, which is then used as the GC detector.

High-resolution mass spectrometry is particularly useful in analyzing the high-molecular-weight ranges of petroleum. Table 11-2 shows a typical group-type analysis of a high boiling petroleum fraction by mass spectrometry (Gallegos et al. 1967). Further characterization of an oil fraction can be made by the MS analysis of individual structures within any one group. An excellent review of high-mass spectrometers and their application to problems in organic geochemistry is given by Burlingame and Schnoes (1969).

An isotope-ratio mass spectrometer operates on the same principle as a conventional MS. However it is designed to compare the ion masses of compounds containing isotopes such as $^{12}CO_2$ and $^{13}CO_2$. Scalan and Morgan (1970) describe a high-precision isotope-ratio MS that can analyze CO_2 and SO_2 samples interchangeably for carbon and sulfur isotopes.

Table 11-2

Group Type Analyses of 650–800°F (343–427°C) Fraction of Crude Oils
by High-Resolution Mass Spectrometry

Compound group	Liquid volume percent of crude		
	Arabia	California	Louisiana
Saturate Hydrocarbons			
Paraffins	21	2	4
Naphthenes			
Noncondensed	7	9	11
Condensed			
Two-ring	4	9	11
Three-ring	2	8	9
Four-ring	3	10	11
Five-ring	4	4	8
Six-ring	0	—	—
Total Saturates	41	46	57
Aromatic Hydrocarbons			
Alkylbenzenes	8	7	5
Benzocycloparaffins	2	7	4
Benzodicycloparaffins	2	5	4
Naphthalenes	0.6	2	2
Acenaphthenes	1	5	2
Fluorenes	4	6	5
Phenanthrenes	10	12	13
Pyrenes	5	6	4
Chrysenes	2	0	0
Total Aromatics	35	50	39
Sulfur Compounds			
Benzothiophenes	5	2	0.6
Dibenzothiophenes	19	2	3
Naphthobenzothiophenes	0	0	0
Total Sulfur Compounds	24	4	4

Other Methods

A nuclear magnetic resonance (NMR) spectrometer can be used to identify
the structures of hydrocarbon molecules. It defines the environment of
hydrogen atoms on the molecule. The total hydrogen content and the
distribution of hydrogen among the functional groups in a liquid sample is
determined nondestructively For example, the position of the CH, CH_2, and
CH_3 groups on chains and rings can be determined. NMR does require that
the sample be separated into a saturate fraction free of olefins and aromatics

or an aromatic fraction free of saturates and olefins The most detailed analyses are made on very narrow boiling fractions of a single compound class. It is customary to combine GCMS with NMR for the most complete structure analyses.

Infrared absorption spectrometry also is used for molecular-type analysis, but to a large extent it has been replaced by the GC–MS–NMR spectrometer combinations.

Ultraviolet (UV) and visible absorption spectrometry are widely used as a fast, inexpensive method for monitoring the purification of samples separated by liquid chromatography. Ultraviolet is used to determine the quantities of molecular structures that absorb UV light, such as porphyrin derivatives, in chromatographic column eluants.

Other analytical equipment for characterizing crude oils includes atomic adsorption and neutron activation to measure trace elements and photoelectric polarimeters to measure the optical activity of oil fractions in solution.

CORRELATION OF RESERVOIR OILS

In crude oil correlation, genetically related oils are differentiated from unrelated oils on the assumption that the same source material and environment of deposition will produce the same oil. A biological marker compound dominant in the source rock would be expected to be in the oils it generated. A particular ratio of two hydrocarbons not affected by extraneous factors would be expected to be the same in oils generated from the same rock. The correlation problem is more complex, because the same source rock may generate different oils at different times in its history. Also, the oils, once generated, could be altered by maturation, gravity segregation, water washing, biodegradation, and migration. In actual practice, a series of parameters are picked that are least affected by nongenetic factors for grouping the oils. Depending on the sophistication of the analysis, the oils fall into a few or many groups with scattered abnormal samples. The geochemist then correlates the groups with the geological data to arrive at an inferred genetic relationship.

Gross properties of oils are not useful for correlation since they are readily changed by nongenetic factors. An example is the oil produced from the steeply dipping Velma sand of Pennsylvanian age in Stephens County, Oklahoma. The data for oils from two wells producing from the updip (1) and downdip (2) edges of the same pool are shown in Table 11-3. The first five properties are different for this oil primarily because of gravity segregation within the pool. However, in terms of hydrocarbon ratios, C_{15+} n-paraffins, vanadium/nickel ratios, and correlation indices of the high boiling fractions, the oils are very similar. Consequently, these latter properties can be used for genetic classification.

Table 11-3

Properties of Velma Oil Samples

	No. 1	No. 2
Producing depth, ft	4,350	5,850
Gravity, °API	37	23
Percent boiling to 200°C	42	14
Percent asphaltenes	1.3	8
Percent sulfur	0.95	1.4
Vanadium + nickel in ppm	41	259
Vanadium/nickel ratio	1.7	2.1
Percent n-paraffins in C_{15+} range	5.6	5.1
Correlation indices (compare with Figure 11-2)		
Hempel fraction 12	42	43
13	44	46
14	48	49
15	50	50

Note: No. 1 is the updip end, 2 the downdip end of the same accumulation.

The Williston Basin oils were grouped into three major types by Williams (1974) on the basis of GC analysis, n-paraffin distribution, and carbon isotope values. No grouping was possible from gross properties such as API gravity and percent sulfur. Erdman and Morris (1974) showed a genetic relationship between 18 oils of the Interior Salt Dome Basin of Mississippi from the close similarity of carbon isotope values, n-paraffin distribution, and odd–even predominance (OEP) of the topped crudes. The same oils showed no correlation in °API gravity and in percentage of sulfur, nitrogen, saturates, aromatics, asphaltics, and asphaltenes. All of these latter parameters are altered more extensively by nongenetic factors.

The lithology of the reservoir rock can cause nongenetic differences. Metwalli (1970) noted that the Hungarian oils in limestone reservoirs or in sandstones with a calcareous matrix were heavier than oils in sands with a clay matrix.

Chemical and physical degradation processes that change crude oils in the reservoir were discussed in Chapter 8. A prerequisite to selecting suitable correlation parameters is an understanding of the effects of these processes on such parameters. For example, the absolute concentration of vanadium and nickel in crude oil varies with the weathering of the oil but the ratio of vanadium to nickel is not appreciably altered by weathering.

Crude oil correlation is not routine, in that each problem may require a different approach. The correlation of two high-gravity oils might involve a comparison of hydrocarbon distributions in the gasoline range, whereas two heavy oils with no gasoline could be correlated on the basis of n-paraffin distributions or vanadium/nickel ratios. At least three correlation parameters that are not altered appreciably by nongenetic factors should be used in any field problem.

Trace Elements

The chlorophyll molecule loses its magnesium at the time of deposition. During diagenesis, both vanadium and nickel become complexed to the porphyrin. As the porphyrin is introduced into a crude oil from its source rock, it carries the imprint of vanadium–nickel distribution with it. Many trace elements in crude oils are simply a reflection of those picked up during migration or from the reservoir, so they have limited value in exploration. Vanadium and nickel differ in having a genetic origin as a metal–organic complex. The concentration of vanadium and nickel in oil is several thousand times what would be expected if these elements were not complexed to organic structures.

Figure 11-3 shows the vanadium–nickel contents of crude oils from four different areas plotted against °API gravity. The lines shown are best-fit lines calculated by the least-squares method. Two lines were calculated for each metal, the vanadium splitting at 30 ppm, the nickel at 3 ppm. The correlation coefficients were about 0.9 for the nickel data and 0.7 for the vanadium data. The increasing vanadium and nickel content with decreasing gravity is a reflection of the increase in heavy ends containing the porphyrin structures in the lower-gravity oils. Oils in general contain more vanadium than nickel.

No differences are evident in the crude oils of these four areas based on trace element content. Differences are recognizable by using vanadium/nickel ratios as shown in Figure 11-4. Each of the four groups of oils is concentrated in a specific range of vanadium/nickel ratios. Hodgson (1954) noted that the vanadium/nickel ratio of crude oils appears to decrease with age. Rosscup and Bowman (1967) measured the thermal stabilities of vanadium and nickel petroporphyrins and found that the vanadium is less stable than the nickel. As shown in Figure 11-4, the youngest group of oils has the highest ratios and the oldest group the lowest, in accordance with the observed stabilities of these metal–porphyrin complexes.

Biological Markers

The porphyrin structures themselves may be used for crude oil correlation, because most oils contain a variety of porphyrin structures derived from the original chlorophyll molecule. Distribution patterns of high-molecular-weight biological markers, such as porphyrins, steranes, and triterpanes, are

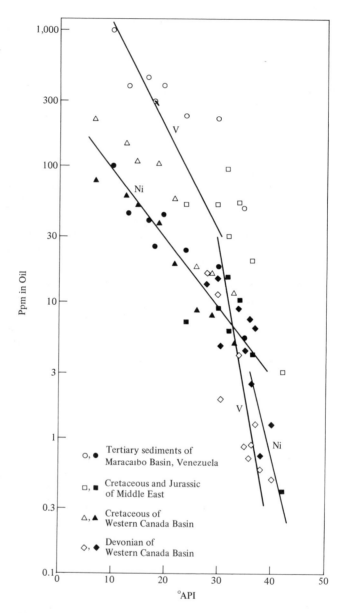

Figure 11-3
Parts per million by weight of vanadium (open symbols) and nickel (solid symbols) in oil vs. °API gravity for crudes of four areas. [Baker 1964; Hodgson 1954]

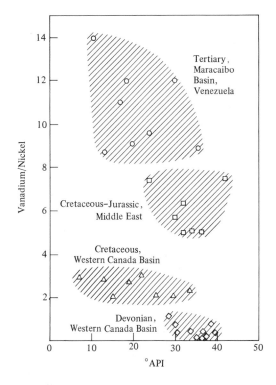

Figure 11-4
Vanadium/nickel ratios vs.
gravity for the same oils shown in
Figure 11-3.

particularly useful in crude oil–seep and oil–source rock correlations, because they are relatively unaffected by weathering and loss of light ends.

Pym et al. (1975) noted significant differences in the distribution patterns of eight hopanes in crude oils from different geographical areas. Hopanes are derivatives of the basic hopene structure (Figure 4-8) that is widely distributed in plants. Distribution patterns for four of the oils examined by Pym et al. are shown in Figure 11-5. The H, D, and G hopanes with molecular formulas $C_{27}H_{46}$, $C_{29}H_{50}$, and $C_{30}H_{52}$ appear to be the most diagnostic for differentiating these oils. The D triterpane peak is the most prominent in all of the Middle East oils examined by Pym et al. The G hopane is the most prominent in Nigerian petroleum.

The naphthene rings of steranes and triterpanes gradually aromatize in very old crude oils so the derivatives with one or more aromatic rings may be used for correlation. Tissot, Espitalié, et al. (1974) was able to classify the oils of the Illizi Basin of Algeria into three groups based on the distribution of monoaromatic steroid derivatives of the formula C_nH_{2n-12}. The largest differences between the oils occurred in the range where $n = 17$ through 22.

Seifert and Moldowan (1978) defined several specific molecular ratios of terpanes and steranes as useful in distinguishing differences in crude oils due

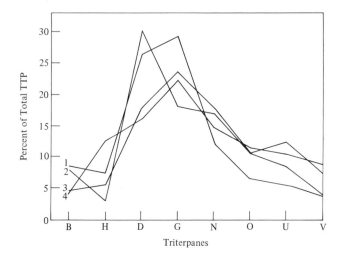

Figure 11-5
Distribution pattern for eight hopane series triterpanes in crude oils
from (1) Nigeria, (2) Kuwait, (3) Prudhoe Bay, Alaska, (4) North Sea.
Letters refer to eight hopane structures defined by Pym et al. (1975).

to source material, maturation, and migration. For example, internal ratios of
5α-steranes are considered to be a source indicator independent of migration
and maturation. Two homologous series of monoaromatized steranes characterized by molecular ion values of 239 and 253 also are good source
indicators. The ratio of primary (C_{29} + C_{30}) to secondary (C_{27} + C_{28})
terpanes is considered to be a source plus maturation parameter. Seifert and
Moldowan (1978) use parent and fragment ion maps for correlating oils with
biological markers. These maps display any series of mass chromatograms
requested by the analyst from a particular GCMS run. They provide a quick
detailed survey of any compound class in a complex mixture. Seifert and
Moldowan (1979) noted that several biological markers can be used to
correlate biodegraded crude oils (see p. 383).

Stable Isotopes

The stable isotopes of carbon have been used for correlating both crude oils
and natural gases (Fuex 1977; Stahl 1977). Sulfur isotopes have been useful in
following biodegradation, maturation, and sulfurization–desulfurization reactions of petroleum (Orr 1974; Krouse 1977), but they have not been used
extensively in crude oil correlation. Hydrogen isotopes are being used to
correlate gases, and a few studies have been made of oxygen and nitrogen
isotopes in the CO_2 and N_2, respectively, of reservoir gas (Stahl 1977).

Most carbon isotope work is reported in conjunction with other geochemical and geological data. Results are usually published in terms of ^{13}C content in parts per thousand relative to the PDB standard (see Chapter 2). The range of $\delta^{13}C_{PDB}$ values for crude oils is from $-18\ \%_{00}$ to $-34\ \%_{00}$. In other words, oils contain between 18 and 34 parts per thousand less ^{13}C than the PDB standard. The n-paraffins in an oil are the most deficient in ^{13}C, while aromatics and NSO compounds are the most enriched in ^{13}C. However, the different compound classes in the C_{15+} fraction of the same oil rarely vary by more than 2 or 3 $\%_{00}$. Differences of several ppt occur in the gas–gasoline–kerosine range of crude oil, and this fact has been used occasionally for correlation. For example, the three oils producing from the Cretaceous Blairmore, Viking, and Devonian Nisku reservoirs of the Joffre field of the Western Canada Basin are all different in this range. The Viking oils from the Bentley and Joffre fields are the same. Correlations in this range are not possible with oils subjected to water washing or biodegradation. Microbial destruction of n-paraffins leaves an oil more enriched in ^{13}C. Also, formation of an asphalt layer by water washing at the oil–water contact tends to cause ^{13}C enrichment.

Most carbon isotope correlations have been made on the C_{15+} fraction of crude oil, because it is less affected by degradation processes than the light fractions. Fuex (1977) reported that the unaltered crude oils of the Mackenzie Delta were correlated chemically into two groups by heptane isomers, but the biodegraded crudes had no heptanes to correlate. The carbon isotope analysis was able to correlate the C_{15+} fraction of all the crudes into the aforementioned two groups irrespective of the degree of biodegradation. Fuex emphasizes that valid oil–oil correlations with carbon isotopes can only be made on identical fractions of the oils. He also recommends that $\delta^{13}C$ values be determined for each compound class (n-paraffins, naphthenes, aromatics) within each fraction.

On a total crude, $\delta^{13}C$ values largely reflect the distribution of paraffins, naphthenes, and aromatics in the crude, so they are of little value compared to the $\delta^{13}C$ of fractions. Figure 11-6 shows the relationship between correlation indicies (CI) and $\delta^{13}C$ for some total crudes for which both data were available. The more elaborate $\delta^{13}C$ measurement is providing the same information as the simple CI analysis in terms of correlation.

The most useful applications of $\delta^{13}C$ analyses in exploration have been in correlating gases. As discussed in Chapter 5, biogenic methane has $\delta^{13}C$ values from about -55 to $-75\ \%_{00}$. Gas formed thermally from liptinitic kerogen, which is deficient in ^{13}C, ranges from about -40 to $-58\ \%_{00}$. Coal gas and deep dry gas, which forms thermally from vitrinitic kerogen, range from about -25 to $-40\ \%_{00}$ (Fuex 1977). These ranges are approximate and the limits can vary as much as $\pm 5\ \%_{00}$ because of differences in the parent organic matter.

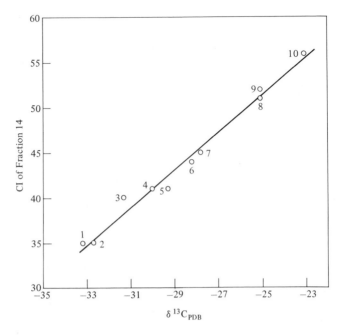

Figure 11-6
Correlation between $\delta^{13}C$ of whole crude oils and the correlation index of the 14th distillation fraction boiling between 250° and 275°C at 40 mm pressure. This fraction is least affected by crude oil degradation and by handling procedures. *Field, Age, Locality:* (1) Dollarhide, Silurian, Texas; (2) Magutex, Ordovician, Texas; (3) Red Wash, Eocene, Utah; (4) Dollarhide, Devonian, Texas; (5) E. Dollarhide, Devonian, Texas; (6) University Land, Permian, Texas; (7) Ashley Valley, Pennsylvanian, Utah; (8) Yellow Creek, Cretaceous, Mississippi; (9) Heidelberg, Cretaceous, Mississippi; (10) Oxnard, Oligocene, California. Samples 8 and 9 represent the C_{11+} fraction of the crude. [Data from Silverman and Epstein 1958; Kvenvolden and Squires 1967; Erdman and Morris 1974; McKinney et al. 1966]

Gases associated with oil in a reservoir generally have from 10 ‰ to 22 ‰ less ^{13}C than the oil. If the gases are formed from the thermal cracking of the oil in the reservoir they tend to increase in $\delta^{13}C$ values with time, because the earliest formed gases are the most deficient in ^{13}C. Also, as reservoir temperatures increase with deeper burial, there is less fractionation of the carbon isotopes during cracking of the oil.

Fuex (1977) and Stahl (1977) both cite several examples of carbon isotope studies of natural gases. Most of these were directed toward evaluating their origin and migration history.

Source rock–crude oil correlations have been attempted by recognizing the progressive change in $\delta^{13}C$ values of components of the oil and rock (Stahl

1977). Source rock extracts usually are isotopically heavier than the corresponding crude oil but are lighter than the asphaltenes of the oil and the kerogen of the rock. In both oils and extracts, the compound classes become isotopically heavier in the order n-paraffins, branched paraffins, naphthenes, single-ring aromatics, multi-ring aromatics, and NSO compounds.

Field Examples

Cretaceous Oils of Central Gulf Coast

Koons et al. (1974) classified the oils in the Lower Tuscaloosa Cretaceous reservoirs of southern Mississippi and Alabama into two distinct groups based on the series of parameters shown in Figure 11-7. The first type of oils contained about twice the n-paraffin content and two per mil less ^{13}C in the saturate fraction than did the second type of oils. These are genetic differences, because, as previously discussed, terrestrial source materials contain more n-paraffin waxes and have less of the ^{13}C isotope than do marine-source compounds. The Type I oils also contain the most steranes, with C_{28} steranes dominant, whereas C_{27} was the most prominent in Type II. The fourth parameter, the ratio of cyclopentanes to n-paraffins in the C_4–C_7 range, was much higher in the first compared to the second type of oils.

The most interesting observation of Koons et al. was that all of the Type I oils occurred in unfaulted structural and stratigraphic traps, whereas most of the Type II oils occurred in faulted structures. The Type I oils were

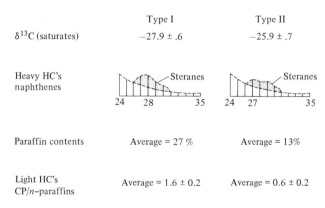

Figure 11-7
Correlation parameters used to classify Lower Tuscaloosa oils of southern Mississippi and Alabama into two types. CP = cyclopentanes. [Koons et al. 1974]

formed in fine-grained shales adjacent to the producing sands, whereas the Type II oils were situated where younger or older source rocks had been brought into contact with the reservoir sands. There also was the possibility that oils generated in deeper formations migrated upward along the faults to form the Type II accumulations.

Williston Basin

Williams (1974) used a variety of correlation parameters to classify over 100 oils from the Williston Basin into three major types. This included correlation index curves, carbon isotope ratios, optical rotation, hydrocarbon type distribution in the C_4–C_7 fraction, and the C_{15+} n-paraffin distribution. The optical rotation and CI curves showed too much overlap for characterization. Initial separation of the oils into two distinct groups was apparent from carbon isotope values of the saturate fractions. The breakdown into three groups showed up in the hydrocarbon type distribution of the C_4–C_7 fraction (Figure 11-8) and the C_{15+} n-paraffin distribution (Figure 11-9). The position for each oil in Figure 11-8 was determined from measuring 11 hydrocarbons in the C_4–C_7 range. The Type I oils contained the most n-paraffins, Type II the most naphthenes, and Type III were intermediate.

The n-paraffin distribution of the Type I oils shown in Figure 11-9 indicates a marine source, from the dominance of the C_{17} and C_{19} peaks.

Williams also made the interesting observation that the Weldon field, lying along the Weldon fault at the western edge of the basin, appeared to be a mixture of Type II and Type I oils based on the carbon isotopes and

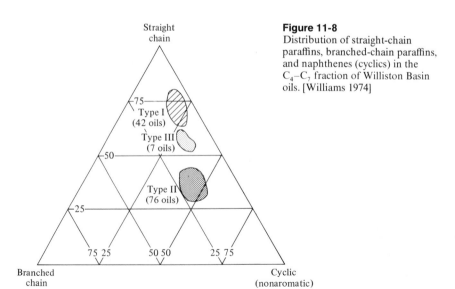

Figure 11-8
Distribution of straight-chain paraffins, branched-chain paraffins, and naphthenes (cyclics) in the C_4–C_7 fraction of Williston Basin oils. [Williams 1974]

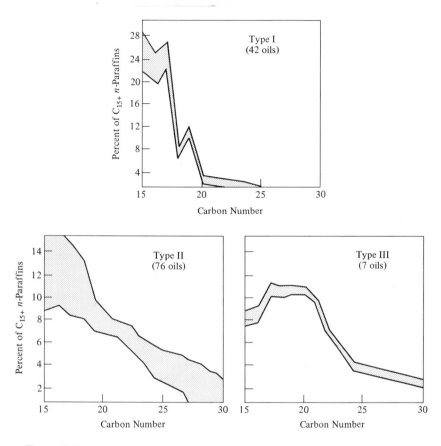

Figure 11-9
Range of C_{15+} n-paraffin distributions in Williston Basin oils. All oil samples of each type occur within shaded areas. [Williams 1974]

hydrocarbon distribution data. He postulated that the reservoir was initially partially filled with Type II oil from nearby source rocks and was later filled with Type I oil through migration along the fault from deeper, older accumulations. The observation of a different oil associated with faulted reservoirs in both the Williston and Tuscaloosa examples tends to confirm the previously mentioned concept that hydrocarbons do intermittently move up some faults to fill overlying reservoirs.

North Sea

Light crude oils and condensates will undergo compositional changes in individual hydrocarbons during the life of the reservoir. Also, they may change during sampling. Erdman and Morris (1974) used ratios of hydrocarbons

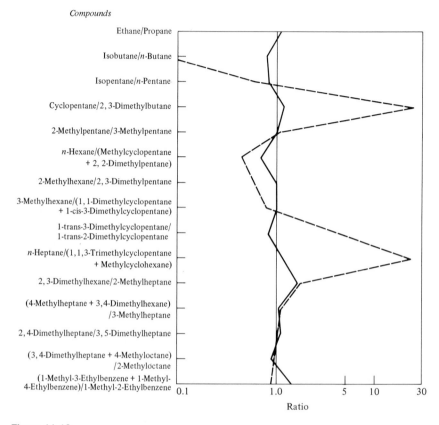

Figure 11-10
Correlation of petroleum based on ratios of hydrocarbons in the C_2–C_{10} range. Solid line, comparison of two North Sea crudes; dashed line, comparison of a North Sea and a Utah, United States, crude. [Erdman and Morris 1974]

for correlation that were similar in chemical structure and boiling point to minimize the effects of these variables. They calculated 16 ratios from the GC analyses of the C_2–C_{10} fraction of crude oils. Figure 11-10 shows a comparison of the ratios of concentrations of one North Sea crude against another and also against a Utah, United States, crude of similar chemical and physical properties. The ratio in Figure 11-10 was calculated from the following equation:

$$\text{Ratio} = \frac{CA_a/CA_b}{CB_a/CB_b}$$

where the numerator is the ratio of concentration of hydrocarbons a and b in Oil A, and the denominator is the ratio of concentrations of a and b in Oil B.

When two oils match perfectly, the ratio (R) = 1. For example, the ratio of cyclopentane to 2,3-dimethylbutane for the two North Sea oils (solid line) is slightly more than 1. The ratio for one North Sea oil and the Utah oil (dashed line) is about 25.

The close correlation between the two North Sea oils having the same source rock is shown by the solid line in the vicinity of the R = 1 line. The Utah crude is considerably different from the North Sea oil, as shown by the dashed-line ratios much higher and lower than 1.

Western Canada Basin

Deroo et al. (1977) correlated over 100 crude oils from the Western Canada Basin into three groups, based on

1. Percentage composition of alkanes plus distillate below 210°C, aromatics, and resins plus asphaltenes

2. Percentage composition of the 210°C+ fraction in terms of normal and branched alkanes, cycloalkanes, and aromatics

3. Distribution of cycloalkanes with one or two rings compared to those with three or more rings

4. Distribution of isoprenoids relative to n-alkanes

5. Thiophenic component distribution

6. Sulfur content.

Group I oils were in Upper and Lower Cretaceous reservoirs; Group II in lowermost Cretaceous, Jurassic, and Mississippian; and Group III in Upper and Middle Devonian. Group II oils are the richest in aromatics, asphaltics, thiophenes, and sulfur and are the poorest in alkanes. Also, the ratio of isoprenoids to n-alkanes is lowest for this group. Group I oils are the richest in alkanes and have the lowest aromatic hydrocarbon and sulfur content of the three groups. Group I can be distinguished from Group III by its lower content of one- and two-ring cycloalkanes and its lower benzothiophene content. Some exceptions to these classifications were noted.

The Western Canada Basin oils also can be correlated in the same three groups based on individual hydrocarbon ratios (Table 11-4). Group I Viking oils have a much higher toluene/benzene ratio (ratio 5 in Table 11-4) than the other groups. Also, Viking oils have fewer 4- to 6-ring polycondensed naphthenes (cycloalkanes) relative to one-ring naphthenes (ratio 3) than the other oils. The Group II Mannville oils can be distinguished from the Group III Devonian oils based on their higher ratio of n-heptane to 1,2-trans-dimethylcyclopentane (ratio 2) and on their lower ratio of methylcyclopentane to methylcyclohexane (ratio 4).

The fact that the ratios correlate by formation indicates that each formation has its own characteristic source rocks that are producing its own unique oil.

Table 11-4
Hydrocarbon Ratios of Oils from the Western Canada Basin

Formation	Field	Hydrocarbon ratios[a]				
		1	2	3	4	5
Group I, Lower Cretaceous						
Viking	Joffre	0.64	3.63	0.56	0.38	4.97
Viking	Joffre	0.68	3.89	0.59	0.37	4.75
Viking	Gilby	0.62	3.42	0.51	0.39	4.76
Viking	Joarcam	0.62	2.44	0.57	0.40	4.7
Group II, Lower Cretaceous (lowermost)						
Mannville	Joffre	0.65	3.88	1.74	0.36	3.54
Mannville	Medicine River	0.71	4.34	1.85	0.38	3.15
Group III, Devonian						
Nisku	Fenn–Big Valley	1.09	0.98	1.12	0.81	2.81
Nisku	Stettler	1.01	0.92	1.27	1.05	2.14
Woodbend	Sturgeon Lake	0.87	0.94	0.68	1.13	2.18
Woodbend	Normandville	0.84	1.20	0.67	0.69	1.84
Beaverhill Lake	Springburn	0.78	1.17	1.03	0.62	2.52
Beaverhill Lake	Snipe Lake	0.77	0.90	0.83	0.90	2.93

[a] 1 = n-heptane/methylcyclohexane; 2 = n-heptane/1,2-trans-dimethylcyclopentane; 3 = sum of polycondensed naphthenes containing four, five, and six rings/one-ring naphthenes; 4 = methylcyclopentane/methylcyclohexane; 5 = toluene/benzene.

Hydrocarbon ratios also are affected by maturation, and in some cases this outweighs the effect of source material. Also, because maturation can occur in the source rock, it is difficult to distinguish whether trends with depth are occurring in the reservoir or from changes in the composition of the oil fed to the reservoir from the source rock over geologic time. Ratio 1 is affected primarily by variations in source material. Ratio 2 decreases in the older formations, because these contain larger quantities of the dimethylcyclopentanes. Ratio 3 is normally expected to decrease with burial, because the polycondensed naphthenes become aromatized, while single-ring naphthenes are formed in greater amounts. In the Paris Basin, Tissot et al. (1971) showed this ratio to decrease appreciably through the depth interval from about 750 to 2,500 m (2,460 to 8,200 ft). This is not the case in the Western Canada Basin, where the highest ratio is in the lowermost Cretaceous Mannville Formation, while the Devonian Beaverhill Lake Formation has ratios higher than the Cretaceous Viking. This is probably due to differences in source material. Ratio 4 is known to decrease with depth of burial in areas such as the Gulf Coast, but in western Canada the higher values are in the deeper Paleozoic formations. The decrease in the toluene/benzene ratio (ratio 5) is mainly caused by an increase in the benzene content of the older, more mature oils.

The ratios of hydrocarbon components in the light gasoline–kerosine range are particularly effective for correlation when combined with carbon isotope data, as previously discussed.

The hydrocarbon ratios given in these examples are only a few of the many used by geochemists in crude oil correlation. Some ratios are strongly source dependent and others temperature dependent, while others, such as 2-methylpentane/3-methylpentane show no large differences with origin or burial. Some biological marker ratios, such as pristane/phytane, increase with temperature to a maximum and then decrease. Changes with temperature (depth) can be overshadowed by large differences in source material. To some extent, the use of hydrocarbon ratios in crude oil correlation is a trial-and-error method where the geochemist utilizes past experience to determine the best ratios for a particular problem. Any two oil samples can be shown to differ in composition if enough parameters are examined. The logical approach is to define the chemical parameters that are most affected by the geological information needed, whether it be the identification of a source rock, or the evaluation of changes with depth.

SOURCE ROCK–CRUDE OIL CORRELATIONS

Oil–source rock correlations are more difficult than oil–oil correlations, because many problems are involved both in sampling and interpreting the data. Removing oil from a fine-grained sediment by conventional extraction loses the hydrocarbons present up into the $C_{12}–C_{15}$ range. To compare this with a reservoir oil sample requires evaporating the latter to constant weight at a temperature such as 45°C (113°F) in order to remove the same range of volatile hydrocarbons. Comparison of the $C_1–C_{15}$ range can be made by low-temperature heating of the source rock and GCMS comparison of the product with the associated reservoir oil.

These comparisons are analytically straightforward, but the interpretations are difficult. A source rock oil is usually not similar in composition to its corresponding reservoir oil for several reasons. First, there is evidence the oil fractionates during the process of leaving the source and migrating to the reservoir accumulation. Second, source rocks do not yield oils of the same composition throughout their generation history. As the rocks are buried to greater depths and higher temperatures, the reactions of higher activation energy are initiated, yielding different products. Third, degradation processes can affect the reservoir oil, as previously discussed, and also may affect a source rock oil by processes such as mineral recrystallization, or uplift and erosion followed by reburial. All of these problems require either that the correlation be made by parameters that are unchanged by the preceding factors or that each factor be studied in sufficient detail to be able to predict the composition of a reservoir oil from its composition in the source rock.

Correlation Problems

Weathering and Subsurface Oxidation

Sometimes it is desirable to compare an oil or seep with a rock outcrop. Weathering greatly reduces both the hydrocarbon and organic carbon content of all types of outcrops. The process is complex, because it depends on the porosity and permeability of the rock, the climate, the extent of surface fracturing, and microbiological activity. Surface weathering is visible to the field geologist as a rind of gray or white on black, organic-rich rocks—or yellow, if iron minerals are present. Organic weathering goes deeper than the visible layer. Leythaeuser (1973) reported that two samples of the Mancos Shale of Utah were weathered to a depth of about 3 m (10 ft) below the visible weathering zone. If outcrops are to be used, the hydrocarbon data must be interpreted with caution. One weathering criterion is the presence or absence of pyrite. If a rock contains pyrite, its disappearance during weathering often coincides with the largest decrease in hydrocarbon and organic carbon content. Consequently, if pyrite is present the organic data can be used with more confidence.

Recrystallization of carbonates, which is known to age spores and pollen, would be expected to alter the source rock oil.

Contamination

Source rock extracts in the C_{15+} range are more easily contaminated than the crude oils obtained for correlation. Side wall cores and cuttings are both subject to contamination from the mud and from handling procedures in the drilling operation. As previously mentioned, cores are usually more heavily contaminated because they are exposed to the drilling mud until the drilling is terminated. Deroo et al. (1977) observed contamination in the C_{18} range in rock extracts of the Western Canada Basin. Mass spectrometric analyses indicated that it was diesel oil. Deroo et al. also observed contamination by iso- and normal alkanes in the C_{20} range of 60 out of 110 cores from the central part of the basin. The exterior of the cores showed far more contamination than the interior. The distribution of naphthenes, aromatics, and asphaltenes was not affected. The source of this contamination was not identified.

Absorption-Migration Effects

In 1958, Brenneman and Smith reported that the oils from source rocks contain more NSO compounds and fewer paraffins than the oils from associated reservoirs. Their data for four oil–source rock pairs is shown in

Table 11-5
Data on Source Rock Extracts and Associated Reservoir Oils

Sample	Paraffin–naphthenes	Aromatics	NSO compounds	Empirical formula
Phosphoria Limestone extract, Wyo.	45	30	25	$C_{30}H_{51}O_{0.8}N_{0.1}S_{0.4}$
Tensleep oil, Wyo.	32	49	19	$C_{30}H_{51}O_{0.1}N_{0.1}S_{0.4}$
Frontier Shale extract, Wyo.	49	31	20	$C_{30}H_{50}O_{0.9}N_{0.2}S_{0.2}$
Wall Creek oil, Wyo.	60	31	9	$C_{30}H_{55}O_{0.15}N_{0.06}S_{0.03}$
Hoxbar Shale extract, Okla.	59	24	17	
Hewitt oil, Okla.	77	21	2	
Woodford Shale extract, Okla.	44	41	15	
Misener oil, Okla.	73	25	2	

Note: Values in weight percent of oil or extract normalized to 100 percent, excluding material retained during chromatography on alumina. Empirical formulas determined on samples prior to chromatography.
Source: Data from Brenneman and Smith 1958.

Table 11-5. These pairs were selected by petroleum geologists familiar with the sample areas. The fine-grained source rocks were taken as near as possible to the oil reservoirs, generally within 12 miles (19.3 km) laterally and 200 ft (61 m) stratigraphically. The crude oils were topped to an initial boiling point of about 200°C (392°F) to correspond to the source rock extracts. The Phosphoria Limestone was the only carbonate source rock, the others being shales.

In all of the shale–sand pairs, the reservoir oils contained more paraffin and fewer NSO compounds than the source rock extracts. In the Phosphoria Limestone–Tensleep Sandstone pair, there were more aromatics and fewer paraffins and NSO compounds in the reservoir oil. This suggests some process of fractionation during desorption or migration from source rock to accumulation. If absorption is primarily on the organic matter, as indicated by Young and McIver (1977), then a difference would be expected between carbonate and shale source beds, because the former contains more marine amorphous organic matter than the latter.

The empirical formulas for extracts and oils indicate that there is strong retention of the oxygenated compounds in the source rock. Retention of the nitrogen and sulfur compounds also occurs in the shale but is barely evident in the carbonate.

A ring analysis of several oil–source rock pairs using the refractive index–density–molecular weight (n–d–M) technique showed more naphthene rings within the source rock hydrocarbons than the reservoir oil

Table 11-6

Naphthene Ring Distribution and Optical Activity for Source Rock Extracts and Associated Reservoir Oils

Sample	Weight percent carbon in naphthene rings[a]	Optical activity, α_D^t
Phosphoria Limestone extract, Wyo.	40	1.17
Tensleep oil, Wyo.	35	0.55
Hoxbar Shale extract, Okla.	42	0.77
Hewitt oil, Okla.	39	0.28
Woodford Shale extract, Okla.	43	0.38
Misener oil, Okla.	35	0.13
Eocene B-7 Shale extract, Venezuela	52	0.72
B-6 sand oil, Venezuela	47	0.28
La Rosa Shale extract, Venezuela	45	1.17
La Rosa sand oil, Venezuela	45	0.49
Ryzdvyanoe Shale extract, USSR	58	–
Achikulak oil, USSR	24	–
Datykh Shale extract, USSR	52	–
Datykh oil, USSR	30	–

[a]Determined by n–d–M method of Waterman. Values for optical activity measured on 345–400°C fraction were all plus.
Source: Data from Brenneman and Smith 1958; Gimplevitch 1960.

hydrocarbons (Table 11-6). This includes data on oil–source rock pairs reported by Gimplevitch (1960). The corresponding optical activity data indicate that the tetra- and pentacyclic triterpanes (Figure 8-18) are two to three times more concentrated in the source rock extracts than the reservoir oils. These polycondensed structures caused more carbon atoms to be in rings in the extracts but the total naphthene content, which includes mostly monocyclic naphthenes is nearly equal or sometimes greater in the reservoir oil. Figure 11-11 shows the hydrocarbon type distribution for two crude oil–shale extract pairs reported by Hunt (1961). One of these pairs also was analyzed by Brenneman and Smith (1958). The reservoir naphthenes are greater than the shale extracts in one case and about the same in the other.

Saturate/aromatic ratios are almost always higher in the reservoir oil than in rock extracts. The shale extracts in Table 11-5 and Figure 11-11 all have a lower saturate/aromatic hydrocarbon ratio than do the corresponding crude oils. Baker (1962) reported saturate/aromatic ratios of 3.1 and 2.1 for the Burbank and Thrall oils, compared to about 1.3 and 0.8 for the corresponding shale extracts. Gimplevitch (1960) also reported a higher saturate/aromatic ratio in the reservoir oils of his Russian oil–source rock pairs. He also found

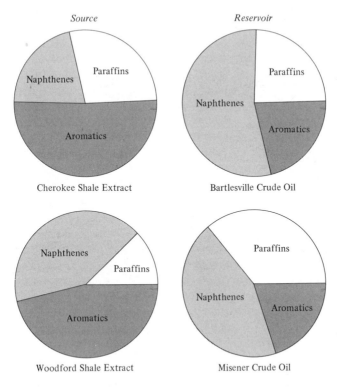

Figure 11-11
Hydrocarbon type distribution as percent of the total in two source
rock extract–crude oil pairs. [Hunt 1961]

perylene, a five-ring condensed aromatic hydrocarbon to be present in the
source rock extracts but not in the oils. Bray and Evans (1965) found that the
extracts from 76 shales ranging from Ordovician to Cretaceous in age from
eight geological provinces in the United States averaged 22 percent paraffins
in the saturated fraction. In contrast, 215 crude oils from the same general
areas averaged 38 percent paraffins in the saturates. No attempt was made to
relate oils to particular shales, but only shales with odd/even n-paraffin (CPI)
values less than 1.2 were analyzed. These would be the more probable source
rocks.

Where does this fractionation of hydrocarbons occur? There is evidence
that it occurs during primary migration from source to reservoir. Baker
(1962), in his detailed study of the Cherokee Group, noted that the largest
organic changes occurred at lithologic boundaries. He did not observe any
progressive change in going from one shale to another. However, in going
from an overlying shale to a sand beneath, there was a distinct decrease in the

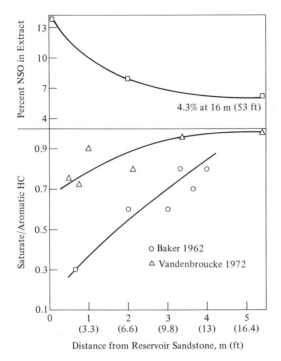

Figure 11-12
Changes in the organic matter of shales approaching reservoir sands. Top: Increase in nitrogen, sulfur, and oxygen content of shale extract. [Vyshemirsky et al. 1974] Bottom: Decrease in saturate/aromatic hydrocarbon ratio.

saturate/aromatic ratio. This is shown in Figure 11-12, along with similar data from Vandenbroucke (1972). In both cases, the most significant shale change is within 3 m (9.8 ft) of the sand. Saturate/aromatic ratios in the sand were 1.8 for Vandenbroucke's data and 3.1 for Baker's data. The largest change occurs at the shale–sand contact as concluded by Baker. However, there is a small progressive change in the shale prior to contact.

Vandenbroucke also reported that the asphaltene content of the shale extracts doubled from background levels within 1 m (3.3 ft) of the reservoir sand. Asphaltenes would be high in NSO compounds. Comparable results were noted by Vyshemirsky et al. (1974), who compared the NSO content of the extracts of Jurassic shales of the West Siberian lowland. The largest increase in NSO compounds occurred in the 2 m (6.6 ft) next to the reservoir sand (Figure 11-12).

The question of hydrocarbon fractionation during secondary and tertiary migration in the reservoir rock is more controversial. Although some geochemists have postulated that chromatographic fractionation occurs during reservoir rock migration, most of the evidence can be explained by other processes, such as degradation by meteoric water or gravity segregation. If chromatographic effects occur by absorption on organic matter (Young and McIver 1977), then the low organic content of most reservoir rocks would preclude any significant changes during secondary or tertiary migration. This may explain the difficulties of investigators such as Kawai and Totani (1971) in finding any clear evidence of changes in crude oils of the Niigata Basin during migration.

Field Examples

Uinta Basin

A well-defined chemical relationship between bitumens in source rocks and in reservoirs was demonstrated in the Uinta Basin from the analysis of several hundred outcrop and subsurface mine samples (Hunt et al. 1954). Four distinctly different bitumens—ozocerite, albertite, gilsonite, and wurtzilite—were correlated with their source rocks, based on similarities in the source rock extracts. The analytical techniques available at the time were mainly infrared analysis, liquid chromatography, refractive index, and elemental analysis. Sometime later, mass spectrometry analyses were carried out, and the results confirmed the earlier conclusions.

The correlations were possible because the bitumens varied widely in composition, and most of them were in close proximity to their source rocks. The largest difference was between the naphthenic–asphaltic bitumen formed from algae remains deposited in the open lacustrine facies of the Green River Formation and the paraffin waxes from land sources deposited in the alluvial shore line facies of the Wasatch Formation. Differences in the paraffin–naphthene, aromatic, and NSO content were previously shown in Table 8-7. Fine-grained rock samples representing over 1,000 ft (305 m) of the Wasatch Formation all contained anisotropic needles of wax in their extracts. These waxes were identical to ozocerite in physical and chemical properties. In contrast, no wax was found in fine-grained rocks representing over 1,000 ft of the Green River Formation. Instead, its extract was high in naphthene and aromatic hydrocarbons and in NSO compounds.

The predictive capability of geochemical research was evident in this early study, which was completed in late 1949 prior to any oil discoveries in the Uinta Basin. At the time, it was predicted in company reports that any oil found in the Wasatch or basal Green River section would be paraffinic, whereas any discovered in the middle and upper Green River would be asphaltic. Subsequent drilling proved this to be true. All the oils produced from the Wasatch were highly paraffinic, and the oil shows encountered in the upper Green River were asphaltic (Bell and Hunt 1963, p. 340).

The Uinta Basin study was carried out on sediments in an early stage of maturation where temperatures were not high enough to form a typical crude oil. Uinta oils have been discovered only in the deeper parts of the basin where the sediments are well into the catagenetic stage. This indicates that correlations can sometimes be made on outcrops that are not at the same stage of maturation as the prospective horizons.

Cherokee Basin

Baker (1962) made a detailed comparison of the extracts from individual lithologic beds of the Cherokee Group in eastern Kansas and Oklahoma. The study was significant in emphasizing that fine-grained rocks of several different lithologies can contribute oil in varying amounts to a reservoir. The Burbank and Thrall producing sands are overlain by a shoreline facies blending into a predominately marine cyclic deposition. A cyclothem of greenish gray, gray, and black shales, plus limestones, underclays, and coal, cover vertical intervals as short as 5 m (16 ft) but have lateral continuity of several kilometers.

The composition of the extracts is shown in Table 11-7. The high hydrocarbon yields relative to organic carbon for the limestone and black shale (column 3) are typical of rocks containing mostly marine amorphous organic matter. Yields from the other samples that have substantial amounts of land-derived material are much lower. How much hydrocarbon each of these lithologies contributed to the reservoirs would be difficult to determine without some detailed comparison of the chemical and isotopic compositions of extracts from each bed. Baker only reported hydrocarbon-type composition, but modern analytical techniques could do much more in interpreting this oil–source rock relationship. It is doubtful if the underclays and greenish gray shales were significant contributors, and the coal will yield mainly paraffin waxes.

In any correlation problem of source rock and crude oil, the maximum core section available next to the reservoir should be analyzed, irrespective of lithology. If the field geologist picks only those samples that appear to be

Table 11-7
Composition of Cherokee Group Rock Extracts

Rock type	Hydrocarbon (HC), ppm	Organic carbon (C_{org}), wt. %	$HC/C_{org} \times 10^{-2}$	Saturate/ aromatic HC
Underclay (9)	19	0.34	1.06	1.28
Greenish-gray shale (43)	31	0.31	1.26	1.64
Limestone (11)	91	0.19	4.72	1.79
Gray shale (37)	129	1.52	0.92	0.62
Black shale (19)	2920	7.94	3.88	1.60
Coal (2)	6900	68.6	1.01	0.28

Note: Number of samples in parentheses. All data are averages. Saturate/aromatic hydrocarbon ratios for Burbank and Thrall oils were 3.1 and 2.1, respectively (Baker 1962).

good source rocks, the ultimate interpretation could be misleading. The black shale in Table 11-7 has a high hydrocarbon content, but it represents less than 5 percent of the lithologic section, and the nearest sample is about 30 m (98 ft) above the producing sands.

Williston Basin

Dow (1974) used lithofacies maps combined with organic geochemistry data and thermal history to outline effective oil source areas for the three types of Williston Basin oils defined by Williams (1974) and discussed earlier in this chapter. In this basin, the source rocks of the three oils are separated by the Devonian Prairie and the Mississippian Charles salts, which prevent mixing except along fault and fracture zones and beyond the limits of the evaporites. The source rocks were defined as those containing adequate oil-generating matter, buried deeper than 7,000 ft, or 2,134 m (5,000 ft, or 1,524 m, below sea level datum). This depth was defined by Nixon (1973) as the minimum required for oil generation in this area.

Figure 11-13 shows the stratigraphic relationships of the three oil systems. Nearly all of the Type I oil production and all of the Winnipeg and Red River oil shows in the basin are found within the limit of the Winnipeg Shale, where it is buried deeper than 5,000 ft below sea level. Type I oil is produced in the Red River Formation and is accumulated in post-Red River reservoirs beyond the depositional edge of thin anhydrites in the upper Red River. No Type I oil occurs above the Devonian Prairie evaporite except in the Richey area, where vertical migration has occurred through fracture zones associated with the Weldon–Brockton fault system.

Williams (1974) compared the extracts of a series of potential source rocks from the section in Figure 11-13 with the three crude oil types, using the techniques previously discussed. Carbon isotope ratios, hydrocarbon type distribution in the C_4–C_7 fraction, and distribution of C_{15+} n-paraffins for the Winnipeg Shale extracts, all correlated with the type I oils. Data for the latter two parameters are shown in Figures 11-14 and 11-15. The characteristic predominance of C_{15} and C_{17} n-paraffins in the Type I oils was also evident in the Winnipeg Shale.

In the Mississippian section, the Lodgepole and Bakken (Madison) shales were possible sources for the Type II oil. For Lodgepole extracts, $\delta^{13}C$ values were around -27 and -28, whereas Bakken extracts and Type II oils were both in the -29 to -30 range. This indicated the Bakken Shale to be the Type II source. Other similarities between Bakken extracts and Type II oils are shown in Figures 11-14 and 11-15.

The Bakken Shale averages about 60 ft thick and contains over 10 weight percent organic carbon near its depocenter east of the Nesson anticline. It contains mostly marine amorphous kerogen with a high oil yield, as indicated in Figure 11-13. Production from fractured Bakken Shale reservoirs is

Period	Formation			Estimated Amount of Oil Expelled 10^6 bbl	Major Oil System
Cretaceous					
Jurassic and Triassic	Spearfish		Salt		
Permian and Pennsylvanian	Tyler		Source rock Type III oil	300	Tyler System
Mississippian	Charles		Salt		
	Mission Canyon				Bakken–Madison System
	Lodgepole				
	Bakken		Source rock Type II oil	10,000	
Devonian	Prairie		Salt		
	Winnipegosis				Winnipeg–Red River System
Silurian					
Ordovician	Red River		Source rock Type I oil	600	
Cambrian	Winnipeg				
Precambrian					

Figure 11-13
Schematic columnar section in Williston Basin, showing probable location of source rocks of three oil types, and showing evaporite barriers. [Dow 1974]

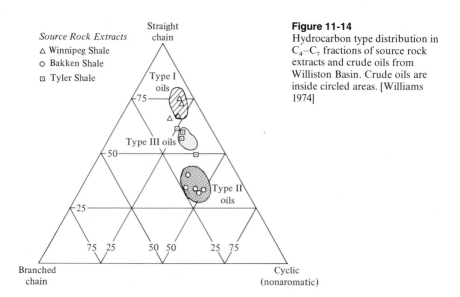

Source Rock Extracts
△ Winnipeg Shale
○ Bakken Shale
▣ Tyler Shale

Straight chain

Type I oils

Type III oils

Type II oils

Branched chain

Cyclic (nonaromatic)

Figure 11-14
Hydrocarbon type distribution in C_4–C_7 fractions of source rock extracts and crude oils from Williston Basin. Crude oils are inside circled areas. [Williams 1974]

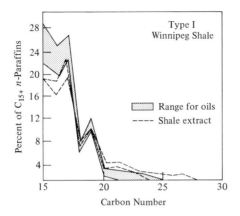

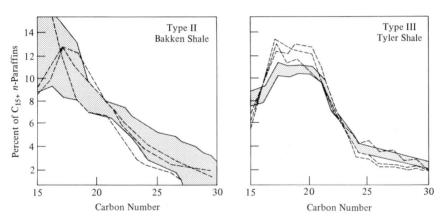

Figure 11-15
Comparison of C_{15+} n-paraffin distribution of source rock extracts and crude oils from
Williston Basin. [Williams 1974]

restricted to the area within the 5,000 ft (1,524 m) subsea structure contour.
Some Type II oil in underlying Devonian carbonates (Figure 11-13) occurs
where the uppermost Devonian reservoirs are brought into contact with
Bakken source rocks by faults. No Type II Bakken (Madison) oil is known to
penetrate the Charles evaporites. Post-Madison production occurs only
beyond the depositional edge of these salts in areas associated with the
regional post-Mississippian unconformity. Dow (1974) believes (see Chapter
6) that oil from Bakken source rocks migrated vertically along fracture zones
on the Nesson anticline, filling Mississippian Mission Canyon and Charles
reservoirs to closure and then spilling northward up the axis of the
south-plunging structure (Figure 6-35). Some oil was trapped along porosity
pinchouts in the Madison, but most of it accumulated where truncated

porosity zones were sealed by Jurassic red shale. Dow concluded that migration along the unconformity extended nearly 100 miles (161 km) beyond the limits of effective Bakken source rocks.

The similarity of Pennsylvanian Tyler Shale extracts and Type III oils is evident in Figures 11-14 and 11-15. The only other potential post-Madison source rock was the Heath Shale. Its C_4–C_7 hydrocarbons have a high percentage of cyclics, placing it in the lower right corner of Figure 11-14, out of the range of the Type III oils.

Dow pointed out that most of the Tyler Shale in the basin is oxidized to red, orange, and varicolored sediments. Type III oil production is limited to a small area of unoxidized Tyler Shale below the 5,000 ft (1,524 m) subsea contour in southwestern North Dakota.

This source rock–crude oil study was able to separate the oils of the Williston Basin into three genetic types, relate each type to a specific source sequence, estimate the oil generated by the sources, and map the lateral limits of the effective source rocks. It is a good example of the kind of information that can be obtained by combining geological, geophysical, and geochemical data.

Southern Mid-Continent, United States

The Arbuckle and Ellenburger groups are a series of Cambro-Ordovician carbonate formations that have produced oil and gas for over 50 years in Kansas, Oklahoma, and Texas. The formations are underlain by volcanics and igneous rocks so the only possible source rocks are either within the Arbuckle and Ellenburger groups or in the overlying Pennsylvanian shales. Cardwell (1977) compared oil samples from eight reservoirs in four fields with extracts of several Arbuckle rocks in terms of gross composition of the C_{15+} fraction, distribution of C_4–C_7 hydrocarbons, and C_{15+} n-paraffin distributions. The extracts were distinctly different than the oils. The differences were greater than would be expected from absorption-migration effects. Cardwell (1977) also compared the Arbuckle oils with some younger Pennsylvanian oils and found them to be remarkably similar in terms of gross composition, C_4–C_7 hydrocarbon distributions, and eight hydrocarbon ratios from among those (Figure 11-10) used by Erdman and Morris (1974). Cardwell (1977) concluded that the Arbuckle and Pennsylvanian oils had the same source—probably Pennsylvanian shales.

Alaska to Germany

Welte et al. (1975) examined 20 sediment–crude oil pairs, some of which had a known genetic relationship, while others were unknown. They concentrated on biological marker compounds as correlation parameters. This included tetracyclic and pentacyclic steroid and triterpenoid hydrocarbons, plus pristane, phytane, and norpristane. In addition, they looked at more conventional parameters, such as carbon isotope ratios and distribution of the C_{15+} n-paraffins. For 6 oil–source rock pairs, there was a good degree of correlation, indicating a true genetic relationship. For 10 pairs, the correlation was moderate, although still indicating a genetic association, and 4 pairs were considered unrelated. Petrographic examination of the source rocks showed those with the best correlation to contain the most bitumenite that fluoresced under UV light.

The highest degree of correlation resulted from the GCMS analysis of the C_{27+} cyclics, the steroids and triterpenoids. From mass spectra plots of key ions, Welte et al. (1975) determined that the following compounds with the molecular ion values shown in parentheses were most useful: cholestane (372), ergostane (386), adiantane (398), sitostane (400), hopane–friedelane (412), and possibly lanostane (414). Correlations were best where the reservoir rock was also the source rock, or where the source rock was in close proximity to the reservoir rock. The carbon isotope data and distribution of C_{15+} n-paraffins did not show as good a correlation. The isoprenoid compound ratios involving pristane, norpristane, and phytane appeared to have little value for correlating these samples.

Subsequently, Leythaeuser et al. (1977) extended this study to additional crude oil–source rock pairs, including an oil field in southern Germany that had four possible source beds extending from Middle Jurassic to Oligocene in age. The compositional data indicated the Jurassic to be the source. The authors emphasized, however, that such conclusions should be supported by additional correlation parameters chemically independent of the C_{27+} cyclics.

Seifert (1977) was able to differentiate between crude oils producing from three formations in the McKittrick area of the San Joaquin Basin of California on the basis of sterane and terpane content plus other parameters, such as carbon isotopes and the sum of branched paraffins over n-paraffins in the C_{14}–C_{30} range. He also compared presumed source rock–crude oil pairs from several areas in terms of the cis- and trans-stereoisomers of the steranes. Later, Seifert (1978) pyrolyzed previously extracted shale samples and was able to show correlations based on the steranes and terpanes cracked from the kerogen matrix. He heated 10 to 300 g of shale in a sodium chloride bath at 2–$3°C/min$, up to $550°C$. The saturates of the trapped pyrolyzate were analyzed by GCMS. The release of steranes and terpanes started around $375°C$, peaked at about $500°C$, and was completed around $550°C$. Seifert compared crude oil from the Bluebell (Eocene) field in Utah with the extract of associated lower Green River shales and the pyrolyzate of the same shales after they had been extracted. The GCMS pattern of all three were generally similar in the detailed terpane profiles and in having low quantities of steranes. Seifert did observe differences in hydrocarbon ratios in the pyrolyzate compared to the bitumen extract of the same source rock, such as a higher ratio of moretanes to hopanes in the pyrolyzate. All biomarker steranes and terpanes in the bitumen however, were also in the pyrolyzate, so this technique can be used for correlation as long as the procedure is standardized. Tricyclic diterpanes and the $5\alpha/5\beta$ stereochemistry of the common steranes are largely preserved in the pyrolyzate. Seifert used ratios such as hopane/moretane and $17\alpha(H)/17\beta(H)$-trisnorhopane to distinguish pyrolyzates of shales of different maturation stages.

The correlation of kerogens of different rocks by pyrolysis–GCMS appears promising as a rapid definitive technique for source rock–crude oil correlations. Although the sterane and terpane fingerprint patterns of the source rock pyrolyzates will contain products not in the oils, it should be possible to distinguish consistent differences between oils and rock extracts that can be recognized in standard pairs and applied to unknown pairs.

Petrov et al. (in press) used pentacyclic triterpanes of the $17\alpha H$ series to recognize genetic differences in oils of a given region. They also noted that ratios of the stereoisomers of steranes and triterpanes can serve as indicators of the aging of oils with increasing time and temperature.

All of these studies are significant in showing the potential of using MS spectra in selective molecular weight ranges for correlation purposes. Similar data on such structures as porphyrins might also be useful for correlation.

Akita Basin, Japan

Taguchi (1975) compared the extracts of 250 Tertiary shale and mudstone samples with 19 Miocene crude oils to identify the source rocks of the oils. CPI values for the Miocene Funakawa Formation were 2 to 4 and for the Miocene Onnagawa Formation, 0.8 to 1.5. This indicated that samples of Funakawa age and younger were too immature to generate economic accumulations of oil. Nickel/vanadium ratios of the oils were in about the same range as the Miocene extracts of both formations, but outside of the range of the Pliocene extracts. Taguchi also noted that there were no five-ring polycyclic aromatic hydrocarbons in the 19 oils from different fields and producing horizons. Polycyclic aromatics were present in the Funakawa and younger formations in concentrations about 30 times that of the Onnagawa and older formations. He concluded that the Onnagawa and possibly the lower part of the Funakawa Formation were the source rocks for the Tertiary Japanese oils.

Cretaceous Oils of Central Gulf Coast

The most critical problem in crude oil–source rock correlation is trying to evaluate how the oil changes when it leaves the source rock and accumulates in the reservoir. The fact that the saturate fraction preferentially accumulates in the reservoir, leaving the aromatics and NSO compounds preferentially accumulated in the source rock, implies that some kind of adsorption or absorption phenomena is taking place, because aromatic–NSO compounds are more polar. Young and McIver (1977) measured the absorbing tendencies of hydrocarbons in the source rocks and in the oil. They calculated the distribution equilibrium of individual hydrocarbons between source and reservoir based on the absorbing tendencies. From this, it was possible to calculate the composition of a hypothetical oil that would be released by a specific source rock and to compare it with the composition of the real oil. Their results showed that it was possible to predict the composition of crude oils genetically related to source rocks, particularly in the older, more deeply buried rocks. Predictions in some young and shallow oil–source rock pairs were less satisfactory. In some of these young examples, it was believed that the reservoirs were filled by oils migrating from older and deeper horizons. It was also suspected that distribution equilibrium between hydrocarbons in the reservoir and source rock is not reached except in the older and deeper sediments. This is a reasonable postulate, because a reservoir oil is constantly changing in composition as long as hydrocarbons are being fed to it from source rocks.

An example of the application of this technique is shown in Figure 11-16 for an oil–source rock pair from the Lower Tuscaloosa Cretaceous formation previously discussed. Ratios of C_{15+} hydrocarbons from a shale extract and its associated reservoir oil (Conerly) are plotted. The ratio of normal plus branched paraffins (P) plus naphthenes (N) to aromatics (Ar) is plotted on the ordinate, and the ratios of paraffins to aromatics (P/Ar) and naphthenes to aromatics (N/Ar) on the abscissa. These ratios were measured by column chromatography of the C_{15+} fractions of the oil and the shale extract. The ratio of $(P+N)/Ar$ for the crude oil is about 4, and for the hydrocarbons

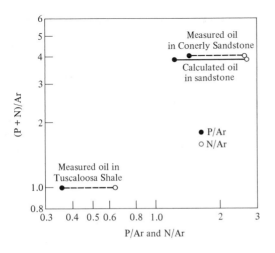

Figure 11-16
Comparison of measured ratios of C_{15+} paraffin (P), naphthene (N), and aromatic (Ar) hydrocarbons in Tuscaloosa Conerly oil with same ratios in oil extracted from Tuscaloosa Shale. Also shown are ratios calculated for a reservoir oil in distribution equilibrium with the source rock oil, taking into consideration the absorbing tendencies of the hydrocarbons. [Young and McIver 1977]

extracted from the shale it is about 1 (Figure 11-16). Again, this shows the higher concentration of aromatics in source rocks compared to reservoir oils (see Figure 11-11). This higher concentration is because of the greater tendency of the aromatics to be absorbed by the source rock. Young and McIver (1977) were able to calculate the composition (P, N, and Ar hydrocarbons) in the oil leaving the source rock from their measurements of the absorbing tendencies of this particular Tuscaloosa shale for paraffin, naphthene, and aromatic hydrocarbons. The absorption measurements were actually made on light gasoline hydrocarbons in the C_5–C_7 range, but it was assumed that the absorbing tendencies of the C_5–C_7 paraffin, naphthene, and aromatic hydrocarbons would be approximately proportional to the absorbing tendencies of the corresponding C_{15+} heavier homologs. The calculated compositions resulted in the ratios also shown in Figure 11-16. It is evident that the calculated and measured values are very similar. This indicates that the reservoir oil is genetically related to the source rock oil, the differences being caused by absorption on the shale organic matter.

Koons et al. (1974) noted that in the two types of Tuscaloosa oils the second type was formed from a higher percentage of marine-derived organic matter. This is evident in Figure 11-7, where the Type II oils have a higher ^{13}C

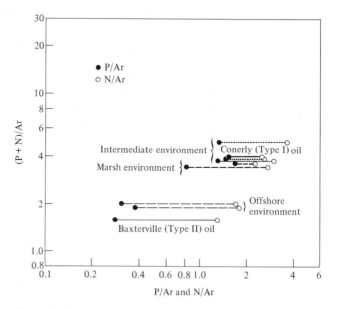

Figure 11-17
Comparison of measured ratios of C_{15+} hydrocarbons in Tuscaloosa Conerly (Type I) oil and Tuscaloosa Baxterville (Type II) oil, with calculated ratios of oils formed from seven Tuscaloosa shales from three depositional environments. [Young and McIver 1977]

content and a lower *n*-paraffin (wax) content. Comparison of the hydrocarbon type ratios in the C_{15+} fraction by the method of Young and McIver leads to the same conclusion. Their ratios of paraffins and naphthenes to aromatics for the Conerly oil (Type I) are compared with calculated ratios of hypothetical oils in distribution equilibrium with fine-grained Tuscaloosa sediment samples from three different environments (Figure 11-17). Comparison indicates that the Conerly oil (I) is genetically related to the marsh environment source rock.

A similar calculation for a Baxterville oil (Type II) also is shown in Figure 11-17. In this case, the ratios of the oils calculated to come from the offshore environment most closely resemble the Baxterville oil. The conclusion from Figure 11-17 is that the Type II oils were formed from a higher proportion of marine source materials than the Type I. Thus, the calculations based on hydrocarbon absorption by the sediments agree with the chemical data of Koons et al.

The interpretation of crude oil–source rock correlations is difficult because of the unknown degree of fractionation in hydrocarbon components when the oil leaves the source and migrates to the reservoir. Young and McIver (1977) have shown that good correlations can be made with parameters that change such as the saturate/aromatic ratio if the effect of absorption on the organic matter of the source rock is taken into account. In some respects, this is a more powerful correlation technique than the others, because it recognizes the degree of fractionation that occurs when the source rock releases the hydrocarbons to the reservoir. This means that the measurement of hydrocarbon contents and absorbing tendencies of potential source rocks in wildcat wells will enable the detailed composition of a crude oil to be predicted prior to its discovery.

SUMMARY

1. Useful crude oil correlation analytical techniques include the Hempel distillation of the U.S. Bureau of Mines; gas chromatography; column chromatography; mass spectrometry, including isotope and high resolution techniques; nuclear magnetic resonance spectrometry; and ultraviolet and visible absorption spectrometry and polarimetry.

2. Gross properties such as gravity and asphaltene content are not useful for crude oil correlation, because the original genetic composition of an oil changes by processes such as gravity segregation, water washing, biodegradation, maturation, and deasphalting.

3. Parameters that have been successfully used in crude oil correlation include the correlation index (U.S. Bureau of Mines); C_4–C_7 and C_{15+}

hydrocarbon type distributions; C_2–C_{10} hydrocarbon ratios; distribution patterns of *n*-paraffins, steranes, and triterpanes and their aromatized derivatives; ^{13}C content of oil fractions; optical rotation; and vanadium/nickel ratios.

4. Crude oil–source rock correlations are affected by changes in the composition of oil as it leaves the source rock, by changes in reservoir composition as new oil is added over geologic time, and by biological and chemical alteration of reservoir oil.

5. Rock outcrops and subsurface samples near unconformities may be used for source rock correlation if there is adequate evidence, such as the presence of pyrite, that the organic matter has not been appreciably altered by weathering.

6. Sampling for source rock–crude oil correlation should include all fine-grained rocks, regardless of lithology, in close proximity to the coarse-grained formations in the drainage area leading to the accumulation.

7. When crude oil leaves the source rock, the saturate hydrocarbons are preferentially released, while the aromatic and NSO compounds are preferentially retained.

8. The distribution equilibrium of hydrocarbons between source and reservoir rocks may be calculated from data on the absorption of those hydrocarbons on the organic matter of the source rocks. From the equilibrium data, it is possible to predict the composition of the oils formed by each of the potential source rocks. Comparison of these calculated oils with the known oils enables a correlation to be made on parameters such as the saturate/aromatic ratio, because it corrects for the degree of fractionation when the oil leaves the source rock.

12

Prospect Evaluation

Petroleum exploration in the United States is carried out by companies that historically have been classified into two groups, the independents and the majors. The majors are defined as companies having a total refining capacity in excess of 175,000 bbl per day and a crude oil self-sufficiency greater than 30 percent. All other exploratory well operators are considered to be independents. The number of majors and independents are listed in Table 12-1, along with the number of wildcat wells they drilled from 1969 through 1974 (Jackson 1977).

During this period, the success ratio for the majors was about twice that of the independents, but, because the latter drilled eight times as many wells, they discovered 75 percent of the new fields or about 52 percent of the total new reserves. The majors found larger fields, but this was because they drilled more deep wells and more offshore wells where large structures are still untested.

About 59 percent of the total new-field wildcat discoveries were gas, which reflects the deeper drilling. As drilling depth increases linearly, the cost increases exponentially, so gas discoveries inevitably cost more than oil discoveries.

The fact that only 1 wildcat well in 50 makes a significant new discovery means that we still need a better understanding of why a structure or trap is barren. Geochemistry can help improve this understanding by providing information on dry holes that will assist in leading to new discoveries. This chapter includes a review of the kinds of information that should be obtained when drilling a wildcat well, as well as some case histories and a discussion of the role of geochemistry in future exploration.

Table 12-1
Exploratory Successes in United States, 1969 through 1974

	16 Major oil companies	5,819 Independent oil companies
Wildcat wells drilled	3,565	28,634
Significant oil and gas discoveries (1 × 10^6 bbl recoverable oil or 6 × 10^9 ft^3 of natural gas)	177	537
Success ratio	5 %	1.9 %
Average depth, ft (m)	8,986 (2,739)	5,625 (1,715)
New reserves in 10^9 bbl of oil equivalent (6,000 ft^3 gas = 1 bbl oil)	2.6	2.9

ANALYSIS AND INTERPRETATION OF GEOCHEMICAL DATA

Many major oil companies have geochemical laboratories equipped to carry out the source rock analyses and crude oil correlation techniques discussed in this book. Other companies may prefer to hire a geochemical consultant and farm out their analyses. Any company that has several geologists on its staff should either have a consultant on a continuing basis or one or more geochemists on its permanent staff. Geologists and geophysicists frequently have an interest but have limited knowledge of geochemistry, so they may misinterpret geochemical data. Also, the number and types of analyses to be run can best be determined by a geochemist working in a close, continuous relationship with the geologists. Organizationally, this is most effective if a geochemical section is formed within the geological division.

The more effective geochemists are usually those with either an undergraduate or graduate degree in chemistry, with a major in one of the earth sciences. It is important to have a strong background in chemistry, because this is the discipline usually lacking in an exploration department.

Geochemical techniques discussed in the preceding chapters are listed in Table 12-2. These include techniques for source rock analysis, crude oil correlation, and seep analysis. Only a few of these analyses need to be run to answer any one question by the geologist. Small companies may obtain the most benefits by setting up only two or three of the most frequently used techniques while farming out the others.

Every geological problem, whether it be evaluation of source sediments in a new area or correlation of crudes in an old area, requires a thorough understanding of the geological variables involved in the problem. A geologist and geochemist working together can best determine the number of samples and the types of analyses that should be done.

Table 12-2
Summary of Geochemical Techniques

1. Gas logging (mud logging)
2. Organic carbon
3. Pyrolysis–FID (Flame Ionization Detector)
4. Pyrolysis–Gas Chromatography–Mass Spectrometry (P–GCMS)
5. Pyrolysis–fluorescence
6. C_1–C_7 hydrocarbon extraction
   a. C_4–C_7 gas chromatography analysis
7. C_{15+} hydrocarbon extraction
   a. Carbon Preference Index (CPI), Odd–Even Predominance (OEP)
   b. n-paraffin distribution
   c. Saturate–aromatic distribution
   d. Mass spectrometry on fractions (polycondensed naphthenes, aromatics, and so forth)
   e. Optical activity on fractions
   f. Carbon isotopes on fractions
8. Vitrinite reflectance
9. Kerogen color
10. Elemental analysis of kerogen and solid bitumens
11. C_4–C_{10} gas chromatography analysis of oil
12. GCMS analysis of oil (terpane distributions, C_{27+} hydrocarbons, and so forth)
13. Hempel analysis
14. Carbon, oxygen, hydrogen, and sulfur isotopes of gases and oil fractions
15. Vanadium/nickel ratios

Rank Wildcat Areas

Reconnaissance

Geological and geophysical surveys can outline structures, potential traps, and areas of good source–reservoir facies. They also can delineate subsurface faults and fractures that may influence hydrocarbon migration. These kinds of information are essential in order to make effective use of any geochemical studies.

Petroleum geochemistry is not a remote sensing technique. It is simply a hydrocarbon survey, similar to the lithologic survey of the geologist and the structural survey of the geophysicist. The difference is that the geochemistry defines the subsurface zones and areas of probable hydrocarbon generation and of hydrocarbon buildup.

The three important factors in a hydrocarbon survey are (1) the quantity of organic matter—organic richness, (2) the quality of organic matter—oil or gas source, and (3) the maturation level—immature, mature, or metamorphosed. These three important parameters can be determined from only two analyses in Table 12-2; namely, pyrolysis–FID (No. 3) and vitrinite reflectance (No. 8). If these two analyses were run on unweathered outcrop samples of all available formations, they would provide an indication of the relative hydrocarbon variations that could be present in the subsurface. The vitrinite reflectance would be useful in indicating how deep the samples were buried in the geologic past. Both the organic richness and type could be estimated from the pyrolysis. The latter also indicates the level of maturation, which may be checked against the vitrinite readings. These two methods are suggested for a new laboratory being set up for geochemical surveys, because they give the maximum amount of data for the time and effort involved. If a laboratory has already accumulated a background of data and experience with other methods, it will naturally continue in this direction. New methods must be introduced on a comparative basis; otherwise, the results cannot be integrated with earlier data.

Organic carbon analyses represent the least expensive way of monitoring organic richness throughout the reconnaissance area. Organic carbon is frequently run on all available samples as a guide to selecting a more limited number for pyrolysis–vitrinite studies. Another advantage of these three techniques over others in Table 12-2 is that they can be run on old cuttings samples from test wells that may have been drilled in the reconnaissance area decades ago.

Sometimes geophysical shot holes drilled in a reconnaissance area offer a means for obtaining samples below the weathered layer.

If some formations in the reconnaissance area produce oil in other provinces, a detailed hydrocarbon analysis should be made on the rock extracts (Method 7, Table 12-2). A comparison of the extracts in the reconnaissance area with the oils and extracts of the productive province will show similarities or differences in comparing source facies.

Any solid bitumen seeps can be characterized by elemental analysis, extraction and pyrolysis. Oil seeps will probably be biodegraded, so techniques such as gas chromatography–mass spectrometry and vanadium/nickel ratios are in order. Carbon and hydrogen isotopes should be determined on the individual hydrocarbon components of gas seeps to evaluate sources and migratory pathways.

First Well

Prior to drilling, a decision should be made, based on the outcrop analyses, as to which subsurface formation should be sampled in greater detail. A mud logger is advisable for the first well in rank wildcat areas, to detemine if the formations are alive or dead with respect to hydrocarbons. If oil is used in the drilling mud, steam or CO_2 is required to drive the gases out of the mud trap. Mud logging gives an instant well-site reading but the analysis of canned cuttings gives more detail. Cuttings should be kept wet at the well site, canned under water and analyzed by pyrolysis–FID for light hydrocarbons (Method 3) or by the C_1–C_7 extraction and GC analysis (Method 6). The more detailed analyses such as those under the C_{15+} extraction (Method 7 in Table 12-2) can be run on selected samples, depending on the results of the initial survey. Vitrinite reflectance should be run on the cuttings, because it gives not only maturation data but also the paleothermal history of the sediments.

The fluids on all drill-stem tests should be analyzed, water, gas, or oil, even if the latter are present in only trace amounts. Such analyses may indicate that a preexisting oil field was destroyed by microbial degradation, especially if the water is of low salinity.

A temperature log should be obtained on all rank wildcats. The importance of temperature in hydrocarbon generation has been emphasized many times. A comparison of the geothermal gradient data with vitrinite reflectance and any information on high-pressure zones is most useful in interpreting possible changes in source potential through geologic time.

If oil is discovered, it should be analyzed by some of the methods 11 through 15 in Table 12-2. Source rock–crude oil correlations should be carried out, as discussed in Chapter 11. Also, comparison of the oil with local seeps and with oils from nearby provinces would be useful.

An example of the application of isotopes to gas origin problems is in the Williston Basin, as discussed in Chapter 11. Dow (1974) concluded that the Cretaceous sediments in this basin were too immature to generate oil or gas, yet gas is produced from the Cretaceous Eagle Sandstone in Montana. Rice (1975) showed that this gas has a $^{13}C_{PDB}$ value of $-70\%_0$ indicating it to be biogenetic in origin.

Any discoveries related to fault–fracture systems or unconformities must recognize that the source beds may be some distance away, either vertically or horizontally, from the reservoir accumulation. Source–reservoir facies are frequently, but not always, in close juxtaposition.

Developed Areas

Geochemical techniques are important even in old areas where major source–reservoir facies have been worked out. Too often, the existence of production in some formations encourages a prejudice against the others. When the first well was drilled by a major oil company over the Scurry Reef of Texas, it had been predetermined that the deeper Ellenburger Limestone was the only horizon worth testing. The rotary drill mudded off the 2 billion bbls of oil in the Scurry Reef, and the test in the Ellenburger was dry, so the well was abandoned. Almost any hydrocarbon monitoring would have spotted the Scurry production.

In many sedimentary basins, there are formations and areas that have not been adequately tested. Source facies can vary widely in the same formation in different parts of a basin, because of different geothermal histories and sources of organic matter. Oil accumulations may be destroyed by meteoric water infiltration or seepage in one part of a basin and retained in another. Geochemical surveys of well samples are relatively inexpensive compared to the cost of drilling the well, and they can provide a wealth of information on the subsurface distribution of hydrocarbons. In all these studies, there must be a close relationship between the geologist, who thoroughly understands the area, and the geochemist, who understands the implications of the chemical analyses.

In developed areas, it is generally necessary to use more of the analytical techniques in Table 12-2 because crude oil correlation and source rock–oil correlation is more useful where oil and gas samples are available from several horizons. At the same time, it is possible to limit the analyses to those that can best answer specific questions posed by the geologist rather than making a routine survey of the entire well, as is typical with a rank wildcat.

Case Histories

Douala Basin, Cameroon

A good example of the application of some of these techniques is the detailed study of the Douala Basin carried out by Albrecht et al. (1976). This basin is located southeast of the present Niger River Delta. The Upper Cretaceous shales are 70 percent clay minerals interbedded with sandy lenses of limited lateral extension. Depositional rates were high and the kerogen is almost entirely the herbaceous, woody type derived from the continent. Production from associated Douala reservoirs is mostly gas, as would be expected, because most of the kerogen has an H/C ratio less than 1. Data on the hydrocarbon extracts and the kerogen composition for the gas source rocks are shown in Figure 12-1. Most of the hydrocarbons are generated in the depth range between about 1,500 and 2,600 m (4,921 and 8,530 ft), equivalent to a vitrinite reflectance R_o value between 0.7 and 1.5 percent. During this interval, the odd/even ratio of the normal paraffins CPI (carbon preference index) drops from around 2 to 1. The hydrogen content of the kerogen decreases from about 5 to 3.5 and the H/C ratio decreases from 0.8 to 0.5. No more hydrocarbons are being generated after the hydrogen content falls below about 3.4 percent. Laboratory thermal evolution experiments performed on some of the samples confirm the maturation changes observed in

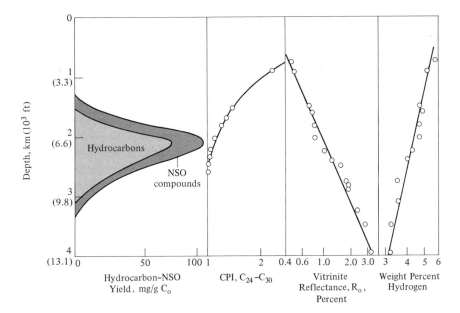

Figure 12-1
Geochemical data on the Upper Cretaceous Logbaba shales of the Douala Basin, Cameroon: yield of hydrocarbon and nitrogen, sulfur, and oxygen (NSO) compounds; carbon preference index (CPI) of n-paraffins; vitrinite reflectance; weight percent hydrogen in kerogen. [Data from Albrecht et al. 1976; Durand and Espitalié 1976]

the natural environment. All of the data shown in Figure 12-1 can be obtained on well cuttings. This type of information can be most valuable in delineating the oil and gas source rocks and the barren zones in wildcat wells drilled in new areas.

Coastal Offshore Stratigraphic Tests (COST Wells)

A stratigraphic test to 11,000 ft (3,353 m) was drilled on the northeast flank of Cortes Bank on the southern California Outer Continental Shelf in 1975. The objective was to acquire information on the stratigraphy, structure, and geochemistry of this unexplored area of the southern California borderland. The geothermal gradient in the well was calculated from wire-line log data as averaging 1.8°F/100 ft (3.2°C/100 m) to about 10,000 ft (3,048 m) and slightly higher at greater depths. The true geothermal gradient is probably somewhat higher than the measured value, because circulation times did not exceed two hours prior to logging (OCS-CAL 75-70 # 1—Paul et al. 1976).

The mud log showed only methane gas in low amounts. No liquid oil shows were reported in the cores, side wall samples, or cuttings. Solid hydrocarbon chips found in the cuttings at 4,020 ft (1,225 m) were identified as gilsonite used in the casing cement at 1,549 ft (472 m). This was knocked loose while setting a deeper casing string at 3,891 ft (1,186 m). Solid hydrocarbon chips were also encountered in a 6-inch sandy siltstone at the top of a graded turbidite sandstone bed at 8,734 ft (2,662 m). It appeared to be a pyrobitumen, which may have been recycled with the turbidite material.

Ten core samples representing the depth interval from 3,307 to 10,920 ft (1,008 to 3,328 m) were analyzed for organic carbon, C_{15+} hydrocarbon extraction, and by pyrolysis–FID–items 2, 3, and 7 in Table 12-1 (Claypool et al. 1976). The three critical questions concerning organic richness, organic quality (oil, gas, or nonsource), and maturation level were answered from these three analyses.

Part of the data is shown in Figures 12-2, 12-3, and 12-4. The organic carbon (Figure 12-2a) was high in the Oligocene and Upper Eocene but very low in all deeper sections. As these are clastic sediments with appreciable land input, it is probable that part of the organic carbon is recycled and of no value as a petroleum source. This is verified by the pyrolysis data (Figure 12-2c,d), which show both the free and generated hydrocarbons to be low at depths greater than 5,000 ft (1,524 m). The only rocks capable of generating economic accumulations of oil are in the intervals shallower than 5,000 feet. This section, however, is too immature. The HC/non-HC ratio (Figure 12-2b) is less than 1 above 5,000 ft, which indicates immaturity (compare to Table 10-4). Also the weight of C_{15+} HC is less than 1 percent of the organic carbon (Figure 12-3) for all samples above 7,000 ft (2,134 m). This chart (compare Figure 7-1) usually indicates immaturity for oil-generating samples plotted on the lower left side. It does not apply to gas generation, and some immature oil shales plot in the mature range above 1 percent, but it is useful as a rough indicator. Further verification of immaturity is the fact that the pyrolysis peak in the upper section is around 475–490°C (887–914°F), which Claypool et al. (1976) interpreted as being in the immature range, by their technique. The pyrolysis peaks in this COST well showed a general increase to 520°C (968°F) at 10,920 ft (3,328 m). The transition from immature to mature facies occurred in the 8,000–9,000 ft (2,438–2,743 m) range, based on the pyrolysis peak.

Gas chromatograms of the saturate hydrocarbon extractions shown in Figure 12-4 also verify that the shallow samples are too immature. Figure 12-4a has the typical bimodal distribution pattern of an immature biogenetic hydrocarbon assemblage. The steranes and triterpanes are dominant in the second mode. For comparison, see Figure 7-22 at 1,500 m (4,921 ft) of immature, shallow Douala Basin samples and Figure 7-27 of the immature Irati oil shale. Figure 12-4a also shows pristane and phytane plus other isoprenoids to be dominant.

In Figure 12-4b, there is too high an odd/even ratio (CPI–OEP) in the C_{25}–C_{32} range for this to qualify as a mature source rock. The sample in

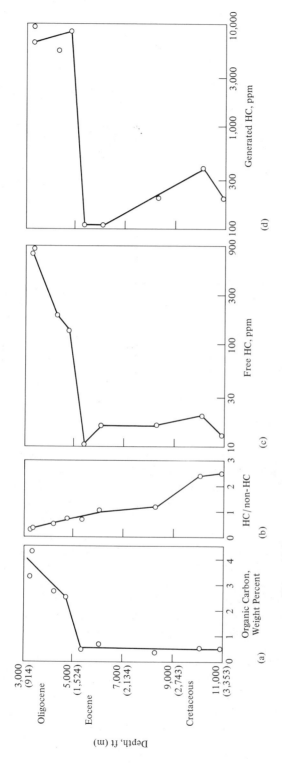

Figure 12-2
Variation in hydrocarbon and organic carbon contents with depth in COST well OCS-CAL 75-70 No. 1 off California. [Claypool et al. 1976]

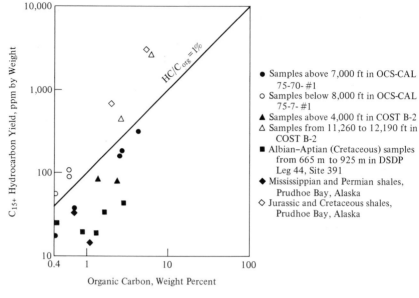

Figure 12-3

Plot of hydrocarbon yield vs. organic carbon. Samples plotting below diagonal line are usually immature or are continental deposits, including coals. Some immature oil shales plot above the line, along with typical oil source rocks. [Data from Claypool et al. 1976; Scholle 1977; Deroo et al. 1978; Morgridge and Smith 1972]

Figure 12-4c is well into the mature range, as indicated by the relatively uniform odd–even, *n*-paraffin peaks.

The conclusion from this study is that the Oligocene–Upper Eocene rocks are organic rich but immature, while the section below about 7,000 ft (2,134 m) is mature but organic lean. Significant discoveries of oil in this general location are unlikely. Gas is possible in the deeper sections, but the low pyrolysis yield and the low yields on the mud logs make even gas a poor prospect. Further evaluation of the deeper, mature section could be made by microscopic examination of the kerogen and CO_2 yield on pyrolysis to determine proportions of land-derived and recycled organic matter in the rocks. Vitrinite reflectance would give the most information on recycled material.

This study does indicate that future exploration should move to where the Oligocene and Upper Eocene are more deeply buried. The richness of these rocks indicates that they will be prolific petroleum producers when they reach the maturation stage.

This study also shows that significant geochemical interpretations can be made from relatively few samples. As previously discussed, it is important to take samples every 50 to 100 ft (15 to 30 m) in a wildcat well with additional sampling across important lithologic boundaries. Only 30 to 40 samples need to be run in a 10,000 ft (3,048 m) well, however, to obtain an initial picture of

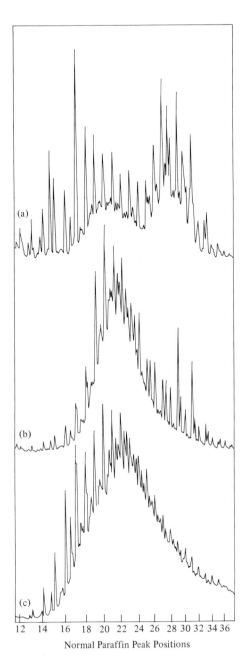

12 14   16   18 20 22   24 26 28 30 32 34 36

Normal Paraffin Peak Positions

Figure 12-4
Gas chromatograms of saturated
hydrocarbons extracted from OCS-
CAL 75-70 No. 1 core samples:
(a) 4,778–4,784 ft (1,456–1,458 m),
(b) 6,008–6,014 ft (1,831–1,833 m),
(c) 10,696–10,702 ft (3,260–3,262 m).
[Claypool et al. 1976]

the changes with depth. Additional analyses can be done as needed when the
major geochemical breaks are defined. Thus, in this study, it would be useful

to analyze additional samples in the critical 6,000–9,000 ft (1,829–2,743 m) range where the change from immaturity to maturity takes place.

Another stratigraphic test (COST B-2) was drilled to 4,890 m (16,043 ft) on the U.S. Atlantic Outer Continental Shelf (Scholle 1977). The present-day geothermal gradient is 2.3°C/100 m (1.3°F/100 ft) with a bottom hole temperature of about 135°C (275°F). This is an interesting case study, because data were obtained on several parameters by several laboratories.

The oldest sediments drilled were Early Cretaceous, which on the Connan chart (Figure 4-17) at 135°C (275°F) would predict the hole is well into the oil-generating range. This assumes there has been a uniform continuous rate of deposition, which is not necessarily the case. The bottom hole vitrinite reflectance values around 0.75 percent support the concept that bottom temperatures are well into peak oil generation. A plot of hydrocarbon yield versus organic carbon indicates samples above 4,000 ft (1,219 m) to be immature and below 11,000 ft (3,353 m) mature (Figure 12-3).

Table 12-3 contains additional geochemical data. The kerogen is the amorphous oil-forming type above 6,000 ft and the gas-forming herbaceous, woody, coaly type at all depths below 6,000 ft. This is verified by the H/C ratio, which is above 1 in the upper section but falls below 1 in the 5,000–6,000 ft (1,524–1,829 m) interval, and below 0.6 in the bottom 2,000 ft (610 m). The yield of C_4–C_7 hydrocarbons is negligible in the 0–9,000 and 14,000–16,000 ft (4,267–4,877 m) intervals but is relatively high in the 9,000–14,000 ft (2,743–4,267 m) interval. Several samples in this range also were high in C_{15+} hydrocarbons, but the detailed gas chromatography analysis indicated a high percentage of biogenic rather than thermally formed hydrocarbons. Also, the gas chromatographic analysis of the C_4–C_7 hydro-

Table 12-3

Geochemical Data on COST B-2 Well Baltimore Canyon, East Coast, United States

Depth, ft (m)	Kerogen Type	TAI[a]	H/C ratio	Yield of C_4–C_7 HC, ppm	Pyrolysis–FID maximum yield temperature, °C
0–6,000 (0–1,829)	20–70 % amorphous	1 to 2	1 to 1.4	60	460
6,000–9,000 (1,829–2,743)	30–50 % herbaceous	2 to 3	0.6 to 0.95	100	470
9,000–14,000 (2,743–4,267)	30–60 % woody, coaly			1,000	480
14,000–16,000 (4,267–4,877)	80 % woody, coaly 20 % herbaceous	2½ to 3	Below 0.6	150	485

[a]Thermal alteration index: 1 = immature, 2 = mature (oils), 3 = condensate, gas.
Source: Data from Scholle 1977.

carbons showed a high percentage of cyclics, which is typical of immaturity. The maximum yield temperature by pyrolysis–FID never exceeded 488°C (910°F), which is only on the verge of the mature zone, by the Claypool et al. (1976) technique. Interpretation of maximum pyrolysis temperatures, however, depend on the procedure. The Espitalié et al. (1977) technique uses 400–435°C (742–815°F) as the maximum yield temperature for the immature zone, 435–460°C (815–860°F) for the mature oil zone, and above 460° C (860° F) for mature gas.

Other data cited as indications of immaturity were a CPI of 1.5 at 12,000 ft (3,658 m) and abundant pristane in the sample. Pristane, however, is more an indicator of land source material than of maturation (Table 7-5).

Considering all the geochemical and maturity data in the COST B-2 report (Scholle 1977), it appears that the section above 9,000 ft (2,743 m) is too immature (by all parameters) to generate commercial accumulations of oil. There could be biogenic accumulations of methane, because none of these techniques can monitor the generation of biogenic gas.

Below 9,000 ft to a total depth of 16,000 ft (2,743 to 4,877 m) more than 80 percent of the organic matter is land derived, with an H/C ratio that is too low for generating significant quantities of oil. It could generate gas or condensate.

The critical question remaining is where will gas or condensate be generated? The conclusions of the COST B-2 report were that this would be deeper than the well depth. This was based on the assumption that peak generation would have been in the 11,000–16,000 ft (3,353–4,877 m) range, and condensate and gas would peak successively deeper.

The general concept of condensate peaking below oil was discussed in Chapter 7, but there are exceptions. Snowdon (1977) has pointed out that tree resins yield naphthenic condensates at vitrinite reflectance levels (R_o) of 0.45 percent. He believes that both gas and condensate may be forming at low R_o and high CPI values indicative of early maturity levels in the Sverdrup Basin of Canada. Tree resins have not been studied in detail as a source of petroleum hydrocarbons, but they are known to yield diterpenes that will form cyclic (naphthenic and aromatic) hydrocarbons in the gasoline–kerosine range on maturation. Trees produce resins in the heartwood, bark, and leaves (Thomas 1969), and the diterpane derivatives are common in extracts of lignite and sub-bituminous coal. Tree resins could be the source of the cyclic light hydrocarbons found in land-derived organic matter in the 9,000–14,000 ft (2,743–4,267 m) interval of the B-2 well.

Just how early a recoverable oil in commercial quantities can form has not yet been satisfactorily answered. Kerosines, condensates, and some waxy oils have been encountered in drill-stem tests in many areas where associated shales have CPI values above 1.3 and R_o values as low as 0.4 percent, such as Cameroon (Africa), the McKenzie Delta (Canada), and the Arctic Isles.

Considerable research still needs to be done to define the immature–mature boundary with different types of both marine- and land-derived organic matter. Until more is known about this, it will have to be concluded that the B-2 well below 9,000 ft has the potential for gas or condensate, assuming there is enough organic matter. In the interval between 14,500 and 16,000 ft (4,420 and 4,877 m), the organic carbon content is generally too low for any hydrocarbon potential (average less than 0.2 percent).

Western Atlantic Outer Continental Margin

The Deep Sea Drilling Project of the National Science Foundation drilled Site 388 on the lower continental rise off Cape Hatteras and Sites 389 through 391 in the Blake-Bahama Basin. These sites were of considerable interest as future areas for offshore drilling. Four oil company laboratories, three academic institutions, the U.S. Geological Survey, and the Institut Français du Pétrole participated in geochemical studies of the samples. All source rock techniques, Nos. 2 through 10 in Table 12-2 were used. Four samples from Site 388A to a depth of 332 m (1,089 ft) and about 15 from Sites 391A and 391C to a depth of 1,393 m (4,570 ft) were analyzed by more than one laboratory. The other sites had too little sample available for analysis because of poor recoveries. Details of the geochemical studies are published in Volume 44 of the *Initial Reports of the Deep Sea Drilling Project*, edited by Benson and Sheridan (1978).

Gas was detected in very few cores during drilling. The concentration of C_2–C_5 hydrocarbons in the total gas was about 1/100th of the amounts found in other areas such as the Black Sea and the Angola and Cape basins off southern West Africa.

The major conclusions of the various geochemical investigations were similar with some minor differences. The quantity of organic matter in the Miocene and Cretaceous is adequate to generate oil or gas, except in the Lower Cretaceous section below about 1,200 m (3,937 ft). The quality of the organic matter at Site 391 is largely land derived, of the type that will generate primarily gas. The maturation level at Site 391 is immature throughout the section drilled, although different types of data indicated that the deepest samples were near the threshold of the oil-generating zone.

Figure 12-5 shows some of the geochemical data for Site 391. There is a bulge in the organic carbon content, peaking in the Late Albian. Visual kerogen inspection indicates the Miocene contains 75 percent or more of amorphous marine organic matter, the remainder being humic material from the continent. The amorphous material decreases in the Cretaceous samples with increasing concentrations of humic and recycled material. Dow (1978) identified two populations of recycled vitrinite in the Cretaceous and Jurassic.

The immaturity of the sediments was clearly evident from the kerogen color, pyrolysis data, the GCMS analysis of the extracts, the HC/non-HC

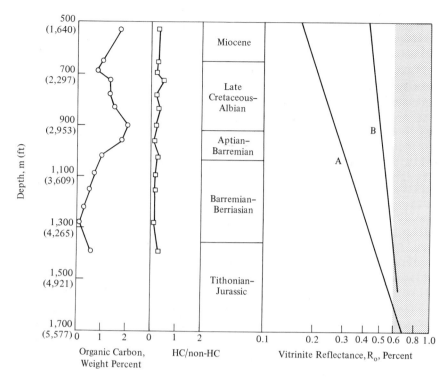

Figure 12-5
Geochemical data obtained on cores from DSDP Leg 44, Site 391. [Vitrinite line A from Cardoso et al. 1978, and line B from Dow 1978]

ratio, and vitrinite reflectance. The latter data are shown in Figure 12-5. The HC/non-HC ratio is comparable to young, Recent sediments throughout the section analyzed (compare with Figure 12-2 and Table 10-4). The vitrinite reflectance data indicated that the oil generation range was being approached in the 1,400–1,600 m (4,593–5,249 ft) depth interval. The discrepancy here is due to the fact that there were only six to eight primary vitrinite data points for constructing these lines. This is inadequate, considering that a slight change in slope results in a large change in the depth interpretation. The kerogen color was yellow with a maturation index of 1+ in all samples from both Sites 388 and 391 except for 391C-52-2 at a depth of 1,393 m (4,570 ft). This kerogen was orange with an index of 2+, indicating it is entering the catagenetic oil-generating stage.

These maturation data were further verified by pyrolysis analyses by two separate laboratories, both of whom concluded from the temperature of the maximum hydrocarbon generation peak that the samples were immature at least to 959 m (3,146 ft), which was the deepest sample pyrolyzed.

Other indicators of immaturity included an abundance of isoprenoids, steranes, triterpanes, odd n-paraffins, and free base chlorins. Thermal stress usually converts chlorins to metalloporphyrins.

Analyses for the quality of the organic matter indicated that if these sediments reach maturation, they will generate primarily gas with only trace amounts of oil. The pyrolysis experiments yielded mostly gas with very little C_{15+} hydrocarbons. The pyrolysis–fluorescence values, which can range up to several thousand units in excellent source rocks, had values from 0 to 3 in the Cretaceous section and up to 12 in the Miocene. This indicates the Cretaceous rocks could not generate oil even if they were mature.

The C_{15+} hydrocarbon yields and organic carbon contents are plotted on Figure 12-3 for Albian—Aptian (Cretaceous) samples (Deroo et al. 1978). All of these are in the immature zone below the diagonal line. The H/C and O/C ratios were calculated from the elemental analyses for carbon, hydrogen, and oxygen (Deroo et al. 1978) in the kerogen and plotted on a kerogen maturation diagram (Figure 12-6). The base of this is from Figure 7-47, which shows the maturation curves for liptinite, vitrinite, and inertinite along with the approximate positions of the vitrinite reflectance levels. The data for eight Site 391 samples in the depth range from 665 m to 959 m (2,182 to 3,146 ft) are plotted on Figure 12-6. Also shown, for comparison, are samples of the kerogen of Silurian–Devonian source rocks from the Sahara in Libya. These are liptinitic in origin and show the typical drop in H/C ratio that occurs with oil and gas generation. In contrast, the Leg 44, Site 391, samples are vitrinitic in origin and are only on the verge of the hydrocarbon-generating phase. The H/C atomic ratio of the Site 391 samples ranges between 0.55 and 0.84, which is in the gas generation range for kerogen.

Information on source quantity and quality also is available from the analysis of free C_4–C_7 hydrocarbons formed in the sediments. In Figure 12-7, the yield of these hydrocarbons in nanograms per gram of organic carbon is compared with similar data from the Black Sea, Angola Basin, and Cape Basin. All of the samples in this figure are in the immature diagenetic stage, yet they will generate small amounts of C_4–C_7 hydrocarbons if they have the potential for forming oil. The data indicate that the samples from Sites 361, 364, and 380 have about 100 times the oil-forming potential of the samples at Site 391 (Hunt and Whelan 1978b).

The limited data obtained at Site 388 indicated that the Miocene claystones have an adequate amount of the amorphous type of organic matter to form oil but they are very immature.

In the Leg 44 investigation, several different laboratories, using a variety of geochemical techniques, arrived at essentially the same conclusions; namely, that the Cretaceous of this part of the western Atlantic has the capability of generating gas with little or no oil, if it is found in a higher temperature regime. It should be emphasized that this conclusion is based on very few samples, largely from one site. Depositional environments can change over

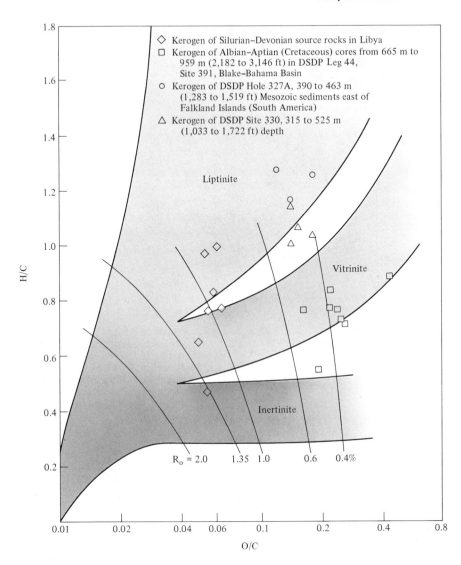

Figure 12-6
Maturation pathways of the dominant kerogen types in terms of their H/C–O/C ratios. R_o is the approximate location of vitrinite reflectance values. [Data from Espitalié et al. 1977; Deroo et al. 1978; Comer and Littlejohn 1976]

relatively short distances, so more analyses at additional sites are needed to verify this initial conclusion.

Samples from DSDP Leg 36, Hole 327A, and Site 330 east of the Falkland Isles (South America) were analyzed by Comer and Littlejohn (1976) for organic matter content and maturity. The kerogen was visually described as amorphous and mixed type. A relatively high palynomorph translucency and

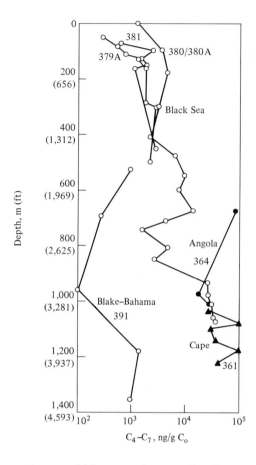

Figure 12-7
Concentration of free C_4–C_7 hydrocarbons in DSDP samples from Sites 379A, 380/380A, and 381 in the Black Sea, 361 in the Cape Basin, 364 in Angola Basin, and 391 in the Blake–Bahama Basin. [Hunt and Whelan 1978b]

a strong odd/even carbon predominance in the extracted hydrocarbons suggested immaturity. The H/C and O/C ratios obtained from elemental analysis of the kerogen are plotted in Figure 12-6. The data support the immaturity conclusion, because most points are in the vitrinite reflectance range of $R_o = 0.4$ to 0.5 percent. The samples from both Hole 327A and Site 330 have a greater oil-forming potential than do the Cretaceous cores of Site 391 in the Blake–Bahama Basin. The kerogen of Hole 327A is particularly rich in liptinite and plots close to the sapropelic kerogens of the Toarcian shales in the Paris Basin. As such, it suggests good prospects for oil if found deeper in the basin.

Prudhoe Bay, Barrow Arch Area, Alaska

The Prudhoe Bay Field on the eastern Arctic slope of Alaska is considered the largest field on the North American continent, with recoverable reserves of about 20 billion bbl of oil and 35 trillion ft³ of gas. Oil is produced from the

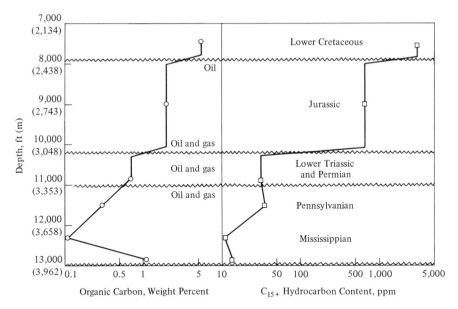

Figure 12-8
Variation in organic carbon and C_{15+} hydrocarbons in Mississippian through Lower Cretaceous sediments of Prudhoe Bay, Alaska.

Kuparuk River Formation just below the Lower Cretaceous–Jurassic unconformity. Both oil and gas are produced from several formations extending from the Pennsylvanian to Lower Jurassic.

Two geochemical analyses were run on the Prudhoe Bay sediments in order to determine organic richness and oil-generating potential (Morgridge and Smith 1972); these were weight percent organic carbon and C_{15+} hydrocarbon content respectively. These data are plotted in Figures 12-3 and 12-8. The organic carbon in the basal Mississippian is high enough to generate oil or gas, but the very low C_{15+} yields indicate it cannot yield oil, and probably not much gas. The Mississippian red bed sequence at about 12,400 ft (3,780 m) is low in both organic carbon and hydrocarbons. The Pennsylvanian carbonates and basal Permian shale have too low a C_{15+} content to be an oil source, but their organic carbon content is high enough to qualify as a gas source. The best source rocks in the area based on these two parameters are clearly the Lower Cretaceous and Jurassic sediments. The organic carbon contents are well above the worldwide mean of about 1 percent, and the C_{15+} hydrocarbons are in the "good to excellent" source rock category, as defined in Table 7-1. In Figure 12-3, both the Cretaceous and Jurassic samples plot well into the mature oil-generating range.

The geological study indicates that the Lower Cretaceous marine shale is the probable source of Prudhoe Bay oil, because it is the only shale in contact with all of the major oil reservoirs in the field. Each of the separate reservoirs shown in Figure 12-8 is truncated at the eastern updip edge by the unconformity at the base of the Cretaceous and is in contact with the overlying Cretaceous marine shale. However, Young et al. (1977) list an age of 87 × 10⁶ years for oil in a Prudhoe Bay Cretaceous reservoir and an age of 218 × 10⁶ years for oil in an early Triassic reservoir. It appears that there are at least two ages of oil sources.

The gas in the deeper sections would be expected since both source and reservoir maturation are greater at the greater depths. The geothermal gradient in this area is about 2.4°F /100 ft (4.4°C/100 m), which indicates a reservoir temperature of about 212°F (100°C) for the oil at 8,000 ft (2,438 m). The deeper pay horizon starting at about 10,000 ft (3,048 m) would have temperatures in excess of 260°F (127°C). Plotting these temperatures on the diagram in Figure 4-17 shows the 8,000 ft (2,438 m) reservoir in the oil zone and the deeper reservoirs in the zone where the oil phases out to gas. Vitrinite reflectance measurements in this well would provide a more accurate reading of the depth changeover from oil to gas. This technique is particularly useful in areas such as Prudhoe Bay that have many unconformities. Construction of a burial history diagram showing temperature changes for each formation over geologic time (Lopatin 1971, 1976) in the Barrow Arch area would clarify the relationship of the Prudhoe Bay field to other potential oil and gas accumulations.

THE ROLE OF GEOCHEMISTRY IN FUTURE EXPLORATION

Geochemistry is the last of the three disciplines—geology, geophysics, and geochemistry—that are beginning to be an integral part of all exploration programs. The slowness in utilizing geochemical concepts has been partly caused by the fact that the science itself is relatively young. Although research in subsurface petroleum geochemistry has been going on for about 30 years, it is only in the last decade that geochemical techniques have been successfully applied to exploration problems on a widespread scale. During this period of gestation, geochemists not only had to show what their techniques could do but they also had to convince the traditionally conservative operating managers that these techniques really paid off in finding new fields and avoiding barren areas.

Geochemistry also was handicapped in some companies because it started in the research laboratory instead of in the field, and there often was a lack of rapport between the research geochemist and the operating geologist. To some extent, the required use of maturation techniques has bridged this gap. The paleontologist and petrographer who studies kerogen maturation can more easily be exchanged with field personnel than can the geochemist. The companies that utilize geochemistry most successfully have geochemical seminars to train new geologists and geophysicists, and they rotate geochemists and geologists between operations and the laboratory.

The sophistication of geochemical techniques requires that many of them be done in the laboratory. It is essential that the laboratory does not become an ivory tower but, rather, an open line of communications and contact with operating problems.

New and Deeper Pay Horizons

Dry holes are caused by a variety of circumstances, including sometimes simply a near miss. Among the stories that used to circulate in the oil companies is the one about the driller who was trucking a rig to the drilling site, which had been carefully picked out by a thorough evaluation of all the geophysical and geological data. The truck broke down two miles from the site, but the driller set up his rig anyway and commenced drilling. A new field was discovered, and its limits did not extend to the original drill site.

Pools have been discovered by geologists reinterpreting old data and opening and retesting dry holes. They have also been discovered by persistent testing during drilling. The Louden field of Illinois, which sustained the old Carter Oil Company through the pre-World War II period, was discovered because the driller kept baling for three days and finally produced some oil. The geochemist can help in the decision as to whether to drill deeper or to try again by emphasizing when the hydrocarbon indicators are positive and when they are negative.

There is still an enormous potential for new oil and gas pools on the continents of the world if the political, economic, and environmental problems can be reconciled. In the United States, where anyone with sufficient capital can go out and drill a well, there were about 23,000 oil and gas fields discovered as of January 1, 1968, in a sedimentary basin area of about 6.5 million km^2 or 2.5 million miles^2 (Halbouty 1970). This calculates to around 3,500 fields per million km^2 (9,200 fields/10^6 miles^2). In contrast, the USSR had 1,118 fields in a sedimentary basin area equivalent to 14 million km^2 (5.4 $\times$ 10^6 miles^2) which is about 80 fields per million km^2 (207 fields/10^6 miles^2). If the USSR were drilled up as much as the United States, it could conceivably have 40 times its number of present fields. Similar calculations can be made for the rest of the world. Clearly, the geochemist can play an important role in minimizing the number of dry holes drilled in order to find these new fields.

Geochemical monitoring of maturation levels is particularly important in deep wells. In 1975, 413 wells were drilled deeper than 15,000 ft (4,572 m) in the United States. About 130 of these went below 18,000 ft (5,486 m). This was only 6 and 2 percent respectively of the total 7,026 wildcats drilled. The potential for hydrocarbons is high in these intervals, because the temperatures are usually well into the oil- and gas-generating range. Oil would be found in the areas of low geothermal gradients and gas in areas of high gradients. Wilson (1973) has outlined several areas in the continental United States and Alaska favorable to deep drilling. The continental shelf of Alaska alone is about one-fourth the size of the conterminous United States, and much of it has petroleum potential.

More deep wells are needed, but there is the hazard that deep drilling can go beyond the greatest depths of hydrocarbon generation, accumulation, and retention. Although methane is thermally indestructible at sedimentary basin temperatures, it can be lost through diffusion, migration, or chemical reaction, as discussed in Chapter 8. Any well drilled beyond 20,000 ft (6,096 m) should continuously monitor the organic maturation very closely. The cost of this is trivial compared to the cost of deep drilling. At present, no economic accumulation of hydrocarbons is known where the kerogen of the associated fine-grained rocks has a vitrinite reflectance above 3.5 and a hydrogen content below 3 percent. Such a field may some day be found, but the odds are against it. The geochemist's job is to keep the well geologist advised as to whether the odds are getting better or worse as the drill goes deeper.

Giants in the Earth

Some subsurface conditions are such that huge source rock drainage areas contribute hydrocarbons to a large, favorably situated, tightly sealed trap, and a giant is formed. To discover a giant is the dream of all wildcatters, but only a few areas in the United States, such as the overthrust belt along the western Wyoming border, northern Alaska, and the U.S. offshore probably still contain giants.

Few laypersons realize the enormous effect a few giants have on the economics of oil. Only 8 fields contain 56 percent of all proved recoverable oil in the world outside the Soviet Bloc (Halbouty 1970). Six of these fields are in the Middle East. In the USSR in 1961, only 8 percent of the oil fields held 80 percent of Russia's proved oil reserves. In the United States, only 46 of 23,000 fields contain 36 percent of the oil reserves. Gas is similarly distributed. One field, Panhandle–Hugoton, will produce about 10 percent of the gas ultimately recovered in the United States. Eleven Canadian giants produced 58 percent of Canada's gas in 1965.

Halbouty el al. (1970) defined a giant oil field as having more than 500×10^6 bbl of recoverable oil with present technology and a giant gas field as having more than 3.5 trillion ft^3 of recoverable gas. As of 1970, there were 187 giant oil fields and 79 giant gas fields. The five sedimentary basins containing the most giants were (1) Arabian Platform and Iranian Basin, 56; (2) West Siberian Basin, USSR, 29; (3) Sirte Basin, Libya, 13; (4) Volga–Ural Basin, USSR, 10; and (5) Gulf Coast Basin, United States, 10. It is a quirk of nature that the United States, which leads the world in petroleum technology and in the number of wells drilled, has only one basin in the top five with the most giants.

Some basins with potential giant fields have barely been explored. The Campeche area in southern Mexico may contain more oil and gas than the entire Middle East, based on early drilling results along the Yucatan Barrier Reef. Oil columns over a kilometer in height and fields over 200 km^2 in area have been discovered to date, and more than 200 seismic structures remain to be tested.

All giants have in common large traps with great vertical thickness or wide lateral extent or both. Structure is not the only criterion, however, as is evident from the dry Destin anticline off Florida and the small hydrocarbon accumulations in the giant Porcupine dome of central Montana. The presence of source rocks and the time of maximum generation and migration of hydrocarbons relative to the availability of porosity and impervious caps are equally important. Petroleum geochemists could make a significant contribution to their profession by studying the whole process of origin, migration, and accumulation for a few of these giant fields. There may be some geochemical parameters unique to the giants. Even the origin of the oil in that giant of giants, Ghawar (Saudi Arabia), is not clearly defined. Young et al. (1977) calculate the age of the Ghawar oil as 166 million years, Middle to Upper Jurassic (see Figure 8-26). Although the Jubaila Limestone (Jurassic) is a shallow-water facies in central Saudi Arabia, it becomes marine in the east, where the basal part is a black argillaceous limestone with varying amounts of black calcareous shale (Steineke et al. 1958). Dunnington (1967b) mentions extensive Upper Jurassic and Lower and Middle Cretaceous shales and shaley limestones deposited in euxinic basins, but there are no published geochemical studies on these presumed source rocks. To what temperatures were these beds buried and is the volume of potential source rocks enough to explain the observed accumulations? A critical question is: How far to the northeast in the Persian Gulf synclinorium does the drainage area of the source beds extend? If there is continuity of permeability through this Jurassic carbonate sequence, the source bed drainage area may extend more than 100 miles (161 km) into the folded belt of the Zagros Mountains. Dickinson (1974) has postulated that partial subduction of the Arabian plate in its collision with the Eurasian plate has expedited the migration of oil along rifted-margin sediments from source beds beneath the Zagros fold–thrust belt. Certainly the high temperatures associated with the subduction zone would cause maximum generation of hydrocarbons. The question as to the continuity of more than 100 miles of permeability could be clarified by source rock– crude oil comparisons extending from the folded belt out to the platform area.

Halbouty et al. (1970) outlined the geological factors concerning the world's giant oil and gas fields. Similar information is needed on the geochemical factors.

The Continental Shelves and Beyond

Known world reserves of oil, natural gas liquids, and gas equivalent to oil, on January 1, l975, were 150 billion MT, or 1,125 billion bbl (McCaslin 1975). Undiscovered resources on the continents and continental shelves are estimated to be 350 billion MT (2,625 billion bbl). At least half of this undiscovered petroleum is believed to be on the continental shelves. Undiscovered resources are really an educated guess at the quantity of hydrocarbons remaining in the earth that are recoverable by present technology. It does not represent oil that we can count on, because much of it will be difficult to find, and some of it may never be discovered. Politics, economics, and environmental restrictions are three factors that could considerably reduce the quantity of world resources that will eventually become proved reserves.

Even when the economics and other factors are favorable, there can be a time lag of several decades between initial discoveries in a basin and the recognition of its true potential. The first commercial field in the Middle East was proven in 1908. During World War II, 35 years later, published estimates of Middle East oil reserves were between 5 and 10 billion bbl (0.7 and 1.3 billion MT). In 1978, 70 years after the first discovery, the oil and gas reserves were estimated to exceed 450 billion bbl, or 60 billion MT (as oil). New basins are proven up more rapidly today, but it will still probably be another 20 years or more before we know the full potential of areas such as the North Sea, or the Colville Basin of Alaska, or the Gulf of Campeche.

The estimated world resources of undiscovered petroleum do not include oceanic areas beyond the continental shelves. Table 12-4 lists the organic carbon content of these deep ocean sediments as summarized from the *Initial Reports of the Deep Sea Drilling Project* (Emery 1975). Abyssal plains can be written off as having too low an organic carbon content and too thin a sedimentary section to develop the temperatures required for generating petroleum. The continental slopes, which extend to depths of 1,500 to 3,500 m (4,921 to 11,483 ft) may be suitable source beds but they appear to contain few layers of coarse-grained reservoir beds required to accumulate the oil.

Table 12-4
Organic Carbon Content of Deep Sea Drilling Project Samples

	No. of holes	No. of samples	Average wt.% organic carbon
Continental slopes	7	239	0.50
Continental rises	18	637	0.60
Closed basins	1	58	0.58
Deep ocean floor	158	4,612	0.19

Source: Emery 1975.

Coarse-grained sediments carried seaward by turbidity currents go beyond the slope to the continental rise, usually via submarine canyons. The rises are gently sloping topographic features that extend from the bottom of the slopes to depths of 3,500 to 5,500 m (11,483 to 18,044 ft). Emery and Skinner (1977) estimated the total area of continental rise and deep marginal basin sediments to be 62 × 10⁶ km² (24 × 10⁶ miles²), which is equal to the combined areas of continental shelves and slopes. The rises are at least 5 km (16,400 ft) thick off eastern North and South America and western Africa. Emery estimates that the rises contain more than half the world's total volume of sediments. They must be considered potential major sources of oil and gas, because at least the upper sections contain coarse-grained turbidite layers interbedded with fine-grained pelagic sediments. This combination of source and reservoir facies in thick sediments with adequate organic matter is highly prospective. Emery and Skinner (1977) estimate the average geothermal gradient of continental rise sediments to be 3.5°C/100 m (1.9°F/100 ft). This means that at a sediment depth of 2,500 m (8,202 ft) on the rise the temperature would be about 90°C (194°F), approaching the oil generating stage. Several rise sediments, such as those in the Gulf of Mexico, contain salt domes, which can create traps and also act as a conduit for heat from deep-seated sources.

The rises occur where the continental margin is within a crustal plate in areas of divergence of sea floor plates. The deep marginal basins are in areas of convergence of sea floor crustal plates with continental plates. The marginal basins of economic interest are inshore of and shallower than the deep trenches along the underthrust belt.

It is impossible to predict the oil and gas potential of the rises and deep marginal basins because no one has drilled deeply enough in these areas to encounter the hydrocarbon-generating zone. The technology to drill and complete wells in 10,000 ft (3,048 m) of water is essentially developed. In 1976, a well was completed in 3,461 ft (1,055 m) of water off Thailand. When that first deep well is drilled on a continental rise or deep marginal basin, it will be most important that geochemists continuously monitor the hydrocarbons in the cores and cuttings by the best techniques available.

SUMMARY

1. The three important factors in a hydrocarbon survey are (1) the quantity of organic matter—organic richness, (2) the quality of organic matter—oil or gas source, and (3) the maturation level—immature, mature, or metamorphosed. They can be determined from organic carbon analysis, pyrolysis–FID, and vitrinite reflectance.

2. In rank wildcat areas, these three parameters should be determined on unweathered outcrop samples, well cuttings of former dry holes, and selected cores or cuttings from nearby provinces with similar formations. Also, any bitumen samples or oil or gas seeps should be analyzed and correlated with the nearest productive province.

3. The first well in a wildcat area should have a mud logger capable of reading shale gas, or well cuttings should be taken at intervals of 15–30 m (50–100 ft) and canned at the well site. The initial hydrocarbon survey can be made on a small selection from this group. A temperature log should be obtained and fluids from drill-stem tests should be analyzed for traces of hydrocarbons.

4. If oil is discovered, correlation with extracts of the presumed source rocks and correlation with oils from nearby provinces is important.

5. In developed areas, the shallower and deeper horizons from known pay zones as well as extensions into areas of complex tectonics should have geochemical monitoring. Variations in source facies, geothermal gradients, and tectonic history can cause wide differences in hydrocarbon prospects for different parts of a basin.

6. Significant geochemical interpretations can be made from analyzing relatively few samples in a wildcat well as long as major geochemical changes are clearly defined.

7. Published case histories of geochemical surveys involving many different tests by several independent laboratories have shown good agreement in interpretations.

8. The successful application of geochemical techniques to petroleum exploration requires that the geochemist and operating geologist work in close cooperation.

9. Geochemical monitoring of maturation levels is particularly important in wells drilled deeper than 15,000 ft (4,572 m), because the drill can go beyond the maximum depth of hydrocarbon generation, accumulation, and retention.

10. There is still an enormous potential for new oil and gas fields both on the continents of the world and offshore. More than half of the petroleum yet to be discovered is believed to be beneath the world's oceans. The prime exploration areas are the continental shelves, continental rises, and deep

marginal basins. The rises are believed to contain more than half the world's total volume of sediments. The political, economic, and engineering problems involved in drilling them will be prodigious. The exploration will require the best combined expertise of geologists, geophysicists, and geochemists to define where the oil and gas have formed and accumulated.

APPENDIX:

Units of Measurement

The units of measurement used in this book include both those customarily used in the petroleum industry and the international system of units (known as SI). A few examples of SI units and symbols are given below. For more details, see the guide published by the American Society for Testing and Materials (1974).

Quantity	Unit	SI Symbol	Formula
area	square meter		m^2
density	kilogram per cubic meter		kg/m^3
force	newton	N	$kg \times m/s^2$
length	meter	m	
mass	kilogram	kg	
pressure	pascal	Pa	N/m^2
time	second	s (or sec)	
volume	cubic meter		m^3

SI PREFIXES

Multiple and submultiple prefixes are used to indicate orders of magnitude thus eliminating insignificant digits and decimals. It is preferable to use them in steps of 1,000, as shown on the following page.

Multiplication Factor	Prefix	SI Symbol
10^{12}	tera	T
10^9	giga	G
10^6	mega	M
10^3	kilo	k
10^{-3}	milli	m
10^{-6}	micro	μ
10^{-9}	nano	n
10^{-12}	pico	p

CONVERSIONS

1,000 ft³ = 28.3 m³

1 ft³ gas/barrel water = 0.1812 m³/kiloliter

1 metric ton oil = 680 m³ gas (chemical conversion basis)

1 barrel oil = 3,200 ft³ gas (chemical conversion basis)

1 metric ton oil = 1,280 m³ gas (BTU basis)

1 barrel oil = 6,040 ft³ gas (BTU basis)

1 barrel oil = 42 gallons = 0.159 m³

1 acre–foot = 1,233.5 m³

1 atmosphere = 101.3 kilopascals

pound–force/inch² (psi) = 6.895 kilopascals

kilogram–force/m² = 9.807 pascals

Density and Specific Volumes of Petroleum at 15.6°C (60°F)

°API gravity	Specific gravity	Barrels per metric ton
0	1.076	5.86
10	1.000	6.30
15	0.9659	6.53
20	0.9340	6.75
26	0.8984	7.02
30	0.8762	7.19
36	0.8448	7.46
40	0.8251	7.64
46	0.7972	7.91
50	0.7796	8.09
60	0.7389	8.53

Note: Gravity, $°API = \dfrac{141.5}{\text{specific gravity } 60°/60°F} - 131.5$

Source: Data from Levorsen 1967, p. 687. Adapted from *Facts and Figures*, 9th ed., American Petroleum Institute.

Glossary

Abnormal Pressure   Any departure from hydrostatic pressure. Overpressures generally range above 12 kPa/m (0.53 psi/ft) and underpressures range below 9.8 kPa/m (0.43 psi/ft).

Activation Energy (E)   The energy that must be absorbed by a molecule, or molecular complex, to break the bonds and form new products.

Aromatic (Arene) (Ar)   Hydrocarbons containing one or more benzene rings. Monoaromatics have the molecular formula C_nH_{2n-6}. Typical arenes are benzene, toluene, and the xylenes. Polycyclic aromatic hydrocarbons (PAH) contain several rings with two or more carbon atoms shared between rings.

Asphalt   Black to dark brown solid or semisolid bitumens that gradually liquefy when heated, composed principally of the elements carbon and hydrogen but containing appreciable quantities of nitrogen, sulfur, and oxygen and largely soluble in carbon disulfide.

Asphaltenes   Asphaltic constituents of crude oil that are soluble in carbon disulfide but insoluble in petroleum ether or *n*-pentane.

Asphaltite   Black to dark brown, comparatively hard, solid bitumens that soften at temperatures above 110°C (230°F) and are largely soluble in carbon disulfide. Examples are gilsonite, glance pitch, and grahamite.

Base of Crude Oil   The "base" of a crude oil is descriptive of the chemical nature of its main constituents. A paraffin-base oil contains predominantly paraffinic hydrocarbons. An intermediate- or mixed-base crude contains roughly equivalent mixtures of paraffins and naphthenes (cycloparaffins). A naphthenic-base crude contains predominantly naphthene hydrocarbons. An asphalt-base crude is one containing a relatively high proportion of nonhydrocarbon constituents such as nitrogen, sulfur, and oxygen compounds. The term aromatic base is not used because there are no oils known to contain predominantly aromatic hydrocarbons.

Biological Markers   Organic compounds whose carbon structure, or skeleton, is formed by living organisms and is sufficiently stable to be recognized in crude oil or the organic matter of ancient sediments. Typical markers are the porphyrins, pristane, phytane, steranes, carotanes, and pentacyclic triterpanes.

Bitumens   Native substances of variable color, hardness, and volatility, composed principally of the elements carbon and hydrogen and sometimes associated with mineral matter, the nonmineral constituents being largely soluble in carbon disulfide.

Carbohydrates   Organic compounds with the approximate general formula $(C \cdot H_2O)_n$, where n is equal to or greater than 4. Sucrose (table sugar), glucose, starch, and cellulose are typical carbohydrates.

Carbon Preference Index (CPI)   The ratio of odd- to even-chain-length normal paraffins in a specific molecular weight range. Originally, Bray and Evans (1961, p. 9) calculated this as the ratio of the sum of the mole percentages of $C_{25}-C_{33}$ odd-carbon n-paraffins to the sum of $C_{26}-C_{34}$ even-carbon n-paraffins. Slightly modified ratios are used today, such as the ratio of twice the $C_{25}-C_{33}$ odd-carbon n-paraffins to the $C_{26}-C_{34}$ plus the $C_{24}-C_{32}$ even-carbon n-paraffins.

Carbon Ratio Theory   The theory that the API gravity of oil increases (specific gravity decreases) as the carbon ratio of coals in the same area increases. As coals increase in rank, oils become lighter eventually changing to gas. The carbon ratio is the ratio of fixed carbon to total carbon (fixed plus volatile) in a coal.

Carotenoids   Plant pigments composed of mono- and dicyclic tetraterpenes. Carotenes are precursors to vitamin A. The yellow, brown, and red colors of leaves in the fall are caused by carotenoids. Carotanes are the alkanes derived from carotenoids.

Casing Head Gasoline   The liquid hydrocarbon recovered from casing head gas by adsorption, compression, or refrigeration. Casing head gas is the gas recovered at the surface from an oil well.

Catagenesis   The process by which organic material in sediments is thermally altered by increasing temperature. Catagenesis covers the temperature range between diagenesis and rock metamorphism, approximately $50-200°C$ ($122-392°F$).

Chitin A polysaccharide (carbohydrate polymer) with a structure similar to cellulose, except that two OH groups are replaced by CH_3ONH groups in each $(C \cdot H_2O)_6$ unit. Chitin forms the horny, hard, outer cover of insects, crustaceans, and parts of some other invertebrates.

Coal A readily combustible rock containing more than 50 percent by weight, and more than 70 percent by volume, of organic material formed from the compaction or induration of variously altered plant remains. Humic coals form from plant cell and wall material deposited under aerobic conditions, while sapropelic coals form from spores, pollen, and algae deposited under anerobic conditions.

Coal Maceral Microscopically recognizable constituents of coal that can be differentiated by their morphology. Macerals are analogous to the minerals of inorganic rocks but differ in having less uniform chemical composition and physical properties. The carbon content of macerals increases with increasing temperature. The major maceral groups are vitrinite, liptinite, and inertinite.

Condensate A hydrocarbon mixture that is gaseous in the ground but condenses into liquid when produced. Its gravity ranges from 60° API upward.

Connate Water Fossil reservoir water that has not been in contact with the atmosphere since deposition. It is high in chloride and calcium and frequently contains more than 100,000 ppm dissolved total solids.

Correlation Index (CI) The correlation index was developed empirically by H. M. Smith of the U.S. Bureau of Mines. He set up an equation such that the addition of the reciprocal absolute boiling point and the specific gravity with appropriate constants was equal to 0 for the normal paraffins and to 100 for the aromatic hydrocarbon benzene.

The equation is

$$CI = \frac{48640}{K} + 473.7\,G - 456.8$$

where CI is the correlation index
K is the average boiling point of the fraction in degrees Kelvin
and G is the specific gravity of the fraction at 60°F (16°C).

Crude Oil A petroleum that is removed from the earth in liquid state or is capable of being so removed.

Crude Oil Fractions

  Gasoline—The fraction of crude petroleum boiling between about 15 and 200°C (60 and 392°F).

  Kerosine—The fraction of crude petroleum boiling between about 200 and 260°C (392 and 500°F).

Gas Oil—The fraction of crude petroleum having a viscosity less than 50 seconds S.U. at 38°C (100°F) and a boiling range between about 260 and 332°C (500 and 630°F).

Lubricating Oil—The fraction of crude petroleum having a viscosity above 50 seconds S.U. at 38°C (100°F) and a boiling range between about 332 and 421°C (630 and 790°F).

Residuum—The residue obtained from the distillation of crude oil after all fractions, including lubricating oils, have been taken off.

Diagenesis   The process involving biological, physical, and chemical alteration of the organic debris in sediments without a pronounced effect from rising temperature.

Distillate   The hydrocarbon fluid produced from natural gas processing. It is denser than condensate and generally is run into the tanks with the crude oil. Gravities range from 50°API upward.

Drill-Stem Test (DST)   A test of the productive capacity of a well while it is still full of drilling mud. The testing tool is lowered into the hole attached to the drill pipe and placed opposite the formation to be tested. Packers are set to shut off the weight of the drilling mud, and the tool is opened to permit the flow of any formation fluid into the drill pipe, where the flow is measured.

Dry Gas   Natural gas consisting principally of methane and devoid of readily condensable constitutents such as gasoline. Dry gas contains less than 0.1 gallon natural gas liquid vapors per 1,000 ft^3 (1.3 liters per 100 m^3).

Gas Hydrates   Crystalline compounds in which the ice lattice of H_2O expands to form cages that contain the gas molecules. Methane hydrates can hold 8 methane molecules in each 46 H_2O molecules, a formula of $CH_4 \cdot 5.75\ H_2O$.

Hempel Distillation   A method of distilling an oil into 15 fractions and the residuum. The first fraction consists of everything distilling up to 50°C (122°F). The next nine fractions are taken at intervals of 25°C (45°F) up to 275°C (527°F) under atmospheric pressure. The last five fractions are distilled under vacuum at 40 mm pressure. The eleventh fraction consists of all materials boiling up to 200°C (392°F) and the succeeding four fractions are taken at intervals of 25°C to 300°C (572°F) at 40 mm. The base of the crude and the quantities of gasoline and other constituents in the crude can be calculated from the distillation data.

Heterocyclic Compounds   Ring compounds in which one or more of the carbon atoms in the ring are replaced by an atom of another element such as nitrogen, sulfur, or oxygen.

Hilt's Law   The volatile carbon in coal decreases proportionally with depth of the coal in normal stratigraphic sequences. The nonvolatile or fixed carbon of coals increases with increasing depth and temperature.

Humic Organic Matter   The decomposition and polymerization products of the lignin, tannins, and cellulose of plant cell and wall material plus carbonized organic matter deposited in swamps and soils under aerobic conditions with partial restriction of oxygen.

Hydrocarbon   A compound composed only of the elements hydrogen and carbon. Bitumens such as petroleum are composed principally, but not only, of hydrocarbons.

Hydrostatic Gradient   The pressure increase with depth of a liquid in contact with the surface. The gradient for freshwater is 9.8 kPa/m (0.433 psi/ft).

Inertinite   The coal maceral group that shows little or no reaction during the coking process. Inertinite includes fusinite (fossil charcoal), sclerotinite derived from fungal remains, and other high-carbon materials with a low-hydrogen content and a high reflectance.

Isomers   Molecules that have the same number and kinds of atoms but are different substances. Structural isomers differ in the way atoms are linked together; for example, n-butane and isobutane. Stereoisomers differ in the spatial arrangements of groups; for example, cis- and trans- (boat and chair) isomers of 1,2-dimethylcyclopropane. Optical isomers are nonidentical mirror images comparable to right- and left-handed gloves; for example, D- and L-lactic acid (see Chapter 4).

Isoprenoid   A hydrocarbon whose molecular structure contains the basic unit isoprene consisting of five carbon atoms with a branch at the second atom as shown.

$$
\begin{array}{c}
\text{C} \\
| \\
\text{C}=\text{C}-\text{C}=\text{C}
\end{array}
$$

This is the basic building block of many natural products such as terpenoids, steroids, carotenoids, pristane, and phytane. Many petroleum hydrocarbons are diagenetic derivatives of isoprenoid polymers. The designation ip-19 and ip-20, as in Figure 7-27, refers to isoprenoids with 19 and 20 carbon atoms; in this case, pristane and phytane, respectively.

Isotopes   Atoms whose nuclei contain the same number of protons but a different number of neutrons. All carbon atoms have six protons, but there are three carbon isotopes containing 6, 7, and 8 neutrons giving atomic masses respectively of 12, 13, and 14 (written as ^{12}C, ^{13}C, ^{14}C).

Kerogen   The disseminated organic matter of sedimentary rocks that is insoluble in nonoxidizing acids, bases, and organic solvents. The organic matter initially deposited with unconsolidated sediments is not kerogen but a precursor that is converted to kerogen during diagenesis. Sapropelic kerogens yield oil and gas on heating, while humic kerogens yield mainly gas. Kerogen includes both marine- and land-derived organic matter, the latter being the same as the components of coal.

Lipids   A broad term that includes all oil-soluble, water-insoluble substances such as fats, waxes, fatty acids, sterols, pigments, and terpenoids.

Liptinite (Exinite)   A coal maceral group that is the dominant organic constituent of boghead coals. Liptinite macerals include sporinite, cutinite, resinite, and alginite, which are derived from spores and pollen, cuticles, resins, and algae respectively. Bituminite is an amorphous liptinite maceral. Liptinite is widely disseminated in sediments and is an important source of crude oil. The kerogen of oil shales is mostly of liptinitic origin.

Lithostatic Gradient   The total pressure increase with depth caused by rock grains and water. It averages about 24.4 kPa/m (1.08 psi/ft).

Mercaptans   Compounds containing the sulfhydryl (—SH) group. They are sulfur analogs of the alcohols in that the oxygen in the alcohol (—OH) group is replaced by sulfur.

Metamorphism   The transformation of preexisting rocks into new types by the action of heat, pressure, stress, and chemically active migrating fluids. Metamorphism usually begins at temperatures above 200°C (392°F). At such temperatures, the organic matter is already reduced to a low-hydrogen carbon residue capable of yielding only small amounts of gas.

Meteoric Water   Fresh surface water entering a subsurface sedimentary section through permeable outcrops. It is high in sodium, bicarbonate, and sulfate and usually contains less than 10,000 ppm dissolved total solids.

Mineral Wax   A species of bitumen having a characteristic luster and unctuous feel, composed principally of saturated hydrocarbons and containing considerable crystallizable paraffins, the nonmineral constituents being soluble in carbon disulfide. An example is ozocerite.

Naphthene (Cycloalkane, Cycloparaffin)   A hydrocarbon ring with the molecular formula C_nH_{2n}. Cyclopentane (C_5) and cyclohexane (C_6) ring structures are the most common in petroleum. Condensed or polycyclic naphthenes contain rings in which two or more carbon atoms are shared. Tetracyclic and pentacyclic naphthenes contain four and five rings, respectively, fused together.

Naphthenic Acids Petroleum acids containing a naphthene or cycloparaffin structure. The most common naphthenic acids contain a cyclopentane (five-carbon atom) ring.

Natural Gas A petroleum consisting of varying proportions of gaseous hydrocarbons such as methane, ethane, propane, isobutane, and occasionally containing liquid hydrocarbons such as pentanes and hexanes, and nonhydrocarbon gases such as carbon dioxide, hydrogen sulfide, nitrogen, hydrogen, and helium.

Odd/Even Predominance (OEP) The ratio of odd- to even-chain-length organic compounds in a specific molecular weight range. If C_i is the relative weight percent of an *n*-alkane containing *i* carbon atoms per molecule, then

$$OEP = \left| \frac{C_i + 6C_{i+2} + C_{i+4}}{4C_{i+1} + 4C_{i+3}} \right|^{(-1)^{i+1}}$$

For further details, see Scalan and Smith (1970).

Oil Shale A compact rock of sedimentary origin with an ash content of more than 33 percent and containing organic matter that yields oil when destructively distilled but not appreciably when extracted with petroleum solvents.

Paraffin (Alkane) A hydrocarbon with the molecular formula C_nH_{2n+2}. It includes normal straight-chain paraffins and branched alkanes, such as methane, ethane, propane and isobutane.

Paraffin Dirt A yellow-brown, gummy, soil organic matter composed of nitrogenous–humic–cellulosic remains of plant material heavily impregnated with fungi, yeasts, actinomyces, and bacteria. The word *paraffin* is a misnomer, since this substance contains less than 3 percent lipid material. The waxy appearance is caused by living and dead microbial cells. Paraffin dirt tends to accumulate in moist soils near hydrocarbon seeps.

Petroleum A species of bitumen composed principally of hydrocarbons and existing in the gaseous or liquid state in its natural reservoir.

Phytane A saturated isoprenoid containing 20 carbon atoms.

Porphyrins Complex compounds originating in living materials and having a basic structure consisting of four interconnected rings, each ring containing four carbon atoms and one nitrogen atom. The red hemoglobin in blood and the green chlorophyll in plants are porphyrins.

Pour Point The temperature at which crude oil will not flow when a tube containing it is first heated in a bath to dissolve all wax and then cooled slowly. At cooling intervals of 3°C (5°F), the tube is held horizontal until there is no flow for 5 seconds.

Primary Migration   The movement of oil and gas out of the fine-grained source rocks into the permeable reservoir rocks.

Pristane   A saturated isoprenoid containing 19 carbon atoms.

Proteins   High-molecular-weight polymers of amino acids that constitute more than 50 percent of the dry weight of animals. The organic nitrogen and sulfur of living organisms is concentrated in the protein fraction. Gelatin, albumin, collagen (connective tissues), keratin (hair, hoofs, nails), and serum globulins are typical proteins.

Pyrobitumen   Black to dark brown, hard bitumens that are infusible and relatively insoluble in carbon disulfide. Typical pyrobitumens are albertite, wurtzilite, and impsonite.

Resin   Petroleum resins are the fraction of residuum that is insoluble in liquid propane but soluble in normal pentane. Plant resins are terpenoids ranging in molecular size from sesquiterpenes (C_{15}) to tetraterpenes (C_{40}). They contain the olefinic double bonds of the isoprene building block that causes the liquids to polymerize and oxidize to hard resins on exposure to air. Balsam and mastic are plant resins.

Resinite   A coal maceral in the liptinite group. Resinites are fossil tree resins derived from balsam, mastic, latex, and some plant gums, waxes, oils, and fats. Amber and copal are resinites.

Sapropelic Organic Matter   The decomposition and polymerization products of high-lipid organic materials, such as spores and planktonic algae deposited in subaquatic muds (marine or lacustrine) under predominately anaerobic conditions.

Saturates (Saturated Hydrocarbons)   A general term including *n*- and branched alkanes and cycloalkanes (paraffins and naphthenes). Saturates are the nonaromatic hydrocarbon fraction of an oil.

Secondary Migration   The movement of fluids within the permeable rocks that eventually leads to the segregation of oil and gas into accumulations in certain parts of these rocks.

Sour Oil   An oil containing noticeable quantities of noxious sulfur compounds such as hydrogen sulfide and mercaptans. A sweet oil does not contain these compounds.

Source Rock   A fine-grained sediment that in its natural setting has generated and released enough hydrocarbons to form a commercial accumulation of oil or gas.

Tar   A thick, black or dark brown, viscous liquid obtained by the destructive distillation of coal, wood, or peat. Tar is not a natural product, and it is a misnomer to refer to asphalt deposits or seeps as *tars*.

Terpenoid   An isoprenoid polymer usually containing 10, 15, 20, 30, or 40 carbon atoms (mono-, sesqui-, di-, tri-, and tetra-, respectively). Terpenoids include hydrocarbons, alcohols, and acids. Among the common terpanes (hydrocarbons) are the hopanes (pentacyclic triterpanes—see Figure 4-8).

Tertiary Migration   The movement of an oil or gas accumulation to a different part of the reservoir rock.

Thermal Alteration Index (TAI)   A maturation color index for the particulate organic matter of sedimentary rocks. The index indicates the degree of thermal alteration that the organic matter has undergone. As proposed by Staplin (1969), the index numbers 1 to 5 include color changes from yellow to brown to black, representing immature, mature, and metamorphosed facies of organic matter.

Viscosity Index (VI)   A series of numbers ranging from 0 to 100 that indicate the rate of change of viscosity with temperature. A VI of 100 indicates a small change in viscosity between temperatures of 38 and 99°C (100 and 210°F), whereas a VI of 0 indicates a large change.

Vitrinite   A coal maceral group that is the dominant organic constituent of humic coals. Vitrinite forms the familiar brilliant black bands of coal. Macerals in the vitrinite group include telinite derived from plant cell walls and collinite from the cell filling. Vitrinite particles are found in about 80 percent of the clays and sands of sedimentary basins.

Wax   *See* Mineral Wax.

Wet Gas   Natural gas consisting of methane and heavier hydrocarbons, the natural gas liquid vapors amount to 4 liters or more per 100 m³ (0.3 or more gallons per 1,000 ft³).

References

Abelson, P. H. 1956. Paleobiochemistry: Inorganic synthesis of amino acids. *Carnegie Inst. Wash. Yearbook, 55,* 171–174.

Abraham, H. 1960. *Asphalts and allied substances.* Vol. 1: *Historical review and natural raw materials.* Princeton, N.J.: Van Nostrand. 370 p.

Aizenshtat, Z. 1973. Perylene and its geochemical significance. *Geochim. Cosmochim. Acta, 37,* 559–567.

Akishev, I. M., R. Kh. Muslimov, N. P. Lebedev, and V. I. Troyepol'skiy. 1974. Bitumen deposits of the Permian sediments of Tataria and prospects for their exploration. *Geol. Nefti Gaza, 12* (3), 23–28.

Al'bov, S. V. 1971. The Kerch—Taman hydrogeochemical and mud volcano region. *Dokl. Akad. Nauk SSSR, 197* (1), 175–177.

Albrecht, P. 1969. Constituants organiques des roches sédimentaires: Étude de la diagénèse dans une serie sedimentarie epaisse. Doctoral thesis, University of Strasbourg, France.

Albrecht, P. 1970. Étude de constituants organiques des séries sédimentaires de Logbaba et de Messel. Transformations diagénétiques. Université de Strasbourg, Mémoires du Service de la Carte Géologique d'Alsace et de Lorraine, No. 32. 119 p.

Albrecht, P., and G. Ourisson. 1971. Biogenic substances in sediments and fossils. *Angew. Chem., Internat. Ed., 10* (4), 209–286.

Albrecht, P., M. Vandenbroucke, and M. Mandengué. 1976. Geochemical studies on the organic matter from the Douala Basin (Cameroon). Part 1: Evolution of the extractable organic matter and the formation of petroleum. *Geochim. Cosmochim. Acta, 40* (7), 791–799.

Aliev, A. G., G. A. Aliyeva, and D. G. Osika. 1966. Distribution of I_2, Br_2, NH_4 and B in rock solutions and stratal water of oil deposits in North Dagestan. *Geokhim.,* no. 12, pp. 1497–1502.

Alliquander, O. 1973. High pressures, temperatures plague deep drilling in Hungary. *Oil and Gas J., 71,* 97–100.

Almon, W. R., and W. D. Johns. 1977. Petroleum-forming reactions. The mechanisms and rate of clay catalyzed fatty acid decarboxylation. In R. Campos and J. Goñi (eds.), *Advances in organic geochemistry, 1975.* S.A. Madrid: Nacional Adaro de Investigaciones Mineras, pp. 157–172.

American Society for Testing and Materials. 1974. *Standard metric practice guide.* E380-74. American Society for Testing and Materials, 1916 Race St., Philadelphia, Pa. 19103. 34 p.

Ammosov, I. I., and Tan Syu-i. 1961. *Stages of alteration of coals and the paragenetic relation of fossil fuels.* Moscow: Izd. Akad. Nauk SSSR. 117 p.

Amosov, G. A. 1951. Optical rotation of petroleums. Geokhimicheskii Sbornik, Trudy VNIGRI. New series, No. 5, No. 2–3. English translation (in *Contributions to geochemistry*) by Israel Program for Scientific Translations, Jerusalem, 1965, pp. 225–233.

Amosov, G. A., and T. A. Kozina. 1966. Interaction of crude oils and water in Sakhalin oil fields. *Internat. Geol. Rev., 9* (7), 883–889.

Anders, D. E., and W. E. Robinson. 1971. Cycloalkane constituents of the bitumen from Green River Shale. *Geochim. Cosmochim. Acta, 35* (6), 661–678.

Anders, D. E., and W. E. Robinson. 1973. Geochemical aspects of the saturated hydrocarbon constituents of Green River oil shale—Colorado No. 1 core. U.S. Bureau of Mines, Report of Investigations 7737. Washington, D.C.: U.S. Bureau of Mines. 23 p.

Anderson, D. M., and P. F. Low. 1958. The density of water adsorbed by lithium-, sodium-, and potassium-bentonite. *Soil Sci. Soc. Amer. Proc., 22* (2), 99–103.

Andreev, P. F., A. I. Bogomolov, A. F. Dobryanskii, and A. A. Kartsev. 1968. *Transformation of petroleum in nature.* London: Pergamon Press. 466 p.

Andrews, E. B. 1861. Rock oil, its geological relations and distribution. *Amer. J. Sci.,* second series, *32* (94), 85–93.

Antonov, P. L. 1953. On the measurement of the diffusion parameters of several rocks. *Sbornik Geokhim. Metody Poiskov Nefti i Gaza,* no. 1.

Antonov, P. L., G. A. Gladysheva, and V. P. Kozlov. 1958. Diffusion of carbon dioxide across rock salt. *Petrol. Geol., 2* (2b), 175–178.

Arnold, R. 1959. Pseudo-evidences of oil and gas. *AAPG Bull., 43* (5), 1058–1079.

Athy, L. F. 1930. Compaction and oil migration. *AAPG Bull., 14* (1), 25–36.

Attaway, D. H., P. L. Parker, and J. A. Mears. 1970. Normal alkanes of five coastal spermatophytes. *Contributions in Marine Science.* Vol. 15. pp. 13–19.

Avrov, V. P., and E. M. Galimov. 1968. Microbiologic nature of a methane pool detected at considerable depth (based on isotopic analysis). *Dokl. Akad. Nauk SSSR, 179* (5), 201–202.

Baedecker, M. J., R. I. Ikan, R. Ishiwatari, and I. R. Kaplan. 1977. Thermal alteration experiments on organic matter in Recent marine sediments as a model for petroleum genesis. In T. F. Yen (ed.), *Chemistry of marine sediments.* Ann Arbor, Mich.: Science Publishers, pp. 55–72.

Bailey, N. J. L., C. R. Evans, and C. W. D. Milner. 1974. Applying petroleum geochemistry to search for oil: Examples from Western Canada Basin. *AAPG Bull., 58* (11), 2284–2294.

Bailey, N. J. L., A. M. Jobson, and M. A. Rogers. 1973. Bacterial degradation of crude oil: Comparison of field and experimental data. *Chem. Geol., 11,* 203–221.

Bailey, T. L. 1947. Origin and migration of oil into Sespe redbeds, California. *AAPG Bull., 31* (11), 1913–1915.

Baker, D. A., and P. T. Lucas. 1972. Strat trap production may cover 280 + square miles. *World Oil, 180* (3), 65–68.

Baker, D. R. 1962. Organic geochemistry of Cherokee Group in southeastern Kansas and northeastern Oklahoma. *AAPG Bull., 46* (9), 1621–1642.

Baker, D. R. 1972. Organic geochemistry and geological interpretations. *J. Geol. Ed., 20* (5), 221–234.

Baker, D. R., and G. E. Claypool. 1970. Effects of incipient metamorphism on organic matter in mudrock. *AAPG Bull., 54* (3), 456–468.

Baker, E. G. 1967. A geochemical evaluation of petroleum migration and accumulation. In B. Nagy and U. Columbo (eds.), *Fundamental aspects of petroleum geochemistry.* Amsterdam: Elsevier, pp. 299–329.

Baker, E. W. 1964. Vanadium and nickel in crude petroleum of South America and Middle East origin. *J. Chem. and Eng. News, 42* (15), 307–308.

Baker, E. W., and G. Eglinton. 1978. Organic geochemistry related developments in the Deep Sea Drilling Project. Paper presented at Tenth Annual Offshore Technology Conference, May 8–11, 1978, Houston, Texas. OTC 3050, pp. 65–70.

Baker, E. W., and G. D. Smith. 1973. Chlorophyll derivatives in sediments, site 147. In B. C. Heezen and I. D. MacGregor (eds.), *Initial reports of the Deep Sea Drilling Project.* Vol. 20. Washington, D.C.: U.S. Government Printing Office, pp. 193–194.

Baker, E. W., and G. D. Smith. 1974. Pleistocene changes in chlorophyll pigments. In B. Tissot and F. Bienner (eds.), *Advances in organic geochemistry, 1973.* Paris: Éditions Technip, pp. 649–660.

Barker, C. 1972. Aquathermal pressuring—role of temperature in development of abnormal pressure zones. *AAPG Bull., 56* (10), 2068–2071.

Barker, C. 1974a. Pyrolysis techniques for source-rock evaluation. *AAPG Bull., 58* (11), 2349–2361.

Barker, C. 1974b. Some thoughts on primary migration of hydrocarbons. *AAPG-SEPM Annu. Mtg. Abstr., 1* (April), 4.

Barker, C., and M. P. Kemp. 1977. Floor for gas. Paper presented at AAPG-SEPM Annual Convention, June 12–16, 1977, Washington, D.C.

Bars, E. A., and L. N. Nosova. 1962. The problem of dissolved organic matter in waters of the Cretaceous and Jurassic deposits of the Ob'-Irtysh Basin. In *Geokhimiya Nefti i Neftyanykh Mestorzhdenii.* Moscow: Akad. Nauk SSSR, pp. 181–198. Translated by Israel Program for Scientific Translations, Jerusalem, 1964.

Bass, N. W. 1963. Composition of crude oils in northwestern Colorado and northeastern Utah suggests local sources. *AAPG Bull., 47* (12), 2039–2064.

Beebe, B. W., and B. F. Curtis. 1968. Natural gases of North America: A summary. In B. W. Beebe and B. F. Curtis (eds.), *Natural gases of North America.* AAPG Memoir 9. Vol. 2. Tulsa, Okla.: American Association of Petroleum Geologists, pp. 2245–2355.

Bell, K. G., and J. M. Hunt. 1963. Native bitumens associated with oil shales. In I. A. Breger (ed.), *Organic geochemistry.* New York: Pergamon Press.

Belyayeva, L. S. 1968. Organic matter in dolerite from Tungska test hole. *Dokl. Akad. Nauk SSSR, 182* (3), 655–658.

Bendoraitis, J. G. 1974. Hydrocarbons of biogenic origin in petroleum-aromatic triterpenes and bicyclic sesquiterpenes. In B. Tissot and F. Bienner (eds.), *Advances in organic geochemistry, 1973.* Paris: Éditions Technip, pp. 209–224.

Benson, W. E., and R. E. Sheridan (eds.). 1978. *Initial reports of the Deep Sea Drilling Project.* Vol. 44. Washington, D.C.: U.S. Government Printing Office. 1005 p.

Bernard, B. B., J. M. Brooks, and W. M. Sackett. 1976. Natural gas seepage in the Gulf of Mexico. *Earth and Planet. Sci. Letters, 31*, 48–54.

Bernard, B. B., J. M. Brooks, and W. M. Sackett. 1977. A geochemical model for characterization of hydrocarbon gas sources in marine sediments. In *Offshore Technology Conference* (May 1977). OTC 2934, pp. 435–438.

Berry, F. A. F. 1969. Relative factors influencing membrane filtration effects in geologic environments. *Chem. Geol., 4*, 295–301.

Bestougeff, M. A. 1967. Petroleum hydrocarbons. In B. Nagy and U. Columbo (eds.), *Fundamental aspects of petroleum geochemistry.* Amsterdam: Elsevier, pp. 73–108.

Biju-Duval, B., J. Letouzey, L. Montadert, P. Courrier, J. F. Mugniot, and J. Sancho. 1974. Geology of the Mediterranean Sea basins. In C. A. Burk and C. L. Drake (eds.), *The geology of continental margins.* New York: Springer-Verlag, pp. 695–721.

Bikkenina, D. A., and A. I. Shapiro. 1969. Hydrocarbons in organic matter from rocks of the Kuonam Suite on the south bank of the Anabar anticline. *Dokl. Akad. Nauk SSSR, 185* (4), 171–174.

Bily, C., and J. W. L. Dick. 1974. Naturally occurring gas hydrates in the Mackenzie Delta, N.W.T. *Canadian Petrol. Geol. Bull., 22*, 340–352.

Blake, W. P. 1855. Preliminary geological report of the U.S. Pacific Railroad survey, 1853. *Amer. J. Sci., 19*, 433–434.

Blasko, D. P., W. J. Wenger, and J. C. Morris. 1972. *Oilfields and crude oil characteristics: Cook Inlet Basin, Alaska.* U.S. Department of Interior, U.S. Bureau of Mines, Report 7688. Washington, D.C.: U.S. Government Printing Office. 44 p.

Blumer, M. 1961. Benzpyrenes in soil. *Science, 134*, 474–475.

Blumer, M. 1967. Hydrocarbons in digestive tract and liver of a basking shark. *Science, 156*, 390–391.

Blumer, M. 1976. Polycyclic aromatic compounds in nature. *Sci. Amer., 234* (3), 35–45.

Blumer, M., and G. Souza. 1965. "Zamene," isomeric C_{19} mono-olefins from marine zooplankton, fishes, and mammals. *Science, 148*, 370–371.

Blumer, M., R. R. L. Guillard, and T. Chase. 1971. Hydrocarbons of marine phytoplankton. *Mar. Biol., 8* (3), 183–189.

Blumer, M., M. M. Mullin, and R. R. L. Guillard. 1970. A polyunsaturated hydrocarbon (3-, 6-, 9-, 12-, 15-, 18-heneicosahexaene) in the marine food web. *Mar. Biol., 6* (3), 226–235.

Blumer, M., M. M. Mullin, and D. W. Thomas. 1963. Pristane in zooplankton. *Science, 140* (3570), 974.

Blumer, M., M. M. Mullin, and D. W. Thomas. 1964. Pristane in marine environment. *Helgol. Wiss. Meeresunters, 10,* 187–201.

Blytas, G. C., and D. L. Peterson. 1967. Determination of kerosine-range *n*-paraffins by a molecular-sieve, gas-liquid chromatography method. *Anal. Chem., 39* (12), 1434–1437.

Bogomolov, A. I., and K. I. Panina. 1961. Low-temperature catalytic conversions of organic compounds under clay. *Geokhim. Sbornik, 7* (174).

Bogomolov, G. V., A. V. Kudel'skiy, and M. F. Kozlov. 1970. The ammonium ion as an indicator of oil and gas. *Dokl. Akad. Nauk SSSR, 195* (4), 938–940.

Bordenave, M., A. Combaz, and A. Giraud. 1970. Influence de l'origine des matières organiques de leur degre d'évolution sur les produits de pyrolyse du kérogène. In G. D. Hobson and G. C. Speers (eds.), *Advances in organic geochemistry, 1966.* London: Pergamon, pp. 389–405.

Bordovskiy, O. K. 1965. Accumulation and transformation of organic substance in marine sediments. *Mar. Geol., 3* (3–4), 1–114.

Borneff, J., F. Selenka, H. Kunte, and A. Maximos. 1968. Experimental studies on the formation of polycyclic aromatic hydrocarbons in plants. *Envir. Res., 2,* 22–29.

Borst, R. L., and W. D. Keller. 1969. Scanning electron micrographs of API reference clay minerals and other selected samples. In *International Clay Conference, Tokyo, Proc.* Vol. 1. Jerusalem: Israel University Press, pp. 871–901.

Borst, R. L., and F. J. Shell. 1971. The effects of thinners on the fabric of clay muds and gels. *J. Petrol. Tech.,* October, pp. 1193–1201.

Bostick, N. 1973. Time as a factor in thermal metamorphism of phytoclasts (coaly particles). In *Congrès Internat. Strat. Geol. Carbonif., Compte Rendue.* Septième, Krefeld, August 23–28. Vol. 2. Maastricht, Netherlands: E. Van Aelst, pp. 183–193.

Bray, E. E., and E. D. Evans. 1961. Distribution of *n*-paraffins as a clue to recognition of source beds. *Geochim. Cosmochim. Acta, 22,* 2–15.

Bray, E. E., and E. D. Evans. 1965. Hydrocarbons in non-reservoir-rock source beds: Part 1. *AAPG Bull., 49* (3), 248–257.

Brenneman, M. C., and P. V. Smith, Jr. 1958. The chemical relationships between crude oils and their source rocks. In L. G. Weeks (ed.), *Habitat of oil: A symposium.* Tulsa, Okla.: American Association of Petroleum Geologists, pp. 818–849.

Brooks, B. T. 1948. Active-surface catalysts in formation of petroleum. *AAPG Bull., 32* (12), 2269–2286.

Brooks, J. D. 1970. The use of coals as indicators of the occurrence of oil and gas. *APEA Journal* (Australia), *10* (2), 35–50.

Brooks, J. D., K. Gould, and J. W. Smith. 1969. Isoprenoid hydrocarbons in coal and petroleum. *Nature, 222,* 257–259.

Brooks, J. M., and W. M. Sackett. 1972. Light hydrocarbon concentrations in the Gulf of Mexico. Paper presented at a joint meeting of the American Geophysical

Union and the American Meteorological Society, August 15–17, St. Petersburg Beach, Florida.

Brooks, J. M., J. R. Gormly, and W. M. Sackett. 1974. Molecular and isotopic composition of two seep gases from the Gulf of Mexico. *Geophys. Res. Letters, 1* (5), 213–216.

Buckley, S. E., C. R. Hocott, and M. S. Taggart, Jr., 1958. Distribution of dissolved hydrocarbons in subsurface waters. In L. G. Weeks (ed.), *Habitat of oil: A symposium.* Tulsa, Okla.: American Association of Petroleum Geologists, pp. 850–882.

Burgess, J. D. 1974. Microscopic examination of kerogen (dispersed organic matter) in petroleum exploration. Geological Society of America, Special Paper 153, pp. 19–30.

Burgess, J. D. 1975. Historical review and methods for determining thermal alteration of organic materials. In *Palynology: Proceedings of the eighth annual meeting of the American Association of Stratigraphic Palynologists.* Vol. 1. Houston, Texas: American Association of Stratigraphic Palynologists, pp. 1–7.

Burlingame, A. L., and H. K. Schnoes. 1969. Mass spectrometry in organic geochemistry. In G. Eglinton and M. T. J. Murphy (eds.), *Organic geochemistry: Methods and results.* New York: Springer-Verlag, pp. 89–149.

Burst, J. F. 1969. Diagenesis of Gulf Coast clayey sediments and its possible relation to petroleum migration. *AAPG Bull., 53* (1), 73–93.

Butler, B. S., and W. S. Burbank. 1929. The copper deposits of Michigan. U.S. Geological Survey, Professional Paper 144, pp. 169–172.

Byramjee, R. 1967. Problème géochimique de la genèse et de la migration du pétrole. *Rev. de L'Inst. Français Pètrol.* et *Ann. Combust. Liq., 22* (5), 791–796. Ref. 14:475.

Cardoso, J. N., A. M. K. Wardroper, C. D. Watts, P. J. Barnes, J. R. Maxwell, G. Eglinton, D. G. Mound, and G. C. Speers. 1978. Preliminary organic geochemical analyses, Site 391, Leg 44 of the Deep Sea Drilling Project. In W. E. Benson and R. E. Sheridan (eds.), *Initial reports of the Deep Sea Drilling Project.* Vol. 44. Washington, D.C.: U.S. Government Printing Office, pp. 617–624.

Cardwell, A. L. 1977. Petroleum source-rock potential of Arbuckle and Ellenburger groups, southern mid-continent, United States. *Quarterly of the Colorado School of Mines, 72* (3), 1–134.

Carlisle, C. T., G. S. Bayliss, and D. G. Van Delinder. 1975. Distribution of light hydrocarbon in sea floor sediments: Correlations between geochemistry seismic structure and possible reservoired oil and gas. In *1975 Offshore Technology Conference, Proceedings.* Vol. 3. Dallas, Texas, pp. 65–70.

Cartmill, J. C. 1976. Obscure nature of petroleum migration and entrapment. *AAPG Bull., 60* (9), 1520–1530.

Cartmill, J. C., and P. A. Dickey. 1970. Flow of a disperse emulsion of crude oil in water through porous media. *AAPG Bull., 54,* 2438–2447.

Cernock, P. J., and G. S. Bayliss. 1977. Organic geochemical analyses of drill cuttings from KCM No. 1 Forest Federal Well Hidalgo Country, New Mexico. New Mexico Bureau of Mines and Mineral Resources, Circular 152, pp. 37–47.

Chapman, R. E. 1972. Primary migration of petroleum from clay source rocks. *AAPG Bull.*, *56* (11), 2185–2191.

Chapman, R. E. 1973. *Petroleum geology: A concise study.* Amsterdam: Elsevier. 304 p.

Chibnall, A. C., S. H. Piper, A. Pollard, E. F. Williams, and P. N. Sahai. 1934. Constitution of primary alcohols, fatty acids, and paraffins present in plant and insect waxes. *Biochem. J.*, *28*, 2189.

Chilingar, G. V., H. J. Bissell, and K. H. Wolf. 1967. Diagenesis of carbonate rocks. In G. Larsen and G. V. Chilingar (eds.), *Diagenesis in sediments.* Amsterdam: Elsevier, pp. 179–322.

Clark, R. C., Jr., 1966. Saturated hydrocarbons in marine plants and sediments. Master's thesis, M.I.T.–W.H.O.I. Joint Program.

Clark, R. C., Jr., and M. Blumer. 1967. Distribution of *n*-paraffins in marine organisms and sediment. *Limnol. and Oceanogr.*, *12* (1), 79–87.

Claypool, G. E. 1974. Anoxic diagenesis and bacterial methane production in deep sea sediments. Doctoral thesis, University of California at Los Angeles.

Claypool, G. E. 1976. Review of shipboard gas analysis, DSDP legs 10–31. *JOIDES Journal, 1* (Special Issue No. 4), c6–c9.

Claypool, G. E., and J. P. Baysinger. 1978. Thermal analysis/pyrolysis of Cretaceous sapropels, DSDP Leg 44, Hole 391 C, Blake–Bahama Basin. In W. E. Benson and R. E. Sheridan (eds.), *Initial reports of the Deep Sea Drilling Project.* Vol. 44. Washington, D.C.: U.S. Government Printing Office, pp. 635–638.

Claypool, G. E., and I. Kaplan. 1974. The origin and distribution of methane in marine sediments. In I. Kaplan (ed.), *Natural gases in marine sediments.* New York: Plenum Press, pp. 99–139.

Claypool, G. E., and P. R. Reed. 1976. Thermal-analysis technique for source rock evaluation: Quantitative estimate of organic richness and effects of lithologic variation. *AAPG Bull.*, *60* (4), 608–626.

Claypool, G. E., A. H. Love, and E. K. Maughan. 1978. Organic geochemistry, incipient metamorphism, and oil generation in black shale members of Phosphoria Formation, western interior United States. *AAPG Bull.*, *62* (1), 98–120.

Claypool, G. E., C. M. Lubeck, J. M. Patterson, and J. P. Baysinger. 1976. Organic geochemical analyses of cores. U.S. Department of the Interior Geological Survey, Open-File Report 76–232. Washington, D.C.: U.S. Government Printing Office, pp. 45–56.

Clayton, J. L., and P. J. Swetland. 1978. Subaerial weathering of sedimentary organic matter. *Geochim. Cosmochim. Acta 42* (3), 305–312.

Clayton, R. B. 1965. Biogenesis of cholesterol and the fundamental steps in terpenoid biosynthesis. *Quart. Rev. London Chem. Soc., 19,* 168–230.

Coleman, D. D. 1976. The origin of drift gas deposits as determined by radiocarbon dating of methane. Paper presented at the Ninth International Radiocarbon Conference, June 20–26, 1976, University of California, Los Angeles, and San Diego.

Coleman, H. J., J. E. Dooley, D. E. Hirsch, and C. J. Thompson. 1973. Compositional studies of a high-boiling 370–535° distillate from Prudhoe Bay, Alaska crude oil. *Anal. Chem., 45* (9), 1724–1737.

Collins, A. G. 1975. *Geochemistry of oilfield waters: Developments in petroleum science.* Amsterdam: Elsevier. 496 p.

Columbo, U., and G. Sironi. 1959. Geochemical analysis of Italian oils and asphalts. In *5th. World Petroleum Congress, Proc.* Vol. 1. New York. pp. 177–205.

Columbo, U., F. Gazzarrini, R. Gonfiantini, E. Tongiorgi, and L. Caflisch. 1969. Carbon isotopic study of hydrocarbons in Italian natural gases. In P. A. Schenck and I. Havenaar (eds.), *Advances in organic geochemistry, 1968.* New York: Pergamon Press, pp. 499–516.

Combaz, A. 1975. Essai de classification des roches carbonees et des constituants organiques des roches sédimentaires. In B. Alpern (ed.), *Pétrographie organique et potentiel pétrolier.* Paris: Centre National de la Recherche Scientifique, pp. 93–101.

Comer, J. B., and R. Littlejohn. 1976. Content, composition, and thermal history of organic matter in Mesozoic sediments, Falkland Plateau. In P. F. Barker and I. W. D. Dalziel (eds.), *Initial reports of the Deep Sea Drilling Project, Leg 36.* Washington, D.C.: U.S. Government Printing Office, pp. 941–944.

Connan, J. 1967. Geochemical significance of the extraction of amino acids from sediments. *Bull. Centre Rech. Pau-SNPA, 1* (1), 165–171.

Connan, J. 1974. Time–temperature relation in oil genesis. *AAPG Bull., 58,* 2516–2521.

Connan, J. 1976. Time–temperature relation in oil genesis: Reply. *AAPG Bull., 60* (5), 885–887.

Connan, J., K. Le Tran, and B. Van Der Weide. 1975. Alteration of petroleum in reservoirs. In *9th World Petroleum Congress, Proc.* Vol. 2. London: Applied Science Publishers, pp. 171–178.

Conybeare, C. E. B. 1965. Hydrocarbon-generation potential and hydrocarbon-yield capacity of sedimentary basins. *Canadian Petrol. Geol. Bull., 13* (4), 509–528.

Cordell, R. J. 1972. Depths of oil origin and primary migration: A review and critique. *AAPG Bull., 56* (10), 2029–2067.

Correia, M. 1969. Contribution a la recherche de zones favorables a la genèse du pétrole par l'observation microscopique de la matière organique figuree. *Rev. de l'Inst. Français Pétrol., 24,* 1417–1454.

Coustau, H. 1977. Formation waters and hydrodynamics. *J. Geochem. Explor., 7,* 213–241.

Craig, H. 1953. The geochemistry of the stable carbon isotopes. *Geochem. Cosmochim. Acta, 3,* 53–92.

Craig, H. 1969. Geochemistry and origin of the Red Sea brines. In E. T. Degens and D. A. Ross (eds.), *Hot brines and recent heavy metal deposits in the Red Sea.* New York: Springer-Verlag, pp. 208–250.

Culberson, O. L., and J. J. McKetta, Jr. 1951. Phase equilibria in hydrocarbon–water systems. Part 3: The solubility of methane in water at pressures to 10,000 psia. *Petroleum Transactions, AIME, 192,* 223–226.

Datsko, V. G. 1959. *Organic matter in Soviet southern waters.* Moscow: Izd. Akad. Nauk SSSR.

Davis, J. B. 1967. *Petroleum microbiology.* Amsterdam: Elsevier. 604 p.

Davis, J. B., and E. E. Bray. 1969. Analyses of oil and cap rock from Challenger (Sigsbee) Knoll. In M. Ewing and J. L. Worzel (eds.), *Initial reports of the Deep Sea Drilling Project, Leg 1.* Washington, D.C.: U.S. Government Printing Office, pp. 415–426.

Davis, J. B., and R. M. Squires. 1954. Detection of microbially produced gaseous hydrocarbons other than methane. *Science, 119* (3090), 381–382.

Davis, J. C. 1970. Petrology of Cretaceous Mowry shales of Wyoming. *AAPG Bull., 54* (3), 487–502.

Davis, T. L. 1972. Velocity variations around Leduc Reefs, Alberta. Geophys., *17* (4), 548–604.

Day, W. C., and J. G. Erdman. 1963. Ionene: A thermal degradation product of β-carotene. *Science, 141*, 808.

Degens, E. T. 1969. Biogeochemistry of stable carbon isotopes. In G. Eglinton and M. T. J. Murphy (eds.), *Organic geochemistry: Methods and results.* New York: Springer-Verlag, pp. 304–329.

Degens, E. T., G. V. Chilingar, and W. D. Pierce. 1964. On the origin of petroleum inside freshwater carbonate concretians of Miocene age. In U. Columbo and G. D. Hobson (eds.), *Advances in organic geochemistry, 1962.* New York: Macmillan, pp. 149–164.

Demaison, G. J., and G. T. Moore. 1978. Anoxic conditions and organic carbon enrichment in modern aquatic sediments. Paper presented at the 10th International Congress on Sedimentology, July 9–14, 1978, Jerusalem, Israel.

Dembicki, H., Jr., W. G. Meinschein, and D. E. Hatton. 1975. Possible ecological and environmental significance of the predominance of even-carbon number C_{20}–C_{30} n-alkanes. *Geochim. Cosmochim. Acta, 39,* 203–208.

Deroo, G., T. G. Powell, B. Tissot, and R. G. McCrossan, with contributions by P. A. Hacquebard. 1977. *The origin and migration of petroleum in the western Canadian sedimentary basin, Alberta: A geochemical and thermal maturation study.* Geological Survey of Canada Bull. 262. Ottawa: Geological Survey of Canada. 136 p.

Deroo, G., B. Tissot, R. G. McCrossan, and F. Der. 1974. *Geochemistry of heavy oils of Alberta.* Canadian Soc. Petrol. Geol. Memoir 3. Calgary, Canada: Stacs Data Services, pp. 148–167.

Deroo, G., J. P. Herbin, J. Rouchaché, B. Tissot, P. Albrecht, and M. Dastillung. 1978. Organic geochemistry of some Cretaceous claystones from Site 391, Leg 44, western North Atlantic. In W. E. Benson and R. E. Sheridan (eds.), *Initial reports of the Deep Sea Drilling Project.* Vol. 44. Washington, D.C.: U.S. Government Printing Office, pp. 593–598.

Desouza, N. J., and W. R. Nes. 1968. Steroids: Isolation from a blue-green alga. *Science, 162,* 363.

Deuser, W. G., and E. T. Degens. 1967. Carbon isotope fractionation in the system CO_2 (gas)–CO_2 (aqueous)–HCO_3^- (aqueous). *Nature, 215,* 1033.

Deuser, W. G., E. T. Degens, and R. R. L. Guillard. 1968. Carbon isotope relationships between plankton and seawater. *Geochim. Cosmochim. Acta, 32,* 657–660.

Deuser, W. G., E. T. Degens, G. R. Harvey, and M. Rubin. 1973. Methane in Lake Kivu: New data bearing on its origin. *Science, 181,* 51–54.

Dickey, P. A. 1969. Increasing concentration of subsurface brines with depth. *Chem. Geol., 4,* 361–370.

Dickey, P. A. 1975. Possible primary migration of oil from source rock in oil phase. *AAPG Bull., 59* (2), 337–345.

Dickey, P. A., and A. Baharlou. 1973. Chemical composition of pore water in some Paleozoic shales from Oklahoma. Paper presented at Fourteenth Annual Logging Symposium of Society of Professional Well Log Analysts, May 6–9, 1973. 7 p.

Dickey, P. A., and J. M. Hunt. 1972. Geochemical and hydrogeologic methods of prospecting for stratigraphic traps. In R. E. King (ed.), *Stratigraphic oil and gas fields.* AAPG Memoir 16. Tulsa, Okla.: American Association of Petroleum Geologists, pp. 136–137.

Dickey, P. A., and C. Soto. 1974. Chemical composition of deep subsurface waters of the western Anadarko Basin. SPE-AIME Paper No. SPE 5178. Dallas: Society of Petroleum Engineers and American Institute of Mining, Metallurgical, and Petroleum Engineers. 17 p.

Dickey, P. A., A. G. Collins, and I. Fajardo M. 1972. Chemical composition of deep formation waters in southwestern Louisiana. *AAPG Bull., 56* (8), 1530–1533.

Dickey, P. A., R. E. Sherrill, and L. S. Matteson. 1943. *Oil and gas geology of the city quadrangle, Pennsylvania.* Harrisburg: Pennsylvania Geological Survey, 4th series, Bull. M25. 201 p.

Dickinson, G. 1953. Geological aspects of abnormal reservoir pressures in Gulf Coast Louisiana. *AAPG Bull., 37* (2), 410–432.

Dickinson, W. R. 1974. Subduction and oil migration. *Geology, 2,* 421–424.

Dooley, J. E., D. E. Hirsch, H. J. Coleman, and C. J. Thompson. 1973. *Compound-type separation and characterization studies for a 370° to 535° C distillate of Swan Hills (south field), Alberta, Canada, crude oil.* U.S. Department of the Interior, Bureau of Mines, Report 7821. Washington, D.C.: U.S. Department of the Interior, Bureau of Mines. 30 p.

Dorsey, N. E. 1940. *Properties of ordinary water-substance.* New York: Reinhold.

Dostalek, M., M. Staud, and A. Rosypalova. 1957. The action of microorganisms on petroleum hydrocarbons. *Cesk. Mikrobiol., 2* (1), 43–47.

Douglas, A. G. 1969. Gas chromatography. In G. Eglinton and M. T. J. Murphy (eds.), *Organic geochemistry: Methods and results.* New York: Springer-Verlag, pp. 161–180.

Douglas, A. G., and B. J. Mair. 1965. Sulfur: Role in genesis of petroleum. *Science, 147,* 499–501.

Dow, W. G. 1974. Application of oil-correlation and source-rock data to exploration in Williston Basin. *AAPG Bull., 58* (7), 1253–1262.

Dow, W. G. 1977a. Kerogen studies and geological interpretations. *J. Geochem. Explor., 7* (2), 77–79.

Dow, W. G. 1977b. Petroleum source beds on continental slopes and rises. In *Geology of continental margins.* AAPG Continuing Education Course Note Series 5, AAPG Continuing Education Program. Tulsa, Okla.: American Association of Petroleum Geologists. 37 p.

Dow, W. G. 1978. Geochemical analysis of samples from Holes 391A and 391C, Leg 44, Blake–Bahama Basin. In W. E. Benson and R. E. Sheridan (eds.), *Initial reports of the Deep Sea Drilling Project.* Vol. 44. Washington, D.C.: U.S. Government Printing Office, pp. 625–634.

Dow, W. G., and D. B. Pearson. 1975. Organic matter in Gulf Coast sediments. In *Preprints, Offshore Technology Conference, Houston, Texas,* pp. 85–94.

Drost-Hansen, W. 1969. Structure of water near solid interfaces. *I. & E. Chem., 61* (11), 10–46.

Drozdova, T. V., and A. V. Kochenov. 1960. Organic material of fossil fish bone. *Geochem.,* no. 8, pp. 900–903.

Dubrova, N. V., and Z. N. Nesmelova. 1968. Carbon isotope composition of natural methane. *Geokhim.,* no. 9, pp. 1066–1071.

Dufour, J. 1957. On regional migration and alteration of petroleum in South Sumatra. *Geol. en Mijnbouw,* new series, *19,* 172–181.

Dungworth, G., and A. W. Schwartz. 1974. Organic matter and trace elements in Pre-Cambrian rocks from South Africa. *Chem. Geol., 14* (3), 167–172.

Dunlap, H. F., and C. A. Hutchinson, Jr. 1961. Marine seep detection. *Offshore, 21,* 11–12.

Dunnington, H. V. 1958. Generation, migration, accumulation and dissipation of oil in northern Iraq. In L. G. Weeks (ed.), *Habitat of oil: A symposium.* Tulsa, Okla.: American Association of Petroleum Geologists, pp. 1194–1251.

Dunnington, H. V. 1967a. Aspects of diagenesis and shape change in stylolitic limestone reservoirs. In *7th World Petroleum Congress, Proc.* Vol. 2. London: Elsevier, pp. 339–352.

Dunnington, H. V. 1967b. Stratigraphical distribution of oil fields in the Iraq-Iran-Arabian Basin. *J. Inst. Petrol., 53* (520), 129–161.

Dunton, M. L., and J. M. Hunt. 1962. Distribution of low-molecular-weight hydrocarbons in Recent and ancient sediments. *AAPG Bull., 46* (12), 2246–2248.

Durand, B., and J. Espitalié. 1972. Formation and evolution of C_1 to C_1 hydrocarbons and permanent gases in the Toarcian clays of the Paris Basin. In H. R. v. Gaertner and H. Wehner (eds.), *Advances in organic geochemistry, 1971.* New York: Pergamon Press, pp. 455–468.

Durand, B., and J. Espitalié. 1976. Geochemical studies on the organic matter from the Douala Basin (Cameroon). Part 2: Evolution of kerogen. *Geochim. Cosmochim. Acta, 40,* 801–808.

Durmish'yan, A. G. 1973. The compaction of clay rocks. *Izd. Akad. Nauk SSSR,* Ser. Geol. (in Russian), no. 8, pp. 85–89. Translated in *Internat. Geol. Rev.,* 1973, *16* (6), 650–653.

Dzhangir'yantz, D. A. 1965. Geothermal characteristics of the Emba region. *Petrol. Geol., 9* (1), 55–61.

Eichmann, R., and M. Schidlowski. 1975. Isotopic fractionation between coexisting organic carbon–carbonate pairs in Precambrian sediments. *Geochim. Cosmochim. Acta. 39,* 585–595.

Eisma, E., and J. W. Jurg. 1967. Fundamental aspects of the diagenesis of organic matter and the formation of hydrocarbons. In *7th World Petroleum Congress, Proc.* Vol. 2. London: Elsevier, pp. 61–72.

Emery, K. O. 1960. *The sea off southern California.* New York: Wiley. 366 p.

Emery, K. O. 1975. Fossil fuels in the ocean. In *A joint meeting of National Academy of Sciences—National Academy of Engineering on national materials policy, Proc.* October 25–26, 1973, Washington, D.C. Washington, D.C.: U.S. Government Printing Office, pp. 79–86.

Emery, K. O., and D. Hoggan. 1958. Gases in marine sediments. *AAPG Bull., 42* (9), 2174–2188.

Emery, K. O., and B. J. Skinner. 1977. Mineral deposits of the deep-ocean floor. *Mar. Min., 1* (1–2), 1–71.

Engler, C. 1911–1912. Die Bildung der Hauptbestandteile des Erdöls. *Petrol, Zhurnal, 7,* 399.

Epstein, A. G., J. B. Epstein, and L. D. Harris. 1975. *Conodont color alteration—An index to organic metamorphism.* U.S. Department of the Interior, Geological Survey, Open-File Report 75-379. 54 p.

Erdman, J. G., 1962. Oxygen, nitrogen, and sulfur in asphalts. *API Division of Science and Technology Summer Symposium, 42* (8), 33–40.

Erdman, J. G. 1975. Geochemical formation of oil. In A. G. Fischer and S. Judson (eds.), *Petroleum and global tectonics.* Princeton, N.J.: Princeton University Press, pp. 225–248.

Erdman, J. G., and D. A. Morris. 1974. Geochemical correlation of petroleum. *AAPG Bull., 58* (11), 2326–2337.

Erdman, J. G., R. L. Borst, and W. J. Hines. 1969. Composition of gas sample 1 (cores) by components (1.5.1). In M. Ewing and J. L. Worzel (eds.), *Initial reports of the Deep Sea Drilling Project, Leg 1.* Vol. 1. Washington, D.C.: U.S. Government Printing Office, pp. 461–463.

Eremenko, N. A. 1961. *Geologia Nefti i Gaza* [*Geology of Oil and Gas*]. Leningrad: Gostoptechizdat. 372 p.

Eremenko, N. A., and I. M. Michailov. 1974. Hydrodynamic pools at faults. *Canadian Petrol. Geol. Bull., 22* (2), 106–118.

Espitalié, J., J. L. LaPorte, M. Madec, F. Marquis, P. Leplat, J. Paulet, and A. Boutefeu. 1977. Méthode rapide de caractérisation des roches mères de leur potentiel pétrolier et de leur degré d'evolution. *Rev. de l'Inst. Français Pétrol., 32* (1), 23–42.

Evans, C. R., and F. L. Staplin. 1971. Regional facies of organic metamorphism in geochemical exploration. In *3rd Internat. Geochemical Exploration Symposium, Proc.* Canadian Institute of Mining and Metallurgy, Special Vol. 11. Montreal: Canadian Institute of Mining and Metallurgy, pp. 517–520.

Evans, C. R., D. K. McIvor, and K. Magara. 1975. Organic matter, compaction history and hydrocarbon occurrence—MacKenzie Delta, Canada. In *9th World Petroleum Congress, Proc.* Vol. 2. London: Applied Science Publishers, pp. 149–157.

Farmer, R. E. 1965. Genesis of subsurface carbon dioxide. In A. Young and J. E. Galley (eds.), *Fluids in subsurface environments.* AAPG Memoir 4. Tulsa, Okla.: American Association of Petroleum Geologists, pp. 378–385.

Fenske, M. R., F. L. Carnahan, J. N. Breston, A. H. Caser, and A. R. Rescorla. 1942. Optical rotation of petroleum fractions. *I & E Chem., 34,* 638–646.

Fertl, W. H. 1972. Worldwide occurrence of abnormal formation pressures. Part 1. American Institute of Mining, Metallurgical, and Petroleum Engineers, Paper No. SPE 3844. Dallas: American Institute of Mining, Metallurgical, and Petroleum Engineers.

Fertl, W. H. 1976. *Abnormal formation pressures: Implications to exploration, drilling, and production of oil and gas resources.* Developments in Petroleum Science No. 2. Amsterdam: Elsevier. 382 p.

Fertl, W. H., and D. J. Timko. 1972. How downhole temperatures affect drilling pressures, an engineering practices report. Part 3. *World Oil, 179,* 36–39, 66.

Forgotson, J. M., Sr. 1969. Indication of proximity of high-pressure fluid reservoir, Louisiana and Texas Gulf Coast. *AAPG Bull., 53* (1), 171–173.

Forsman, J. P., and J. M. Hunt. 1958. Insoluble organic matter (kerogen) in sedimentary rocks of marine origin. In L. G. Weeks (ed.), *Habitat of oil: A symposium.* Tulsa, Okla.: American Association of Petroleum Geologists, pp. 747–778.

Frank, D. J., J. R. Gormly, and W. M. Sackett. 1974. Revaluation of carbon isotope compositions of natural methanes. *AAPG Bull., 58* (11), 2319–2325.

Friedman, G. M. 1975. The making and unmaking of limestones or the downs and ups of porosity. *J. Sed. Petrol., 45* (2), 379–398.

Frost, A. V. 1945. The role of clays in the formation of petroleum in the crust. *Progress in Chemistry, 14* (6), 1–20.

Fuex, A. N. 1977. The use of stable carbon isotopes in hydrocarbon exploration. *J. Geochem. Explor., 7* (2), 155–188.

Gabrielyan, A. G. 1962. On the formation of oil and gas pools in the Volgograd–Volga region. *Geologicheskoe Stroenie i Neftegazonosnost' Volgogradskoi Oblasti, 1,* 248–273.

Gagosian, R. B. 1976. A detailed vertical profile of sterols in the Sargasso Sea. *Limnol. and Oceanogr., 21* (5), 702–710.

Galimov, E. M. 1968. Isotopic composition of carbon in gases of the crust. *Internat. Geol. Rev., 11* (10), 1092–1104.

Galimov, E. M., N. G. Kuznetsova, and V. S. Prokhorov. 1968. The composition of the former atmosphere of the earth as indicated by isotopic analysis of Pre-Cambrian carbonates. *Geokhim.,* no. 11, pp. 1376–1381.

Gallegos, E. J. 1973. Identification of phenylcycloparaffin alkanes and other monoaromatics in Green River Shale by gas chromatography–mass spectrometry. *Anal. Chem., 45* (8), 1399–1403.

Gallegos, E. J. 1975. Terpane–sterane release from kerogen by pyrolysis gas chromatography–mass spectrometry. *Anal. Chem., 47* (9), 1524–1528.

Gallegos, E. J., J. W. Green, L. P. Lindeman, R. L. LeTourneau, and R. M. Teeter. 1967. Petroleum group-type analysis by high-resolution mass spectrometry. *Anal. Chem., 39* (14), 1833–1838.

Galwey, A. K. 1972. The rate of hydrocarbon desorption from mineral surfaces and the contribution of heterogeneous catalytic-type processes to petroleum genesis. *Geochim. Cosmochim. Acta., 36* (10), 1115–1130.

Gardner, F. J. 1971. Here are the whys, wherefores of elephant hunting. *Oil and Gas J., 69,* 108–110.

Gavrilov, A. Ya., and V. S. Dragunskaya. 1963. Aromatic condensates of Eastern Turkmen, SSR. *Akad. Nauk Turkm. SSR*, Ser. F12. *Khm. Geol. Nauk, 3,* 111–113.

Gehman, H. M., Jr. 1962. Organic matter in limestones. *Geochim. Cosmochim. Acta, 26,* 885–897.

Gelpi, E., H. Schneider, J. Mann, and J. Oro. 1970. Hydrocarbons of geochemical significance in microscopic algae. *Phytochem., 9,* 603–612.

Geodekian, A. A., and V. A. Stroganov. 1973. Geochemical prospecting for oil and gas using reference horizons. *J. Geochem. Explor., 2,* 1–9.

Gerarde, H. W., and D. F. Gerarde. 1961–1962. *The ubiquitous hydrocarbons.* Vols. 25, 26. Washington, D.C.: Association of Food and Drug Officials of the United States, pp. 1–47.

Germann, F. E. E., and W. Ayres. 1942. The origin of underground carbon dioxide. *J. Phys. Chem., 46,* 61–68.

Germanov, A. I. 1965. Geochemical significance of organic matter in the hydrothermal process. *Geokhim.,* no. 7, pp. 834–843.

Getz, F. A. 1977. Molecular nitrogen: Clue in coal-derived-methane hunt. *Oil and Gas J., 75,* 220–221.

Gimplevich, E. D. 1960. Similarities and differences in the chemical composition of oil and bituminous hydrocarbons from scattered organic substances. In *Problema Proiskhozdeniya Nefti i Gaza i Usloviya Forirovaniya ikh Zalezheli.* Moscow: Trudy Vesesoyuznogo Soveshchaniya, pp. 455–459.

Giraud, A. 1970. Application of pyrolysis and gas chromatography to geochemical characterization of kerogen in sedimentary rock. *AAPG Bull., 54* (3), 439–455.

Gnilovskaja, M. G. 1971. The oldest aquatic plants of the Wendian of the Russian Platform (late Pre-Cambrian). *J. Paleontol., 5* (3), 372–378.

Goldberg, I. S. 1973. Solid bitumens in petroleum deposits of the Baltic region as indicators of stages in the migration of petroleum. *Dokl. Akad. Nauk SSSR, 209* (2), 462–465.

Goldberg, I. S., and K. A. Chernikov. 1968. The nature of bitumens in the alkalic and ultramafic rocks on the northern part of the Siberian platform. *Geochem.,* no. 4, pp. 476–482.

Goldstein, T. P. In press. Geocatalytic reactions in the formation of oil. *AAPG Bull.*

Goncharov, V. S., E. S. Goncharov, V. F. Perepelichenko, and V. G. Khel'kvist. 1973. Possible reserves of hydrogen sulfide from connate fluids. *Petrol. Geol., 3,* 445–452.

Gorokhov, S. S., N. I. Petrova, and V. S. Kovalenko. 1973. Experimental study of the alteration of biogenic carbon at high temperatures and pressures. *Dokl. Akad. Nauk SSSR, 209,* 194–196.

Gorskaya, A. I. 1950. Investigations of the organic matter of Recent sediments. In *Recent analogs of petroliferous facies* (Symposium). Moscow: Gostoptekhizdat.

Govett, G. J. S. 1966. Origin of banded iron formations. *Geol. Soc. Amer. Bull., 77,* 1191–1212.

Gransch, J. A., and E. Eisma. 1970. Characterization of the insoluble organic matter of sediments by pyrolysis. In G. D. Hobson and G. C. Speers (eds.), *Advances in organic geochemistry, 1966.* New York: Pergamon, pp. 407–426.

Gray, J., and A. J. Boucot. 1975. Color changes in pollen and spores: A review. *Geol. Soc. Amer. Bull., 86,* 1019–1033.

Grayson, J., and R. E. La Plante. 1973. Estimated temperature history in the lower part of Hole 181 from carbonization measurements. In L. D. Kulm and R. von Huene (eds.), *Initial reports of the Deep Sea Drilling Project, Leg 18.* Washington, D.C.: U.S. Government Printing Office. 1077 p.

Greensfelder, B. S., H. H. Voge, and G. M. Good. 1949. Catalytic and thermal cracking of pure hydrocarbons. *I. & E. Chem., 41*, 2573–2584.

Gretener, P. E. 1976. *Pore pressure: Fundamentals, general ramifications and implications for structural geology.* AAPG Continuing Education Program Course Note Series No. 4. Tulsa, Okla.: American Association of Petroleum Geologists. 87 p.

Grob, K. 1975. The glass capillary column in gas chromatography: A tool and a technique. *Chromatographia, 8* (9), 423–433.

Grob, K., K. Grob, Jr., and G. Grob. 1975. Organic substances in potable water and its precursor. Part 3: The closed-loop stripping procedure compared with rapid liquid extraction. *J. Chromatog., 106*, 299–315.

Gurevich, M. S. 1954. *Materials for oil prospecting hydrogeology of the southern part of the West Siberian lowland.* Vol. 1. Trudy VSEGEI. Moscow: Gosgeoltekhizdat.

Gurevich, M. S. 1956. *West Siberian Artesian Basin, supplement to hydrochemical map of Siberia and the Far East.* Trudy VSEGEI. Moscow: Gosgeoltekhizdat.

Guseva, A. N., and L. A. Fayngersh. 1973. Conditions of accumulation of nitrogen in natural gases as illustrated by the central European and Chu-Sarysu oil–gas basins. *Dokl. Akad. Nauk SSSR, 209* (2), 210–212.

Gutjahr, C. C. M. 1966. Carbonization measurements of pollen grains and spores and their application. *Leidse Geol. Mededelingen, 38*, 1–29.

Halbouty, M. T. 1970. Geology of giant petroleum fields: Introduction. In M. T. Halbouty (ed.), *Geology of giant petroleum fields.* AAPG Memoir 14. Tulsa, Okla.: American Association of Petroleum Geologists, pp. 1–7.

Halbouty, M. T., A. A. Meyerhoff, R. E. King, R. H. Dott, Sr., H. D. Klemme, and T. Shabad. 1970. World's giant oil and gas fields. In M. T. Halbouty (ed.), *Geology of giant petroleum fields.* AAPG Memoir 14. Tulsa, Okla.: American Association of Petroleum Geologists, pp. 502–528.

Hammond, D. E. 1974. Dissolved gases in Cariaco Trench sediments: Anaerobic diagenesis. In I. R. Kaplan (ed.), *Natural gases in marine sediments.* New York: Plenum Press, pp. 71–89.

Han, J., and M. Calvin. 1969. Hydrocarbon distribution of algae and bacteria and microbiological activity in sediments. *Proc. Nat. Acad. Sci. U.S.A., 64* (2), 436–443.

Hanshaw, B. B., and T. B. Coplen. 1973. Ultra-filtration by a compacted clay membrane. Part 2: Sodium ion exclusion at various ionic strengths: *Geochim. Cosmochim. Acta, 37* (10), 2311–2327.

Hare, P. E. 1969. Geochemistry of peptides and amino acids. In G. Eglinton and M. T. J. Murphy (eds.), *Organic geochemistry: Methods and results.* New York: Springer-Verlag, pp. 438–463.

Hare, P. E. 1973. Amino acids, amino sugars, and ammonia in sediments from the Cariaco Trench. In B. C. Heezen and I. D. MacGregor (eds.), *Initial reports of the Deep Sea Drilling Project.* Vol. 20. Washington, D.C.: U.S. Government Printing Office, pp. 941–942.

Harwood, R. J. 1973. Biodegradation of oil. In *The geology of fluids and organic matter in sediments*. National Conference on Earth Science, April 30–May 4, 1973, Banff, Alberta, Canada. Edmonton: University of Alberta, pp. 149–156.

Harwood, R. J. 1977. Oil and gas generation by laboratory pyrolysis of kerogen. *AAPG Bull., 61* (12), 2082–2102.

Hase, A., and R. A. Hites. 1976. On the origin of polycyclic aromatic hydrocarbons in Recent sediments: Biosynthesis by anaerobic bacteria. *Geochim. Cosmochim. Acta, 40*, 1141–1143.

Heacock, R. L., and A. Hood. 1970. Process for measuring the live carbon content of organic samples. U.S. Patent 3.508.877. Patent Trademark Office, Washington, D.C. 20231.

Hedberg, H. D. 1926. The effect of gravitational compaction on the structure of sedimentary rocks. *AAPG Bull., 10* (11), 1035–1073.

Hedberg, H. D. 1936. Gravitational compaction of clays and shales. *Am. J. Sci.*, fifth series, *31* (184), 241–287.

Hedberg, H. D. 1954. World oil prospects from a geological viewpoint. *AAPG Bull., 38*, 1724.

Hedberg, H. D. 1964. Geological aspects of origin of petroleum. *AAPG Bull., 48* (11), 1755–1803.

Hedberg, H. D. 1968. Significance of high-wax oils with respect to genesis of petroleum. *AAPG Bull., 52* (5), 736–750.

Hedberg, H. D. 1974. Relation of methane generation to undercompacted shales, shale diapirs, and mud volcanoes. *AAPG Bull., 58* (4), 661–673.

Hedberg, H. D., L. C. Sass, and H. J. Funkhouser. 1947. Oil fields of Greater Oficina area, Central Anzoategui, Venezuela. *AAPG Bull., 31* (12), 2089–2149.

Hedberg, W. H. 1967. Pore water chlorinities of subsurface shales. Doctoral dissertation, University of Wisconsin. Ann Arbor, Mich.: University Microfilms, No. 67–12. 121 p.

Heling, D. 1971. Alteration of smectitic mixed-layered minerals to illite controlled by temperature in Tertiary shales of the Rhein Graben. Paper presented at 8th International IAS Sedimentological Congress Program, July 1971, Heidelberg, Germany.

Hilt, C. 1873. Die Beziehungen zwischen der Zusammensetzung und den technischen Eigenschaften der Steinkohle. *Sitzber. Aachener Bezirksvereinigung VDI, 4.*

Hitchon, B. 1963. Geochemical studies of natural gas. *J. Canadian Petrol. Tech., 2* (3), 100–116; (4), 165–174.

Hitchon, B. 1974a. Application of geochemistry to the search for crude oil and natural gas. In A. A. Levinson (ed.), *Introduction to exploration geochemistry*. Calgary, Canada: Applied Publishing, pp. 509–545.

Hitchon, B., 1974b. Occurrence of natural gas hydrates in sedimentary basins. In I. R. Kaplan (ed.), *Natural gases in marine sediments*. Vol. 3. New York: Plenum Press, pp. 195–225.

Hitchon, B., and M. Gawlak. 1972. Low molecular weight aromatic carbons in gas condensates from Alberta, Canada. *Geochim. Cosmochim. Acta., 36*, 1043–1059.

Hitchon, B., and M. K. Horn. 1974. Petroleum indicators in formation waters from Alberta, Canada. *AAPG Bull., 58* (3), 464–473.

Hobson, G. D., and E. N. Tiratsoo. 1975. *Introduction to petroleum geology.* Beaconsfield, England: Scientific Press. 300 p.

Hodgson, G. W. 1954. Vanadium, nickel and iron trace metals in crude oils of western Canada. *AAPG Bull., 38* (12), 2537–2554.

Hoefs, J. 1969. Carbon abundance in common igneous and metamorphic rock types. In K. H. Wedepohl (ed.), *Handbook of geochemistry.* New York: Springer-Verlag, pp. 6E1–6, 6M–1.

Holder, G. D., D. L. Katz, and J. H. Hand. 1976. Hydrate formation in subsurface environments. *AAPG Bull., 60,* (6), 981–988.

Holland, H. D. 1962. Model for the evolution of the earth's atmosphere. In A. E. J. Engel, H. L. James, and B. F. Leonard (eds.), *Petrologic studies: A volume to honor A. F. Buddington.* Geological Society of America, pp. 447–477.

Holland, H. D. 1973. Ocean water, nutrients and atmospheric oxygen. In *Proc. Symposium on Hydrogeochemistry and Biochemistry.* Vol. 1. Washington, D.C.: Clark, pp. 68–81.

Holland, H. D. 1975. The evolution of seawater. Paper presented at the meeting on early Earth, Spring 1975, Leicester, England. 12 p.

Holmquest, H. J. 1965. Deep pays in Delaware and Val Verde basins. In A. Young and J. E. Galley (eds.), *Fluids in subsurface environments.* AAPG Memoir 4. Tulsa, Okla.: American Association of Petroleum Geologists, pp. 257–279.

Hood, A., and J. R. Castaño. 1974. Organic metamorphism: Its relationship to petroleum generation and application to studies of authigenic minerals. *United Nations ESCAP CCOP* [Committee for Coordination of Joint Prospecting for Mineral Resources in Asian Offshore Areas] *Technical Bull., 8,* 85–118.

Hood, A., C. C. M. Gutjaha, and R. L. Heacock. 1975. Organic metamorphism and the generation of petroleum. *AAPG Bull., 59,* 986–996.

Hubbert, M. K. 1953. Entrapment of petroleum under hydrodynamic conditions. *AAPG Bull., 37,* 1954–2026.

Hubbert, M. K., and W. W. Rubey. 1959. Role of fluid pressure in mechanics of overthrust faulting. Part 1: Mechanics of fluid-filled porous solids and its application to overthrust faulting. *AAPG Bull., 70,* 115–165.

Huc, A. Y., B. Durand, and J. C. Monin. 1978. Humic compounds and kerogens in cores from Black Sea sediments, Leg 42B, Holes 379A, B, and 380A. In D. A. Ross and Y. P. Neprochov (eds.), *Initial reports of the Deep Sea Drilling Project, Leg 42B.* Vol. 42, Part 2. Washington, D.C.: U.S. Government Printing Office, pp. 737–748.

Hunt, J. M. 1953. Composition of crude oil and its relation to stratigraphy in Wyoming. *AAPG Bull., 37,* 1837–1872.

Hunt, J. M. 1954. Chemistry applied to exploration. In *Preprints, 10th Southwest Regional ACS Mtg.,* Fort Worth, Texas, December 2–4, 1954. Washington, D.C.: American Chemical Society. 6 p.

Hunt, J. M. 1961. Distribution of hydrocarbons in sedimentary rocks. *Geochim. Cosmochim. Acta, 22,* 37–49.

Hunt, J. M. 1963a. Composition and origin of the Uinta Basin bitumens. In A. L. Crawford (ed.), *Oil and gas possibilities of Utah, re-evaluated.* Utah Geological and Mineralogical Survey, Bull. 54, Paper 24. Salt Lake City: Utah Geological Society, pp. 249–273.

Hunt, J. M. 1963b. Geochemical data on organic matter in sediments. In V. Bese (ed.), *3rd International Scientific Conference on Geochemistry, Microbiology and Petroleum Chemistry, Proc.* Vol. 2. Budapest, October 8–13. Budapest: KULTURA, pp. 394–412.

Hunt, J. M. 1967. The origin of petroleum in carbonate rocks. In G. V. Chilingar, H. J. Bissell, and R. W. Fairbridge (eds.), *Carbonate rocks.* New York: Elsevier, pp. 225–251.

Hunt, J. M. 1968. How gas and oil form and migrate. *World Oil, 167,* 140–150.

Hunt, J. M. 1970. The significance of carbon isotope variations in marine sediments. In G. D. Hobson and G. C. Speers (eds.), *Advances in organic geochemistry, 1966.* New York: Pergamon Press, pp. 27–35.

Hunt, J. M. 1972. Distribution of carbon in the crust of the earth. *AAPG Bull., 56* (11), 2273–2277.

Hunt, J. M. 1973. An examination of petroleum migration processes. *Doklady Po Geokhimicheskim i Khimiko-Fizicheskim Voprosam Razvedki i Dobychi Nefti i Gaza, 1,* 219–229.

Hunt, J. M. 1975a. Is there a geochemical depth limit for hydrocarbons? *Petrol. Eng., 47* (3), 112–127.

Hunt, J. M. 1975b. Origin of gasoline range alkanes in the deep sea. *Nature, 254* (5499), 411–413.

Hunt, J. M. 1976. Origin of Athabaska oil. *AAPG Bull., 60,* 1112.

Hunt, J. M. 1977a. Distribution of carbon as hydrocarbons and asphaltic compounds in sedimentary rocks. *AAPG Bull., 61* (1), 100–104.

Hunt, J. M. 1977b. Ratio of petroleum to water during primary migration in Western Canada Basin. *AAPG Bull., 61* (3), 434–435.

Hunt, J. M. 1978. Characterization of bitumens and coals. *AAPG Bull., 62* (2), 301–303.

Hunt, J. M. 1979. Hydrocarbon studies in deep ocean sediments. In E. W. Baker (ed.), *Symposium on organic geochemistry of Deep Sea Drilling Project cores.* Princeton, N.J.: Science Press.

Hunt, J. M., and G. W. Jamieson. 1956. Oil and organic matter in source rocks of petroleum. *AAPG Bull., 40* (3), 477–488.

Hunt, J. M., and A. L. Kidwell. 1958. Oil migration in recent sediments. *World Oil, 147,* 79–83.

Hunt, J. M., and R. N. Meinert. 1954. Petroleum prospecting. U.S. Patent 2.854.396. Patent Trademark Office, Washington, D.C. 20231.

Hunt, J. M., and J. K. Whelan. 1978a. Dissolved gases in Black Sea sediments. In D. A. Ross and Y. P. Neprochov (eds.), *Initial reports of Deep Sea Drilling Project, Leg 42B.* Vol. 42, Part 2. Washington, D.C.: U.S. Government Printing Office, pp. 661–665.

Hunt, J. M., and J. K. Whelan. 1978b. Light hydrocarbons in sediments of DSDP Leg 44 holes. In W. E. Benson and R. E. Sheridan (eds.), *Initial reports of the Deep Sea Drilling Project.* Vol. 44. Washington, D.C.: U.S. Government Printing Office, pp. 651–652.

Hunt, J. M., F. Stewart, and P. A. Dickey. 1954. Origin of hydrocarbons of Uinta Basin, Utah. *AAPG Bull., 38* (8), 1671–1698.

Hunt, T. S. 1861. Notes on the history of petroleum or rock oil. *Canadian Natur. and Geol.*, *6* (4), 241–255.

Hunt, T. S. 1863. Report on the geology of Canada. In *Canadian Geological Survey Report: Progress to 1863*. Canadian Geological Survey. 983 p.

Illing, V. C. 1933. Migration of oil and natural gas. *J. Inst. Petrol.*, *19* (114), 229–274.

Initial reports of the Deep Sea Drilling Project. 1971–1978. Washington, D.C.: U.S. Government Printing Office.

International petroleum encyclopedia. (J. McCaslin, ed.). 1975. Tulsa, Okla.: Petroleum Publishing. 480 p.

Jackson, J. R., Jr. 1977. Independents-majors: Their exploratory role. *Oil and Gas J.*, *75*, 95–96.

Jeffrey, D. A., and W. M. Zarella. 1970. Geochemical prospecting at sea. *AAPG Bull.*, *54*, (5), 583–584.

Jewell, D. M., J. H. Weber, J. W. Bunger, H. Plancher, and D. R. Lantham. 1972. Ion-exchange, coordination and adsorption chromatographic separation of heavy-end petroleum distillates. *Anal. Chem.*, *44* (8), 1391–1395.

Jobson, A., F. D. Cook, and D. W. S. Westlake. 1972. Microbial utilization of crude oil. *Applied Microbiol.*, *23*, (6), 1082–1089.

Johns, W. D., and A. Shimoyama. 1972. Clay minerals and petroleum-forming reactions during burial and diagenesis. *AAPG Bull.*, *56*, 2160–2167.

Jones, P. H., and R. H. Wallace, Jr. 1974. Hydrogeologic aspects of structural deformation in the northern Gulf of Mexico Basin. *J. Res. U.S. Geol. Survey, 2* (5), 511–517.

Jones, R. W. 1978. Some mass balance and geological restraints on migration mechanisms. In *Physical and chemical constraints on petroleum migration*. Vol. 2. Notes for AAPG short course, April 9, 1978, AAPG National Meeting, Oklahoma City. 43 p.

Jüntgen, H., and J. Klein. 1975. Origin of natural gas from coaly sediments. *Erdöl und Kohle, 28* (2), 65–73.

Juranek, J. 1958a. A contribution to the problem of the origin of C_1–C_5 hydrocarbons in samples of soil air in gas survey work. *Czech. Inst. Petrol. Res., Transactions, 9*, 57–79.

Juranek, J. 1958b. Das Verhältnis der Oberflächen-Anomalien hangender Kennzeichen des Bitumengehaltes in Gesteinen zu den Bitumenlagersätten. *Chemie der Erde, 19* (4), 450–453.

Kalomazov, R. U., and M. A. Vakhitov. 1975. Appearance and nature of anomalously high formation pressures in the Kuyab Mega syncline of the Tadzhik depression. *Nefti Gazovaya Geolog. Geofiz.*, no. 10, pp. 3–6.

Kaneda, T. 1969. Hydrocarbons in spinach: Two distinctive carbon ranges of aliphatic hydrocarbons. *Phytochem., 8*, 2039–3044.

Kanehara, K., K. Motojima, and Y. Ishiwada. 1958. Natural gas. Tokyo: Asakurashoten. Cited in R. Sugisaki, 1964, *AAPG Bull., 48* (1), 85–101.

Kartsev, A. A. 1963. *Hydrogeology of oil and gas fields.* Moscow: Gosudarstvennoe Nauchno-Tekhnicheskoe Izdatel'stvo Neftyanoi y Gorno-Toplivnoi Literatury. 353 p.

Kartsev, A. A. 1964. Geochemical transformation of petroleum. In U. Columbo and G. D. Hobson (eds.), *Advances in organic geochemistry, 1962.* New York: Pergamon Press, pp. 11–14.

Kartsev, A. A., N. B. Vassoevich, A. A. Geodekian, S. G. Neruchev, and V. A. Sokolov. 1971. The principal stage in the formation of petroleum. In *8th World Petrol. Congr., Proc.* Vol. 2. London: Applied Science Publishers, pp. 3–11.

Karweil, J. 1969. Aktuelle Probleme der Geochemie der Kohle. In P. A. Schenk and I. Havenaar (eds.), *Advances in organic geochemistry, 1968.* Oxford, England: Pergamon Press, pp. 59–84.

Katz, D. L. 1971. Depths to which frozen gas fields (gas hydrates) may be expected. *J. Petrol. Tech., 23,* 419–558.

Katz, D. L., D. Cornell, R. Kobayashi, F. H. Poettmann, J. A. Vary, J. R. Elenbaas, and C. F. Weinaug. 1959. Water-hydrocarbon systems. In D. L. Katz et al. (eds.), *Handbook of natural gas engineering.* New York: McGraw-Hill, pp. 189–221.

Kawai, K. 1963. Some considerations concerning the genesis of natural gas deposits of dissolved-in-water type in the Kuzasa Group, with special reference to diffusion of methane from gas reservoirs. *J. Japanese Assoc. Petrol. Tech., 28* (1), 6–15.

Kawai, K., and S. Totani. 1971. Relationship between crude-oil properties and geology in some gas fields in the Nigata Basin, Japan. *Chem. Geol., 8,* 219–246.

Kendrick, J. W., A. Hood, and J. R. Castaño. 1978a. Petroleum-generating potential of sediments from Leg 41, Deep Sea Drilling Project. In J. Gardner and H. Herring (eds.), *Initial reports of the Deep Sea Drilling Project.* Vol. 41. Washington, D.C.: U.S. Government Printing Office, pp. 817–820.

Kendrick, J. W., A. Hood, and J. R. Castaño. 1978b. Petroleum-generating potential of sediments from the Eastern Mediterranean and Black Seas. In J. I. Usher and P. Supko (eds.), *Initial reports of the Deep Sea Drilling project.* Vol. 42. Washington, D.C.: U.S. Government Printing Office, pp. 729–735.

Kennedy, W. A. 1963. Solubilization of hydrocarbons as a process of formation of petroleum deposits. Doctoral thesis, University of Texas at Austin.

Kharaka, Y. K., and W. C. Smalley. 1976. Flow of water and solutes through compacted clays. *AAPG Bull., 60* (6), 973–980.

Khitarov, N. I., and V. A. Pugin. 1966. Behavior of montmorillonite under elevated temperatures and pressures. *Geokhim.,* no. 7, pp. 790–795. English translation, *Geochemistry,* no. 3, pp. 621–626.

Kidwell, A. L., and J. M. Hunt. 1958. Migration of oil in Recent sediments of Pedernales, Venezuela. In L. G. Weeks. (ed.), *Habitat of oil: A symposium.* Tulsa, Okla.: American Association of Petroleum Geologists, pp. 790–817.

Kim, A. G. 1973. *The composition of coalbed gas.* U.S. Bureau of Mines, Report of Investigations, RI7762. Washington, D.C.: U.S. Department of Interior, Bureau of Mines. 9 p.

King, L. H. 1963. *On the origin of anthraxolite and impsonite.* Mines Branch Research Report R116. Ottawa: Department of Mines and Technical Surveys. 9 p.

King, L. H., F. E. Goodspeed, and D. S. Montgomery. 1963. *A study of sedimented organic matter and its natural derivatives.* Mines Branch Research Report R114. Ottawa: Department of Mines and Technical Surveys. 68 p.

Kissin, I. G., and S. I. Pakhomov. 1967. The possibility of carbon dioxide generation at depth at moderately high temperatures. *Dokl. Akad. Nauk SSSR, 174,* 181–183.

Klein, J., and H. Jüntgen. 1972. Studies on the emission of elemental nitrogen from coals of different rank and its release under geochemical conditions. In H. R. v. Gaertner and H. Wehner (eds.), *Advances in organic geochemistry, 1971.* New York: Pergamon Press, pp. 647–656.

Klemme, H. D. 1972. Heat influences size of oil giants. Parts 1 and 2. *Oil and Gas J., 70* (29, 30), 76–78, 134–144.

Klemme, H. D. 1975. Geothermal gradients, heat flow and hydrocarbon recovery. In A. G. Fischer and S. Judson (eds.), *Petroleum and global tectonics.* Princeton, N.J.: Princeton University Press, pp. 251–306.

Knebel, G. M., and G. Rodriguez-Eraso. 1956. Habitat of some oil. *AAPG Bull., 40* (4), 547–561.

Knoche, H., and G. Ourisson. 1967. Organic compounds in fossil plants (*Equisetum,* horsetails). *Angew. Chem., Internat. Ed., 6,* 1085.

Komae, H. 1960. Catalytic reaction of Japanese acid clay with 1,8-cineole and with limonene. *J. Sci. Hiroshima Univ.,* Series A, *24,* 699–731.

Kononova, M. M. 1966. *Soil organic matter.* New York: Pergamon Press.

Konyukhov, A. I., and G. I. Teodorovich. 1969. Optimum depths for the formation of oil in the terrigenous Jurassic in eastern Ciscaucasia. *Dokl. Akad. Nauk SSSR, 188* (2), 50–53.

Koons, C. B., J. G. Bond, and F. L. Peirce. 1974. Effects of depositional environment and postdepositional history on chemical composition of Lower Tuscaloosa oils. *AAPG Bull., 58* (7), 1272–1280.

Koons, C. B., G. W. Jamieson, and L. S. Ciereszko. 1965. Normal alkane distributions in marine organisms: Possible significance to petroleum origin. *AAPG Bull., 49* (3), 301–316.

Kortsenshteyn, V. N. 1968. Comparison of the benzene content of stratal water in Mesozoic complexes of southern Mangyshlak and eastern Ciscaucasia. *Dokl. Akad. Nauk SSSR, 180* (3), 697–699.

Kotel'nikov, D. D., and T. V. Florenskaya. 1974. Clay minerals in Upper Riphean–Wendian terrigenous rocks and in the base of the Markovsk deposit. *Internat. Geol. Rev., 16* (2), 194–201.

Kozlov, V. P. 1960. The heavy hydrocarbons in the gases of coals of the Donets Basin. *Petrol. Geol., 4* (6A), 363–365.

Kravets, V. V. 1974. Temperature field of the Dnieper–Donets depression and distribution of oil and gas pools within it. *Geologiya i Geokhimiya Goryuchikh Iskopayemykh,* no. 39, pp. 58–66. English translation, *Petrol. Geol., 12* (6), 1975, 275–278.

Krouse, H. R. 1977. Sulfur isotope studies and their role in petroleum exploration. *J. Geochem. Explor., 7* (2), 189–212.

Kruglikov, N. M. 1964. *Hydrogeology of the northwest border of the West Siberian Artesian Basin.* Moscow: Nedra.

Kudryatseva, Y. I., A. A. Andreyeva, and O. I. Suprunenko. 1974. Discovery of natural kerosene in southwestern Kamchatka. *Dokl. Akad. Nauk SSSR, 216* (2), 418–421.

Kudryavtsev, N. A. 1959. *Oil, gas, and solid bitumen in igneous and metamorphic rocks.* Vses. Nauchno-Issledovatel'skogo, Geologo Razvedchnogo Instituta, no. 142. 263 p.

Kushnareva, T. I., 1971. Evidence of hydrothermal activity in oil-bearing carbonate rocks of the Upper Devonian of the Pechora Ridge. *Dokl. Akad. Nauk SSSR, 198* (1), 175–177.

Kvenvolden, K. A. 1972. Organic geochemistry of early Precambrian sediments. In *24th International Geological Congress, Proc.* Montreal, Quebec, Sec. 1. Ottawa: International Geological Congress; Gardenvale, Pa.: Harpell's Press Cooperative, pp. 31–41.

Kvenvolden, K. A., and R. M. Squires. 1967. Carbon isotopic composition of crude oils from Ellenburger Group (Lower Ordovician), Permian Basin, West Texas and eastern New Mexico. *AAPG Bull., 51* (7), 1293–1303.

Landa, St., Sl. Hala. 1958. Über die Bestimmung von Adamantan in Rohölen durch Bildung von Addukten mit Thioharnstoff. *Erdöl und Kohle, 11,* 698–700.

Landes, K. K. 1967. Eometamorphism and oil and gas in time and space. Part 1. *AAPG Bull., 51* (6), 828–841.

Landes, K. K. 1973. Mother Nature as an oil polluter. *AAPG Bull., 57* (4), 637–641.

LaPlante, R. E. 1974. Hydrocarbon generation in Gulf Coast Tertiary sediments. *AAPG Bull., 58* (7), 1281–1289.

Larskaya, Ye. S., and D. V. Zhabrev. 1964. Effects of stratal temperatures and pressures on the composition of dispersed organic matter (From the example of the Mesozoic-Cenozoic deposits of the western Ciscaspian region): *Dokl. Akad. Nauk SSSR, 157* (4), 135–139.

Le Tran, K. 1972. Geochemical study of hydrogen sulfide absorbed in sediments. In H. R. v. Gaertner and H. Wehner (eds.), *Advances in organic geochemistry, 1971.* New York: Pergamon Press, pp. 717–726.

Le Tran, K., J. Connan, and B. Van Der Weide. 1974. Problems relatifs à la formation d'hydrocarbures et d'hydrogène sulfuré dans Le Bassin Sud-ouest Aquitain. In B. Tissot and F. Bienner (eds.), *Advances in organic geochemistry, 1973.* Paris: Éditions Technip, pp. 761–789.

Leventhal, J. S. 1976. Stepwise pyrolysis–gas chromatography of kerogen in sedimentary rocks. *Chem. Geol., 18,* 5–20.

Levorsen, A. I. 1967. *Geology of petroleum.* (2nd ed.). San Francisco: W. H. Freeman. 724 p.

Levy, E. J., R. R. Doyle, R. A. Brown, and F. W. Melpolder. 1961. Identification of components in paraffin wax by high-temperature gas chromatography and mass spectrometry. *Anal. Chem., 33* (6), 698–704.

Lewis, C. R., and S. C. Rose. 1970. A theory relating high temperatures and overpressures. *J. Petrol. Tech., 22,* 11–16.

Leythaeuser, D. 1973. Effects of weathering on organic matter in shales. *Geochim. Cosmochim. Acta, 37,* 113–120.

Leythaeuser, D., and D. H. Welte. 1969. Relation between distribution of heavy *n*-paraffins and coalification in carboniferous coals from the Saar district, Germany. In P. A. Schenck and I. Havenaar (eds.), *Advances in organic geochemistry, 1968.* New York: Pergamon Press, pp. 429–442.

Leythaeuser, D., A. Hollerbach, and H. W. Hagemann. 1977. Source rock/crude oil correlation based on distribution of C_{27+} cyclic hydrocarbons. In R. Campos and J. Goñi (eds.), *Advances in organic geochemistry, 1975.* S.A. Madrid: Empresa Nacional Adaro de Investigaciones Mineras, pp. 3–20.

Lijmbach, G. M. G. 1975. On the origin of petroleum. In *9th World Petroleum Congress, Proc.* Vol. 2. London: Applied Science Publishers, pp. 357–369.

Lindblom, G. P., and M. D. Lupton. 1961. Microbiological aspects of organic geochemistry. *Develop. Ind. Microbiol., 2,* 9–22.

Link, W. K. 1952. Significance of oil and gas seeps in world oil exploration. *AAPG Bull., 36* (8), 1505–1540.

Livingston, H. K. 1951. Knock resistance of pure hydrocarbons in correlations with chemical structure. *I. & E. Chem., 43* (12), 2834–2840.

London, Ye. E. 1964. Degree of saturation of formation waters by dissolved hydrocarbons and sulfates as an exploration indicator in oil–gas evaluation. *Geolog. Nefti Gaza, 8,* 643–649.

Lopatin, N. V. 1971. Temperature and geological time as factors of carbonification. Akad. Nauk SSSR. Izv. Ser. Geol., No. 3, pp. 95–106.

Lopatin, N. V. 1976. The determination of the influence of temperature and geologic time on the catagenic processes of coalification and oil–gas formation. In *Issledovaniya organicheskogo veshchestva sovremennykh i iskopaemykh osadkov [Research on organic matter of modern and fossil deposits].* Moscow: Akad. Nauk SSSR, Izdatel'stvo, "Nauka," pp. 361–366.

Louis, M. C. 1966. Etudes géochimiques sur les "Schistes cartons" du Toarcien du Bassin de Paris. In G. D. Hobson and M. C. Louis (eds.), *Advances in organic geochemistry, 1964,* pp. 85–94.

Louis, M. C., and B. P. Tissot. 1967. Influence de la temperature et de la pression sur la formation des hydrocarbures dans les argiles a kerogen. In *7th World Petroleum Congress, Proc.* (Mexico). Vol. 2. London: Elsevier, pp. 47–60.

Low, P. F. 1976. Viscosity of interlayer water in montmorillonite. *Soil Sci. Soc. Amer. Proc., 40* (4), 500–505.

Lutz, M., J. P. H. Kaasschieter, and D. H. Van Wijhe. 1975. Geological factors controlling Rotliegend gas accumulations in the mid-European Basin: In *9th World Petroleum Congress, Proc.* Vol. 2. London: Applied Science Publishers, pp. 93–103.

McAuliffe, C. D. 1966. Solubility in water of paraffin, cycloparaffin, olefin, acetylene, cyclo-olefin, and aromatic hydrocarbons. *J. Phys. Chem., 70* (4), 1267–1275.

McAuliffe, C. D. 1969. Determination of dissolved hydrocarbons in subsurface brines. *Chem. Geol., 4,* 225–233.

McAuliffe, C. D. 1978. Chemical and physical constraints on petroleum migration with emphasis of hydrocarbon solubilities in water. In *Physical and chemical constraints on petroleum migration.* Vol. 2. Notes for AAPG short course, April 9, 1978, AAPG National Meeting, Oklahoma City. 39 p.

MacCarthy, G. R. 1926. Colors produced by iron in minerals and the sediments. *Amer. J. Sci., 12,* 17–36.

McCaslin, J. (ed.). 1975. *International petroleum encyclopedia.* Tulsa, Okla.: Petroleum Publishing. 480 p.

McCrossan, R. G. 1961. Resistivity mapping and petrophysical study of Upper Devonian inter-reef calcareous shales of central Alberta, Canada. *AAPG Bull., 45* (4), 441–470.

McIver, R. D. 1967. Composition of kerogen—clue to its role in the origin of petroleum. In *7th World Petroleum Congress, Proc.,* Mexico City. Vol. 2. London: Elsevier, pp. 26–36.

McKay, J. F., and D. R. Latham. 1973. Polyaromatic hydrocarbons in high-boiling petroleum distillates: Isolation by gel permeation chromatography and identification by fluorescence spectrometry. *Anal. Chem., 45* (7), 1050–1055.

McKenna, E. J., and R. E. Kallio. 1965. The biology of hydrocarbons. *Ann. Rev. Microbiol., 19,* 183–208.

McKinney, C. M., E. P. Ferrero, and W. J. Wenger. 1966. *Analyses of crude oils from 546 important oilfields in the United States.* U.S. Bureau of Mines Report of Investigations 6819. Washington, D.C.: U.S. Department of Interior, Bureau of Mines. 345 p.

McNair, H. M., and E. J. Bonelli. 1969. *Basic gas chromatography.* Madison, N.J.: Gow Mac Instrument Co. 306 p.

Magara, K. 1968. Compaction and migration of fluids in Miocene mudstone, Nagoaka Plain, Japan. *AAPG Bull., 52,* 2466–2501.

Magara, K. 1971. Permeability considerations in generation of abnormal pressures. *J. Soc. Petrol. Eng.,* pp. 236–242.

Magara, K. 1973. Compaction and fluid migration in Cretaceous shales of western Canada. Canadian Geological Survey Paper No. 72-18. Ottawa: Canadian Geological Survey. 90 p.

Magara, K. 1976a. Water expulsion from clastic sediments during compaction— directions and volumes. *AAPG Bull., 60* (4), 543–553.

Magara, K. 1976b. Thickness of removed sedimentary rocks, paleopore pressure, and paleotemperature, southwestern part of Western Canada Basin. *AAPG Bull., 60* (4), 554–565.

Mair, B. J. 1967. *Annual report for the year ending June 30, 1967.* American Petroleum Institute Research Project 6. Pittsburg, Pa.: Carnegie Institute of Technology.

Makarenko, F. A., and S. I. Sergiyenko. 1970. Geothermal zoning of the composition of oil in eastern Ciscaucasia. *Dokl. Akad. Nauk SSSR, 210,* 207–209.

Makarenko, F. A., and S. I. Sergiyenko, 1974. Heat flow in oil, gas and gas-condensate fields of continents. *Dokl. Akad. Nauk SSSR, 214* (1), 45–47.

Makogon, Y. F., V. P. Tsarev, and N. V. Cherskiy. 1972. Formation of large natural gas fields in zones of permanently low temperatures. *Dokl. Akad. Nauk SSSR, 205* (3), 215–218.

Makogon, Y. F., F. A. Trebin, A. A. Trofimuk, V. P. Tsarev, and N. V. Cherskiy. 1971. Detection of a pool of natural gas in a solid (hydrated gas) state. *Dokl. Akad. Nauk SSSR, 196,* 197–200.

Malyshek, V. T., P. A. Shoikhet, M. V. Gasanov, and S. K. Sal'miev. 1962. The biogenic formation of the higher gaseous hydrocarbons in bottom sediments.

Izv. Akad. Nauk Azerbaidzhanian SSR, Ser. Geolog.-Geograf. Nauk i Nefti, no. 1, pp. 63–72.

Manheim, F. T., and M. K. Horn. 1968. Composition of deeper subsurface waters along Atlantic continental margin. *Southeastern Geol., 9* (4), 215–236.

Marsden, S. S., and K. Kawai. 1965. "Suiyōsei-Ten'nengasu," a special type of Japanese natural gas deposit. Part 1. *AAPG Bull., 49* (3), 286–295.

Martin, R. L., and J. C. Winters. 1963. Determination of hydrocarbons in crude oil by capillary-column gas chromatography. Paper delivered at American Chemical Society Meeting, Division of Petroleum Chemistry, September 1963. 5 p.

Martin, R. L., J. C. Winters, and J. A. Williams. 1963. Distributions of *n*-paraffins in crude oils and their implications to origin of petroleum. *Nature, 199,* 110–113.

Matviyenko, V. N. 1975. Comparative characteristics of geothermal conditions in some fields of West Siberia. *Nefti Gazovaya Geolog. Geofiz.,* no. 10, pp. 12–14.

Maxwell, J. R., A. G. Douglas, G. Eglinton, and A. McCormick. 1968. The botryococcenes—hydrocarbons of novel structure from alga *Botryococcus braunii* Kutzing. *Phytochem., 7,* 2157–2171.

Meinschein, W. G. 1969. Hydrocarbons—saturated, unsaturated and aromatic. In G. Eglinton and M. T. J. Murphy (eds.), *Organic geochemistry: Methods and results.* New York: Springer-Verlag, pp. 330–356.

Meissner, F. F. 1978a. Patterns of source-rock maturity in non-marine source-rocks of some typical western interior basins. In *Non-marine Tertiary and Upper Cretaceous source rocks and the occurrence of oil and gas in west central U.S.* Rocky Mountain Association of Geologists (RMAG) Continuing Education Lecture Series, January 16–18, 1978, Denver, Colorado, pp. 1–43.

Meissner, F. F. 1978b. Petroleum geology of the Bakken Formation, Williston Basin, North Dakota and Montana. *Proceedings of 1978 Williston Basin Symposium, "The Economic Geology of the Williston Basin,"* September 24–27, 1978, Montana Geological Society, Billings, pp. 207–227.

Merchant, P., Jr. 1968. Resolution of C_4 to C_{12} petroleum mixtures by capillary gas chromatography. *Anal. Chem., 12* (14), 2153–2167.

Metwalli, M. H. 1970. Investigations on the relation between oil gravity, depth, environment of the oil-bearing rock and the composition of the crudes. In *7th Arab Petroleum Congress, Proc.* Kuwait, March 1970. Paper No. 50 (B-3). 20 p.

Miller, C. J. 1938. Carbon dioxide accumulations in geologic structures. *Amer. Inst. Min. Met. Eng., 129,* 439–468.

Mogilevskii, G. A. 1963. Sovremennoe sostoyanie i perspektivy razvitiya kompleska gazobiokhimicheskikh metodov poiskov Nefti i Gaza v SSSR. In V. Bese (ed.), *3rd International Scientific Conference on Geochemistry, Microbiology, and Petroleum Chemistry, Proc.* Budapest, October 8–13, 1962. Vol. 2, Part 1, pp. 579–623.

Mold, J. D., R. K. Stevens, R. E. Means, and J. M. Ruth. 1963. The paraffin hydrocarbons of tobacco: Normal, iso-, and anteiso-homologs. *Biochem. J., 2,* 605–610.

Momper, J. A. 1978. Oil migration limitations suggested by geological and geochemical considerations. In *Physical and chemical constraints on petroleum migration.* Vol. 1. Notes for AAPG short course, April 9, 1978, AAPG National Meeting, Oklahoma City. 60 p.

Moody, J. D. 1975. Distribution and geological characteristics of giant oil fields. In A. G. Fischer and S. Judson (eds.), *Petroleum and global tectonics.* Princeton, N.J.: Princeton University Press, pp. 307–320.

Moody, J. D., J. W. Mooney, and J. Spivak. 1970. Giant oil fields of North America. In M. T. Halbouty (ed.), *Geology of giant petroleum fields.* AAPG Memoir 14. Tulsa, Okla.: American Association of Petroleum Geologists, pp. 8–16.

Morgridge, D. L., and W. B. Smith, Jr. 1972. Geology and discovery of Prudhoe Bay field, eastern Arctic slope, Alaska. In R. E. King (ed.), *Stratigraphic oil and gas fields—classification, exploration methods, and case histories.* AAPG Memoir 16. Tulsa, Okla.: American Association of Petroleum Geologists, pp. 489–501.

Mukhin, Yu. V. 1966. Evaluation of oil and gas prospects in ancient basins from gases in solution in formation waters. *Sovietskaya Geol.,* no 6, pp. 20–33. Translation, *Internat. Geol. Rev.,* 1968, *10,* (1), 68–78.

Müller, E. P., K. Goldbecher, and T. Abotnewa. 1973. Zur Geochemie und Genese Stickstoffreicher Erdgase. *Zeitschrift für Angewandte Geologie, 19* (10), 494–499.

Muller, J. 1964. Palynological contributions to the history of Tertiary vegetation in N.W. Borneo. In D. Murchison and T. S. Westoll (eds.), *Coal and coal-bearing strata.* New York: Elsevier, pp. 39–40.

Munn, M. J. 1909. Studies in the application of anticlinal theory of oil and gas accumulation. *Econ. Geol., 4,* 141.

Murchison, D. 1969. Some recent advances in coal petrology. *Congrès Internat. Strat. Géol. Carbonif., Compte Rendu,* Sheffield. Vol. 1. Maastricht, Netherlands: E. van Aelst, pp. 351–368.

Murray, G. 1965. Indigenous Pre-Cambrian petroleum. *AAPG Bull., 49* (1), 3–21.

Nakai, N. 1960. Carbon isotope fractionation of natural gas in Japan. *J. Earth Sci., Nagoya Univ., 8,* 174–180.

Nakai, N. 1962. Geochemical studies on the formation of natural gases. *J. Earth Sci., Nagoya Univ., 10,* 71–111.

Nechayeva, O. L. 1968. Hydrogen in gases dissolved in water of the West Siberian Plain. *Dokl. Akad. Nauk SSSR, 179* (4), 961–962.

Neglia, S. 1979. Migration of fluids in sedimentary basins. *AAPG Bull., 63* (4), 573–597.

Nelson, W. L. 1962. Wax and lube content of crude oil—an approximation. *Oil and Gas J., 60,* 122–124.

Nelson, W. L. 1972. What's the average sulfur content vs. gravity? *Oil and Gas J., 70* (5), 59.

Nelson, W. L. 1974. What are the amounts of nitrogen and oxygen in U.S. products? *Oil and Gas J., 72* (5), 112–114.

Newberry, J. S. 1860. *The rock oils of Ohio.* Ohio Agricultural Report for 1859. 14 p.

Nixon, R. P. 1973. Oil source beds in Cretaceous Mowry shales of northwest interior United States. *AAPG Bull., 57* (1), 136–161.

Oakwood, T. S., D. S. Shriver, H. H. Fall, W. J. McAleer, and P. R. Wunz. 1952. Optical activity of petroleum. *I. & E. Chem., 44,* (11), 2568–2570.

O'Connor, J. G., F. H. Burow, and M. S. Norris. 1962. Determination of normal paraffins in C_{20} to C_{32} paraffin waxes by molecular sieve adsorption: Molecular weight distribution by gas-liquid chromatography. *Anal. Chem., 34* (1), 82–85.

Odom, I. F. 1967. Clay fabric and its relation to structural properties in mid-continent Pennsylvanian sediments. *J. Sed. Petrol., 37,* 610–623.

Oro, J., and D. W. Nooner. 1967. Aliphatic hydrocarbons in Precambrian rocks. *Nature, 213,* 1082–1085.

Oro, J., T. G. Tornabene, D. W. Nooner, and E. Gelpi. 1967. Aliphatic hydrocarbons and fatty acids of some marine and freshwater microorganisms. *J. Bacteriol., 93,* 1811–1818.

Orr, W. L. 1974. Changes in sulfur content and isotopic ratios of sulfur during petroleum maturation—study of Big Horn Basin Paleozoic oils. Part 1. *AAPG Bull., 58* (11), 2295–2318.

Owen, E. W. 1975. Trek of the oil finders. In *A history of exploration for petroleum.* AAPG Memoir 6. Tulsa, Okla.: American Association of Petroleum Geologists, pp. 1–4.

Pakhomov, S. I., and I. G. Kissin. 1968. More information about the geochemistry of carbon dioxide in subsurface zones of the hydrosphere. *Dokl. Akad. Nauk SSSR, 180,* 194–197.

Park, R., and S. Epstein. 1960. ^{13}C in lake waters and its possible bearing on paleolimnology. *Amer. J. Sci., 258,* 253–272.

Parker, C. A. 1973. Geopressures in the deep Smackover of Mississippi. *J. Petrol. Tech., 28* (8), 971–979.

Patterson, C. 1956. Age of meteorites and the earth. *Geochim. Cosmochim. Acta, 10,* 230–237.

Paul, R. G., R. E. Arnal, J. P. Baysinger, G. E. Claypool, J. L. Holte, C. M. Lubeck, J. M. Patterson, R. Z. Poore, R. L. Slettene, W. V. Sliter, J. C. Taylor, R. B. Tudor, and F. L. Webster. 1976. *Geological and operational summary, Southern California Deep Stratigraphic Test OCS-CAL 75-70 No. 1, Cortes Bank area offshore southern California.* U.S. Department of the Interior Geological Survey, Open-File Report 76-232. Washington, D.C.: U.S. Department of the Interior Geological Survey. 65 p.

Perry, E. A., Jr., and J. Hower. 1972. Late-stage dehydration in deeply buried pelitic sediments. *AAPG Bull., 56,* 2013–2021.

Petersil'ye, I. A., Ye. K. Kozlov, K. D. Belyayev, V. V. Sholokhnev, and V. S. Dokuchayeva. 1970. Nitrogen and hydrocarbon gases in ultramafic rocks of the sopcha stock of the Monchegorsk pluton, Kola Peninsula. *Dokl. Akad. Nauk SSSR, 194,* 200–203.

Petrov, A. A., D. S. Pustilnikova, and N. N. Abriutina. 1977. Sterane hydrocarbons of oils and their geochemical importance. In N. B. Vassoevich et al (eds.), *8th International Congress on Organic Geochemistry, Abstracts of Reports.* Moscow, May 10–13, 1977. Vol. 1, pp. 162–164.

Petrov, A. A., T. V. Tichomolova, and S. D. Pustilnikova. 1969. The distribution of hydrocarbons in the gasoline fraction obtained upon thermocatalysis of fatty acids. In P. A. Schenck and I. Havenaar (eds.), *Advances in organic geochemistry, 1968.* New York: Pergamon Press, pp. 401–405.

Philippi, G. T. 1957. Identification of oil-source beds by chemical means. In *20th International Geological Congress, Proc.* Mexico City, 1956, Sec. 3, pp. 25–38.

Philippi, G. T. 1965. On the depth, time and mechanism of petroleum generation. *Geochim. Cosmochim. Acta, 29,* 1021–1049.

Philippi, G. T. 1974a. Depth of oil origin and primary migration: A review and critique: Discussion. *AAPG Bull., 58* (1), 149–154.

Philippi, G. T. 1974b. The influence of marine and terrestrial source material on the composition of petroleum. *Geochim. Cosmochim. Acta, 38,* 947–966.

Philippi, G. T. 1975. The deep subsurface temperature controlled origin of the gaseous and gasoline-range hydrocarbons of petroleum. *Geochim. Cosmochim. Acta, 39,* 1353–1373.

Philippi, G. T. 1977. On the depth, time and mechanism of origin of the heavy- to medium-gravity naphthenic crude oils. *Geochim. Cosmochim. Acta, 41* (1), 33–52.

Pixler, B. O. 1969. Formation evaluation by analysis of hydrocarbon ratios. *J. Petrol. Tech., 24,* 665–670.

Polak, L. S. 1954. *Studies in the physical properties of rocks of the Ural-Emba region carried out during 1953.* MNP SSSR Glavnefte-Geofizika, Kazakhstanskaya Geofizicheskaya Kontora.

Polivanova, A. N. 1977. The relationship between the carbon isotopic composition of methane and hydrogen sulphide and saliferous occurrences. In N. B. Vassoevich et al (eds.), *8th International Congress on Organic Geochemistry, Abstracts of Reports.* Moscow, May 10–13, 1977. Vol. 2, pp. 164–166.

Polonskaya, B. Ya., Ye. D. Rozonova, and A. G. Andrianova. 1974. Redistribution of organic matter in carbonate sediments. *Geol. Nefti Gaza,* no. 2, pp. 42–45.

Pomeyrol, R., F. Bienner, and M. Louis. 1961. Exemple de prospection géochimique par l'analyse des gaz adsorbés en surface dans le bassin de Fort Polignac. *Rev. de l'Inst. Français Pétrol. et Ann. Combust. Liq., 16* (7–8), 868–874.

Potonie, H. 1908. Die rezenten Kaustobiolithe und ihre Lagerstatten: Die Sapropeliten. *Abh. Kgl. Preuss. Geol. Landesanstalt,* new series, *1* (55).

Powell, T. G., and D. M. McKirdy. 1973. The effect of source material, rock type and diagenesis on the *n*-alkane content of sediments. *Geochim. Cosmochim. Acta, 37,* 623–633.

Powell, T. G., P. J. Cook, and D. M. McKirdy. 1975. Organic geochemistry of phosphorites: Relevance to petroleum genesis. *AAPG Bull., 4,* 618–632.

Pratt, W. E. 1943. *Oil in the earth.* Lawrence: University of Kansas Press. 106 p.

Press, F., and R. Siever. 1974. *Earth.* San Francisco: W. H. Freeman. 945 p.

Price, L. C. 1973. The solubility of hydrocarbons and petroleum in water as applied to the primary migration of petroleum. Doctoral thesis, University of California, Riverside.

Price, L. C. 1976. Aqueous solubility of petroleum as applied to its origins and primary migration. *AAPG Bull., 60* (2), 213–244.

Proshlyakov, B. K. 1960. Reservoir properties of rocks as a function of their depth and lithology. *Geol. Nefti Gaza, 4* (12), 24–29. Assoc. Tech. Services Translation RO3421.

Pusey, W. C. III. 1973. The ESR-kerogen method: A new technique of estimating the organic maturity of sedimentary rocks. *Petrol. Times, 77,* 21–26.

Pustilnikova, S. D., A. L. Tsedilina, M. J. Krasavchenko, and A. A. Petrov. 1973. Isoprene petroleum hydrocarbon formation from phytane. *Geol. Nefti Gaza*, no. 12, pp. 56–59.

Pym, J. G., J. E. Ray, G. W. Smith, and E. V. Whitehead. 1975. Petroleum triterpane fingerprinting of crude oils. *Anal. Chem., 47* (9), 1617–1622.

Radchenko, O. A., I. P. Karpova, and A. S. Chernysheva. 1951. *A geochemical investigation of weathered and highly altered mineral fuels from South Fergana.* Trudy VNIGRI, new series No. 5, Contributions to Geochemistry, No. 2–3, pp. 180–202. Translated by Israel Program for Scientific Translations, Jerusalem, 1965.

Rall, H. T., C. J. Thompson, H. J. Coleman, and R. L. Hopkins. 1972. *Sulfur compounds in crude oil.* U.S. Bureau of Mines Bull. No. 659. 193 p.

Rankama, K. 1954. Early Pre-Cambrian carbon of biogenic origin from the Canadian shield. *Science, 119,* 506–507.

Reerink, H., and J. Lijzenga. 1973. Molecular weight distributions of Kuwait asphaltenes as determined by ultracentrifugation: Relation with viscosity of solutions. *J. Inst. Petrol., 59* (569), 211–222.

Regnault, V. 1837. *Ann. Mines, 3* (12), 161.

Rehm, B. 1972. Worldwide occurrence of abnormal pressures. Part 2. Paper No. SPE 3845. Dallas: American Institute of Mining, Metallurgical, and Petroleum Engineers. 8 p.

Reznikov, A. N. 1967. The geochemical conversion of oils and condensates in the zone of catagenesis. *Geolog. Nefti Gaza,* no. 5, pp. 24–28.

Rhead, M. M., G. Eglinton, and G. H. Draffan. 1971. Hydrocarbons produced by the thermal alteration of cholesterol under conditions simulating the maturation of sediments. *Chem. Geol., 8* (4), 277–297.

Rice, D. D. 1975. Origin of and conditions for shallow accumulations of natural gas. In *27th Annual Field Conference, Wyoming Geological Association Guidebook.* Casper: Wyoming Geological Association, pp. 267–272.

Rittenhouse, G. 1973. Pore-space reduction in sandstones—controlling factors and some engineering implications. *Offshore Technology Conference,* Paper No. OTC 1806. Dallas. 10 p.

Robinson, C. J. 1971. Low-resolution mass spectrometric determination of aromatics and saturates in petroleum fractions. *Anal. Chem., 43* (11), 1425–1434.

Rodionova, K. F., and O. P. Chetverikova. 1962. On the composition of residual organic matter in the Paleozoic rocks of the middle Volga region. *Geochem.,* no. 10, pp. 1024–1029.

Rogers, M. A., J. D. McAlary, and N. J. L. Bailey. 1974. Significance of reservoir bitumens to thermal-maturation studies, Western Canada Basin. *AAPG Bull., 58* (9), 1806–1824.

Ronov, A. B. 1958. Organic carbon in sedimentary rocks (in relation to the presence of petroleum). *Geochem.,* no. 5, pp. 497–509.

Ronov, A. B., and A. A. Yaroshevsky. 1969. Chemical composition of the earth's crust. *The earth's crust and upper mantle. American Geological Union Monograph 13.* Washington, D.C.: American Geological Union, pp. 37–57.

Ronov, A. B., A. A. Migdisov, and A. A. Yaroshevsky. 1973. The main stages of the geochemical history of the outer shells of the earth. In *Proc. Symposium on Hydrogeochemistry and Biochemistry*. Vol. 1. Washington, D.C.: Clark, pp. 40–53.

Rosaire, E. E. 1938. Shallow stratigraphic variations over Gulf Coast structures. *Geophys., 3* (3), 96–115.

Rosscup, R. J., and J. Bowman. 1967. Thermal stabilities of vanadium and petro porphyrins. *Preprints of the Division of Petroleum Chemistry, Amer. Chem. Soc.* Vol. 12. Washington, D.C.: American Chemical Society, p. 77.

Rossini, F. D. 1960. Hydrocarbons in petroleum. *J. Chem. Ed., 37* (11), 554–561.

Rozin, A. A. 1966. Gas hydrochemical zonality of subsurface waters of the Mesozoic sediments of the south, central and east parts of the West Siberian Basin. In *Problems of oil and gas of the south part of the West Siberian Lowland.* Tomsk University.

Rubinstein, I., O. P. Strausz, C. Spyckerelle, R. J. Crawford, and D. W. S. Westlake. 1977. The origin of the oil sand bitumens of Alberta: A chemical and microbiological simulation study. *Geochim. Cosmochim. Acta, 41,* 1341–1353.

Rumeau, J.-L., and C. Sourisse. 1973. Un exemple de migration primaire en phase gazeuse. *Bull. Centre Rech. Pau-SNPA, 7* (1), 53–67.

Ryther, J. 1970. Is the world's oxygen supply threatened? *Nature, 227,* 374–375.

Rzasa, M. J., and D. L. Katz. 1950. The coexistence of liquid and vapor phases at pressures above 10,000 psi. *Transactions, AIME, 189,* 119.

Sachanen, A. N. 1945. *Chemical constituents of petroleum.* New York: Reinhold. 451 p.

Sackett, W. M. 1978. Carbon and hydrogen isotope effects during the thermocatalytic production of hydrocarbons in laboratory simulation experiments, *Geochim. Cosmochim. Acta, 42* (6), 571–580.

Sackett, W. M., B. Bernard, and J. Brooks. 1975. Isotopic and molecular compositions of seep gases in the Gulf of Mexico. Paper presented at the Southern Regional Organic Geochemists' Meeting, November 1975, Port Aransas, Texas.

Sackett, W. M., W. R. Eckelmann, M. L. Bender, and A. W. H. Be. 1965. Temperature dependence of carbon isotope composition in marine plankton and sediments. *Science, 148,* 235–237.

Sackett, W. M., S. Nakaparksin, and D. Dalrymple. 1966. Carbon isotope effects in methane production by thermal cracking. In G. D. Hobson and G. C. Speers (eds.), *Advances in organic chemistry, 1964.* New York: Pergamon Press, pp. 37–53.

Saraf, D. N. 1970. A mathematical model for the vertical migration of underground reservoir fluids. *Indian J. Tech., 8,* 237–242.

Sayles, F. L., and F. T. Manheim. 1975. Interstitial solutions and diagenesis in deeply buried marine sediments: Results from the Deep Sea Drilling Project. *Geochim. Cosmochim. Acta, 39,* 103–127.

Scalan, R. S., and T. D. Morgan. 1970. Isotope ratio mass spectrometry instrumentation and application to organic matter contained in Recent sediments. *Int. J. Mass. Spectrom. Ion Phys., 4,* 267–281.

Scalan, R. S., and J. E. Smith. 1970. An improved measure of the odd-even predominance in the normal alkanes of sediment extracts and petroleum. *Geochim. Cosmochim. Acta, 34* (6), 611–620.

Schiessler, R. W., and D. Flitter. 1952. Urea and thiourea adduction of C_5–C_{42}-hydrocarbons. *J. Amer. Chem. Soc., 74,* 1720–1723.

Schmidt, G. W. 1973. Interstitial water composition and geochemistry of deep Gulf Coast shales and sands. *AAPG Bull., 57* (2), 321–337.

Scholle, P. A. 1975. Chalk diagenesis. In A. W. Woodland (ed.), *Petroleum and the continental shelf of north-west Europe.* Vol. 1. New York: Wiley, pp. 420–427.

Scholle, P. A. (ed.). 1977. *Geological studies on the COST No. B-2 Well, U.S. Mid-Atlantic outer continental shelf area:* U.S. Geological Survey Circular 750. Arlington, Va.: U.S. Geological Survey. 71 p.

Schopf, J. M. 1956. A definition of coal. *Econ. Geol., 51* (6), 521–527.

Schopf, J. W. 1970. Pre-Cambrian microorganisms and evolutionary events prior to the origin of vascular plants. *Biol. Rev., 45,* 319–352.

Schopf, J. W. 1975. The age of microscopic life. *Endeavour, 34* (122), 51–58.

Schopf, J. W., B. N. Haugh, R. E. Molnar, and D. F. Satterthwait. 1973. On the development of metaphytes and metazoans. *J. Paleontol., 47* (1), 1–9.

Schrayer, G. J., and W. M. Zarella. 1963. Organic geochemistry of shales. Part 1: Distribution of organic matter in siliceous Mowry Shale of Wyoming. *Geochim. Cosmochim. Acta, 27,* 1033–1046.

Schrayer, G. J., and W. M. Zarella. 1966. Organic geochemistry of shales. Part 2: Distribution of extractable organic matter in the siliceous Mowry Shale of Wyoming. *Geochim. Cosmochim. Acta, 30,* 415–434.

Secor, D. T., Jr. 1965. Role of fluid pressure in jointing. *Amer. J. Sci., 263,* 633–646.

Seifert, W. K. 1977. Source rock-oil correlations by C_{27}–C_{30} biological marker hydrocarbons. In R. Campos and J. Goñi (eds.), *Advances in organic geochemistry, 1975.* S.A. Madrid: Empresa Nacional Adaro de Investigaciones Mineras, pp. 21–44.

Seifert, W. K. 1978. Steranes and terpanes in kerogen pyrolysis for correlation of oils and source rocks. *Geochim. Cosmochim. Acta, 42* (5), 473–484.

Seifert, W. K., and J. M. Moldowan. 1978. Applications of steranes, terpanes and monoaromatics to the maturation, migration and source of crude oils. *Geochim. Cosmochim. Acta, 42* (1), 77–95.

Seifert, W. K., and J. M. Moldowan. 1979. The effect of biodegradation on steranes and terpanes in crude oils. *Geochim. Cosmochim. Acta, 43* (1), 111–126.

Seifert, W. K., and R. M. Teeter. 1970. Identification of polycyclic aromatic and heterocyclic crude oil carboxylic acids. *Anal. Chem., 42* (7), 750–758.

Sengupta, S. 1975. Experimental alterations of the spores of lycopodium clavatum as related to diagenesis. *Rev. Paleobot. and Palynol., 19,* 173–192.

Sergienko, S. R. 1964. *High molecular compounds in petroleum.* (2nd ed.). *Vysokomolekularnye Soedineniya, Nefti,* Izd. Moscow: Khimiya. Translated by Israel Program for Scientific Translations, Jerusalem, 1965. 440 p.

Sever, J. R. 1970. The organic geochemistry of hydrocarbons in coastal environments. Doctoral dissertation, University of Texas at Austin. University Microfilm No. 72-2418; at University of Texas at Austin.

Shabarova, N. T. 1955. The characteristics of nitrogenous substances in the sedimentary deposits of ancient Caspian and Apsheron formations. *Dokl. Akad. Nauk SSSR, 105* (4), 774–776.

Shaulov, M. A. 1973. Problem of the existence of gas pools feeding mud volcanism in Taman. *Petrol. Geol., 7,* 98–100.

Shibaoka, M., A. J. R. Bennett, and K. W. Gould. 1973. Diagenesis of organic matter and occurrence of hydrocarbons in some Australian sedimentary basins. *APEA Journal* (Australia), *13,* 73–80.

Shimoyama, A., and W. D. Johns. 1971. Catalytic conversion of fatty acids to petroleum-like paraffins and their maturation. *Nature Phys. Sci., 232,* 140–144.

Shtogrin, O. D. 1974. Nitrogen and its compounds in subsurface waters of the north Crimean downward and their value to oil–gas exploration. *Geologiya i Geokhimiya Goryuchikh Iskopaymykh,* no. 37, pp. 78–82.

Sidorenko, A. V., Y. A. Borshchevskiy, S. A. Sidorenko, V. A. Ustinov, and N. K. Popova. 1972. Isotopic composition of elementary carbon from Pre-Cambrian metamorphic rocks. *Dokl. Akad. Nauk SSSR, 206* (2), 463–466.

Sieck, H. C. 1973. Gas-charged sediment cones pose possible hazard to offshore drilling. *Oil and Gas J., 71,* 148, 150, 155, 163.

Sikka, D. B. 1959. Radiometric survey of Redwater Oilfield, Alberta, Canada. Paper presented at Symposium on Petroleum Geochemistry, June 4, Fordham University, New York City. 8 p.

Siller, C. W., G. E. Murray, R. M. Hopkins, and D. A. McNaughton. 1963. Aussie strike attracts interest. *Oil and Gas J., 61,* 189–191.

Silverman, S. R. 1964. Investigations of petroleum origin and mechanisms by carbon isotope studies. In S. L. Miller and G. J. Wasserburg (eds.), *Isotopic and cosmic chemistry.* Amsterdam: North-Holland Publishing, pp. 92–102.

Silverman, S. R., and S. Epstein. 1958. Carbon isotopic composition of petroleums and other sedimentary organic materials. *AAPG Bull., 42* (5), 998–1012.

Simoneit, B. R. 1975. Sources of organic matter in oceanic sediments. Doctoral thesis, University of Bristol, England.

Simoneit, B. R., W. G. Howells, and A. L. Burlingame. 1973. Preliminary organic geochemical analyses of the Cariaco Trench Site 147 Deep Sea Drilling Project, Leg 15. In B. C. Heezen and I. G. MacGregor (eds.), *Initial reports of Deep Sea Drilling Project.* Vol. 20. Washington, D.C.: U.S. Government Printing Office, pp. 157–184.

Simonenko, V. F. 1974. Possible participation of anomalous water of clay in processes of hydrocarbon migration. *Geol. Nefti Gaza,* no. 2, pp. 42–45.

Skrigan, A. I., 1951. Composition of turpentine from a swamp rosin 1,000 years old. *Dokl. Akad. Nauk SSSR, 80,* 607–609.

Skrigan, A. I. 1964. Preparation and utilization of fichtelite and retene. *Tr. Vses. Nauchn-Tekhn. Sovisch. Gorki,* pp. 108–115.

Smith, H. M. 1940. Correlation index to aid in interpreting crude-oil analyses. U.S. Bureau of Mines Tech. Paper 610. Washington, D.C.: U.S. Department of Interior, Bureau of Mines. 34 p.

Smith, J. E. 1971a. Relationship between petroleum reservoir rocks and source rocks. *Math. Geol., 3* (2), 183–192.

Smith, J. E. 1971b. Shale compaction. Society of Petroleum Engineers of AIME.

Paper No. SPE 3633. Dallas: American Institute of Mining, Metallurgical, and Petroleum Engineers. 12 p.

Smith, J. E., J. G. Erdman, and D. A. Morris. 1971. Migration, accumulation and retention of petroleum in the earth. In *8th World Petroleum Congress, Proc.* Moscow, 1971. London: Applied Science Publishers, pp. 13–26.

Smith, J. W., J. W. Schopf, and I. R. Kaplan. 1970. Extractable organic matter in Pre-Cambrian cherts. *Geochim. Cosmochim. Acta, 34,* 659–675.

Smith, P. V., Jr. 1954. Studies on origin of petroleum: Occurrence of hydrocarbons in Recent sediments. *AAPG Bull., 38* (3), 377–404.

Snarskii, A. N. 1964. Relationship between primary migration and compaction of rocks. *Petrol. Geol., 5* (7), 362–364.

Snarskii, A. N. 1970. The nature of primary oil migration. *Izv. Vyssh. Ucheb, Zavedenii, Neft Gaz, 13* (8), 11–15. Translated by Assoc. Tech. Services.

Snider, L. C. 1934. Current ideas regarding source beds for petroleum. In W. E. W. Rather and F. H. Lahee (eds.), *Problems of petroleum geology.* AAPG Memoir l. Tulsa, Okla.: American Association of Petroleum Geologists, pp. 51–66.

Snowdon, L. R. 1977. Organic geochemistry of the Upper Cretaceous/Tertiary delta complexes of the Beaufort-Mackenzie sedimentary basin, northern Canada. Doctoral thesis, Rice University.

Snowdon, L. R. In press. Errors in the extrapolation of experimental kinetic parameters to organic geochemical systems. *AAPG Bull.*

Snowdon, L. R., and K. J. Roy. 1975. Regional organic metamorphism in the mesozoic strata of the Sverdrup Basin. *Canadian Petrol. Geol. Bull., 23* (1), 131–148.

Snyder, L. R. 1969. Nitrogen and oxygen compound types in petroleum: Total analysis of an 850–1000°F distillate from a California crude oil. *Anal. Chem., 41* (8), 1084–1094.

Sokolov, V. A. 1933. *New methods of prospecting for oil and gas deposits.* Trudy Neftyanoi.

Sokolov, V. A. 1960. Chemical processes in the formation of oil and primary migration problems. In *Problema proiskhozhdeniya nefti i gaza i usloviya formirovaniya ikl zalezhei.* Trudy Vsesoyuz, Soveshchaniya, Moscow, October 20–27, 1958. Moscow. pp. 100–109.

Sokolov, V. A., and S. I. Mironov. 1962. On the primary migration of hydrocarbons and other oil components under the action of compressed gases. In *The chemistry of oil and oil deposits.* Acad. Sci. USSR, Inst. Geol. and Exploit. Min. Fuels, pp. 38–91 (in Russian). English translation by Israel Program for Scientific Translation, Jerusalem, 1964.

Sokolov, V. A., and Y. M. Yurovskii. 1961. *The theory and practice of gas logging.* Moscow: Gostoptekhizdat. 335 p.

Sokolov, V. A., Z. A. Buniat-Zade, A. A. Geodekian, and F. D. Dadashev. 1969. The origin of gases of mud volcanoes and the regularities of their powerful eruptions. In P. A. Schenck and I. Havenaar (eds.), *Advances in organic geochemistry, 1968.* Oxford, England: Pergamon Press, pp. 473–484.

Sokolov, V. A., T. V. Tichomolova, and O. A. Chermisinov. 1972. The composition and distribution of gaseous hydrocarbons independently of depth as a consequence of their generation and migration. In H. R. v. Gaertner and H. Wehner (eds.), *Advances in organic geochemistry, 1971.* Oxford, England: Pergamon Press, pp. 479–486.

Sokolov, V. A., T. P. Zhuse, N. B. Vassoevich, P. L. Antonov, G. G. Grigoryev, and V. P. Kozlov. 1963. Migration processes of gas and oil, their intensity and directionality. Paper presented at 6th World Petroleum Congress, June 19–26, Frankfurt, Maine.

Speers, G. C., and E. V. Whitehead. 1969. Crude petroleum. In G. Eglinton and M. T. J. Murphy (eds.), *Organic geochemistry: Methods and results.* New York: Springer-Verlag, pp. 638–667.

Spevak, Yu. A. 1972. Study of the gas saturation of formation waters in connection with temperature conditions of the metamorphism of organic matter. *Geol. Nefti Gaza,* no. 2, pp. 57–63.

Spjeldnaes, N. 1963. A new fossil (*Papillomembrana sp.*) from the upper Pre-Cambrian of Norway. *Nature, 200,* 63–65.

Stach, E., M.-Th. Mackowsky, M. Teichmüller, G. H. Taylor, D. Chandra, and R. Teichmüller. 1975. *Stach's textbook of coal petrology.* (2nd ed.). Berlin: Gebrüder Borntraeger. 428 p.

Stahl, W. J. 1974. Carbon isotope ratios of German natural gases in comparison with isotope data of gaseous hydrocarbons from other parts of the world. In B. Tissot and F. Bienner (eds.), *Advances in geochemistry, 1973.* Paris: Éditions Technip, pp. 453–462.

Stahl, W. J. 1977. Carbon and nitrogen isotopes in hydrocarbon research and exploration. *Chem. Geol., 20,* 121–149.

Staplin, F. L. 1969. Sedimentary organic matter, organic metamorphism, and oil and gas occurrence. *Canadian Petrol. Geol. Bull., 17* (1), 47–66.

Steineke, M., R. A. Bramkamp, and N. J. Sander. 1958. Stratigraphic relations of Arabian Jurassic oil. In L. G. Weeks (ed.), *Habitat of oil: A symposium.* Tulsa, Okla.: American Association of Petroleum Geologists, pp. 1294–1329.

Stevens, N. P. 1956. Origin of petroleum—a review. *AAPG Bull., 40* (1), 51–61.

Stevens, N. P. 1962. Geochemical prospecting in the United States. In V. Bese (ed.), *3rd International Scientific Conference on Geochemistry, Microbiology and Petroleum Chemistry, Proc.* Vol. 2. Budapest, October 8–13. Budapest: KULTURA, pp. 83–97.

Stoll, R. D., J. Ewing, and G. M. Bryan. 1971. Anomalous wave velocities in sediments containing gas hydrates. *J. Geophys. Res., 76* (8), 2090–2094.

Strakhov, N. M. 1974. Volume of sedimentary cover in Russian platform and the global quantitative geochemistry. *Internat. Geol. Rev., 16* (3), 281–289.

Stroganov, V. P., and M. I. Shubbota. 1973. Comparative characteristics of gases dissolved in waters of the Jurassic, Cretaceous and Paleogene sediments of the platform part of western Central Asia. *Petrol. Geol., 7,* 355–358.

Sugisaki, R. 1964. Genetic relation of various types of natural gas deposits in Japan. *AAPG Bull., 48* (1), 85–101.

Sukharev, G. M., Yu. K. Taranukha, and S. P. Vlasova. 1970. Subsurface heat flow in the Bibieybat oil and gas field. *Dokl. Akad. Nauk SSSR, 190* (1), 176–179.

Sulin, V. A. 1946. *Waters of oil reservoirs in the system of natural waters.* In Russian. Moscow: Gostoptekhizdat.

Swinnerton, J. W., and V. L. Linnenbom. 1969. Low molecular weight hydrocarbon analyses of Atlantis II waters. In E. T. Degens and D. A. Ross (eds.), *Hot brines and recent heavy metal deposits in the Red Sea.* New York: Springer-Verlag, pp. 251–260.

Taguchi, K. 1975. Geochemical relationships between Japanese Tertiary oils and their source rocks. In *9th World Petroleum Congress, Proc.* Vol. 2. London: Applied Science Publishers, pp. 193–194.

Tappan, H. 1974. Molecular oxygen and evolution. In O. Hayaishi (ed.), *Molecular oxygen in biology: Topics in molecular oxygen research.* North-Holland, pp. 81–135.

Teichmüller, M. 1958. Métamorphisme du charbon et prospection du pétrole. *Rev. Ind. Minérale,* Numero Special, pp. 1–15.

Teichmüller, M. 1963. Die Kohlenflöze der Bohrung Münsterland. Part 1. *Fortschr. Geol. Rheinld. u. Westf., 11,* 129.

Teichmüller, M. 1974. Generation of petroleum-like substances in coal seams as seen under the microscope. In B. Tissot and F. Bienner (eds.), *Advances in organic geochemistry, 1973.* Paris: Éditions Technip, pp. 379–407.

Teichmüller, M., and R. Teichmüller. 1950. Das Inkohlungsbild des Niedersächsischen Wealden-Beckens. *Z. Deutsch. Geol. Ges., 100,* 498–517.

Teichmüller, M., and R. Teichmüller. 1968. Geological aspects of coal metamorphism. In D. Murchison and T. S. Westoll (eds.), *Coal and coal-bearing strata.* New York: Elsevier, pp. 233–267.

Teslenko, P. F., and B. S. Korotkov. 1966. Effect of arenaceous intercalations in clays on their compaction. *Internat. Geol. Rev., 9* (5), 699–701.

Thode, H. G., R. K. Wanless, and R. Wallouch. 1954. The origin of native sulfur deposits from isotope fractionation studies. *Geochim. Cosmochim. Acta, 5,* 286–298.

Thom, W. T., Jr. 1934. Present status of the carbon-ratio theory. In W. E. W. Rather and F. H. Lahee (eds.), *Problems of petroleum geology:* Tulsa, Okla.: American Association of Petroleum Geologists, pp. 69–95.

Thomas, B. R. 1969. Kauri resins—modern and fossil. In G. Eglinton and M. T. J. Murphy (eds.), *Organic geochemistry: Methods and results.* New York: Springer-Verlag, pp. 599–618.

Thompson, T. L. 1976. Plate tectonics in oil and gas exploration of continental margins. *AAPG Bull., 60* (9), 1463–1501.

Timko, D. J., and W. H. Fertl. 1972. How downhole temperatures, pressures affect drilling. *World Oil, 175,* 79–82.

Ting, T. C. 1975. Reflectivity of disseminated vitrinites in the Gulf Coast region. Pétrographie de la matière organique des sédiments relations avec la paléotempérature et le potentiel pétrolier. Paper presented to the Centre National de la Recherche Scientifique, September 15–17, 1973, Paris.

Tissot, B. 1966. Problèmes géochimiques de la genèse et de la migration du pétrole. *Rev. de l'Inst. Français Pétrol.* et *Ann. Combust. Liq., 21* (11), 1621–1671.

Tissot, B. 1969. Primières donnees sur le mecanismes et la cinétique de la formation du petrole dans les sediments: Simulation d'un schema reactionnel sur ordinateur. *Rev. l'Inst. Français Pétrol., 24* (4), 470–501.

Tissot, B. 1973. Toward a quantitative evaluation of the petroleum formed in sedimentary basins. *Rev. Ass. Franc. Tech. Pétrol., 222,* 27–31.

Tissot, B., and R. Pelet. 1971. Nouvelles donneés sur les méchanismes de genèse et de migration du pétrole simulation mathematique et application à la prospection. In *8th World Petroleum Congress, Proc.* Vol. 2. London: Applied Science Publishers, pp. 35–46.

Tissot, B., G. Deroo, and J. Espitalié. 1975. Etude comparee de l'epoque de formation et d'expulsion du pétrole dans diverses provinces geologiques. In *9th World Petroleum Congress, Proc.* Vol. 2. London: Applied Science Publishers, pp. 159–169.

Tissot, B., J. L. Oudin, and R. Pelet. 1972. Criteria of the origin and evolution of petroleum, application to the geochemical study of sedimentary basins. In H. R. v. Gaertner and H. Wehner (eds.), *Advances in geochemistry, 1971.* New York: Pergamon Press, pp. 113–134.

Tissot, B., Y. Califet-Debyser, G. Deroo, and J. L. Oudin. 1971. Origin and evolution of hydrocarbons in early Toarcian shales, Paris Basin, France. *AAPG Bull.,* 55 (12), 2177–2193.

Tissot, B., B. Durand, J. Espitalié, and A. Combaz. 1974. Influence of nature and diagenesis of organic matter in formation of petroleum. *AAPG Bull.,* 58 (3), 499–506.

Tissot, B., R. Pelet, J. Roucaché, and A. Combaz. 1975. Utilisation des alcanes comme fossiles geochimiques indicateurs des environnements geologiques. Institut Français du Petròle, Ref. 23 440. Paris: Institut Français du Petròle. 26 p.

Tissot, B., J. Espitalié, G. Deroo, C. Tempere, and D. Jonathan. 1974. Origin and migration of hydrocarbons in the eastern Sahara (Algeria). In B. Tissot and F. Bienner (eds.), *Advances in organic geochemistry, 1973.* Paris: Éditions Technip, p. 315.

Toland, W. G. 1960. Oxidation of organic compounds with aqueous sulfate. *J. Amer. Chem. Soc., 82,* 1911–1916.

Torgovanov, V. B., et al. 1960. *Subsurface waters and gases of the West Siberian lowland.* Trudy VNIGRI, No. 197.

Trask, P. D. 1936. Proportion of organic matter converted into oil in Santa Fe Springs field, California. *AAPG Bull., 20,* 245–257.

Trask, P. D., and H. W. Patnode. 1942. *Source beds of petroleum.* Tulsa, Okla.: American Association of Petroleum Geologists. 566 p.

Trask, P. D., H. E. Hammar, and C. C. Wu. 1932. *Origin and environment of source sediments of petroleum.* Houston: Gulf Publishing. 323 p.

Trebin, F. A., V. A. Khoroshilov, and A. V. Demchenko. 1966. Kinetics of hydrate formation of natural gases. *Gazov. Prom., 11* (6), 10.

Trofimuk, A. A., and A. E. Kontorovich. 1977. Forms of primary migration of oil. In N. B. Vassoevich et al. (eds.), *8th International Congress on Organic Geochemistry, Abstracts of Reports.* Moscow, May 10–13, 1977. Vol. 1, pp. 160–162.

Uspenskii, V. A. 1956. *The carbon balance in the biosphere in relation to the distribution of carbon in the earth's crust.* Moscow: Gostoptekhizdat.

Uspenskii, V. A., and A. I. Gorskaya. 1951a. Elaterites as a mineralogical group. *Geokhimicheskii Sbornik,* Trudy VNIGRI. New series, No. 5, No. 2–3. Moscow. English translation (in *Contributions to geochemistry*), by Israel Program for Scientific Translations, Jerusalem, 1965, pp. 203–218.

Uspenskii, V. A., and A. I. Gorskaya. 1951b. Material composition of organic material from the Lower Silurian limestones in the region of the town of Chvdovo. *Geokhimicheskii Sbornik,* Trudy VNIGRI. New series, No. 5, No. 2–3. Moscow. English translation (in *Contributions to geochemistry*) by Israel Program for Scientific Translations, Jerusalem, 1965, pp. 103–114.

Uspenskii, V. A., O. A. Radchenko, Y. A. Glebovskaya, A. P. Shishkova, T. N. Mel'Tsanskaya, and F. B. Indenbom. 1961. *Principal transformations of bitumens in nature and problems of their classification.* Trudy VNIGRI. No. 185. Moscow, pp. 21–167.

Valitov, N. B. 1974. Elemental sulfur as a factor in the generation of hydrogen sulfide in deep-lying carbonate reservoir rocks. *Dokl. Akad. Nauk SSSR, 219* (4), 206–208.

Vandenbrouche, M. 1972. Study of primary migration: Variation of composition of rock extracts in the source rock-reservoir transition. In H. R. v. Gaertner and H. Wehner (eds.), *Advances in organic geochemistry, 1971.* New York: Pergamon Press, pp. 547–565.

Vandenbroucke, M., P. Albrecht, and B. Durand. 1976. Geochemical studies on the organic matter from the Douala Basin (Cameroon). Part 3: Comparison with the Early Toarcian shales, Paris Basin, France. *Geochim. Cosmochim. Acta, 41,* 1241–1249.

Vandenburg, L. E., and E. A. Wilder. 1970. The structural constituents of carnauba wax. *J. Amer. Oil Chem. Soc., 47,* 514–518.

Van Krevelen, D. W. 1961. *Coal.* New York: Elsevier, 514 p.

Van Nes, K., and H. A. Van Westen. 1951. *Aspects of the constitution of mineral oils.* New York: Elsevier, pp. 335–347.

Vassoevich, N. B. 1960. Experiment in constructing typical curve of gravitational compaction of clayey sediments. *Nov. Neft. Tekh. Geol. Ser., 4,* 11–15.

Vassoevich, N. B. 1971. The source of petroleum: A biogenic carboniferous substance. *Priroda,* no. 3, pp. 58–69.

Vassoevich, N. B., Yu. I. Korchagina, N. V. Lopatin, and V. V. Chernischev. 1969. *The main stage of petroleum formation.* Moscow University Vestnik, no. 6, pp. 3–37 (in Russian); English translation: *Int. Geol. Rev.* 1970, *12,* (11), 1276–1296.

Vassoevich, N. B., I. V. Visotskii, A. N. Guseva, and V. B. Olenin. 1967. Hydrocarbons in the sedimentary mantle of the earth. In *7th World Petroleum Congress, Proc.* Vol. 2. London: Elsevier, pp. 37–45.

Vassoevich, N. B., I. V. Visotskii, V. A. Sokolov, and Ye. I. Tatarenko. 1971. Oil–gas potential of late Pre-Cambrian deposits. *Internat. Geol. Rev., 13* (3), 407–418.

Vernadskii, V. I. 1934. *Outlines of geochemistry.* ONTI, Gornogeolog. Neft. Izd., pp. 152–153.

Vernon, J. W., and R. A. Slater. 1963. Submarine tar mounds, Santa Barbara County, California. *AAPG Bull., 47* (8), 1624–1627.

Vyshemirskiy, V. S., and L. S. Yamkovaya. 1970. Dependence of the migration of bitumenoids on the attitude of strata in western Siberia. *Dokl. Akad. Nauk SSSR, 195* (5), 1197–1199.

Vyshemirskiy, V. S., A. A. Trofimuk, A. E. Kontorovich, and S. G. Neruchev. 1974. Bitumoids fractionation in the processes of migration. In B. Tissot and F.

Bienner (eds.), *Advances in organic geochemistry, 1973.* Paris: Editions Technip, p. 359.

Wakeham, S. G. 1977. Synchronous fluorescence spectroscopy and its application to indigenous and petroleum-derived hydrocarbons in lacustrine sediments. *Envir. Sci. and Tech., 11,* 272.

Waldron, J. D., D. S. Gowers, A. C. Chibnall, and S. H. Piper. 1961. Further observations on the paraffins and primary alcohols of plant waxes. *Biochem. J., 78,* 435–442.

Warner, D. L. 1964. An analysis of the influence of physical-chemical factors upon the consolidation of fine-grained elastic sediments. Doctoral dissertation, University of California at Berkeley.

Weaver, C. E., and K. C. Beck. 1969. Changes in clay–water system with depth, temperature, and time. Georgia Institute of Technology Water Resources Center Project. No. A-008-GA. 95 p.

Webb, G. W. 1976. Oklahoma City oil—second crop from preserved subunconformity source rocks. *AAPG Bull., 60* (1), 115–122.

Weber, V. V., and N. M. Turkel'taub. 1960. The formation of gaseous hydrocarbons in Recent marine sediments. *International Congress on Sedimentology, Doklady Sovetskikh Geologov Proc.* pp. 9–16.

Weeks, L. G. (ed.). 1958. *Habitat of oil: A symposium.* Tulsa, Okla.: American Association of Petroleum Geologists. 1384 p.

Weeks, L. G. 1961. Origin, migration, and occurrence of petroleum. In G. B. Moody (ed.), *Petroleum exploration handbook.* New York: McGraw-Hill, chap. 5.

Weiss, A. 1963. Organic Derivatives of mica-type layer-silicates. *Angew. Chem., Internat. Edit., 2* (3), 134–144.

Welte, D. H. 1972. Petroleum exploration and organic geochemistry. *J. Geochem. Explor., 1,* 117–136.

Welte, D. H., H. W. Hagemann, A. Hollerbach, D. Leythaeuser, and W. Stahl. 1975. Correlation between petroleum and source rock. In *9th World Petroleum Congress, Proc.* Vol. 2. London: Applied Science Publishers, pp. 179–191.

Wenger, W. J., and J. C. Morris. 1971. Utah crude oils: Characteristics of 67 samples. Report 7532. Washington, D.C.: U.S. Department of the Interior, Bureau of Mines, 51 p.

Wenger, W. J., M. L. Whisman, W. L. Lanum, and J. S. Ball. 1957. Characteristics and analyses of ninety-two Colorado crude oils. Report 5309. Washington, D.C.: U.S. Department of the Interior, Bureau of Mines. 60 p.

Wetmore, D. E., C. K. Hancock, and R. N. Traxler. 1966. Fractionation and characterization of low molecular weight asphaltic hydrocarbons. *Anal. Chem:, 38* (2), 225–230.

Whelan, J. K. 1979. C_1 to C_7 Hydrocarbons from IPOD Hole 397/397A. In W. B. F. Ryan and Ulrich von Rad (eds.), *Initial reports of the Deep Sea Drilling Project.* Vol. 47. Washington, D.C.: U.S. Government Printing Office.

Whelan, T. 1975. The geochemistry of methane in coastal and nearshore sediments. Paper presented at the Annual Meeting of the Geological Society of America, 1970, Salt Lake City.

White, D. 1915. Geology: Some relations in origin between coal and petroleum. *J. Wash. Acad. Sci., 5* (6), 189–212.

White, D. E. 1965. Saline waters of sedimentary rocks. In A. Young and J. E. Galley (eds.), *Fluids in subsurface environments.* AAPG Memoir 4. Tulsa, Okla.: American Association of Petroleum Geologists, pp. 342–366.

White, S. M. 1975. Interstitial water studies, Leg 31. In D. E. Karig and J. C. Ingle, Jr. (eds.), *Initial Reports of the Deep Sea Drilling Project.* Vol. 31. Washington, D.C.: U.S. Government Printing Office, pp. 639–653.

Whitehead, E. V. 1971. Chemical clues to petroleum origin. *Chem. and Ind.,* no. 27, pp. 1116–1118.

Whitmore, F. C. 1943. Fundamental research on occurrence and recovery of petroleum. In *American Petroleum Institute Progress Report.* American Petroleum Institute, pp. 124–125.

Williams, J. A. 1974. Characterization of oil types in Williston Basin. *AAPG Bull., 58* (7), 1243–1252.

Wilson, H. H. 1975. Time of hydrocarbon expulsion, paradox for geologists and geochemists. *AAPG Bull., 59,* 69–84.

Wilson, J. E. 1973. Deep is the word for tomorrow's oil. Paper presented at the Third Annual Meeting, Division of Production, American Petroleum Institute, April 9–11, 1973, Denver, Colorado.

Winniford, R. S., and M. Bersohn. 1962. The structure of petroleum asphaltenes as indicated by proton magnetic resonance. Paper presented at Symposium on Tars, Pitches, and Asphalts. In *Preprints, American Chemical Society, Division of Fuel Chemistry.* American Chemical Society, pp. 21–32.

Winters, J. C., and J. A. Williams. 1969. Microbiological alteration of crude oil in the reservoir. Paper presented at Symposium on Petroleum Transformation in Geologic Environments, American Chemical Society, Division of Petroleum Chemistry, September 7–12, New York City. Paper PETR 86: E22-E31.

Worzel, J. L., R. Leyden, and M. Ewing. 1968. Newly discovered diapirs in Gulf of Mexico. *AAPG Bull., 52* (7), 1194–1203.

Yagishita, H. 1963. Geochemical prospecting method for petroleum by hydrocarbon analysis of source rocks. Paper presented at the ECAFE Seminar on Geochemical Prospecting Methods, Bangkok, Thailand.

Yarullin, K. S. 1961. Characteristics of the distribution of gas and oil pools in the Cis-Uralian Trough. *Dokl. Akad. Nauk SSSR, 141* (1), 1142–1145.

Yasenev, B. P. 1959. Gas sampling in wells and its value in exploration. *Geol. Nefti Gaza, 3* (2), 36–39.

Yasenev, B. P. 1962. New data on direct geochemical methods in the exploration for oil and gas fields. *Geol. Nefti Gaza, 6* (12), 54–58.

Yen, T. F. 1974. Structure of petroleum asphaltene and its significance. *Energy Sources, 1* (4), 447–463.

Yermakov, V. I., V. S. Lebedev, N. N. Nemchenko, A. S. Rovenskaya, and A. V. Grachev. 1970. Isotopic composition of carbon in natural gases in the northern part of the West Siberian Plain in relation to their origin. *Dokl. Akad. Nauk SSSR, 190* (3), 196–199.

Young, A., and R. D. McIver. 1977. Application of physical chemistry to petroleum geochemistry. Part 2: Distribution of hydrocarbons between oils and associated fine-grained sediments. *AAPG Bull., 61* (9), 1407–1436.

Young, A., P. H. Monaghan, and R. T. Schweisberger. 1977. Calculation of ages of hydrocarbons in oils: Physical chemistry applied to petroleum geochemistry. Part 1. *AAPG Bull., 61* (4), 573–600.

Youngblood, W. W., M. Blumer, R. L. Guillard, and J. Fiore. 1971. Saturated and unsaturated hydrocarbons in marine benthic algae. *Mar. Biol., 8,* 190–201.

Zankl, H. 1969. Structural and textural evidence of early lithification in fine-grained carbonate rocks. *Sedimentology, 12,* 241–256.

Zarella, W. M., R. J. Mousseau, N. D. Coggeshall, M. S. Norris, and G. J. Schrayer. 1967. Analysis and significance of hydrocarbons in subsurface brines. *Geochim. Cosmochim. Acta, 31,* 1155–1166.

Zaritskaya, O. V., and P. V. Zaritskiy. 1962. Nature of solid bitumens in the reservoir rocks of the Shebela gas-condensate deposit. *Dokl. Akad. Nauk SSSR, 143* (2), 39–40.

Zartman, R. E., G. J. Wasserburg, and J. H. Reynolds. 1961. Helium, argon, and carbon in some natural waters. *J. Geophys. Res., 66* (1), 277–305.

Zhuze, T. P., V. I. Sergeyevich, V. F. Burmistrova, and Y. A. Yesakov. 1971. Solubility of hydrocarbons in water under stratal conditions. *Dokl. Akad. Nauk SSSR, 198* (1), 206–209.

Zieglar, D. L., and J. H. Spotts. 1978. Reservoir and source-bed history of Great Valley, California: *AAPG Bull., 62* (5), 813–826.

Zierfuss, H. 1969. Heat conductivity of some carbonate rocks and clayey sandstones. *AAPG Bull., 53* (2), 251–260.

Zimin, Y. G., Y. A. Spevak, and G. P. Nivkov. 1973. Regularities in the variation of the component composition of water-soluble gases of the West Siberian oil–gas basin. *Petrol. Geol., 7,* 290–293.

Zinger, A. S. 1962. Molecular hydrogen in gas dissolved in waters of oil–gas fields, lower Volga region. *Geokhim.,* no. 10, pp. 890–898.

Zinger, A. S., and T. E. Kravchik. 1973. Soluble alcohols in natural waters in the southeastern part of the Russian Platform. *Dokl. Akad. Nauk SSSR, 210* (1), 208–211.

ZoBell, C. E. 1946a. Action of microorganisms on hydrocarbons. *Bacteriol. Rev., 10* (1–2), 1–49.

ZoBell, C. E. 1946b. Studies on redox potential of marine sediments. *AAPG Bull., 30* (4), 477–511.

ZoBell, C. E. 1947. Microbial transformation of molecular hydrogen in marine sediments with particular reference to petroleum. *AAPG Bull., 31,* 1709–1751.

ZoBell, C. E. 1964. Geochemical aspects of the microbial modification of carbon compounds. In U. Columbo and G. D. Hobson (eds.), *Advances in organic geochemistry, 1962.* New York: Macmillan, pp. 339–356.

Zor'kin, L. M., and E. V. Stadnik. 1970. Composition of gases and salts and the age of ground water of the Caspian depression. *Dokl. Akad. Nauk SSSR, 195* (5), 221–223.

Zor'kin, L. M., E. V. Stadnik, V. K. Soshnikov, and G. A. Yurin. 1972. Geochemistry of gases in ground water in Devonian terrigenous rocks of the Russian Platform. *Dokl. Akad. Nauk SSSR, 203,* 200–202.

Zor'kin, L. M., E. V. Stadnik, V. K. Soshnikov, and G. A. Yurin. 1974. Gases in underground waters of Proterozoic and Paleozoic deposits of the Russian Platform. *Byul. Moskov. Obshchest. Ispytatelei Prirody, Otd. Geol.*, *49* (3), 97–108.

Index of Names

Index of Topics

Arbutin, 76
Area, of clay particle surfaces, 203
Arenes. *See* Aromatic hydrocarbons
Aromatic hydrocarbons, 30, 37–38, 82, 545
Aromatic/saturate ratio, change with
    temperature, 366
Arrhenius equation, 127, 131–134
Arthropoda, 74
Ashley Valley, Utah, 231–232
Asmari reservoirs, Iran:
    age of oil in, 395–396
    seepage of oil from, 415–416
    source of oil in, 252–254
Asphalt, 21, 64–65, 545
    elemental composition of, 29
    underwater mounds of, 421
    world deposits of, 402
Asphaltenes, 44, 58–64, 545
    atomic ratios of, 62
    in crude oil, 58
    in gilsonite, 412
    propane extraction of, 59
    structure of, 63–64
    vanadium–nickel content of, 63–64
Asphaltites, 402, 545
Asphalt mat, at oil–water contact, 390
Asphalt reservoir seal, 255, 382, 388–390
Asymmetric carbon atom, 72
Athabasca heavy oil, 223, 401
    biodegradation of, 389–391
Atlantic Outer Continental Shelf, USA, source rock
    potential of, 526–530
Atmosphere, evolution of, 14–17
Atomic adsorption, 483
Atomic number, 23
Atomic weight, 23
Autochthonous organic matter, 398–400
Azulene, 89

Bacteria:
    n-paraffins and isoprenoids in, 96–97
    oil degradation by sulfate-reducing, 386
    pristane, phytane in, 90
    See also Microorganisms
Bactericide, in cuttings analysis, 442
Bakken Formation, 238–239, 252-253, 505–508
Baku, USSR, 3, 8
Ball mill extraction, 450
Baltimore Canyon, East Coast, USA:
    geochemical data in, 526
    source rock potential of sediments in, 527
Barracouta field, Australia, 357
Barrel, U.S., 544
Bartlesville shale, Oklahoma, 192
Base of crude oil, 545
Basins:
    with most giant fields, 537
    seepage along margins of, 417–420
Bathyal sediments, 283
Baxterville oil, Gulf Coast, 511–513
Beaufort Basin, Canada, 180, 443
Beaver Hill Lake Formation, 441
Beeswax, 79
Bell Creek field, Montana, 384–385
Benthos, annual production of, 105
Benzaldehyde, 115
Benzene, 36, 115
    in formation waters, 448–450
    products from, 52
Benzofuran, 40
Benzoic acid, 41
3,4-Benzpyrenes, 36, 99–100
Bering Sea, 103
Bicarbonate ion, 25–26, 167

Bicarbonate–sodium waters, 193
Biodegradation of crude oil:
    in Bell Creek field, Montana, 384–385
    effect on age dating of oil, 397
    effect on ^{13}C content, 489
    mechanism of, 382–384
    by sulfate-reducing bacteria, 386
    temperature limitations on, 385–386
    in various reservoirs, 386–387
Biodegraded crude oil:
    comparison with immature oil, 391
    correlation of, 488
Biogenic methane, 155
    distinguished from petrogenic, 423–425
    isotopic composition of, 176–178
    relation to lithology, 163
Biological markers, 73, 86–99, 546
    in crude oil correlation, 485–488
    in Irati shale, 313–314
    maturation changes in, 301–314, 381
Biomonomers, 101–102
Biopolymers, 101–102
Biosphere, carbon isotopes in, 23
Bisabolene, 88
Bitumen, 28, 108, 546
    from coal, 284–285
    coefficient of, 284–286
    extracted from sediments, 450–451
    along faults and fractures, 233–234
    in igneous rock, 69–73
    quantity generated from heating kerogen, 318
    and threshold of intense generation, 286–287
Bitumens:
    and asphalts and asphaltites, 402–403
    atomic ratios of, 400–401
    origin and classification of, 398–404
    and pyrobitumens, 403–404
    and waxes, 401–402
Black Sea, 107, 117, 154, 532
Black shale, Ohio, 7, 261
Blake–Bahama Basin, geochemical data in, 528–532
Bluebell field, Utah, 509
Blue haze, 89
Boghead coal, 104, 274–275, 278, 341
    Botryococcus braunii in, 93
Bolivar coastal fields, Venezuela, 388–389, 418–419
Borneo, Seria field of, 390
Boscan field, Venezuela, 357
Botryococcus braunii, 91–93, 274, 341
Bradford oil, Pennsylvania, correlation index curve
    of, 475
Brines. *See* Formation waters
British Columbia, 441, 444–445
    migration in Mesozoic formations of, 226–229
Bureau of Mines, U.S., Hempel analysis, 474–477
Burma oil fields, 3
Butylene, 40

Cadalene, 380–381
Calcium, 167
    in pore waters, 190–191
Calcium carbonate, CO_2 from, 167–169
California:
    Cortes Bank, OCS, in, 521
    offshore seeps in, 521
    pyrolysis of Monterey shale in, 455
    San Joaquin Basin in, 388, 509
Cambrian:
    Davis shale kerogen, 175
    giant fields, 357–358
    HC and C_O of rocks, 266–267
    oil and gas in USSR, 18
Cambrian life, 17